MONSOONS

MONSOONS

Edited by

Jay S. Fein
Pamela L. Stephens

National Science Foundation
Washington, DC

A WILEY-INTERSCIENCE PUBLICATION
JOHN WILEY & SONS
New York • Chichester • Brisbane • Toronto • Singapore

Library of Congress Cataloging in Publication Data:
Monsoons.

   "A Wiley-Interscience publication."
   Includes bibliographies and index.
   1. Monsoons.   2. Monsoons—India.   I. Fein, Jay S.,
1937–   II. Stephens, Pamela L., 1949–

QC939.M7M66   1986   551.5′184   86-15835
ISBN 0-471-87416-7

Printed in the United States of America

10 9 8 7 6 5 4 3 2 1

To Sara for her intellectual and emotional support
—JSF

Contributors

Mr. Boon-Khean Cheang
Malaysian Meteorological Service
Selangor, Malaysia

Professor P. K. Das
Director—General
India Meterological Department
  (retired), and Centre for
  Atmospheric Sciences
Indian Institute of Technology
New Delhi, India

Dr. Robert A. Knox
Scripps Institution of Oceanography
University of California, San Diego
La Jolla, California

Professor T. N. Krishnamurti
Department of Meteorology
Florida State University
Tallahassee, Florida

Professor Gisela Kutzbach
Department of General Engineering
University of Wisconsin–Madison
Madison, Wisconsin

Professor John E. Kutzbach
Director
Center for Climatic Research
Institute for Environmental Studies
University of Wisconsin–Madison
Madison, Wisconsin

Professor Takio Murakami
Department of Meteorology
University of Hawaii
Honolulu, Hawaii

Professor Brian J. Murton
Department of Geography
University of Hawaii
Honolulu, Hawaii

Professor Raaj Kumar Sah
Department of Economics, Economic
  Growth Center, and School of
  Organization and Management
Yale University
New Haven, Connecticut

Professor J. Shukla
Director
Center for Ocean–Land–Atmosphere
  Interactions
Department of Meteorology
University of Maryland
College Park, Maryland

Mr. Khushwant Singh
Member of Parliament
New Delhi, India

Dr. M. S. Swaminathan
Director—General
International Rice Research Institute
Manila, Philippines

Dr. Bruce A. Warren
Department of Physical Oceanography
Woods Hole Oceanographic Institution
Woods Hole, Massachusetts

Professor Peter J. Webster
Department of Meteorology
Pennsylvania State University
University Park, Pennsylvania

Professor John A. Young
Department of Meteorology
University of Wisconsin–Madison
Madison, Wisconsin

Dr. Francis Zimmermann
Centre National de l'Homme et de la
   Société
Paris, France

Preface

Monsoon is a word that evokes images of dark clouds, torrential rains, and flooded villages and cities. The monsoon is this and much more. For more than half of the world's people, the rhythms of life are keyed to the coming and going of the monsoon. It is deeply rooted in their culture, religion, and economy.

With this book, we have tried to provide a multifaceted view of the monsoon: its lore, its societal impacts, and its meteorology. There are many excellent books on monsoon meteorology, but we are unaware of any which also consider the influence of the monsoon on the culture, arts, and economies of past and present societies.

The major portion of the book is devoted to the physical aspects of the monsoon. In this, our bias as meteorologists is evident. We have endeavored to bring together experts in the field of monsoon meteorology to provide for the nonspecialist an appreciation of this complex weather system, one which we have yet to fully understand.

The book focuses on the Asian monsoon and, in particular, the Indian summer monsoon. There are other monsoon systems; however, none is so physically powerful or dramatic, so rich in folklore, literature, and history, or has such an enormous societal impact as the Asian monsoon.

Our intended audience includes administrators and policymakers, researchers and students, and interested laypeople. We have aimed for a level that is understandable by the layperson and at the same time of value to the specialist. A familiarity with the physics of fluids is helpful for the meteorology parts of the book (Parts IV–VI). However, for those anxious about this prerequisite, Chapter 1 should provide adequate background. The remainder of the book requires no particular technical background.

Our goal has been to produce a cohesive book in which each chapter contributes to the overall theme of the multifaceted monsoon. We hope the reader will enjoy learning about the monsoon in all its aspects. We have.

Jay S. Fein
Pamela L. Stephens

Washington, DC
December 1986

Contents

19 Prediction and Warning Systems and International, Government, and Public Response: A Problem for the Future

M. S. Swaminathan

MONSOONS

PART I

INTRODUCTION

Why are monsoons of such interest to so many people? Why have scientists from all parts of the world been so intrigued by monsoons? Certainly one reason is that the monsoon appears to be linked to weather and climate in other parts of the globe. Another is the monsoon itself. As one of the most dramatic of all weather events, tantalizingly complex, rich in variations from place to place, year to year, day to day, and difficult to predict, it is intrinsically interesting.

In this first chapter, Peter Webster, professor of meteorology, defines the monsoon and its extent, and examines its underlying causes. The monsoon, as he states, is a circulation system driven, as are all atmospheric circulations, by the energy of the sun. What sets the monsoon apart from other circulations? Why is it so dramatic, so powerful? His answers to these questions allow the reader without a strong scientific background to understand the important concepts. His chapter provides a basic explantion and introduces complexities of the monsoon systems upon which other authors will expand in later chapters.

1

The Elementary Monsoon

Peter J. Webster
Department of Meteorology
Pennsylvania State University
University Park, Pennsylvania

INTRODUCTION

The term "monsoon" appears to have originated from the Arabic word *mausim**
which means season. It is most often applied to the seasonal reversals of wind
direction along the shores of the Indian Ocean, especially in the Arabian Sea, that
blow from the southwest during one half of the year and from the northeast during
the other. As monsoons have come to be better understood, the definition has been
broadened to include almost all of the phenomena associated with the annual weather
cycle within the tropical and subtropical continents of Asia, Australia, and Africa
and the adjacent seas and oceans. It is within these regions that the most vigorous
and dramatic cycles of weather events on the earth take place.

The dominant characteristic of the great monsoon systems, the annual cycle
itself, has led the inhabitants of the monsoon regions to divide their lives, customs,
and economies into two distinct phases: the "wet" and the "dry." The wet, of
course, refers to the rainy season, during which warm, moist, and very disturbed
winds blow inland from the oceans. The dry refers to the other half of the year,
when the wind reverses bringing cool and dry air from the hearts of the winter
continents. In some locations, the cold and dry winter air flows across the equator
toward the hot continents of the summer hemisphere. In this manner, the dry of
the winter monsoon is tied to the wet of the summer monsoon, and vice versa.

To define the geographic extent of the monsoon we must first decide upon a
definition of what constitutes a monsoon climate. A number exist (1), but the most
common definition uses characteristics of the annual variation of both wind and
rainfall. By these criteria, to be a *monsoon* climate, the wind must reverse in

---

* Editors' Note: The original Arabic spelling of monsoon appears to be obscure. Singh (Chapter 2) uses
mausem. Zimmermann (Chapter 3) agrees with Webster, and Warren (Chapter 7) uses *mawsim*.

direction between summer and winter, blowing inland from the cooler oceans toward the warm continent during summer and from the cold continent toward the warm oceans or land areas during winter. The definition usually requires that the summer season must be very wet and the winter season very dry. The region where all of these criteria are met lies within the rectangles shown as dashed lines in Figure 1.1.* Almost all of the Eastern Hemisphere of the tropics is included in the definition although some meteorologists even include the southwestern North American continent as a monsoon climate. Overall, about one half of the tropics or one quarter of the surface area of the entire globe may be defined as a monsoon climate.

Important features of the summer and winter monsoons are also shown in Figure 1.1. Main areas of precipitation are shaded and the principal surface winds systems are denoted by arrows. In each season there are two principal rainfall maxima; one over Africa and the other, the larger, over southern Asia and Australasia. The classic example of the annual cycle occurs over India where, during summer, the surface winds are from the southwest (the Indian Southwest Monsoon: marked as ISWM in Fig. 1.1a). With the coming of winter the winds change to northeasterlies (the NEWM of Fig. 1.1b) and the climate becomes dry.

In Figure 1.1 the regions of maximum surface temperature are cross-hatched. The hottest locations (e.g., the Sahara and the Sahel of northern Africa, Pakistan, and Rajasthan of the Middle East, and central Australia) are all on the *poleward* side of major monsoon precipitation. The coldest land areas are, of course, in the winter hemisphere (indicated by stippling) from which emanate the low-level winds of the monsoon system. The Southern Hemisphere land masses, smaller and generally closer to the equator than their northern counterparts, possess warmer winter climates. As a result, the winter flow out of the Southern Hemisphere continents is not nearly as cold as that from the larger Northern Hemisphere continents.

In this chapter we shall concentrate mainly on the annual cycle of the monsoon. However, we shall see in Chapter 11 that it is incorrect to think of summer and winter phases of the monsoon as just prolonged periods of rain or drought, each of some months duration. There are also significant variations that exist on time scales ranging from days to weeks. Thus, while the monsoon appears to have a well-defined annual cycle, closer inspection shows that the monsoon varies substantially and that within the cycle a significant substructure exists that becomes evident as the intensity of the monsoon rains wax and wane through the wet season.

Short-term variations include the individual weather disturbances (i.e., a period of disturbed weather or storms lasting some days) that occur in rapid succession during the so-called active-monsoon periods. A prolonged period of one to several weeks marked by an absence of weather disturbances is called a break-monsoon or, more correctly, a break in the active monsoon. During an active phase the

---

* Within the rectangle not all regions pass the rather rigid test of wind and rainfall reversibility that defines the monsoon climate. Near the equator where little contrast between seasons exists, not all winters are completely dry and maximum rainfall may not occur during the summer. Some regions even possess a double maximum (1). However all experience reversal of the wind direction between summer and winter and, as we will see, they all fall within the domain of the monsoon.

NORTHERN HEMISPHERE SUMMER MONSOONS

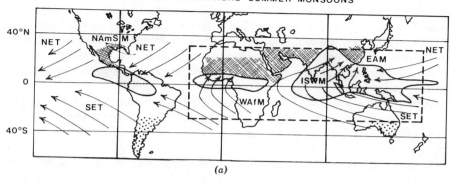

(a)

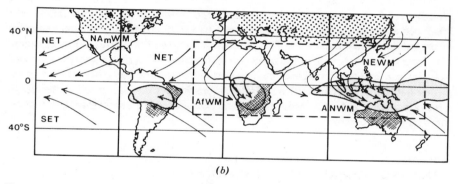

(b)

Figure 1.1. The domains of the principal monsoon systems of the atmosphere during the Northern Hemisphere summer (a) and winter (b). Using the criteria of seasonal wind reversal and distinct wet summers and dry winters, the monsoon region is outlined by the dashed rectangle (1). The main surface wind and areas of maximum seasonal precipitation are indicated as arrows and shaded areas, respectively. The cross-hatching shows the land areas with maximum surface temperatures and the stippling indicates the coldest land surfaces. Key monsoon and tropical systems are indicated. NET and SET refer to the northeast and southeast tradewind regimes. EAM, ISWM, WAfM, and NAmSM indicate, respectively, the East Asia Monsoon, the Indian Southwest monsoon, the West African Monsoon, and the North American Summer Monsoon. NEWM, ANWM, NAmWM, and AfWM show the locations of the northeast Winter Monsoon, the Australian Northwest Monsoon, the North American Winter Monsoon and the African Winter Monsoon. Although often referred to as a monsoon region, the southwestern United States experiences a significant winter rainfall maximum in addition to the summer rainfall. Thus the North American monsoon system fails the precipitation criterion and is included here for interest only.

weather is unstable with frequent storms that produce the rain deluges traditionally associated with the monsoon. But, during a dormant or break phase of the monsoon, the weather is hot, clear, and dry. Monsoon breaks are drought periods and, if prolonged, may cause considerable hardship and even famine in the monsoon lands.

A variable of the monsoon system of considerable importance is the timing of the commencement of the wet. This, the so-called onset of the monsoon, is usually sudden with the weather changing abruptly from the pre-monsoon heat (similar to the torrid climate of the break-monsoon), to the weather disturbances, storms, and intense rainfall of an active period. For a farmer, knowledge of when the onset will occur is critical as with it resides the key to the timing of the planting of his crops. The withdrawal of the monsoon (i.e., the cessation of rainfall over the continents) during the early autumn is a much more gradual transition than the onset.

Added to these subseasonal variations is the interannual variability of the monsoon which modulates the entire annual cycle to produce years where the monsoon is very active or very weak. Such years, of course, are the times of either flood or drought.

Despite the importance of the interannual variability of the monsoon, the difference between the character of the monsoon drought and flood years is very much *smaller* than the difference between the weather in the active and break periods within the annual cycle. The transition of the established monsoon between the active and break periods is discussed at some length in Chapter 11 and the interannual variability of the monsoon in Chapter 14. This chapter defines the basic physical mechanisms that drive the annual cycle of the monsoon and sheds some light on why the monsoon is so intense and why it is restricted to certain regions of the globe. Much of the discussion focuses on the southwest Indian monsoon because it represents the classic monsoon system and it has been studied and documented in a most thorough manner for well over a century.

1 DRIVING MECHANISMS OF THE MONSOON

A practical knowledge of the monsoons played an important role in the social and economic development of the ancient civilizations of the Eastern Hemisphere. Long before the arrival of the Europeans, eastern merchants had plied trade routes between the cities and kingdoms of southern Asia and eastern Africa, successfully adapting their commerce to the distinct seasonal rhythm of the monsoons. After an Arab pilot showed the ancient Portuguese mariner, Vasco da Gama, the secret of navigation between East Africa and India, routes that followed the monsoon winds became the basis of a lucrative European trade and cultural link between eastern and western civilizations. As important as the spices, treasures, and cultural riches that the early explorers and traders carried back to Europe were the fragments of meteorological information secreted away in their ships' logs. Collectively, the logs showed a cycle of alternating winds in the Eastern Hemisphere, southwesterly in summer and northeasterly in winter, of a strength and constancy not experienced in the familiar weather of Europe or the northern Atlantic Ocean. (A detailed discussion of the experiences of the ancient mariners and the scientific inferences made from their records and writings appears in Chapter 7.)

European scholars now had before them observations of low latitude weather which initially stimulated them to wonder about the currents of the atmosphere on a global scale and even to speculate about their physical basis. The collection and deciphering of the logs was not an easy task. Not only were they considered secret (they contained specific information relating to commerce and trading routes), but they were not homogeneous in the quality of their observations or in the quantities that were measured. Two classical studies, perhaps the first in the modern era on the general circulation of the atmosphere, were made in the late seventeenth and early eighteenth centuries by the English scientists Edmund Halley and George Hadley (2, 3).

Halley, a noted astronomer, was intrigued by the work of Sir Issac Newton that suggested that comets travel in parabolic or elliptical orbits. Noting that elliptical orbits meant that the occurrence of comets would be periodic, Halley forecasted in 1682 that a major comet would return in 1758. With this success, of course, was born "Halley's Comet" and a place in history for its discoverer. However, his study of periodic phenomena was not restricted to the heavens. Just four years later, in 1686, Halley hypothesized that the primary cause of the annual cycle of the monsoon circulation was the differential heating between ocean and land caused by the seasonal march of the sun. Differential heating, he reasoned, would cause pressure differences in the atmosphere that could only be equalized by winds blowing from high pressure to low pressure. During the summer months, the winds would blow from the cool ocean to the hot land masses. In winter, as the continents cool to temperatures lower than the adjacent oceans, the winds would reverse, blowing from the land toward the ocean. Thus the first model ever proposed of the monsoon was Halley's planetary scale "sea breeze–land breeze" system (2).

Halley's first chart of winds based on observations and produced in 1688 is shown in Figure 1.2. The diagram shows clearly the northeast trades of the Northern Hemisphere (the NET of Fig. 1.1) and the southeast trades of the Southern Hemisphere (the SET) converging in the equatorial zone. However, in the eastern South Atlantic, the trades swing north and eastward toward the Guinea coast bringing warm ocean air to the hot North African continent. The chart represents the first documentation of the African summer monsoon! The relationship between the trades and the monsoons is discussed in greater detail in Chapter 11.

Hadley (3) noted that the winds of Halley's model monsoon would blow directly toward the coasts (either northerly or southerly, depending on the season) and not obliquely as the southwesterlies or northeasterlies observed by the mariners (see Fig. 1.1) or in Halley's own charts (Fig. 1.2). It was clear that Halley's model was lacking and he argued that the missing physical ingredient was the effect of rotation of the earth about its axis. Because of rotation, moving air is deflected, veering to the right in the Northern Hemisphere and to the left in the Southern Hemisphere. Thus looking at the tropics, Hadley reasoned that as air flowed across the equator toward the heated land mass the earth's rotation would cause the winds to turn to the right, thereby forming the monsoon southwesterlies. His ideas could explain the wind patterns observed at the east–west coastlines of the major continents (Fig. 1.2).

With the work of these two great scientists, theoretical meteorology emerged as a systematic science with strong empirical foundations. But beyond even that

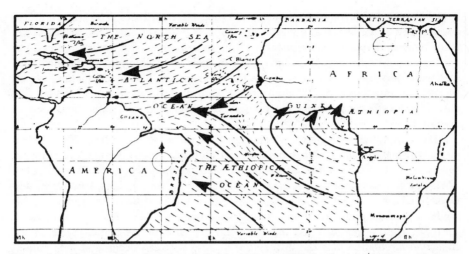

Figure 1.2. The first chart of winds compiled by Halley in 1688 for the Atlantic Ocean (2). Large arrows have been superimposed on Halley's diagram so that the wind direction can be discerned. Notice the flow from the southeast that crosses the equator and recurves to cross the coast of Guinea from the southwest. This is the West African Monsoon (WAfM) shown in Figure 1.1*a*. In the western sections of the North and South Atlantic are the southeasterly and northeasterly trade winds designated as SET and NET in Figure 1.1.

achievement was their identification of two of the most fundamental driving mechanisms of the planetary scale monsoon:

1. *the differential heating of the land and ocean* and the resulting pressure gradient that drives the winds from high pressure to low pressure, and
2. *the swirl introduced to the winds* by the rotation of the earth.

What is amazing is that so much of the essence of the physical nature of the monsoon was understood and explained nearly 300 years ago. Since then, of course, we have increased our understanding of the basic physics of the atmosphere and extended our knowledge of the three-dimensional structure of the monsoon using both observations and theory. It might be argued that compared to the Halley–Hadley quantum jump (although it is interesting to note that it took nearly 60 years for Hadley's work to be appreciated) our advances since have been relatively small. However there is one very important and fundamental aspect of the earth system yet to be considered: the earth is basically an aqueous planet so positioned relative to the sun that the three phases of water (liquid, vapor, and ice) can coexist in the atmosphere. In other words, the average temperature of the earth lies very close to the so-called *triple point* of water.

The triple point of water is the combination of temperature and pressure at which its solid, liquid, and gaseous phases can coexist in equilibrium. Water molecules in an environment close to the triple point can freely move from one state to another. If the transition is to a higher state (e.g., from liquid to vapor), energy must be added to the water molecule. Thus water will not evaporate unless it receives energy, in the case of the atmosphere or the ocean, for example, from the sun. On the other hand, if the transition is to a lower state (e.g., from vapor to liquid) energy is given off by the molecule. Thus the solar energy used to evaporate water over the ocean will be released to the atmosphere when the vapor condenses at some later stage during precipitation. The proximity of the conditions on the earth to the triple point of water determines to a large degree the structure and vigor of monsoons and, more generally, the state of the weather and climate on the planet. The important role that moisture plays in the atmosphere's circulation, making the earth unique among all the planets in our solar system, was not appreciated by either Halley or Hadley.

The major effect of moist processes is to augment the differential heating produced by the land–sea contrasts by providing a mechanism for the heat received from the sun over the vast reaches of the oceans to be collected, stored, and concentrated and later released over the warm oceans and hot continents of the summer hemisphere. Monsoons, if defined only in terms of wind, probably exist on a dry and arid planet like Mars (4, 5). But compared to the earth, the monsoons of Mars would be rather benign affairs since there exists no corresponding mechanism that emulates the earth's phase changes of water that can concentrate the heat from the sun and drive vigorous atmospheric motion.

Thus to the two fundamental physical processes identified by Halley and Hadley we can add a third:

3. *moist processes* that determine the strength, vigor, and location of the major monsoon precipitation by storing, redistributing, and selectively releasing, in the vicinity of the heated continents, the solar energy arriving over most of the tropics and subtropics.

The following sections examine in some detail each of the three driving mechanisms.

2 THE BASIC MECHANISM: DIFFERENTIAL HEATING OF LAND AND SEA

All heating of the earth system (except for very small amounts such as from geothermal heating) must originate from the sun. Consequently all motion in the atmosphere and the ocean is a result of solar heating. Strictly speaking, it is not the heating that forces the motion of the atmosphere and ocean but rather the *distribution* of the heating *and* cooling around the earth. Motion will only occur if one region is heated or cooled more than some other region. This differential heating results from a combination of two factors. These are

1. the variation in space and time of the solar heating of the earth system that results both from the annual cycle of the sun and the character of the earth's surface at a particular location, and
2. the spatial and temporal variation of the cooling of the earth to space.

The most important question with respect to the first factor above is whether the earth's surface is water or land and whether the land is wet or dry. These characteristics not only determine the amount of solar radiation that is reflected back to space but also the magnitude and the way in which heat is stored at the earth's surface. Cooling of the earth, the mechanism by which the earth maintains its heat balance, is fairly independent of the nature of the surface but is a strong function of its temperature.

2.1 The Annual Cycle of Radiative Heating

The monsoon involves the atmospheric reaction to *differential* heating. We recall that heating at a particular location is actually the *net* radiative heating, that is, the sum of heating from the sun and the cooling of the earth to space, rather than just the distribution of solar heating itself.

2.1.1 Gain of Energy by Solar Heating. Because the earth is a sphere and is tilted some 24° to the solar plane, different regions of the globe are subjected to a strong annual cycle in the strength of the incoming solar radiation. In the top panels of Figure 1.3*a* and *b*, the upper curves show the latitudinal distribution of incoming solar energy. They indicate energy arriving at the surface of the earth ignoring variations due to the reflection of solar radiation by clouds and their distribution and the absorption of solar radiation by the atmosphere.

The amount of energy received from the sun at the surface depends on the intensity of the incoming solar radiation and the length of the day, factors that are related directly to the spherical shape of the earth, the tilt of the earth's axis and the rotation of the earth about the sun (5, 6, 7). Thus the long days of the summer polar regions (up to 24 hours of daylight) and the intense solar heating of low latitudes (but with only 12 hours of daylight), combine to produce an almost constant and very high input of solar heat over the entire summer hemisphere. This can be contrasted with a decrease of daylight hours in conjunction with the decreasing intensity of the solar radiation in the winter hemisphere, producing a total solar heating that reduces rapidly to zero inside the winter Antarctic or Arctic Circles. Within these zones there is only the perpetual darkness of the winter polar night.

In Figure 1.3, it has been assumed that cloud amount is constant over the entire globe. Between reflection by clouds and the ground some 30% of the incident solar radiation is reflected back to space. This fraction of the incoming energy cannot heat the atmosphere or the earth's surface. Some of the sun's energy is absorbed directly by the atmosphere but the amount is relatively small and the atmosphere may be considered nearly transparent to incoming solar radiation. Because of this transparency, the solar radiation heats the ground which, in turn, heats the atmosphere from below.

2.1.2 Loss of Energy by Cooling to Space. For the reasons given previously, we cannot discuss solar radiation without considering the heat loss by the earth to space. This heat loss arises because the earth system maintains thermal balance so that, averaged over some period of time, the earth neither heats nor cools.

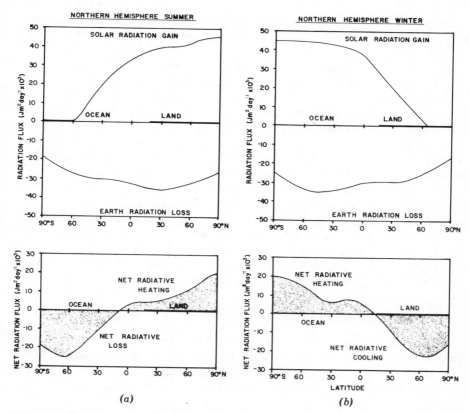

Figure 1.3. The distribution of the incoming solar radiation and the loss of the earth radiation to space as a function of latitude; distributions for the Northern Hemisphere summer (*a*) and winter (*b*). For each season, the upper curve shows the gain of radiation from the sun by the earth system. The radiation loss by the earth system to space is shown as the curve along the negative axis. The lower diagrams show the *net* radiative heating (i.e., the sum of the solar radiation input and the earth radiation loss) gained or lost by the earth system. The winter hemispheres lose more energy to space than they gain from the sun whereas the summer hemispheres gain more than they lose (units: Jm²/day × 10³).

The heat balance of the earth is accomplished by a continuous radiation of heat *from* the earth's surface and the atmosphere to space. But from the perspective of understanding the basic driving force of the monsoons, the important point is that the distribution of heat loss is very different from the heat gain from the sun. This is because the heat loss to space depends on the temperature of the system. Specifically, the loss of heat is proportional to the fourth power of its *absolute temperature* (the absolute temperature is the temperature in degrees Celsius plus 273°). Detailed and elementary descriptions of the earth's radiation balance are given in (4) and (5).

Because of their warmer temperatures, the equatorial regions and the summer hemisphere lose more energy to space than does the winter hemisphere. However the loss varies by less than 30% over the entire globe at any time of the year. The distribution of the loss of earth heat to space (marked as earth radiation loss) is shown in the lower curves on the top panels of Figure 1.3. The radiation loss should be contrasted with the very uneven distribution of incoming solar energy (marked as solar radiation gain).

2.1.3 The Distribution of the Net Radiative Heating.

From the net radiation curves (the sum of the solar radiation gain and the earth radiation loss to space), shown in the bottom panels of Figure 1.3, a clear picture emerges. Broad latitude bands of net radiation loss and net radiation gain show that the winter hemisphere loses far more heat in the net than it gains. The summer hemisphere, on the other hand, accumulates far more heat than it loses. Consequently, a strong *differential radiative heating* exists between the summer and the winter hemispheres.

If the summer hemisphere were to continue heating and/or the winter hemisphere continue to cool at the rates shown in Figure 1.3, the temperatures would be very much hotter or colder than are observed. Such accumulations or losses of heat do not occur because heat is transported from regions of heat abundance (the tropics and the summer hemisphere) to regions of heat deficit (the winter hemisphere). These transports are accomplished by the winds and ocean currents that flow from areas of high pressure to areas of low pressure, as suggested by Halley.

2.2 Land–Ocean Heating Contrasts

Figure 1.3 shows how much heat is available at the earth's surface from net radiative heating. Now we are in a position to ask how the type of surface, land, or ocean, is affected by the distribution of radiative heating.

If the same amount of heat were available to equal areas of ocean and land, the impact on the temperature of each area would be very different. In fact, the rise in land area temperature would be far greater than that over the ocean. The different temperature responses to the same heat input occur because of two different but related physical properties of oceans and land:

1. The *specific heat* (defined as the amount of heat required to raise a given mass of a substance one degree Celsius) of water is *twice* that of *dry* soil. Thus, for the same amount of heat input, the temperature of dry land would increase by twice that of an equal mass of water. Of course, as the land becomes wet and saturated with precipitation, its specific heat will increase toward that of water, and, as a consequence, the temperature increase will be smaller.
2. The *effective heat capacity* (a measure of a system as a whole to store heat) is many orders of magnitude larger for the ocean than for land.

Traditionally, the differential heating between land and oceans is attributed to the first of these two properties, although it is probably the less important of the two. This is especially so once the monsoon precipitation has become established over the continental region because then the soil is moistened and the heat capacity of the ground approaches that of water.

In the overall scheme of the monsoon, it is the second effect, the *difference* in effective heat capacities of land and ocean, that is probably far more important.

The fundamental difference in the manner in which land and ocean react to heating and cooling comes from the way each can transfer heat within itself and store it internally. The crucial aspect is that land is a solid and the oceans are fluid. We may summarize the differences as follows:

1. Both solids and fluids can be heated by solar radiation and both may cool by radiation loss to space. However depending upon its turbidity (or murkiness), a fluid may be fairly transparent to solar heating so that energy can pass through the surface and be absorbed at some depth. Land, on the other hand, will absorb all the incident and nonreflected solar energy within a few microns of the surface.

2. Heat transfer through solids is accomplished by molecule-to-molecule diffusion of heat. This is a very inefficient and slow process that causes the heating of the land's subsurface layers to take a very long time.

3. As in solids, heat transfer through fluids occurs through molecular diffusion but also by the very efficient process of turbulent mixing or stirring. Turbulence may be instigated by mechanical stress applied to the surface of the ocean by the wind or by mixing caused by heavier surface water, which has been cooled by radiative heat loss, sinking to greater depths. Such processes can mix heat absorbed in the surface layers to great depths where it is stored, or bring up warmer water stored at an earlier time, to the surface to replace heat losses occurring there.

Figure 1.4 illustrates the various processes described above by a schematic description of the energy exchanges at the surface of the earth (land in Fig. 1.4*a* and ocean in Fig. 1.4*b*) during summer and winter. The left parts of the figures show the disposition of the radiative heating and cooling. The right parts show a time sequence of the temperature profiles in the column extending from the atmosphere, through the surface and into the subsurface.

The interpretation of the temperature changes over the land areas is rather straight-forward. During early summer, as the net radiation becomes positive, radiative heating of the surface takes place (compare Fig. 1.3) and the land surface heats rapidly; a consequence of the low specific heat of dry land. The rate of heating is very high near the surface because of the slow downward transfer of heat to the cooler subsurface by molecular diffusion. As summer progresses the temperature of the upper surface of the soil increases and the subsurface layers slowly follow suit in the manner described in the sequences I through III in Figure 1.4*a*.

The contact between the hot land surface and the atmosphere allows an upward transfer of sensible heat to the atmosphere which takes place by the *direct* addition of heat at the lower boundary of the atmosphere. A parcel of air, warmed by heat from the surface, will expand and rise, being lighter than its environment and thus buoyant. As the air parcel ascends, it encounters a lower pressure environment and expands, causing the parcel to cool. The parcel will continue to rise through the atmosphere redistributing the heat gained at the surface until it loses its buoyancy. As these hot parcels of air rise up from the earth's surface, cooler parcels from aloft replace them at the surface where they are heated in turn. In this way, in the summer over land, dry buoyant bubbles may transfer heat upward of 5000 m.

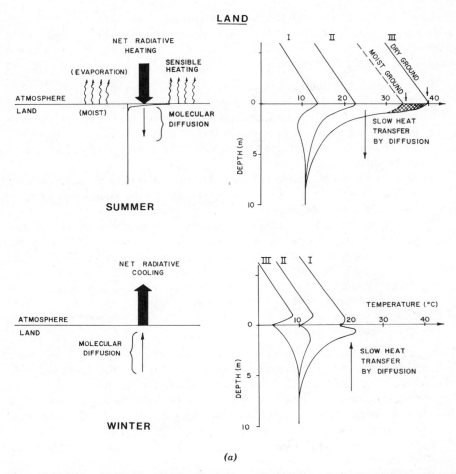

Figure 1.4. Schematic diagrams showing the dominant physical processes involved in heat transport at the land surface (*a*), and the ocean surface (*b*) for summer and winter. The broad arrows represent the net radiative flux at the land or ocean surface (see Fig. 1.3). The net incoming radiation impinging on the land surface is very rapidly attenuated in the first few millimeters of the soil but relatively deeply penetrates the nearly transparent ocean (the attenuation of the subsurface solar radiation curves is shown in the two left panels of *a* and *b*). Wind stress at the ocean surface causes substantial transfers of heat

However, as we shall see later, if the bubbles are moist the transfer of heat may be through the *entire* troposphere or to levels above 12,000 m.

When precipitation occurs, the heating of the surface decreases because the specific heat of the soil increases with the addition of moisture. Furthermore, a considerable portion of the heating from the sun is used in evaporating the moisture in the soil. With ground moisture taken into account the final temperature profile is shown as the dashed curve near the summer curve labeled III in Figure 1.4*a*.

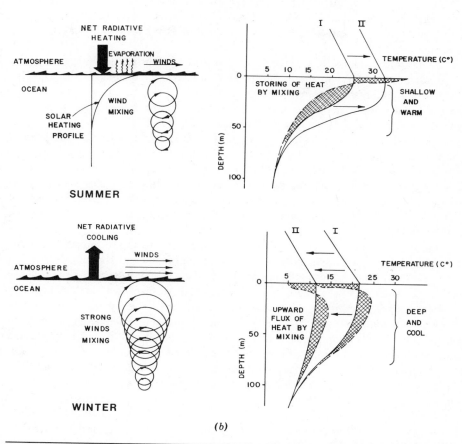

(b)

(upward in winter and downward is summer) between the surface layers and the subsurface ocean by producing turbulent mixing. The right panels show successive temperature profiles (marked I, II, etc.). The cross-hatching in (a) indicates the changes in the subsurface temperature profile that would be expected if the ground were moist. The cross-hatched areas in (b) show the change in the temperature profiles when the heated surface water is mixed downward in the summer case or mixed upward in winter.

The difference in temperature of the uppermost layer of the soil between dry and moist conditions is indicated by the cross-hatched area.

During winter, at the time of net radiative cooling, the reverse sequence occurs but now the subsurface layers of the ground remain warmer than the rapidly cooling surface. The slow upwelling of heat from below does little to mitigate the very large radiative loss and, at the surface, the temperature falls quite rapidly, cooling the air near the ground and rendering the atmosphere very stable. Because it is stable, buoyant bubbles cannot form at the surface and instead heat transfer through the lower part of the atmosphere occurs by molecular diffusion or by the radiation of heat. Since heat diffusion is so slow, the cold surface layer insulates the main body of the atmosphere from the very cold land surface as shown in the right part of Figure 1.4a.

So responsive is the heating or cooling of the land surface to variations in the net radiation that the temperature follows the annual cycle of the sun very closely. Maximum temperatures (often well in excess of 40°C) occur close to the summer solstice and minimum temperatures (dropping to below freezing in central and southern Asia) occur near the winter solstice.

The brisk response of the land surface to the annual variation of radiative heating is very different from that of the ocean which may best be described as sluggish. With the extra process of heat transfer by mixing (either by cold water sinking or by wind stirring) in addition to the radiative heating and cooling at the surface, the temperature of the ocean surface is not in phase with the solar cycle but lags behind by nearly two months. The processes involved in this lagged response are shown in Figure 1.4b.

During summer when the wind is usually light, the stirring of the surface water is minimal and it warms substantially from the solar heating as shown by the ocean temperature profile in the right summer panel. This heated water, being less dense than the water below it, tends to remain near the surface as a shallow layer. A typical temperature profile is given by the dashed line. When wind mixing occurs some warm water is transferred downward and is replaced by the cooler water from below, resulting in a curve labeled I. The cross-hatched area between the dashed line and curve I indicates the exchange of heat in the ocean column with the warm surface water being mixed down and the cooler subsurface water mixed up. The result is a cooler ocean surface and a warmer interior. As summer progresses, the ocean surface continues to heat and the warm layer slowly deepens by the slow stirring induced by the wind. The effect is to store very large quantities of heat to depths of 50 to 100 m below the surface as shown schematically in the upper right part of Figure 1.4b where the subsurface layers have increased in temperature in the sequence I to II.

As winter approaches, the upper layer of the ocean cools very rapidly as the intensity of the incoming solar radiation diminishes and the net radiation at the surface becomes negative (see Fig. 1.3). Due to the surface cooling, the density of the upper layer increases substantially. Under the action of gravity the cool water sinks and is replaced by the lighter and warmer subsurface water. In addition turbulence induced by the strong surface winds of winter mixes the water vertically. The warm subsurface water is, of course, the surface water that was heated and mixed down during the summer and held in storage during the intervening period. Its transport to the surface during winter has the effect of slowing down the winter cooling and is responsible for the sluggish response to the annual cycle of the solar heating. The timing of the maximum and minimum temperatures of the ocean column (shown as the two curves I and II in Fig. 1.4b) do not occur until a full *two months* after the summer and winter solstices, respectively. This is evident from the schematic winter diagram which shows the initial temperature profiles and the final profiles after the heat is mixed up from below. The cross-hatched areas indicate the exchanges of heat that have taken place. In contrast to the summer case, the surface heats at the expense of cooling well below the surface.

The sequences described above for the ocean temperature are very general and assume that the net radiative heating follows the annual cycle suggested in Figure 1.3, and also that the winter winds are stronger than those of summer. In the monsoon regions the latter asumption does not hold as the strongest winds over the ocean occur during summer when the monsoon is fully established. These strong winds increase the turbulent mixing of the ocean to such a degree that the sea surface temperatures tend to *fall* in the monsoon regions *before* the summer solstice! In fact, the maximum sea surface temperatures in the Arabian Sea and the Bay of Bengal occur in April and May. The surface waters cool substantially in June and July as the monsoon winds increase. A second sea surface temperature maximum occurs in September when the monsoon winds diminish in intensity. The effect of the rapid cooling of the surface temperatures in the north Indian Ocean in the early summer tends to enhance the already large temperature gradients across the ocean–continent boundary. In the next section we shall see that the reduction in sea surface temperature near the coast has two offsetting effects: it increases the pressure gradient that drives a stronger monsoon, but it reduces the evaporation which, in turn, reduces the intensity of the monsoon.

3 THE IMPACT OF DIFFERENTIAL HEATING

The distribution of land and ocean and their different effective heat capacities together with the annual cycle of radiative heating all conspire to produce a pattern of atmospheric heating that is variable in both space and time. We shall now examine how heating patterns are converted into the motion field we call the monsoon. The processes necessary for the conversion are:

1. the creation by differential heating of regions with different potential energies;
2. resulting in the creation of a body force, in this case the pressure gradient force; and
3. atmospheric motion produced by work done by the pressure gradient force that converts the potential energy into kinetic energy.

Figure 1.5a provides a simple schematic realization of the sequence described above. Imagine a fluid differentially heated, as described in the preceding sections, so that warm and light air overlies the land area and cooler and denser air (shaded in Fig. 1.5a), lies above the ocean. The greater the differential heating across the fluid, the greater the difference in weight of the adjacent volumes. Suppose that the light and dense fluids were initially separated by a partition (diagram (i) of Figure 1.5a). Since the surface pressure at any point is a measure of the weight of fluid above that point, the surface pressure below the cold air must be greater than

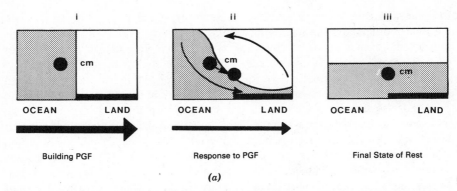

Figure 1.5a. Air motion in a very simple system forced by differential heating between land and ocean which creates adjacent relatively cold and dense (stippled) and warm and light fluid masses. If the two air masses were artificially constrained as in (i), a pressure gradient force (PGF) would act at the surface from the ocean toward the land. If the restraint were removed (ii), the fluid would move under the action of the PGF with the cold dense fluid undercutting the warm light fluid. With the motion of the air the center of mass (cm) of the system falls. Motion continues until the surface pressure is equalized everywhere and the PGF vanishes. At that stage, the fluid is completely at rest with the cm at its lowest possible level (iii).

below the warm air. Thus a *pressure gradient force* exists between the cold fluid and the warm fluid, or between the ocean and the land.

If the partition is removed, the pressure gradient force at the surface of the fluid will cause the cold, dense fluid to move toward and undercut the warm, light air. The arrows in diagram (ii) illustrate the motion that ensues: sinking of cold fluid and motion toward the land at low levels and rising of the warm fluid and motion toward the ocean at upper levels. As the motion continues and the cold fluid further undercuts the warm fluid, the weight of adjacent columns becomes more and more equal. With the undercutting, the center of mass (the solid circle marked "cm" in Fig. 1.5a) of the entire system is lowered. Motion will continue until the pressure gradient force has vanished, which occurs when the weight of all adjacent columns is equal. Then the atmosphere will be at rest as shown in diagram (iii).

An alternative interpretation of the transition between states (i) and (iii) of Figure 1.5a is obtained by noting that motion will continue until all the potential energy available to the system has been exhausted. Potential energy may be defined as the energy a system possesses under the action of gravity by virtue of its location above some level. Changes in potential energy may be measured by noting changes in the location of the center of mass of the simple system of Figure 1.5a through the sequences (i) through (iii). In the initial state (i), the center of mass lies within the heavier fluid and because it is to one side of the geometric center of the fluid it will exert a torque on the entire system causing the counterclockwise motion shown in (ii). Motion will continue until the torque vanishes when the dense fluid resides completely under the less dense fluid. Then the center of mass lies directly below the geometric center of the fluid. At that time the potential energy of the system is at its minimum as shown in (iii).

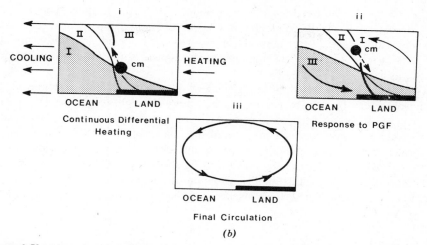

Figure 1.5*b*. Air motion resulting from a continuously maintained differential heating. The differential heating (i) tends to raise the system cm and so increase its potential energy. The effect is to maintain the PGF, recreating state (i) of Figure 1.5*a* by stages I, II, and III (panel i). The PGF causes motion (ii) which reverses the process, lowering the system cm and decreasing its potential energy. This reduces the PGF, recreating state (iii) of Figure 1.5*a* by stages I, II, and III in panel (ii). The combination of *continuous* differential heating (i) and the motion induced by the PGF (ii) creates a continuous monsoon circulation (iii).

Of course the real monsoon, and in fact the entire earth system, is in a state of continual motion with no partitions. So the preceding example is an extremely simple version of the real monsoon. Nevertheless, as the air over the land is being continually heated and the air over the ocean continues to cool, the pressure gradient force between ocean and land is maintained. Such a situation can be seen in diagram (i) of Figure 1.5*b*. Continuous differential heating will effectively maintain a column of denser air over the ocean and a lighter column over the land (as shown in the sequences I, II, and III) and, at the same time, maintain the pressure gradient force (or, alternatively, a counterclockwise torque on the fluid). The final effect is a *continuous* monsoon circulation (diagram iii) that is similar in many aspects to circulations in the real atmosphere. Of course, this circulation is closer to a sea-breeze system, rather than the monsoon. In the next section we shall add other physical features that will bring the model closer to the real monsoon system.

4 A MORE REALISTIC MONSOON MODEL: COMPRESSIBILITY, MOISTURE EFFECTS, AND ROTATION

In the preceding sections we have seen how the differential heating between ocean and land leads to a monsoon circulation. But it is too simplistic to represent the atmosphere by a fluid in a container even though some of the basic physics are quite similar. Some properties are not analogous. Because the atmosphere is made up of gases it is highly compressible and, as such, its temperature and density change with height. In the preceding model we had assumed, implicitly, that the fluids were homogeneous. Furthermore, the atmosphere is moist and the earth rotates about its axis.

Consequences of the vertical variation of the thermodynamic structure in the atmosphere can be seen by examining the observed height of a constant pressure surface as a function of latitude. Pressure is a convenient quantity to discuss because the pressure at any level is a measure of the mass that lies above. For example, assuming that the pressure at the surface of the earth is about 1000 millibars (mb), then half of the atmosphere, by mass, will lie above the 500-mb surface.

Figure 1.6*a* shows the latitudinal variation of the height of the 500-mb surface (solid curve) for the Northern Hemisphere summer between 60°S and 60°N. Plotted

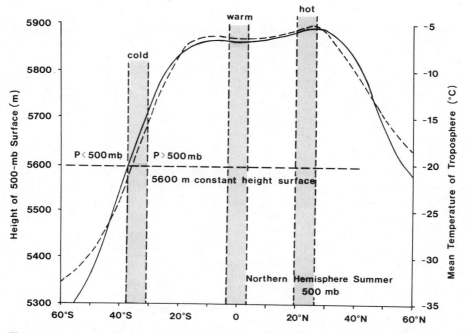

Figure 1.6a. A plot of the observed height of the 500-mb pressure surface (solid line, left scale) and the mean tropospheric temperature (dashed line, right scale) from 60°S to 60°N during the Northern Hemisphere summer. Hot, warm, and cold atmospheric columns are indicated for future reference. Where the height of the 500-mb surface is greatest, the mean tropospheric temperature is warmest. Temperature and height distributions are a result of the differential heating due to land–sea locations and the seasonal variation in the net radiative heating (Fig. 1.3). As the temperature decreases towards the winter (Southern) hemisphere, the height of the 500-mb surface decreases. The 500-mb surface approximately divides the mass of an atmospheric column in half. Thus, as the temperature of the column decreases, the more compressed the column's lower half becomes. As a consequence, a low latitude to high latitude pressure gradient exists along a constant height surface. For example, along the 5600 m constant height surface (marked), the pressure poleward of 35°S is less than 500 mb (in that area the 500-mb surface lies below 5600 m). To the north pressure is higher.

on the same graph (dashed curve) is the mean temperature of the troposphere (i.e., the layer between the surface and 12 km), the distribution of which is dictated, to a large degree, by the distribution of land and ocean and the seasonal variation of the net radiative heating (see Fig. 1.3). Both curves possess a very similar structure. Basically, when the temperature is warm, the 500-mb surface is high. Maximum heights and temperatures are located in the monsoon convective regions of the Northern Hemisphere. Minima in height and temperature occur in the winter hemisphere. That is, the center of mass of a column of the real atmosphere is located at a higher level in the warm air of low latitudes than in the cold air of the winter middle latitudes.

Of course, the fact that the 500-mb surface height and the mean tropospheric temperature have very similar distributions in latitude merely states that the density of air is inversely proportional to its temperature. However the figure also tells something about pressure gradients along constant height surfaces. For example, consider the 5600-m surface, which is shown in Figure 1.6a as the horizontal dashed line intercepting the 500-mb height curve at about 35°S. Poleward of this point, all pressures along the 5600-m surface are less than 500 mb (i.e., the 500-mb surface will be below the 5600-m surface) and all points to the north will have pressures greater than 500 mb. Thus as a consequence of the compressibility of the atmosphere and the distribution of heating, a strong pressure gradient exists between low and high latitudes, in general, and between the monsoon regions of the summer hemisphere and the winter hemisphere mid-latitudes, in particular.

Higher in the troposphere, the effects of compressibility on the pressure gradient between low and high latitudes are even stronger. Figure 1.6b illustrates this point by showing the difference in the heights of the pressure levels between the hot and cold columns of Figure 1.6a (or, in this case, between atmospheric columns located at 25°N and 35°S). The heights of the 500-mb surface differ between the cool and the hot columns by nearly 300m. At 200 mb the height difference between the columns increases by over a factor of two. In the lower troposphere the gradient actually reverses so that, in this example, below about 800 mb (or in the lowest 2 km of the atmosphere) the pressure gradient force is directed to the north or toward the hot column. Above 800 mb, the pressure gradient force is to the south and toward the cold column.

A major constituent of the atmosphere is water vapor, and as we have stated earlier, moist processes have a powerful impact on the character of the monsoon. Their inclusion into the simple system considered in Section 1.3, together with compressibility effects, leads to a much more realistic model of the monsoon circulation. For convenience, we shall consider the effect of compressibility first and then in conjunction with moist processes. Finally, we shall introduce the effects of the earth's rotation.

4.1 The Dry Monsoon: Compressibility Effects

Figure 1.7 describes the driving mechanisms of the monsoon circulation in an atmosphere in the absence of moist processes. Diagram (i) shows the vertical temperature profiles of atmospheric columns located at about 35°S, the equator,

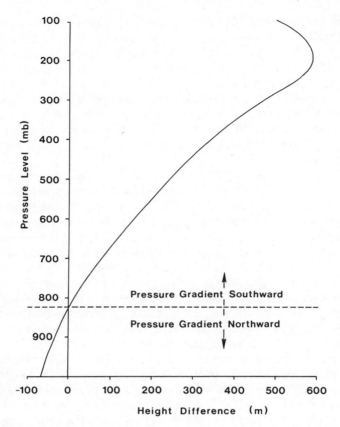

Figure 1.6b. The difference in height of constant pressure levels (left scale) between hot and cold columns. The difference increases substantially with height. Below about 800 mb, the heights of the constant pressure surfaces may be greater in the cold column than in the hot column if the surface pressure is higher in the cold air. This produces a reversal of the pressure gradient with height. Thus, below 800 mb the pressure gradient is toward the north, or toward the hot column. Above 800 mb, the pressure gradient is stronger and directed southward.

and at 25°N during the Northern Hemisphere summer. The profiles are labeled, respectively, cold, warm, and hot and refer to the columns shown in Figure 1.6a. All show a similar decrease of temperature with height. Diagram (ii) depicts the three columns but with each divided into four blocks containing equal masses of air.* Because air density decreases with height, the volume required for the given mass, represented by the size of the blocks, increases with height. Furthermore, since density is inversely proportional to temperature at a given level, the blocks with cold air occupy less volume than those with warm or hot air.

* Although it is a simplification, for the sake of illustration we shall suppose that the four blocks contain the *total* mass of the atmosphere in their respective columns. If this were actually the case, there would be no surface pressure gradient between the columns. In reality there is, but it is relatively weak compared to the upper tropospheric pressure gradient.

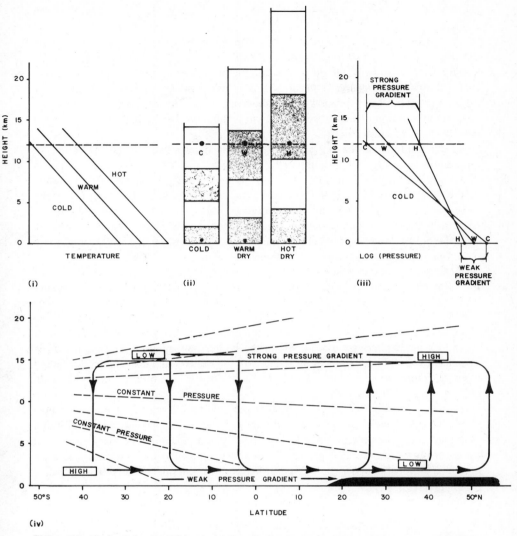

Figure 1.7. Mechanistic view of the production of a dry monsoon where moist processes are ignored. Upper diagrams show the production of the strong upper tropospheric pressure gradient acting from the land *to* the ocean or from the Northern to the Southern Hemisphere. The vertical temperature distribution of three atmospheric columns (cold, warm, and hot are indicative of the winter subtropics, the equatorial region, and the monsoon latitudes, respectively, as shown in Fig. 1.6) decreases at much the same rate with height (i). Diagram (ii) shows the vertical distribution of the mass of the three atmospheric columns. Each has been divided into four blocks of equal mass. As shown in Figure 1.6, the cold column is relatively dense so that the blocks of equal mass of air, denoted by the successive stippled and clear boxes in (ii), have less volume than their warmer counterparts. As a consequence, the pressure at any height, say at the dashed line in (ii), is less in the cold column than in the warm or hot columns. The slower decrease of pressure with height in a warm column compared to the cool column causes a very strong pressure gradient to be produced in the upper troposphere (iii), which is opposite in direction to the lower level pressure gradient force. In the latitude–height domain, the effect is to produce a broad circulation with flow toward the land in the lower troposphere, ascending air over the heated continent and flow toward the cold ocean in the upper troposphere (iv). The circuit is completed by descending air over the oceans.

23

The difference of the density structure with height between hot and cold columns has a great impact on the relative pressure difference at the same level high in the troposphere, as shown in Figure 1.6*a* for the 500-mb surface. Here at the 12-km level, shown as the dashed horizontal line in diagrams (i) through (iii), the same effect is apparent. Since the mass of air above a point is a measure of the atmospheric pressure at that point, the pressure at 12 km is much *higher* in the warm air column than in the cold air column.

The magnitude of the pressure difference is determined by the difference in temperatures through the columns (the density at any level in the atmosphere is directly related to the temperature). From Figure 1.7 (iii), which shows the change of pressure with height for the three columns, it can be seen that even if surface pressure is lower over the land (hot column) than the ocean (cold column), in the high troposphere the pressure over land (hot column) is higher than the pressure over the ocean (cold column). Thus, the pressure gradient reverses with height and increases in magnitude in a manner consistent with Figure 1.6*b*.

Reversing pressure gradients is due to a combination of differential heating of the atmosphere and the effects of compressibility. The impact of the pressure gradient forces is to drive a monsoon circulation that extends between the winter and the summer hemispheres. The surface pressure gradient forces air from the winter hemisphere toward the heated continents. The strong, reversed gradient of the upper troposphere returns the air to the winter hemisphere. The circulation is completed with broad regions of air rising over the heated continents and air subsiding over the cooler oceans. The resulting circulation, the dry monsoon, is shown in diagram (iv) of Figure 1.7.

In many respects, the model monsoon of Figure 1.7 resembles the atmosphere. However it still lacks the form and strength of the real monsoon, principally due to the lack of moist processes. Probably, the circulation is more akin to the seasonal cycle (or monsoon) on dry Mars (4). There are also problems with the vertical scale of our dry monsoon. We shall discuss this point at the end of Section 4 and also in Chapter 11.

4.2 The Moist Monsoon: The Atmospheric Solar Collector

The addition of moisture changes the entire character of the monsoon. Besides being driven by forces produced by temperature differences between land and ocean, the monsoon circulation is also influenced by the vast amounts of solar energy that have been trapped in the tropical oceans. Significant, also, are the continental regions, now moist through precipitation, that act in a manner similar to the oceans so that they are very warm extensions of the oceans. Much of the solar energy incident on the ocean and the moist land is used to evaporate water. This energy is trapped in the water vapor molecules borne along by the trade winds in latent or inaccessible form; trapped energy, that is, until it is released during precipitation. At this point the water vapor condenses to form liquid water and the stored heat is released during the ascent of the buoyant parcels over the heated continent. Thus monsoons become so intense because nearly all of this accumulated heat is released in a geographically limited region over the heated continent. In that sense the heated continent acts as a catalyst, providing an initial region of rising air for the release of the accumulated heat rather than being the primary and final agent for the heating of the atmosphere itself.

Figure 1.8 illustrates the impact of moist processes on the vigor of the monsoon. The diagram follows a sequence identical to that of Figure 1.7. The upper-left panel shows the temperature change of moist air with height. If there is sufficient water vapor and the air is forced to rise over the heated continent, the water vapor condenses into liquid (most of which falls as precipitation), releasing latent heat which adds to the warmth and heat content of the air. The decrease of temperature with height is much less than when the air is dry. Also the warmer the air is, the greater the amount of water vapor it can hold. Thus, hot columns of air have the potential to release more latent heat than cooler columns [compare the differences between the solid and the dashed temperature profiles for the hot and warm moist atmospheres in the upper-left panel (i) of Fig. 1.8].

An important effect of the released latent heat is that the upper tropospheric pressure gradient is greater than that in the dry case, shown when comparing (ii) and (iii) of Figures 1.7 and 1.8. In fact, so intense is the latent heating of the hot moist air over the heated continent that the upper tropospheric temperature over the Northern Hemisphere subtropics (the monsoon region) is far warmer than the air over the equator.

The result of the larger pressure gradient is shown in schematic (iv) of Figure 1.8. As a result of the latent heating, the rising air in the summer hemisphere subtropics is characterized by very deep convective clouds and intense precipitation. The upper-level flow, directed by the pressure gradient force toward the cold air of the Southern Hemisphere, is very strong. Furthermore, because mass is conserved, the strong, upper-level outflow causes the low-level flow toward the heated land mass to be much stronger.

A number of other characteristics of the monsoon flow are affected by moist processes. Besides increasing the intensity of the monsoon circulation, moist processes also provide an upper limit on its intensity. We shall discuss this phenomenon briefly because it provides an excellent example of the many nuances introduced into monsoon meteorology by moist processes.

As far as the land surface is concerned, the major impact of moist processes is to wet the ground. Wetting does two things, each of which leads to the cooling of the land underneath a rather narrow precipitation zone. First, the specific heat of the soil increases so that more heat is necessary to raise its temperature. Second, upon receiving solar radiation, the moisture in the soil evaporates so that a large proportion of the solar energy goes, not into heating the soil, but into changing liquid water to vapor. As these processes occur and the surface temperature drops, the land becomes more "ocean-like" and the region of precipitation moves steadily inland toward the now hotter continent interior. After a number of days the coastal land dries out and warms, and new ascending motion ensues. Moist air is once again forced to rise near the coast and as precipitation takes place, the original precipitating zone, now far inland, is starved of moisture and dies.

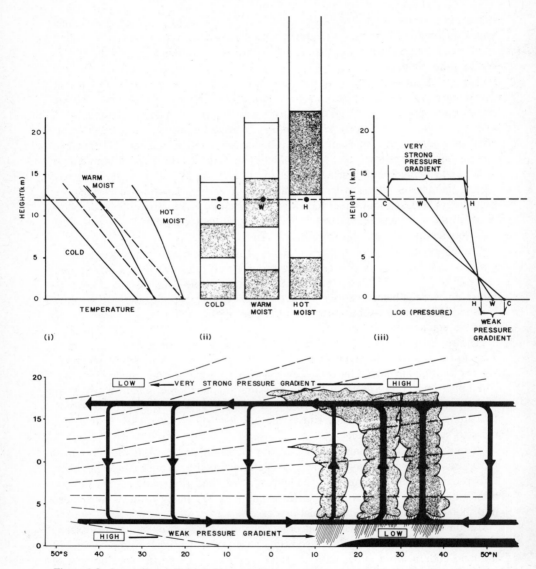

Figure 1.8. Same effect as Figure 1.7 except moist processes are accounted for by the effect of warming the atmospheric column through the latent heat release that accompanies condensation in the rising air. The greater the volume of water vapor, the greater the release of latent heat and, consequently, the smaller the decrease of temperature with height. The result is to produce an even stronger upper-level pressure gradient (ii) and (iii) and an overall stronger monsoon (iv) with stronger upper- and lower-level wind fields and a more vigorous and localized region of rising air over the heated continent. Intense precipitation coincides with the ascending part of the monsoon circulation.

We can now see that moist processes not only increase the vigor of the monsoon but also add a variability on quite another timescale. The migration of the precipitation zone is one of the major subseasonal variations of the monsoon. Viewed locally, these are the active and break periods discussed earlier which are as significant to agriculture as the annual cycle of the monsoon itself. Evidence for the connection between moist processes and the active and break periods will be discussed in detail in Chapter 11.

In the context described above, the monsoon may be thought of as a very efficient solar collector that focuses solar energy collected over a very large area for use in a relatively small geographic region. The total amount of solar energy received from the sun does not change regardless of whether moist processes are involved (except, of course for the higher albedo of the earth because of clouds). But the distribution of the heating is greatly affected by moisture and the regional circulations such as the monsoon are radically different in form and intensity because moist processes provide a means for storing energy for later use.

4.3 The Impact of the Earth's Rotation

Figure 1.9 shows the primary impact of the earth's rotation on a monsoon circulation driven by differential heating where moist processes have been taken into account. In the diagram a continent is depicted as a shaded spherical triangle in the Northern Hemisphere. Figure 1.9a shows the heating distribution consistent with summer net radiative heating (see Fig. 1.3).

Figure 1.9b shows the horizontal air motion on a stationary earth at two levels. H and L indicate regions of relative high and low pressure at these levels. At each level, the arrows depict the wind direction and its strength is denoted by the relative closeness of the arrows to each other. The strongest flow is between the warmest and coldest air columns in the upper troposphere, or identically, in the direction of the strongest pressure gradient force (see Fig. 1.8). At the lower and upper levels, the flow is directly toward or away from the heated continent, respectively.

Now let us consider the same monsoon circulation on the earth rotating from west to east. The impact on the circulation is shown in Figure 1.9c. Rather than a direct flow into or out of the monsoon area, we now see the distinct swirling that characterizes real atmospheric flow. The winds still move into the heated continental region in the lower levels and out of it in the upper troposphere but the path is less direct.

The swirling is, of course, caused by the Coriolis force, a manifestation of the earth's rotation. The Coriolis force deflects motion to the left in the Southern Hemisphere and to the right in the Northern Hemisphere. Its magnitude increases steadily from zero at the equator to a maximum at the poles. Because the Coriolis force is weak in low latitudes, air near the equator under the action of a pressure gradient force will flow almost directly between high and low pressure. Flow at higher latitudes under the action of the same pressure gradient force will deviate or swirl substantially, resulting in the characteristic clockwise flow around low pressure areas and counterclockwise flow around high pressure areas.

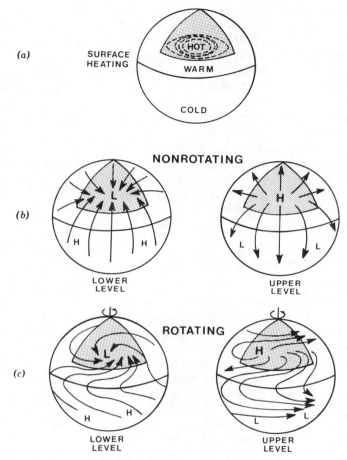

Figure 1.9. The effect of rotation on the monsoon circulation. The figure shows the lower and upper tropospheric response during the Northern Hemisphere summer to thermal forcing produced by differential heating of a continent (shaded area) surrounded by ocean. The middle globes (*b*) represent the response of a nonrotating earth to the heating distribution (*a*). The arrows depict the wind direction, and its strength, at each level, is denoted by the relative closeness of the arrows to each other. The two bottom globes (*c*) show the response to the same forcing of a rotating earth. In both cases the lower tropospheric flow is shown on the left and the upper tropospheric on the right. Relative low and high pressures at each level are shown as H and L.

Figure 1.9 shows the reasons for the lower tropospheric monsoonal flow patterns shown in Figure 1.1. Consider the Asian summer monsoon. From the cold high pressure regions of the winter hemisphere, low-level air moves toward the equator under the action of the northward directed pressure gradient force. The strong middle latitude Coriolis force deflects the flow to the left (i.e., to the west) and it approaches the equator obliquely as southeasterly winds (see Fig. 1.1*a*). As the flow approaches low latitudes, the Coriolis force weakens and the flow crosses the equator almost parallel to the pressure gradient force. Upon crossing the equator the Coriolis force now deflects the flow to the right (i.e., to the east) and it approaches the hot Asian continent as the strong southwesterly monsoon (the ISWN of Fig. 1.1a). The swirl of the low-level flow of the other major monsoon systems can be explained in a similar manner. For example, the NEWM, the ANWM, and the AfWM of Figure 1.1b are merely Northern Hemisphere winter counterparts of the flow shown in Figure 1.9.

Analogous arguments exist for the return upper-level flow. The pressure gradient force is reversed from that in the lower troposphere and the strongest flow is toward the wintertime Southern Hemisphere. After the air crosses the equator, the Coriolis force deflects the wind to the left (eastward), creating extremely strong westerlies in the Southern Hemisphere extratropics. These westerlies form the most intense part of the winter jet stream and play a major role in the Southern Hemisphere winter weather. This jet stream is directly linked to the monsoon heating over Asia and is a good example of the impact the monsoon can have on weather and climate at remote geographical locations. A discussion of the interrelationship of the monsoon with other circulations is given in greater detail in Chapter 11.

The previous discussion shows that when the three principal physical components of the monsoons are included, we can explain a great deal of their observed structure and the variations of their circulation. However, a word of caution is needed. Our considerations have neglected to address the vertical scale of the monsoon and have assumed that the monsoon circulation occupies the entire troposphere. For the moist monsoon (Fig. 1.8), this is a good assumption as the moist processes themselves imply a 15-km vertical scale. This is the height to which a moist parcel will ascend through the release of latent heat in the typical warm tropical atmosphere. However, for the dry monsoon, where the atmosphere is heated at its base (rather than from *within,* as in the moist case), the monsoon circulation would be much shallower with a vertical scale of only a few kilometers.

5 THE ANNUAL CYCLE OF THE MONSOON

With the aid of a schematic diagram and the physical principles described in previous text, we can now discuss the sequence of events that make up the annual cycle of the monsoon. Figure 1.10 shows a sequence of five height and latitude cross sections (similar to the sections shown in Figs. 1.7 and 1.8) along the 90°E meridian for different times of the year. The surface temperature (solid curve) and the precipitation (histograms) are shown beneath each cross section.

The first panel (i) shows the pre-summer (April) flow field. By this time of the year the land areas have started to heat up but can force only weak upward motion. The majority of the precipitation and the strongest rising motion remain close to the equator where the most moist air (shown as the shaded contours) associated with the warmest sea surface temperatures is raised into the middle troposphere by convection in the intertropical convergence zone. Over the continent, weak rising

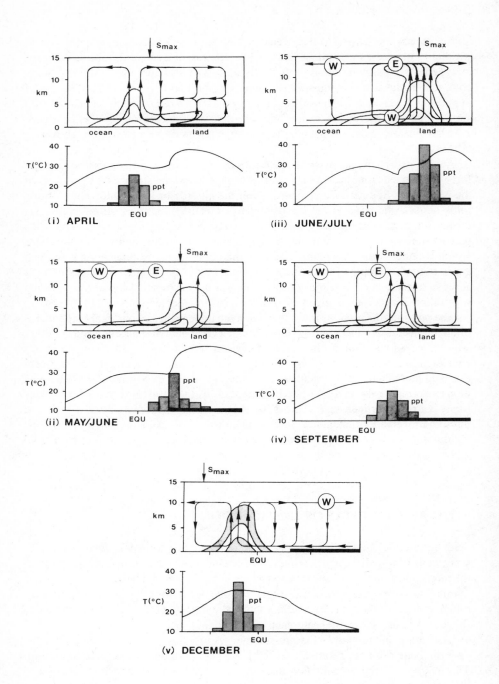

(i) APRIL

(iii) JUNE/JULY

(ii) MAY/JUNE

(iv) SEPTEMBER

(v) DECEMBER

motion is caused by the sensible heat input at the surface. However, this is dominated by subsiding air originating from the equatorial convection so that the vertical extent of the rising motion is limited.

As the sun moves further northward through May and June [panel (ii); the maximum solar input is denoted by S_{max}], the land is heated more intensely producing stronger vertical motion that finally overcomes the subsidence. The moisture content of the troposphere over the land slowly increases as the surface winds turn onshore. At this time, organized precipitation commences and the upper-level return flow becomes sufficiently strong that the deflection by the Coriolis force produces a moderately strong easterly jet stream just to the north of the equator (denoted by Ⓔ) and a westerly jet stream in the winter hemisphere (Ⓦ).

As summer advances and the summer monsoon is established [panel (iii), June/July] the air above the land becomes very moist and warm. At this stage the maximum pressure gradients have been created and the monsoon is in its most intense phase. This stronger circulation is subject to a greater turning by the Coriolis force which produces very strong easterly and westerly jet streams. Even the low-level southwesterly flow over the North Indian Ocean has a strong jetlike core.

Because of precipitation moistening the soil, the surface temperature of the land near the coast decreases. This is the time of maximum monsoon variability as the zones of precipitation successively march inland, away from the cooler, moist coastal land and toward the hot and dry interior. As the locus of maximum insolation moves southward [panel (iv), September], the monsoon loses intensity and eventually the region of maximum precipitation, much diminished, moves over the ocean toward the equator. Finally, with the maximum insolation in the Southern Hemisphere and with the cooling of the continents in the winter hemisphere, convection and precipitation are reestablished in the region of maximum sea surface temperature. The circulation associated with the winter monsoon flow streaming off the cold continents is shown in (v).

6 SOME CONCLUDING REMARKS

The purpose of this chapter has been to dissect the monsoon and display it in its simplest possible physical components. Indeed, if we consider only differential heating, moist processes, and rotation, it appears that the essence of the observed monsoon circulation and structure can be explained.

Figure 1.10. Sequence of the monsoon circulation along the meridian 90°E during the Northern Hemisphere summer in the southern Asia region. Five sections are shown to represent the circulation during (i) April, (ii) late May through early June, (iii) late June through July, (iv) September and (v) December. The position of maximum solar heat input is denoted by the heavy arrow. The position of the maximum easterly and westerly winds in the upper troposphere (the jet streams) are denoted by Ⓔ and Ⓦ. The shaded contours indicate regions of maximum relative humidity. Below each section is plotted the surface temperature curve together with the relative precipitation distribution, shown as bars. Land and ocean areas are marked on the lower axes.

But our description pertains to only the very broad structure of the monsoon and not to the nuances of its character nor to the details of its dramatic variability. Nor does it address the real problem of forecasting monsoon weather or the anticipation of the state of future monsoon seasons. We leave the discussion of such complexities to later chapters. Even so, it is heartening that a basis for understanding the largest, strongest, and from a societal perspective, the most influential weather and climate event on earth can be gained using a few elementary physical principles.

REFERENCES

1. C. S. Ramage, *Monsoon Meteorology*, Academic, New York, 1971, 296 pp.
2. E. Halley, An historical account of the trade winds and the monsoons, observable in the seas between and near the tropicks, with an attempt to assign the phisical cause of the said winds, *Phil. Trans., Roy. Soc. London*, **16**, 153–168 (1686).
3. G. Hadley, Concerning the cause of the general trade winds, *Phil. Trans., Roy. Soc. London*, **39**, 58–62 (1735).
4. P. J. Webster, Low latitude circulation of Mars, *ICARUS*, **30**, 626–649 (1977).
5. P. J. Webster, Monsoons, *Scientific American*, **245**(2), 108–118 (1981).
6. J. C. G. Walker and R. M. Goody, *Atmospheres*, Prentice-Hall, Englewood Cliffs, NJ, 1972, 147 pp.
7. J. M. Wallace and P. V. Hobbs, *Atmospheric Science, An Introductory Survey*, Academic, New York, 1977, 466 pp.

PART II

LITERATURE AND FOLKLORE

The indigenous knowledge of those affected by monsoons has been passed from generation to generation, first orally and later in recorded poetry and prose. Spoken, sung, and written, monsoon folklore accumulated from centuries of observations is rich in imagery and diversity.

In Chapter 2 Khushwant Singh, parliamentarian and journalist, tells us that

> the monsoon is the most memorable experience in the lives of Indians. . . . It has to be a personal experience because nothing short of living through it can fully convey all it means to a people for whom it is not only the source of life, but also their most exciting contact with nature.

As Singh explains, "The Indian's attitude towards [the monsoon] remains fundamentally different from that of the Westerner."

Francis Zimmermann and Brian J. Murton are Western academics. In contrast to Singh, they treat their subjects dispassionately. In Chapter 3 Zimmermann describes traditional ideas about the rainy season in India. These ideas, manifested as rituals, festivals, medical regimen, and poetical and visual symbolism, are products of observations of nature in the context of traditional Hindu religion. In Chapter 4 Murton analyzes proverbs from South India to convey how farmers of that region have coped for thousands of years with the vagaries of the monsoon.

These three chapters convey the enormous influence of the monsoon on virtually every facet of traditional culture in India.

2

The Indian Monsoon in Literature

Khushwant Singh
Member of Parliament
New Delhi, India

INTRODUCTION

Monsoon is not another word for rain. As its original Arabic name (mausem) indicates, it is a season. There is a summer monsoon as well as a winter monsoon, but it is only the nimbus southwest winds of summer that make a "mausem"— the season of rains. The winter monsoon is like a quick shower on a cold and frosty morning. It leaves one chilled and shivering. Although it is good for the crops, people pray for it to end. Fortunately, it does not last very long.

The summer monsoon is quite another affair. It is preceded by several months of working up a thirst so that when the waters come they are drunk deep and with relish. From the end of February, the sun starts getting hotter and spring gives way to summer. Garden flowers wither. Wild flowering trees take their place. First comes the silk cotton, the coral and the flame of the forest, all scarlet and bright orange. They are followed by the firier flamboyant, known in India as the gulmohur. The last of the hot summer's flowering trees is the laburnum which is a bright, golden yellow. Then the trees lose their flowers as well as their leaves. Their bare branches stretch up to the sky as if begging for water, but there is no water. The sun comes up earlier than before and licks up the drops of dew before the fevered earth can moisten its lips. It sears the grass and thorny scrub till they catch fire. The fires spread and dry jungles burn like matchwood.

The sun goes on, day after day, from east to west, scorching relentlessly. The earth cracks and deep fissures open their gaping mouths asking for water, but there is no water—only the shimmering haze at noon making mirage lakes of quicksilver. Poor villagers take their thirsty cattle out to drink; both man and beast are struck down with the heat. The rich wear sunglasses and hide behind curtains of khas fiber on which their servants pour water.

The sun makes an ally of the breeze. It heats the air until it becomes the loo (India's khamsin) and sends it on its errand. Even in the intense heat, the loo's warm caresses are sensuous and pleasant. It brings up prickly heat. It produces a numbness that makes the head nod and the eyes heavy with sleep. It brings on a stroke which takes its victim as gently as the breeze bears a fluff of thistledown.

Then comes a period of false hope. The temperature drops. The air becomes still. From the southern horizon a black wall begins to advance. Hundreds of kites and crows fly ahead. Can it be . . . ? No, it is a dust storm. A fine powder begins to fall. A solid mass of locusts covers the sun. They devour whatever is left on the trees and in the fields. Then comes the storm itself. In furious sweeps it smacks open doors and windows, banging them forward and backward, smashing their glass panes. Thatched roofs and corrugated iron sheets are borne aloft like bits of paper. Trees are torn up by the roots and fall across power lines. The tangled wires electrocute people and set houses afire. The storm carries the flames to other houses till there is a conflagration. All this happens in a few seconds. Before you can say Chakravarti Rajagopalachari, the gale is gone. The dust hanging in the air settles on books, furniture, and food; it gets in the eyes and ears, throat and nose.

Rudyard Kipling has described the pre-monsoon heat of northern India in his story *False Dawn* (1), in which an ardent suitor caught in a dust storm proposed marriage to the wrong sister.

> I had felt that the air was growing hotter and hotter, but nobody seemed to notice it until the moon went out and a burning hot wind began lashing the orange trees with a sound like the noise of the sea. Before we knew where we were the dust-storm was on us, and everything was roaring, whirling darkness.

Again it was Kipling who captured the feeling of listlessness that the months' searing heat produces (2):

> No Hope, no change! The clouds have shut us in,
> And through the cloud the sullen Sun strikes down
> Full on the bosom of the tortured town,
> Till Night falls heavy as remembered sin
> That will not suffer sleep or thought of ease,
> And, hour on hour, the dry-eyed Moon in spite
> Glares through the haze and mocks with watery light
> The torment of the uncomplaining trees.
> Fall off, the Thunder bellows her despair
> To echoing Earth, thrice parched. The lightnings fly
> In vain. No help the heaped-up clouds afford.
> But wearier weight of burdened, burning air,
> What truce with Dawn? Look, from the aching sky
> Day stalks, a tyrant with a flaming sword!

This happens over and over again until the people lose all hope. They are disillusioned, dejected, thirsty, and sweating. The prickly heat on the back of their necks is like emery paper. There is another lull. A hot petrified silence prevails. Then comes the shrill, strange call of a bird. Why has it left its cool bosky shade and come out in the sun? People look up wearily at the lifeless sky. Yes, there it is with its mate! They are like large black-and-white bulbuls with perky crests and long tails. They are pied crested cuckoos (*Clamator Jacobinus*) who have flown all the way from Africa ahead of the monsoon. Isn't there a gentle breeze blowing? And hasn't it a damp smell? And wasn't the rumble which drowned the bird's anguished cry the sound of thunder? The people hurry to the roofs to see. The same ebony wall is coming up from the east. A flock of herons fly across. There is a flash of lightning that outshines the daylight. The wind fills the black sails of the cloud and they billow out across the sun. A profound shadow falls on the earth. There is another clap of thunder. Big drops of rain fall and dry up in the dust. A fragrant smell rises from the earth. Another flash of lightning and another crack of thunder like the roar of a hungry tiger. It has come! Sheets of water, wave after wave. The people lift their faces to the clouds and let the abundance of waters cover them. Schools and offices close. All work stops. Men, women, and children run madly about the streets, waving their arms and shouting "ho, ho"—hosannas to the miracle of the monsoon.

The monsoon is not like ordinary rain, which comes and goes. Once it is on, it stays for three to four months. Its advent is greeted with joy. Parties set out for picnics and litter the countryside with the skins and stones of mangoes. Women and children make swings of branches of trees and spend the day in sport and song. Peacocks spread their tails and strut about with their mates; the woods echo with their shrill cries.

After a few days the flush of enthusiasm is gone. The earth becomes a big stretch of swamp and mud. Wells and lakes fill up and burst their bounds. In towns, gutters get clogged and streets become turbid streams. In villages, mud-walls of huts melt in the water and thatched roofs sag and descend on the inmates. Rivers which keep rising steadily from the time the summer's heat starts melting the snows, suddenly turn to floods as the monsoon spends itself on the mountains. Roads, railway tracks, and bridges go under water. Houses near the river banks are swept down to the sea.

With the monsoon the tempo of life and death increases. Almost overnight grass begins to grow and leafless trees turn green. Snakes, centipedes, and scorpions are born out of nothing. At night, myriads of moths flutter around the lamps. They fall in everybody's food and water. Geckos dart about filling themselves with insects until they get heavy and fall off ceilings. Inside rooms, the hum of mosquitoes is maddening. People spray clouds of insecticide and the floor becomes a layer of wriggling bodies and wings. Next evening, there are many more fluttering around the lampshades and burning themselves in the flames. The monsoon has its own music. Apart from thunder, the rumble of storm clouds and the pitter-patter of rain drops, there is the constant accompaniment of frogs croaking. Aristophanes (3) captured their sound: "*Brek-ek-ek-ex, Koax, Koax! Brekekekex, Koax Koax!*"

While the monsoon lasts, the showers start and stop without warning. The clouds fly across, dropping their rain on the plains as it pleases them, until they reach the

Himalayas. They climb up the mountain sides. Then the cold squeezes the last drops of water out of them. Lightning and thunder never cease. All this happens in late August or early September. Then the season of the rains gives way to autumn.

The monsoon is the most memorable experience in the lives of Indians. Others who wish to know India and her people should also see its impact on the country. It is not enought to read about it in books, or see it on the cinema screen, or hear someone talk about it. It has to be a personal experience because nothing short of living through it can fully convey all it means to a people for whom it is not only the source of life, but also their most exciting contact with nature. What the four seasons of the year mean to the European, the one season of the monsoon means to the Indian. The summer monsoon is preceded by desolation; it brings with it the hopes of spring; it has the fullness of summer and the fulfillment of autumn all in one.

It is not surprising that much of India's art, music, and literature is concerned with the summer monsoon. Innumerable paintings depict people on rooftops looking eagerly at the dark clouds billowing out from over the horizon with flocks of herons flying in front. Of the many melodies of Indian music, *Raga Megha–Malhar* is the most popular because it brings to the mind distant echoes of the sound of thunder and the falling of raindrops. It brings the odor of the earth and of green vegetation to the nostrils; the cry of the peacock and the call of the koel to the ear. There is also the *Raga Desh* and *Hindole*, which invoke scenes of merry-making—of swings in mango groves and the singing and laughter of girls. Most Indian palaces had specially designed balconies from which noblemen could view the monsoon downpour. Here they sat listening to court musicians improvising their own versions of monsoon melodies, sipping wine, and making love to the ladies of their harem. The most common theme in Indian songs is the longing of lovers for each other when the rains are in full swing. There is no joy greater than union during monsoon time; there is no sorrow deeper than separation during the season of the rains.

The Indian attitude toward clouds and rain remains fundamentally different from that of the Westerner. To the one, clouds are symbols of hope; to the other, of despair. The Indian scans the heavens and if nimbus clouds blot out the sun, his heart fills with joy. The Westerner looks up and if there is no silver lining edging the clouds, his depression deepens. The Indian talks of someone he respects and looks up to as a great shadow, like the one cast by the clouds when they cover the sun. The Westerner, on the other hand, looks on a shadow as something evil and refers to people of dubious character as shady types. For him, his beloved is like the sunshine and her smile a sunny smile. An Indian's notion of a beautiful woman is one whose hair is black as monsoon clouds and eyes that flash like lightning. The Westerner escapes clouds and rain whenever he can to seek sunnier climes. An Indian, when the rains come, runs out into the street shouting with joy and lets himself be soaked to the skin.

1 THE MONSOON IN INDIAN LITERATURE

The monsoon has exercised the minds of Indian writers (as well as painters and musicians) over the centuries. Some of the best pieces of descriptive verse were composed by India's classical poets writing in Sanskrit. Amaru (date uncertain, but earlier than ninth century A.D.) describes the heat of the summer and the arrival of the monsoon (4, page 70, verse 68):

> The summer sun, who robbed the pleasant nights,
> And plundered all the water of the rivers,
> And burned the earth, and scorched the forest-trees,
> Is now in hiding; and the autumn clouds,
> Spread thick across the sky to track him down,
> Hunt for the criminal with lightning-flashes.

To be away from one's wife or sweetheart during the season of rains can be a torture (4, page 76, verse 92):

> At night the rain came, and the thunder deep
> Rolled in the distance; and he could not sleep,
> But tossed and turned, with long and frequent sighs,
> And as he listened, tears came to his eyes;
> And thinking of his young wife left alone,
> He sobbed and wept aloud until the dawn.
> And from that time on
> The villagers made it a strict rule that no traveller
> Should be allowed to take a room for the night in the village.

Literary conceit and facetiousness have always been practiced by Indian poets. Thus Sudraka (probably third–fourth century A.D.) has a girl taunt a cloud (4, page 73, verse 81):

> Thundercloud, I think you are wicked.
> You know I'm going to meet my own lover,
> And yet you first scare me with your thunder,
> And now you're trying to caress me
> With your rain-hands!

Bhartrihari (A.D. 500 or a little earlier) went into erotic ecstasies combining descriptions of the monsoon with dalliance (5, page 101, verse 137):

> Flashing streaks of lightning
> Drifting fragrance of tropical pines,
> Thunder sounding from gathering clouds,
> Peacocks crying in amorous tones—
> How will long-lashed maids pass
> These emotion-laden days in their lovers' absence?

He writes of "autumnal rains rousing men's lusts" (5, page 103, verse 140):

> When clouds shade the sky
> And plantain lilies mask the earth,

When winds bear lingering scents
Of fresh verbena and kadamba
And forest retreats rejoice
With the cries of peacock flocks;
Then ardent yearning overpowers
Loved and wretched men alike.

While the downpour lasts there is little that lovers can do besides stay in bed
and make love (5, page 105, verse 142):

Heavy rains keep lovers
Trapped in their mansions;
In the shivering cold a lord
Is embraced by his long-eyed maid,
And winds bearing cool mists
Allay their fatigue after amorous play.
Even a dreary day is fair
For favoured men who nestle in love's arms.

Once the rains have set in good and proper, clouds, lightning, and rain become
a routine affair (4, page 79, verse 102):

Black clouds at midnight;
Deep thunder rolling.
The night has lost the moon:
A cow lowing for her lost calf.

Monsoon is not only trysting time for humans but also for animals and birds,
above all India's national bird, the peacock. Yogesvara (circa A.D. 800) describes
the courtship dance in these beautiful lines (4, page 125, verse 216):

With tail-fans spread, and undulating wings
With whose vibrating pulse the air now sings,
Their voices lifted and their beaks stretched wide,
Treading the rhythmic dance from side to side,
Eying the raincloud's dark, majestic hue,
Richer in color than their own throats' blue,
With necks upraised, to which their tails advance,
Now in the rains the screaming peacocks dance.

Subandhu (late sixth century A.D.) in his *Vasavadatta* (6) is exuberant in his
welcome of the monsoon:

The rainy season had arrived. Rivers overflowed their banks. Peacocks danced at eventide.
The rain quelled the expanse of dust as a great ascetic quells the tide of passion. The
chataka birds were happy. Lightning shown like a bejewelled boat of love in the pleasure-
pool of the sky; it was like a garland for the gate of the palace of paradise; like a lustrous

girdle for some heavenly beauty; like a row of nailmarks left upon the cloud by its lover, the departing day.

The rain was like a chess player, while yellow and green frogs were like chessmen jumping in the enclosures of the irrigated fields. Hailstones flashed like pearls from the necklaces of heavenly birds. By and by, the rainy season yielded to autumn, the season of bright dawns; of parrots rummaging among rice-stalks; of fugitive clouds. In Autumn the lakes echoed with the sound of herons. The frogs were silent and the snakes shrivelled up. At night the stars were unusually bright and the moon was like a pale beauty.

Poet Vidyapati (1352–1448) of Mithila from the eastern part of Bihar used nature to highlight erotic scenes of love-making between Krishna and Radha. Of these many are set in the monsoon (7, page 57, verse 18):

How the rain falls
In deadly darkness!
O gentle girl, the rain
Pours on your path
And roaming spirits straddle the wet night.
She is afraid
Of loving for the first time.
O Madhava,
Cover her with sweetness.
. . . .
How will she cross the fearful river
In her path?
Enraptured with love,
Beloved Radha is careless of the rest.
. . . . . .
Knowing so much,
O shameless one,
How can you be cold towards her?
Whoever saw
Honey fly to the bee?

In another verse Vidyapati describes an empty house during the rainy season (7, page 100, verse 61):

Roaring the clouds break
And rain falls.
The earth becomes a sea
In a far land, my darling
Can think of nothing
But his latest love.
I do not think
That he will now return
The god of love rejects me.
A night of rain,
An empty house

And I a woman and alone
The streams grow to great rivers.
The fields lie deep in water.
Travellers cannot now reach home.
To all, the ways are barred.
May that god without a body
Strip me of my body too.
Says Vidyapati:
When he remembers, Krishna will return.

The prolonged monsoon can become tiresome for some people. Vidyapati writes about their predicament (7, page 110, verse 7):

Clouds break.
Arrows of water fall
Like the last blows
That end the world.
The night is thick
With lamp-black for the eyes.
Who keep so late a tryst?
The earth is a pool of mud
With dreaded snakes at large.
Darkness is everywhere,
Save where your feet
Flash with lightning.

But all said and done, the season of rains is one of exultation (7, page 126, verse 87):

Clouds with lightning,
Lightning with the clouds
Whisper and roar.
Branches in blossom
Shower in joy
And peacocks loudly chant
For both of you.

Another body of literature where many references to monsoons can be found are *bārahmāsā* (12 months) composed by poets of northern India. We are not sure when the tradition of composing *bārahmāsā* came into vogue but by the sixteenth century it had become well established and most poets tried their hand at describing the changing panorama of nature through the year. The Sikh's holy scripture, The Granth Sahib, has two baramahs* (Punjabi version of *bārahmāsā*) of which the one composed by the founder of the faith, Guru Nanak (1469–1539), in *Raga*

* The names of months used in these verses from *Raga Tukhari* are from the old Punjabi. They are similar to the Sanskrit (see Chapter 3) but are pronounced and spelled differently.

Tukhari (8) has some memorable depictions of the weather. Since the monsoons in the Punjab break sometime after mid-July, Nanak first describes the summer's heat in his verse on Asadh (June–July):

In Asadh the sun scorches
Skies are hot
The earth burns like an oven
Waters give up their vapours
It burns and scorches relentlessly
Thus the land fails not
To fulfil its destiny

The sun's chariot passes the mountain tops;
Long shadows stretch across the land
And the cicada calls from the glades.
The beloved seeks the cool of the evening.
If the comfort she seeks be in falsehood,
There will be sorrow in store for her.
If it be in truth,
Hers will be a life of joy everlasting.
My life and its ending depend on the will of the Lord.
To Him says Nanak, I surrendered my soul.

Asadh is followed by Sawan (July–August) when the monsoons break in northern India.

O my heart, rejoice! It's Sawan
The season of nimbus clouds and rain
My body and soul yearn for my Lord.
But my lord is gone to foreign lands.
If he return not, I shall die pining for Him.

The lightning strikes terror in my heart.
I stand all alone in my courtyard,
In solitude and in sorrow.

O mother of mine, I stand on the brink of death.
Without the Lord I have neither hunger nor sleep
I cannot suffer the clothes on my body.
Nanak says, she alone is the true wife
Who loses herself in the Lord.

Since monsoon is trysting time for lovers and thus engrossed they tend to forget their Maker, Nanak admonishes them in his verse on Bhadon (August–September):

In the month of Bhadon
I lose myself in a maze of falsehood
I waste my wanton youth

River and land are on endless expanse of water
For it is the monsoon the season of merry-making.
It rains,
The nights are dark,
What comfort is it to the wife left alone?
Frogs croak
Peacocks scream
The papeeha calls "peeoh, peeoh,"
The fangs of serpents that crawl,
The stings of mosquitoes that fly are full of venom.

The seas have burst their bounds in the ecstasy of fulfillment.
Without the Lord I alone am bereft of joy
Whither shall I go?
Says Nanak, ask the guru the way
He knoweth the path which leads to the Lord.

The poetic tradition has continued to the present time. India's only Nobel Laureate in literature, Rabindra Nath Tagore (1861–1941), has two beautiful pieces in his most celebrated work, *Gitanjali* (9, page 11, verse 18 and page 14, verse 23):

Clouds heap upon clouds and it darkens
Ah, love, why dost thou let me wait outside at the door all alone?
In the busy moments of the noontide work I am with the crowd,
But on this dark lonely day it is only for thee I hope
If thou showest me not thy face, if thou leavest me wholly aside, I know not how I
    am to pass these long, rainy hours.
I keep gazing on the far-away gloom of the sky and my heart wanders wailing with
    the restless wind.

————

Art thou abroad on this stormy night on the journey of love, my friend? The sky
    groans like one in despair.
I have no sleep tonight.
Ever and again I open my door and look out on the darkness, my friend!
I can see nothing before me,
I wonder where lies thy path!
By what dim shore of the ink-black-river, by what far edge of the frowning forest,
Through what mazy depth of gloom art thou threading the course to come to me my
    friend?

These are but a few examples from Sanskrit and languages of northern India illustrating the impact the monsoons make on sensitive minds of poets and men of letters. Similar examples are available from all other languages and dialects spoken in the rest of the country.

2 THE MONSOON DESCRIBED BY ENGLISH POETS AND NOVELISTS

Many foreign writers have given vivid descriptions of the monsoon. Of these there is a memorable one by L. H. Niblett in his *India in Fable, Verse, and Poetry* (10), published in 1938:

> The sky was grey and leaden: the Moon was dull and pale;
> Suspended high, the dust-clouds, in canopying veil,
> O'erlooked wide fields and hamlets of India's arid plains—
> Sun-baked and scorch'd and yellow—athirsting for the Rains.
> The atmosphere was stifling: the air was still as death,
> As the parched jheels emitted their foul and charnel breath.
> Storm-clouded the horizon: a flash across the sky,
> A boom of far-off thunder, and a breeze like a distant sigh:
> 'Tis the dirge of a dying summer: the music of the gods;
> Dead leaves rise up and caper: the Melantolia nods:
> Tall trees to life awaken: the top-most branches sway
> And the long grass is waving along the zephyr way.
> A mantle of red shadow envelops all around—
> The trees, the grass, the hamlets, as the storm-clouds forward bound.
> Of a sudden, comes a whirlwind, dancing, spinning rapidly;
> Then gust on gust bursts quick, incessant, mad, rushing furiously.
> A crash—and the Monsoon's on us, in torrents everywhere,
> With the bellowing roar of thunder, and lightning, flare on flare.
> The tempest's now abated; a hush falls o'er the scene;
> Then myriad birds start chatt'ring and the grass again is green,
> The fields like vast, still mirrors, in sheets of water lie,
> The frogs, in droning chorus, sing hoarse their lullaby,
> Each tankand pool is flooded, great rivers burst their banks,
> King Summer's reign is ended, the Monsoon sovereign ranks.

E. M. Forester, the celebrated author of *A Passage to India*, has an equally vivid portrayal of the rainy season in his *The Hill of Devi* (11):

The first shower was smelly and undramatic. Now there is a new India—damp and grey, and but for the unusual animals I might think myself in England. The full Monsoon broke violently and upon my undefended form. I was under a little shelter in the garden, sowing seeds in boxes with the assistance of two aged men and a little boy. I saw black clouds and felt some spots of rain. This went on for a quarter of an hour, so that I got accustomed to it, and then a wheel of water swept horizontally over the ground. The aged men clung to each other for support, I don't know what happened to the boy. I bowed this way and that as the torrent veered, wet through of course, but anxious not to be blown away like the roof of palm leaves over our head. When the storm decreased or rather became perpendicular, I set out for the Palace, large boats of mud forming on either foot. A rescue expedition, consisting of an umbrella and a servant, set out to meet me, but the umbrella blew inside out and the servant fell down.

Since then there have been some more fine storms, with lightning very ornamental and close. The birds fly about with large pieces of paper in their mouths. They are late, like everyone else, in their preparations against the rough weather, and hope to make a nest straight off, but the wind blows the paper round their heads like a shawl, and they grow alarmed and drop it. The temperature is now variable, becomes very hot between the storms, but on the whole things have improved. I feel much more alert and able to concentrate. The heat made me feel so stupid and sleepy, though I kept perfectly well.

It is strange that the monsoon did not exercise the minds of foreign writers as much as it did of the Indians. A large majority of them were birds of passage who did not stay long enough in the country to share the emotional response of the Indians. During British rule, in most government offices English officers with their families moved up to hill-stations like Simla, Mussoorie, Darjeeling, and Ooty from where they administered the country or went on holidays to Kashmir and so escaped the intense heat that enveloped the plains; consequently they were unable to sense the relief and the joy that came with the monsoon. In any event, they could not rid themselves of their inborn aversion to rain which spoiled their fun at home; for them monsoons were just a succession of rainy days.

3 THE MONSOON IN FOLK LITERATURE

A substantial portion of the folk literature of all of India's 14 languages is devoted to the monsoons. What was observed in the changes of climate, formation of clouds, flora and fauna was put in doggerel or made into proverbs. And every village has its *Sabjantawallah* (Mr. Know-it-all) who could predict when the monsoon would break and how bountiful it would be.

3.1 Portents

In every part of India peasants have their own way of predicting the monsoon. There is a general belief that the more intense the heat during April, May, and June, the heavier will be the rains that follow. In northern India some varieties of thorny bushes like the *karwand* and *heever* break into tiny leaf a month before the rains break. The *papeeha* (hawk cuckoo or the brain-fever bird) is loudest during the hot days and its cry is interpreted in Marathi as "pāos ālā, p̄aos ālā" meaning "the rains are coming." The peasants are also familiar with the monsoon bird (pied crested cuckoo, the *Clamator Jacobinus*), also known as *megha papeeha*—the song-bird of the clouds. Its natural habitat is in East Africa. Taking advantage of the monsoon winds, it flies across the Indian Ocean and the Arabian Sea to arrive on the western coast of India a day or so ahead of the rain bearing clouds. It is rightly regarded as the monsoon herald. It flies at a more leisurely pace inland and is usually sighted in Delhi about 15 days after the monsoon has broken over the western Ghats (12).

3.2 Village Soothsayers

Indians divide the few months of the summer monsoon into eight periods of 13 to 14 days each, depending on the signs of the Zodiac known as *nakṣatras*.

Of the 27 *nakṣatras* the fifteenth known as *svātī* (late October) is considered the most auspicious. According to poets, the mythical bird Chatrik drinks only of the *svātī* rain. And it is only the drops of the *svātī* rain that turn to pearls when they fall into oysters. The *svātī* falling on bamboo trees produces "vanslochan" a precious medicament of Ayurveda, the indigenous Hindu system of medicine.

All Indian languages have innumerable proverbs stressing the importance of rains in their particular regions. For instance, the test of a good monsoon in Maharashtra is when the gunny sacks peasants drape over their heads and shoulders as they go out in the rain remain damp long enough to breed insects. For the Punjab, comprehensive compilation has been made (13). They are largely variations of the single theme "if the rains are good, there will be no famine." There are also proverbs about distribution of rains during the year (13).

> Four months do not need even a rain of gold: *mārgaśīrṣa* (mid-November to mid-December) *caitra* (mid-March to mid-April), *vaiśākha* (mid-April to mid-May) and *jyestha* (mid-May to mid-June). Except for these four months, rain is desirable in all the other months of the year.

For some unknown reason, people expect the monsoons to break over Bombay by the 10th of June. The onset of winter rain is calculated as following 100 days after the end of summer monsoon.

Despite the summer rains being the real monsoon, it is the short winter rains that the Punjabi farmer prizes more. "Winter rain is gold, Hadha (June–July) rains, silver, and Sawan–Bhadon (July–September), mere copper," says a proverb. "If there is a spell of rain in mārgaśīrṣa (mid-November to mid-December) the wheat will have healthy color." There are parallel proverbs instructing farmers what to do during the winter monsoon months (13): "If you do not plough your land in Hadha you will be like a dry Sawan and a child who learnt nothing at school." "If it rains on Diwali (the festival of lamps that falls early in November) the sluggard will be as well off as a conscientious tiller except that the tiller's crop will be more abundant."

3.3 Monsoon Proverbs In Hindi

Most of these are ascribed to Ghagh (seventeenth century) a learned Brahmin poet–astrologer and his even more learned wife, Bhaddari, a low-caste girl he married because of her learning. Says Bhaddari:

> When clouds appear like partridge feathers and are spread across the sky, they will not go without shedding rain.*

* No standard reference.

A similar proverb in Punjabi says exactly the opposite. Ghagh predicts:

> When lightning flashes in the Northern sky and the wind blows from the east, bring oxen under shelter because it is sure to rain.

> When water in the pitcher does not cool, when sparrows bathe in dust and the ants take their eggs to a safer place, you can be sure of a heavy downpour.

> If the southern wind flows in the months of *māgha* and *pauṣa* (i.e., January and February), the summer monsoon is bound to be good.

> Dark clouds in the sky may thunder without shedding a drop; where white clouds may be pregnant with rain.

However, some of their proverbs seemed to have been designed to keep hope alive. "If the clouds appear on Friday and stay till Saturday," Ghagh tells Bhaddari, "be sure that it will rain."

Ghagh–Bhaddari proverbs are on the lips of peasant folk in the Hindi speaking belt stretching from Haryana and Rajasthan across Uttar Pradesh, Madhya Pradesh to eastern boundaries of Bihar. Variations of the same proverbs exist in Bengal and Maharashtra.

4 MONSOON AND THE FUTURE

Since attaining independence in 1947, India has taken enormous strides toward freeing herself from dependence on the vagaries of the monsoons. She has raised enormous dams, laid thousands of miles of irrigation canals, and dug innumerable electrically operated tube-wells to supply water to her farms. As a result, even when the monsoon has been poor, the country has been able to produce enough to feed itself. Nevertheless, even today it is not far wrong to say that "for us in India scarcity is only a missed monsoon away" (14).

However, there is no longer the same agony waiting through long summer months of searing heat to catch a glimpse of the first clouds nor the same ecstasy when they spread across the skies and shed their bounty. The monsoon no longer stirs the imagination of the poet or the novelist with the same intensity it used to. It remains the favorite time of the year for lovers but few now write about it.

REFERENCES

1. C. Carrington, *Rudyard Kipling, His Life and Work*, Macmillan, London, 1955, p. 94.
2. *Rudyard Kipling's Verse 1885–1926*, Hodder and the Staughton, London, 1930, p. 80.
3. Aristophanes, "The Frogs," in *Great Books of the Western World*, Benton, New York, 1952, p. 566.
4. J. Brough, *Poems From the Sanskrit*, Penguin Books Ltd., Harmondsworth, Middlesex, England, 1968.

5. *Bharatrihari: Poems with the Transliterated Sanskrit Text of the Satakatrayam*, translated by B. S. Miller, edited by William Theodore de Bary, Columbia University Press, New York and London, 1967.

6. Subandhu, "Vasavadatta," Sixth century A.D., in B. N. Pandey, Ed., *A Book of India—An Anthology of Prose and Poetry from the Indian Sub-Continent*, Vol. 1, Rupa, New Delhi, 1977, p. 138.

7. W. G. Archer, *Love Songs of Vidyapati*, translated from Maithili by D. Bhattacharya, Allen and Unwin, London, 1963.

8. Guru Nanak, "Ragi Tukhari," 1604, in Guru Arjun Dev, Compiler, *Adi Granth*, Shiromani Gurdwara Prabandhak Committee, Amritsar, India, 1984, pp. 1107–1110. Translated from Gurmukhi to English by K. Singh in *A History of the Sikhs*, Vol. 1, Princeton University Press, Princeton, NJ, 1963, pp. 351–357.

9. R. N. Tagore, *Gitanjali*, Macmillan, New Delhi, 1980.

10. L. H. Niblett, "India in Fable, Verse, and Story," in B. N. Pandey, Ed., *A Book of India—An Anthology of Prose and Poetry from the Indian Sub-Continent*, Vol. 1, Rupa, New Delhi, 1977, p. 138.

11. E. M. Forester, *The Hill of Devi*, Edwin Arnold and Co., London, 1953, p. 93.

12. S. Ali and S. D. Ripley, *Handbook of the Birds of India and Pakistan*, Vol. 3, Exford University Press, Delhi, 1984, pp. 194–200.

13. M. S. Randhawa, Ed., *Agricultural Proverbs of the Punjab*, Director Public Relations, Chandigarh, Punjab, India, 1962.

14. P. Goradia, *So Said Indira Gandhi*, Prafull Goradia, New Delhi, 1983, p. 37.

3

Monsoon In Traditional Culture

Francis Zimmermann
Centre National de la Recherche Scientifique
Sciences de l'Homme et de la Société
Paris, France

INTRODUCTION

The Arabic word *mausim* is used for anything that comes around but once a year, a fixed time, a season. In a derivative sense, it was the name given to the periodic winds of the Indian seas by the Arab pilots. The notion of periodic changes in the direction of the winds is not prevalent in indigenous thought; it is a notion taught by navigators from the West. According to Sanskrit literature as well as to folk knowledge, the monsoon is indeed the result of a periodic reversal, but the realization of this reversal is with the sun and rains, not the winds. In the view of priests, astrologers, and physicians—who are responsible for establishing the cycle of the festivals, the calendar of agricultural activities, the seasonal diets, and regimens—there exists a fundamental alternation between the Dry and the Wet principles. Monsoon comes after the summer solstice, when Agni (the sun) reverses its course and the earth comes under the rule of Soma (the moon), the dispenser of rain. Thus to define monsoon in traditional culture, we have to substitute rains for winds.

"Monsoon" proper was first used by sailors. The voyage to the East Indies, João de Barros wrote in 1553, "has to be made by the prevailing wind, which is called *monção*" (Portuguese form of the word corrupted from the Arabic). This is one of many quotations from the earlier travelers that are found in the *Hobson-Jobson* (1), an ethnographic glossary published in the early 1900s that yields a wealth of information on South Asian culture. The entry for monsoon, however, remains totally external to indigenous ideas. It takes into account only the seafarer's concept of periodic wind reversal and fails to mention that, nowadays in common parlance, monsoon means "the rainy season." In Chapter 7, Warren presents the ancient and medieval reports of monsoon winds and currents; in this chapter, we address the native's viewpoint.

Figure 3.1. Map of South Asia.

It would be unfair to say that an association of the monsoon with reversing winds is totally unknown to local people. For example, the Tamil word *tenṟal*, "south-wind," is used to qualify the damp wind that blows from the south on the Coromandel coast (Fig. 3.1) from July to September. It is the "longshore wind" mentioned in the *Glossary of the Madras Presidency* (2), a wealth of ethnographic information dating back to 1893.

Traditional culture and classical literature originated with rural settlements. Monsoon Asia is an essentially agricultural world; consequently in collective thought, monsoon is viewed in farmer's terms and, in that sense, is linked to the outburst of rains. Of course, the winds might be used in predicting rain. Examples are found in Chapter 4 where Murton traces monsoon concepts in agricultural proverbs. The almanacs, which are printed every year in the various vernaculars for lay audiences, refer to the strength and direction of the winds, but they are overwhelmingly concerned with astrological reflections (and thus are of little value for scientists, although they make fascinating stories for the anthropologist).

While Westerners divide the year into four seasons, the South Asian calendar in its most abstract form consists of a triad: winter, summer, monsoon. The year is also partitioned into six seasons or 12 lunar months. This division of the year into three main phases, Cold, Hot, and the Rains, is quite faithful to local realities. The rains cover a period of four months that, under the Sanskrit name *caturmāsa*,* "the Four Months," reflects the concept of monsoons. One should never forget monsoon in daily prayers: "May the rain come down in the proper time, may the earth yield plenty of corn . . . "(3). The prosperity of a kingdom depends upon the rains, that is why the king in the religious literature is called a "rain-maker" (4). The outburst of the monsoon is of the utmost importance to the man of the soil. One can start plowing as soon as the monsoon has come. It marks the beginning of the agricultural year and it is tinged with drama because of the risk either of floods or of rain failure.

This chapter will examine the role of the rains in the religious Hindu year, stressing the ritual and social significance of monsoon endings. We then proceed to the analysis of indigenous concepts related to the cause and effect of monsoons: the alternating motions of the sun and moon, and the increase and decrease of vegetable saps and bodily humors (the liquids in the body said to determine a person's complexion). The next step is to draw a picture of monsoon symbolism, referring, among other sources, to miniatures (from the Indian schools of portraiture and landscape on vellum) illustrative of seasonal moods and manners. Despite our scanty knowledge of ethnometeorology outside India, we conclude with some remarks on one of the various Hinduized cultures of Southeast Asia to put our Sanskritic bias in context.

It is appropriate to note that much of our material is drawn from Sanskrit texts. We focus on the learned tradition and the way in which the classical Hindu calendar

---

* Regarding the use of diacritics, three cases are distinguished:

1. Proper names—of Sanskrit authors, mythological characters, geographical names—are *without* diacritical marks:

   Kalidasa, Susruta instead of Kālidāsa, Suśruta;
   Vishnu, Krishna instead of Viṣṇu, Kṛṣṇa;
   the Himalayas instead of Himālayas.

2. Technical common names are *with* diacritics and printed in italics:

   varṣā (rains), *bārahmāsā* (twelve months), and so on.

3. Exceptionally, the names of the months and the names of the *nakṣatras* appear in two different forms:

   āṣāḍha, with diacritics, in italics;
   Asadha, in roman, capital A, no diacritics.

These conventions are to simplify and make explanations more readable in those paragraphs where the names of months or of asterisms are frequently repeated. They have been used also by Singh (Chapter 2) and Murton (Chapter 4).

Full diacriticals are used in the list of references, for author's names as well as for Sanskrit titles.

is implemented in the various fields of medicine, astrology, rituals, and the fine arts.

1 THE RAINS IN THE CLASSICAL HINDU CALENDAR

The monsoon raises ambivalent feelings. Preceded by a period of water shortage and physical exhaustion, its advent is a relief and is greeted with great joy, as Singh shows in Chapter 2. But the first rains are followed immediately by a period of anxiety. Hindu religion explains the tenseness of the monsoon time by teaching that god Vishnu goes to sleep during the rainy season, and the earth deprived of its Lord remains in the power of demons. *Caturmāsa*, the Four Months of the monsoon time, are expressly thought of as inauspicious. The gods are absent (holding aloof from human matters), it is the time of *pralaya*, the deluge or world destruction, an ominous time, the ambiguity of which is explained by Judy Pugh (5), an anthropologist studying the North Indian almanacs:

> The monsoon rains constitute the pivot of the agricultural year. Since this is the time for planting rice and other crops which form the major portion of the food supply, the abundance or scarcity of rain during this period determines whether or not the year's crops will be bountiful. . . . In the context of agricultural production, then, temporal and climatic auspiciousness is manifested as "wetness," which ramifies into "fertility," "succulence," and "plenty." "Dryness" during this season is "untimely" and hence connotes "sterility," "harshness," and "scarcity." In an inverted set of significances, the "wetness" of the monsoon rains signifies inauspiciousness for the celebration of marriages and other life-cycle rites, which are proscribed for the four months of the rainy season. Inauspiciousness as "wetness" is manifested as "cosmic deluge," "destruction," "divine absence," "difficulty in travel and communication," "isolation," and "danger from snakes and other creatures."

Hence the outburst of the rains raises expectations and dread simultaneously, which the people try to reconcile through an elaborate cycle of festivals and other observances.

1.1 Festivals and Other Observances

The main calendrical festivals occurring in *caturmāsa* are shown in Figure 3.2.* The Four Months of the rainy season represent one night in the life of god Vishnu.

* *A note on chronology*–The religious calendar is luni–solar. The year is divided into 12 lunar months. One lunar month corresponds to a complete series of lunar phases between two new moons or two full moons. The *amānta* system of reckoning, in which a month is bounded by two new moons, is commonly used in the south, while the *pūrṇimānta* system of reckoning, according to which a month is bounded by two full moons, prevails in North India. The bright fortnight is the 15-day period of the moon's waxing between the new and full moons. The dark fortnight is the period of the moon's waning, between the full moon and the following new moon. The lunar months and the solar months bear the same names. The lunar year is shorter than the solar year by about 10 days. The gap is compensated for by adding one intercalary lunar month every three years. This extra month has no festivals. Interested readers should refer to Merrey (6) for further details.

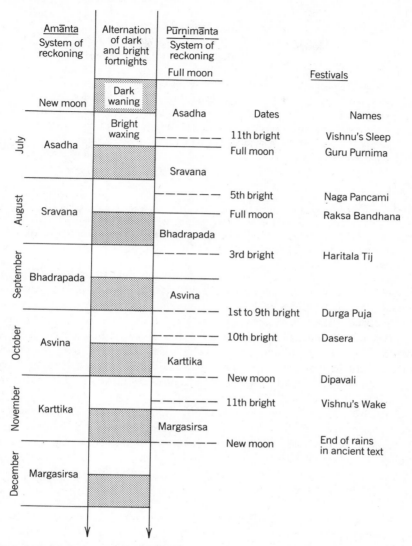

Figure 3.2. Calendar of the monsoon rituals. The names of the months are Sanskrit. The *amānta* system of reckoning, in which a month is bounded by two new moons, is commonly used in the south, while the *pūrṇimānta* system of reckoning, according to which a month is bounded by two full moons, prevails in North India. This calendar gives the position of the main festivals occurring in the Four Months. Approximate equivalents in the Western calendar are indicated on the left.

On the eleventh day of the bright fortnight in the month of *āṣāḍha*, Vishnu is believed to retire below the ocean for his four months' sleep, and on the eleventh day of the bright fortnight of *kārttika* to awake and return. This period covers the last few days of the month of *āṣāḍha*, all of *śrāvaṇa*, *bhādrapada*, and *āśvina*, and almost all of *kārttika*. (These are the Sanskrit names of the months, the full list of which is given in Table 3.5 and Figure 3.6). The Hindi names, the most common

TABLE 3.1. Calendrical Festivals during the Monsoon[a]

Hariśayanī	Vishnu's Sleep	Early July
Gurupūrṇimā	Gurus' Worship	Mid-July
Nāgapañcamī	Serpent Deities' Worship	Early August
Rakṣābandhana	Tying the Amulet	Mid-August
Haritālā tīj	Goddess Parvati's Festival	Early September
Durgāpūjā	Goddess Durga's Nine Days	Early October
Daserā	Warriors' Day	Early October
Dīpāvali	Row of Lights	End of October
Haribodhinī	Vishnu's Wake	Early November

[a] All names are in Sanskrit, except Tij and Dasera (Hindi). It is not possible here to account for regional variations (despite their significance to the anthropologist) such as the duality of Tij in some provinces; the first Tij (Women's Swinging Festival), more exuberant, being followed one month later by the more subdued Haritala Tij.

in India today, are derivatives easily inferred from the Sanskrit originals. The Tamil calendar used in the next chapter (see Table 4.1) follows a different system of reckoning, but the first day of the Tamil month of *āṭi* (July–August) corresponds to the eleventh bright of *āṣādha*, so that the four Tamil months of *āṭi*, *āvaṇi*, *puraṭṭāci*, and *aippaci* correspond to the Sanskrit *caturmāsa*. The Four Months cover most of the agricultural activities which yield the *kharīf* (i.e., the autumn) crop. This crop depends upon the monsoon rains. The Four Months also encompass some of the major religious festivals. The names, definitions, and dates of these festivals are given in Table 3.1.

All rituals are not festive occasions. For example, since Vishnu, who is the special protector of the newly married, retires on the eleventh bright of Asadha, and the shortest wedding celebrations last two days, the latest date on which a wedding ceremony may begin is the ninth day of the bright fortnight of Asadha. After that it would be necessary to postpone the wedding for four months.

When Vishnu goes to sleep:

It is the ritual opening of the monsoon season, a "ticklish" season ritually, for a man's sins may spoil the monsoon crops for others as well as for himself. This is the day on which vows are taken for the term of the monsoon as to what books people will read, how long they will fast, when they will observe silence, whether or not they will eat with the left hand, and so forth. (3)

Conflicting tendencies underlie the tense atmosphere of the period. It is time for us to retreat and take shelter—but it is also a time for renewals and revivals. Certain religious characters, gurus and nagas for instance, clearly show this ambivalence.

1.1.1 The Gurus.

On the full-moon night of Asadha every one should worship the spiritual preceptors (the *gurus*), though nowadays this is not always done. However, laymen and disciples do very often worship Sankara and Vyasa (paragons of wisdom). . . . On this night ascetics of all sects go to the temple of Vishnu in whatever town they happen to be . . . and there take their special vow of not travelling during the monsoon. They . . . are not allowed to travel during this season lest they should commit the sin of killing by inadvertently trampling under foot some of the young life then so abundantly springing into being. (3)

Not only do the flooded rivers of the monsoon months make travel difficult, but the difficulty is ritualized so that the monsoon becomes a season of retreat and religious observances. The temporary settlement of itinerant *sannyāsis* (world-renouncers) is a major cultural feature. The rains bring an afflux of *sādhus* (holy men, ascetics) to pilgrimage cities like Banaras:

> The presence of so many ascetics and renouncers in the city during these months is important to devout Hindu householders. Some of these renouncers will be dependent upon Banaras householders for alms, an act of generosity that benefits the householders as well. Formerly, this was a time when the renouncers, many of whom were learned, would settle among the people for long enough to teach them. In Banaras today something of this tradition lingers on. While the rains pour down in the early evening, certain temples and monasteries will be crowded with eager listeners, both lay people and ascetics, and one of the learned ones will speak. (7)

1.1.2 The Nagas. Deities appearing in the form of snakes, the *nāgas*, are celebrated on the fifth day of the bright fortnight of Sravana:

> These *nāgas* have always been propitiated with special care during the rains, for in this season the floods often force them from their usual habitations in the earth, and they suddenly appear in people's gardens, courtyards, and houses. The *nāgas* are both loved and feared, and on *nāgapañcamī* (the name of their festival) their images are painted on either side of the doorways of houses, and they are propitiated there with offerings of milk and puffed rice.(7)

A symbolical connection is postulated between the serpents, the idea of fertility, and the rains. The snake-deities are propitiated by barren women in want of a son, and fertility associated with the monsoon materializes not only in rice but in sons.

1.1.3 Other Observances. The remaining festivals of the period, *rakṣābandhana*, *tīj*, *durgāpūjā*, and *dīpāvali*, stress the cataclysmic character of the monsoon. They symbolize radical disorder, followed by drastic reordering. Marc Gaborieau (8) has argued that, on a cosmic scale, monsoon is conceived of as a disintegration of the world. Demons are overwhelmingly present in *durgāpūjā* and *dīpāvali* rituals, while the gods have accompanied Vishnu in his retirement. The borders of the living and the dead vanish. The dead are thought to be very close by during the months of Asvina and Karttika; the dark fortnight of Asvina, named Fathers' Fortnight, is devoted to them. *Dīpāvali* also is partly dedicated to the dead and their god Yama. Disorder affects human society itself. In the month of Sravana, low castes become arrogant and ask their high-caste employers for extra wages; in the next month with the beginning of *tīj*, women contest their husbands' and mothers-in-law's authority.

On the other hand, each of these festivals ultimately contributes toward restoring order. At a cosmic level, the gods are gradually called back, until Vishnu during *dīpāvali* eventually defeats the demon Bali, thus putting an end to the demons' rule. The human society also is reconstructed. At *rakṣābandhana*, the twice-born (high-caste Hindus) reassert their superiority over their low-caste employees; *tīj* ends with the women resuming their job of submissive wives.

Not only is *dīpāvali* often regarded as a New Year festival, but particulars of *durgāpūjā* and *daserā* suggest that the end of the monsoon marks the time of new endeavors. For example, *sarasvatīpūjā* (celebration of Sarasvati, the goddess of eloquence and learning) on the ninth day of *durgāpūjā* is traditionally the date when the young brahmins join their guru's house for a new course of studies; from *daserā* onwards, armies can embark on new campaigns, and kings set out for a tour of their territories. Finally the monsoon season, which is widely accepted as June–September on meteorological grounds, has been symbolically protracted in the Sanskrit calendar, to encompass the end of the autumn harvest in October. Thus defined, the rainy season represents a ritual conclusion, a ritual break in the annual cycle, the time of cosmic reversals and commencements afresh.

1.2 The Precession of the Equinoxes

The precession of the equinoxes accounts for a number of discrepancies between the traditionally fixed dates of festivals and the actual timing of monsoons. For example, today the monsoon season ends in October and *dīpāvali* is the festival that marks the end of the monsoon on the new moon of Karttika (end of October). But in ancient times the rainy season ended one month later, accompanied with some other festivals that have now fallen into oblivion. Among various literary sources, the *Bṛhatsaṃhitā* (9), a treatise of astrology, fixes the date for the end of the rainy season on the new moon of Margasirsa (end of November). The calendar set forth in this astrological text of the sixth century A.D. is illustrated in Figure 3.3. According to F. B. J. Kuiper (10), the fact that the rains now end about one month earlier than in the *Mahābhārata* times explains why Karttika was set as the specific time for playing dice in the *Mahābhārata* (a Sanskrit epic dating back to the last few centuries B.C.). Since dice, a demoniac game in which the heroes lose their kingdoms, symbolizes a period of cosmic chaos, Karttika had to be part of the rainy season.

Thus in dealing with classical texts, we have to allow for a time-lag of about one month due to the precession of the equinoxes. The seasons' dates have changed 28 days in 2000 years. Ancient dates for the summer solstice (the outburst of the rains) and for the end of the rainy season four months later can be determined by taking this into account: the summer solstice is on June 21 (today) versus July 19 (2000 years ago), and the end of the rainy season is in October (today) compared to November (2000 years ago). No wonder the very first theatrical performance connected with Indra's banner festival, celebrating god Indra's victory over the demons and symbolizing the re-enactment of the world creation, took place in the month of Margasirsa; it was the end of the rainy season.

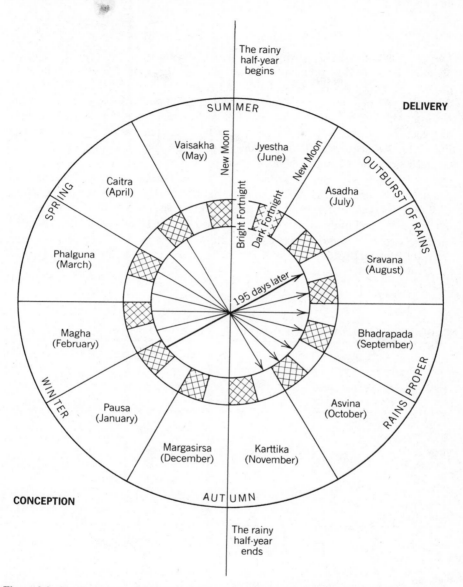

Figure 3.3. The pregnancy of clouds through the cycle of 12 months. From *Bṛhatsaṃhitā* (9), Chapter XXI, verses 9–12. The 12 months are reckoned from new moon to new moon (*amānta* system). The rains pour down during the rainy half-year—*Delivery*, but these rainfalls originated 195 days earlier in the dry half-year—*Conception*.

1.3 The North Indian Context

The geographical setting to which the Hindu calendar best applies is the northern plains of the Indian subcontinent, the Indus and Ganges basins. This is the traditional place of reference for Hindu medicine and astrology. The Delhi Doab constitutes the native soil of Brahminic culture and serves as the point of reference in Sanskrit writing. Although the timings and features of the monsoon differ as one goes from the west coast to the Bay of Bengal, and still more from Assam to Sri Lanka (see Fig. 3.1), the sequence of the seasons remains roughly the same and the ritual calendar is applicable throughout.

Kerala (the southwest coast) more or less fits the northern pattern. Tamilnad (the southeast), by contrast, is said to be deprived of a real monsoon and the sequence of Tamil festivals differs from the Sanskrit one. The *poṅkal* festival in January celebrates the end of the rainy season. However, in the Tamil coastal plains, which produce two crops of rice a year, the paddy-growing cycle is not very different from the classical pattern: summer rice is harvested in September and winter rice

TABLE 3.2. The Cycle of the Seasons in the Dry Zone of Sri Lanka[a]

	Seasons	Climate	Paddy-Growing Cycle	Rites
April	Season between years	Rain	Plowing	New Year
May			Sowing (mid-May)	
June	HOT (*uṣṇa*)			
		Hot		
July				
		Dry		
August				
September			Harvest (September) Plowing	New Rice Festival
October	COOL (*samaśīta*)	Heavy Rains	Sowing (mid-October)	
November				
December				
January		Dry		
February	COLD (*śīta*)			
March			Harvest (March)	

[a] From Leach (11).

in February. This compares with what has been observed by Leach (11) in Sri Lanka. In both cases, most of the annual rainfall is concentrated into two periods: April–May (Sri Lanka) or June–July (Tamilnad), and September–December. The three traditional seasons are here, but their sequence and timings are quite different. The calendar observed in Sri Lanka is outlined in Table 3.2. Significantly, Sanskrit names are given to the seasons. *Uṣṇa*, the hot, the dry season, covers the period of dry season paddy-growing, and harvest. *Samaśīta*, the cool season, is the period of preparing the land and sowing the seed for the monsoon crop. *Śīta*, the cold season, extends over the monsoon paddy-growing period up to the harvest. Similarities to the classical calendar can be recognized: plowing when the rains start and New Year's Day when the rains stop. Although the climate is different, obviously the basic cultural framework remains the same.

2 INDIGENOUS CONCEPTS OF MONSOONS AND MEDICINE

The macrobiotic medicine of India includes a sophisticated doctrine of the relationships between mankind and the environment based on the seasonal pulsation of the sun's track across the sky. Traditionally there is an alternation between *dakṣiṇāyana* (the sun's movement southward) and *uttarāyana* (its movement northward). Consequently, all living things including the vegetable kingdom are affected by the alternative increase and decrease of saps and vital liquids.

2.1 Release and Capture of Saps

Through food, habitat, and physical activities, a person is influenced, penetrated, immersed in the system of humors, flavors and qualities that make up the atmosphere, the climate, and the landscape. *Rasa*, "sap, flavor, chyle, unctuousness," manifests itself as a juice formed in the living body from all the substances assimilated by digestion. It is present in food, drugs, and plants. The sun captures the *rasa* and the moon exudes or frees the *rasa*. The year is divided into a period of *Release* (rains, autumn, winter), when the moon frees all saps, and a period of *Capture* (frosts, spring, summer), when the sun takes all saps back. The first period corresponds to the sun's movement toward the south; it is the *saumya* half-year (ruled by Soma, the moon), which is wet and cold. The second period is marked by the sun's movement northward; it is the *āgneya* half-year (ruled by Agni, the sun), which is hot and dry. In the period when the sun and the wind dominate and together destroy all the unctuousness and softness of the world, human beings lose their strength. This progressive weakening is most extreme during the summer solstice, which also marks the beginning of monsoon time.

2.2 Two Seasonal Cycles

The traditional annual cycle consists of six seasons. Two are needed to cover the Four Months of the monsoon period, namely, the rains and autumn in the more commonly accepted form of the cycle.

However, in the medical treatises we find at least two different forms of the annual cycle, both of which are shown in Figure 3.4. The second form splits the monsoon season into two parts: *prāvṛṣ*, the outburst of the rains, and *varṣā*, the rains proper. Cakrapanidatta (12), a Sanskrit commentator of the eleventh century A.D., offers an explanation for the existence of the two forms:

> In the southern part of the Ganges basin, there is so much of rains (*varṣā*) that they extend farther than the season of their outburst (*prāvṛṣ*); north of the Ganges, there is so much of cold that it extends over the two seasons of winter and frost. Says Kasyapa: "It rains so much, south of the Ganges, that scholars have divided the rainy season into two—*prāvṛṣ* and *varṣā*—, and in the northern part of the Ganges basin, drenched with water coming from the Himalayas (the abode of frosts), there is so much of cold that they have made the cold season twofold—winter and frost."

These alternative arrangements might be useful to the meteorologist who is attempting to identify the connections between what is observed and what is said in the Sanskrit texts. Anyone who has experienced the Indian climate knows that spring and autumn appear as fleeting, almost imperceptible, periods. There are only three true seasons—winter, summer, and the monsoon—which are referred to as *extreme* seasons in the texts. Spring and autumn, the *moderate* seasons, represent

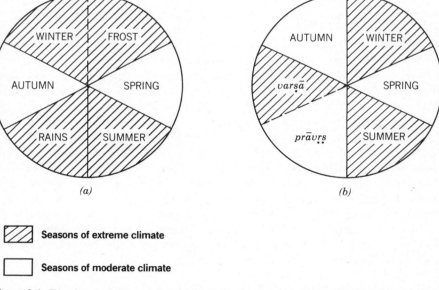

(a)   *(b)*

▨ **Seasons of extreme climate**

☐ **Seasons of moderate climate**

Figure 3.4. Two forms of the annual cycle. Form (*b*) of the annual cycle splits the monsoon season into two: the outburst of the rains (*prāvṛṣ*), and the rains proper (*varṣā*). There are only three true seasons in South Asia—winter, summer, and the rains—referred to as extreme seasons in the texts. Whether the transitional or moderate seasons are two or three in number is questionable. Physiological consequences, and the corresponding Gregorian months for forms (*a*) and (*b*), are indicated in Figures 3.5 and 3.6, respectively.

transitional and fleeting periods between the three strongly defined periods of rigorous cold (in the northern Indian winter), torrid heat, and the monsoon deluges. Some question whether the moderate seasons are two or three in number. No moderate season appears between summer and the monsoon; actually, a clear-cut break is observed that is represented by the summer solstice in the first form of the annual cycle (Fig. 3.4a). However, when *prāvṛṣ* is inserted between summer and the rains proper (Fig. 3.4b), this two-month period (June–August) is referred to as a *moderate* season enjoying mild weather conditions. This seems to be an artificial rearrangement for the sake of symmetry: three hard seasons (winter, summer, monsoon) alternating with three mild ones (spring, *prāvṛṣ*, autumn).

In addition to the cosmic alternation of Capture (by the sun in summer) and Release (by the moon in the rains) of all saps or vital liquids, another mechanism exists that is the succession of the three humors—Wind, Bile, and Phlegm—over the annual cycle. Humors represent organic fluids and pathogenic entities at the same time. Each season, according to its dominant flavor, provokes, in turn, the accumulation, disorders, and pacification of one of the humors. For example, the humor Wind accumulates in summer due to the acrid (Fig. 3.5) or astringent (Fig. 3.6) flavor of the season. Then it is vitiated in the rains (Fig. 3.5) or in *prāvṛṣ* (Fig. 3.6), which brings on fits of rheumatism, and so forth. It is eventually pacified in the autumn (Fig. 3.5) or in the rains proper (Fig. 3.6).

The first form of the seasonal cycle, the relevant portion of which is presented in Figure 3.5, stresses the importance of the summer solstice as a clear-cut dividing line. Before the solstice, dryness predominates and the humor Wind accumulates, leading to rheumatic complaints, paralysis, and nervous diseases. After the solstice, water which is now superabundant turns sour and, consequently, Bile accumulates, which causes fevers. One should also note that the Wind disorders during the monsoon are the delayed result of astringency and acridity suffered in summer. Between the lower part (the cosmic mechanism) and the upper part (the physiological consequences) of Figure 3.5, there is a time lag. Thus weakness, dryness, and Wind disorders (all the same syndrome) in the monsoon time represent the delayed effect of the sun's rule in the preceding seasons. Similarly, the sourness of the rains produces its effect in autumn and later.

Both forms of the annual cycle establish a general framework followed in formulating regimens and treatments, though it remains somewhat artificial and rudimentary. Both arrangements, for example, fail to indicate that Phlegm also accumulates in the rainy season, or that Wind is best pacified by the sweet flavor of the winter season. Of the two, the second form presented in Figure 3.6 is better related to therapeutics, because the three humors are evenly distributed over the six seasons, so that accumulation and pacification, which take place in the three extreme seasons and call for home remedies, alternate with disorders, which take place in the three moderate seasons and call for hospitalization.

2.3 Seasonal Rhythms in Regimen and Therapeutics

The physiological characteristics of each season and the appropriate regimen and diet are described at length in the medical treatises. For example, Caraka (13),

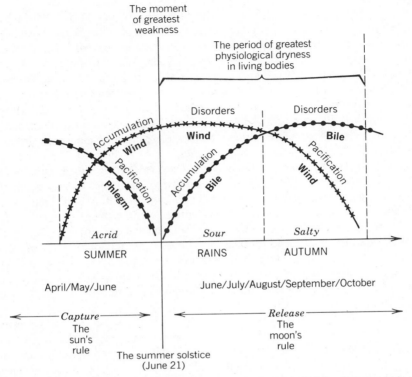

Figure 3.5. Multiple determination of the rainy season. The lower part of the chart summarizes the cosmic alternation of *Capture* (by the sun) and *Release* (by the moon) of all saps or *rasas*. The upper part indicates its physiological effects, namely the predominance of a particular flavor in each season, and the sequence of the three phases—accumulation, disorders, and pacification—of a given humor, which is distributed over three different seasons. The rainy season is characterized by simultaneous disorders of Wind and accumulation of Bile.

Sūtra VI, 33–42 in a medical compendium of the beginning of our era, gives the prescriptions appropriate for the rainy season:

> The sun by capturing the saps has weakened the body, and in the monsoon time the digestive power is low. It is further hindered by the vitiated humors like Wind, for, while the digestive fire dies down, under the effect of moisture oozing from the earth, of precipitations from the sky, and of the tendency of water to turn sour during the monsoon time, the humors, first of all the Wind, get provoked. Accordingly, the general rule that is laid down for the rainy season is moderation. One should avoid gruel diluted to excess, day sleep, chill, river water, physical exercise, sunshine, and sexual intercourse. One should generally use honey in preparing foods and drinks. On very cold days marked by stormy winds and rain, one should, even in the rainy season, take unctuous articles with pronounced sour and salty flavors, to pacify Wind. To maintain the digestive fire, the food should consist of old barley, old wheat and winter rice, pulse soup, and meat from dry land game; drinks should consist of a small quantity of wine or liquor, or of rainwater,

or of water from a well or a tank, if need be, but duly boiled and cooled, and added with honey. Oil massages, dry chafings, baths, perfumes, and garlands are advisable; one should wear light and clean clothes, and reside in a house designated for the monsoon time and free from damp.

Most of these prescriptions are meant to alleviate Wind disorders. However, monsoon is also a time for Bile accumulation. The vitiation of Phlegm is there as well, since moisture and dampness are akin to Phlegm. "The moisture oozing from the earth vitiates all the three humors," a commentator explains in Sanskrit (13), "rainwater provokes Phlegm as well as Wind, and the tendency of water to turn sour provokes Bile and Phlegm." That is why "honey is recommended in small quantity; even if by nature it tends to aggravate Wind disorders; it is meant here to overcome the dampness of the rainy season," and to compensate for the sourness of drinking water. It should be clear that flavors like sourness are not only sensory qualities but abstract entities. They signify more than the trivial fact that "this water tastes sour"; they connote instead a pathogenic property related to the humors.

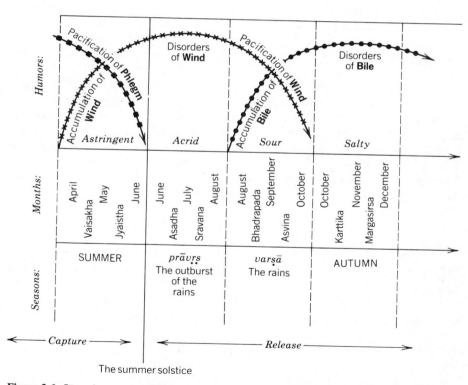

Figure 3.6. The rainy season split into two. This new scheme displays a symmetrical arrangement in the physiology of the three bodily humors, Wind, Bile, and Phlegm. Accumulation and pacification of the humors, which take place in the extreme seasons, alternate with disorders, which take place in the moderate seasons.

Thus sourness will provoke Bile and Phlegm; and Wind disorders are the delayed result of astringency and acridity.

The succession of the flavors through the seasonal cycle induces physicians to formulate particular cycles of diets and other observances. Let us consider, for example, the prescription of unctuous substances like sesame oil and ghee (clarified butter from milk of the buffalo or cow) which are used as vehicles, in the pharmaceutical sense of the word, and are impregnated with saps from the medicinal plants with which they have been boiled. They are administered internally as potions and enemas, and externally through oil baths or frictions. "It is advisable to administer unctuous substances only in the moderate seasons," says Vahata (14), *Sūtra* XVI, 11, "sesame oil in *prāvṛṣ*, ghee in autumn, grease and marrow in spring." This is corroborated by ethnography. For example, the physicians of Kerala, the southwestern province of the subcontinent, where the medicated oils and ghees indicated in the ancient texts are still in vogue today, consider that June–July (i.e., *prāvṛṣ*) and October–November (i.e., the autumn), both of which in Kerala enjoy moderate climatic conditions, are the best months of the year for the prescription of unctuous substances and purifying treatments. These two seasons are when the fashionable nursing homes accommodate a maximum of wealthy guests attracted by the prospective beneficial effects of purifying and rejuvenating therapies.

The alternation of extreme and moderate seasons determines the choice between the two great categories of medical therapy: *śodhana*, the purifying therapy, and *śamana*, the pacifying therapy. This opposition is fairly congruent with that between clinical and pharmaceutical medicine. One case requires hospitalization and the application of emetics, purgatives, and other evacuants. The other case requires a simple prescription of medicines to be taken orally or externally, but always at home. The choice between *śodhana* (hospitalization) and *śamana* (home remedies) according to the season and to the tendencies of the humors is outlined on Table 3.3. In principle, the great cures of Hindu medicine come under the category of *śodhana*, and require hospitalization and the use of complex procedures. They can be undertaken only during the moderate seasons when it is neither too hot nor too cold, too dry nor too wet.

2.4 Qualities of the Winds

Do the classical texts provide practical information about the monsoon? Do they give a faithful account of nature as she is observed? What is said of the prevailing winds in a given season might provide a clue. The few brief remarks we have traced in Sanskrit sources are put together in Table 3.4. In reading these remarks, remember that all observations conform to the religious world view and reflect local conditions in the Delhi Doab (see Fig. 3.1). In this context, not only does the reversal from northerly winds in winter (January) to westerly winds in *prāvṛṣ* (July) undeniably correspond to meteorological realities, but so do the observations of the less persistent southerly winds in spring (March) and easterly winds in autumn (November).

There is symbolism superimposed on these observations. For example, northerlies and southerlies blow from the Himalayas and the Vindhyas, respectively, and these winds are viewed as healthy and fortifying. This is a classical theme which reoccurs

TABLE 3.3. The Alternation of Extreme and Moderate Seasons and Its Medical Consequences

	Summer	*prāvṛṣ*, Outburst of Rains	*varṣā*, Rains Proper	Autumn
Climatic conditions	Extreme	Moderate	Extreme	Moderate
Mode of treatment advisable	Pacifying (*śamana*)	Purifying (*śodhana*)	Pacifying (*śamana*)	Purifying (*śodhana*)
Tendencies of the humors	Pacification of Phlegm Accumulation of Wind	Disorders of Wind	Pacification of Wind Accumulation of Bile	Disorders of Bile
The best pharmaceutical vehicle	(not specified)	Sesame oil	(not specified)	Ghee

in folklore. (In Chapter 4, Murton introduces proverbs about the excellence of northerly and southerly winds.) Specific medical qualities are ascribed to the various quarters of the wind-rose in a passage from Susruta (15), *Sūtra* XX, 23–29 which we may summarize in the following way:

The easterly wind is sweet, unctuous, salty, and heavy. It provokes burning sensation and aggravates *raktapitta* (Bile disorders with hemorrhages), poi-

TABLE 3.4. Reversals in the Directions of the Prevailing Winds, as Noted in the Classics of Hindu Medicine

	Northerly winds prevail in winter (January)	
Westerly winds in *prāvṛṣ* (July) arouse Wind	DELHI DOAB (point of reference)	Easterly winds in autumn (November) arouse Bile
	Southerly winds prevail in spring (March)	

References: Northerlies in winter: Suśruta (15), *Sūtra* VI, 22. Southerlies in spring: Suśruta (15), *Sūtra* VI, 28; Vāhaṭa (14), *Sūtra* III, 23. Westerlies in *prāvṛṣ*: Suśruta (15), *Sūtra* VI, 31. Easterlies in autumn: Caraka (13), *Sūtra* VI, 45 and *Cikitsā* XIV, 19.

sonings, ulcers, and diseases from Phlegm. It is beneficial to the consumptive, and in other diseases from Wind.

The southerly wind is sweet but light, with an astringent aftertaste, and does not give any burning sensation. It is strengthening, good for the eyes, and pacifies *raktapitta* without provoking Wind.

The westerly wind is clear, dry, and piercing. It absorbs all the unctuousness and strength, and by desiccating the Phlegm, it produces consumption and other diseases from Wind.

The northerly wind is sweet, unctuous, and cold, with an astringent aftertaste. It is harmless to all humors. Strengthening, it increases the bodily secretions and cures consumption, cachexia, and poisonings.

If we gather up the threads of our analysis, they display a certain knowledge of the alternation of westerly winds in July and easterly winds in November. In other words, the reversal of the prevailing winds is known to the classic texts. The main contrast is between westerlies from the Indus delta and easterlies from the Bay of Bengal, which corresponds to an antithesis between dry lands (Punjab) and marshy lands (Bengal) in traditional ecology. For humors and diseases, the dry lands of the west suffer from Wind disorders, while the marshy lands of the east suffer from Bile and Phlegm disorders. Comparing the humoral quality of the seasons with that of their prevailing winds:

—*prāvṛṣ*, westerlies, dry lands, disorders of Wind,
—autumn, easterlies, wet lands, disorders of Bile and Phlegm,

we see a somewhat coherent picture emerging of the climatic and physiological characteristics of the monsoon.

3 ASTROLOGICAL AND POETICAL SYMBOLISM

Ours is an easy task in this section on symbolism, since the astrological lore related to the monsoon will be touched on again in Chapter 4, and literary descriptions have been presented more thoroughly in Chapter 2. Here we address the system of signs, of manners, of moods suited to the rainy season, and more specifically, of those that are codified in the learned Hindu tradition. First we shall consider the rules of prognostication formulated in classic texts. Then citing examples from literature and paintings, we shall examine the symbolic connections between animals and birds, and plants and landscapes that characterize the monsoon time. From the far-fetched constructions of astrology to the natural imagery of poems and paintings, we shall find that the same set of symbols relates meteorological realities to human activities.

3.1 The Lunar Mansions or Constellations

The native calendar ties the seasons and climates to one of the most important series of astrological symbols, the 27 *nakṣatras*. The *nakṣatras*—constellations or asterisms—correspond to the lunar mansions, and to fixed periods of the sun's stations on the lunar zodiac. The list of the *nakṣatras* with the dates of the sun's entrance into each of them is given in Table 3.5. The *nakṣatras* define the lunar zodiac, that is, the cycle of the moon's stations. The positions of the sun, in its course over the ecliptic, are also defined by reference to these asterisms resulting in the 27 divisions of the year.

It is this system of reckoning that prevails in proverbs and in the ritual forecasting of rains. The system functions at two levels of language, in the vernacular as well

TABLE 3.5. Asterisms of the Lunar Zodiac and the Corresponding Months (all names in Sanskrit)

Sanskrit Names of the *nakṣatras*	Date of the Sun's Entrance into each *Nakṣatra*	Corresponding Luni–Solar Months in *Pūrṇimānta* Reckoning
1. *Aśvinī*	April 12	*Caitra*
2. *Bharaṇī*	April 26	
3. *Kṛttikā*	May 10	*Vaiśākha*
4. *Rohiṇī*	May 24	
5. *Mṛgaśiras*	June 7	*Jyeṣṭha*
6. *Ārdrā*	June 21	
7. *Punarvasū*	July 5	
8. *Puṣyā*	July 19	*Āṣāḍha*
9. *Āśleṣā*	August 2	
10. *Maghā*	August 16	*Śrāvaṇa*
11. *Pūrvaphalguni*	August 30	
12. *Uttaraphalgunī*	September 12	*Bhādrapada*
13. *Hastā*	September 26	
14. *Citrā*	October 10	*Āśvina*
15. *Svātī*	October 23	
16. *Viśākhā*	November 5	
17. *Anurādhā*	November 19	*Kārttika*
18. *Jyeṣṭhā*	December 2	
19. *Mūlā*	December 15	*Mārgaśīrṣa*
20. *Pūrvāṣāḍhā*	December 27	
21. *Uttarāṣāḍhā*	January 10	*Pauṣa*
22. *Śravaṇā*	January 23	
23. *Dhaniṣṭhā*	February 4	*Māgha*
24. *Śatabhiṣaj*	February 18	
25. *Pūrvabhadrapadā*	March 3	
26. *Uttarabhadrapadā*	March 17	*Phālguna*
27. *Revatī*	March 30	*Caitra*

as in Sanskrit, in popular sayings as well as in learned treatises. This is demonstrated in a comparison of two articles in the *Glossary of the Madras Presidency* (2). One article is devoted to Caurtey, Tamil *kārtti* (see Chapter 4) for Sanskrit *nakṣatra*, which includes a number of agricultural proverbs from South India and represents the more popular level. In contrast, the article devoted to Mazhay, Tamil *malai*, rain, cloud, in which the *Bṛhatsaṃhitā*'s (9) chapters on the pregnancy of clouds and on the connection between rain and the celestial phenomena are printed in full, constitutes a very sophisticated treatment.

Let us consider a few popular sayings from the Madras Glossary about *nakṣatras* which are not considered in Chapter 4. The farmers use these proverbs to forecast the weather and to adjust their cultivation practices accordingly. In Krttika (May 10 to 23), the hottest part of the year, any rains that fall are the showers in advance of the monsoon, and one can start plowing. If there is no rain in Mrgasiras (June 7 to 20), a drought is expected for the next five asterisms, until the middle of August. Conversely, with the first showers in this asterism the sowing of paddy is begun. The most important of the *nakṣatras* is Ardra (June 21 to July 4) because it marks when the monsoon usually begins in full force in North India. This critical asterism is inauspicious for either sowing or transplanting. It is only during Punarvasu (July 5 to 18) that a second sowing of paddy is tried, if the first has failed. The sequence of (a) Mrgasiras, (b) Ardra, and (c) Punarvasu conveys the idea of a dialectic, which translates the weather uncertainty into alternating/alternative prescriptions: sowing is in turn (a) possible, (b) impossible, and (c) then again possible. Proverbs also are formulated as so many alternatives: "If there be thunder in Ardra, the rain will fail for sixty days; but if it rains in Ardra, it will continue for the next six asterisms. . . . " The figures dogmatically set forth ("for sixty days") sum up an immemorial tradition of meteorological observations.

The classical astrological texts establish a scheme for rain forecasting in which the *nakṣatras* do not represent the fixed periods of the sun's stations on the lunar zodiac (as in the proverbs of popular astrology), but the lunar mansions* proper. Rainfall is predicted more than six months in advance by recording the symptoms of the pregnancy of clouds under a particular station of the moon. In principle, the rains pour down during the half-year from Jyestha through Karttika, but these rains originated 195 days before, since the pregnancy of clouds extends over seven sidereal revolutions of the moon (i.e., 195 days). Predictions made in the conception period (December–May) apply to the delivery period (June–November). Figure 3.3 outlines the forecasting system laid down in the *Bṛhatsaṃhitā* (9), Chapter XXI, verses 6–7:

> The symptoms of pregnancy must be detected in the period starting from the day when the moon reaches the asterism Purvasadha during the bright fortnight of the month of Margasirsa. The foetus formed during the moon's stay in a particular asterism will deliver rain 195 days after, when the moon passes through the same asterism according to the laws of its revolution.

* The asterisms through which the moon passes; a different one for each day of the cycle of the moon.

For example, in a given year the moon reaches Purvasadha on the fourth day of the bright fortnight of Margasirsa (i.e., November 26). The symptoms of cloud pregnancy detected on that day are used for predicting rainfall on the third day of the dark fortnight of next Jyestha (June 9), the day when the moon will reach Purvasadha again after seven sidereal revolutions. Now suppose that the symptoms of pregnancy on November 26 were very bad, and those on December 6 (when the moon passes through Krttika) extremely promising, that is, one observes a delightful and cool breeze from the north, the sun and the moon encircled by a glossy halo, a red glow on the horizon at dawn, and so on. One would predict that there is no prospect of rain on June 9, but all signs point to the outburst of the monsoon on June 19 (the moon passing through Krttika again).

At the beginning of the rainy half of the year, short-term predictions are based on the same scheme. The prospect of rainfall in the rainy half-year is determined through the rain that fell in the first few days subsequent to the day when the moon reached the asterism Purvasadha during the dark fortnight of the month of Jyestha. For example, the moon passed through Purvasadha on June 9, but the rains started pouring down on June 19, when the moon reached Krttika. The amount of rain gauged on June 19 and the next few days would enable astrologers to predict the quantity of rainfall for the entire rainy season, and whether the prospects for the autumn harvest are good or poor. The names of the first rain-producing lunar mansions—Krttika (June 19), Rohini (June 20), Mrgasiras (June 21), and so on — have also great predictive value, since the stars that preside over the initial rains are deemed to remain the sources of all rainfall throughout the delivery period (9), Chapter XXIII, verse 5:

> In whichever stars there is rain at the beginning, there will generally be rain once again under the same stars in the delivery season.

Lunar mansions indicate not only the timings but also the quantities of rainfall. For example, Krttika is considered among the less productive stars, so where Krttika (in an arbitrarily given year) plays the role of the first rain-producing lunar mansion, it foretells rains not so abundant that year. A more detailed knowledge of the stars would result in more precise forecasts.

Astrological predictions of rain require a very large amount of information on the individual characters of the stars, and on the shapes and colors of the clouds, the directions of the winds, and the moods of the inhabitants —humans, beasts and birds, herbs and trees— in a given setting, for detecting the symptoms of cloud pregnancy.

3.2 Poems and Miniatures

Besides the medical and astrological fields, two traditional disciplines also use the cycle of the seasons as a symbolical frame for expressing beliefs and emotions. These are the science of love (*kāma*), and the science of poetic moods (*rasa*) and musical modes (*rāga*) which are used in Hindu music, painting, and poetry. The

Hindu treatment of musical *rāgas* and miniature paintings is of a literary kind; they are meant to suit and illustrate the mood of a poem in an attempted union between sounds, images and words. Thus narrative themes, psychological moods, musical modes, and visual imagery are combined, to project the flavor of a few most significant seasons and sceneries. The monsoon is one of these favorite settings.

The *Śṛṅgāraprakāśa* (16), a treatise of poetics of the sixteenth century, associates the monsoon with a series of love festivals and sports which provide a counterpart to ethnographic descriptions presented above. For example, enjoying the sight of the dance of peacocks exhilarated on hearing the rumbling noises of the clouds (a most common symbol in poetry); mock fights with "weapons" made of the soft twigs of the *kadamba* tree which blossoms in the rains; playing on the swing, a sport which is practiced during *tīj*, and illustrated by miniatures of the musical mode Hindola, "Swing."

The symbolism of seasonal plants, beasts and birds, noises and games, and colors in the sky and over the landscape, was first established by Sanskrit poets like Kalidasa (17) in his *Round of the Seasons*:

> Groups of gay amorous peacocks
> Rend the air with jubilant cries
> To hail the friendly rain
> And spreading wide their jewelled trains
> They hold their gorgeous dance parade. . . .

But a poetic genre has developed in the modern vernaculars, as in the Hindi *bārahmāsā*, *Songs of the Twelve Months*, in which climates are used metaphorically to depict situations of distress, grief or suffering. Susan Wadley (18) has shown that the *bārahmāsā* provide a psychological almanac of the meaningful parts of the year:

> The recurrent themes and images are those aspects of the yearly cycle, whether climatically or ritually based, that are in discord with individuals whose personal situations, whatever their cause, are bad. . . . Hence, the *bārahmāsā* provide us with a map or guide to the potentially more difficult times of the year. . . .

The mood usually set in the context of rains is that of *viraha*, the torments of separation, of estrangement, and feverish waits. The mood is symbolized by certain human characters and narrative themes, of which we can give a few examples.

The wife of the traveler is a woman whose husband is away; she does not dress herself the usual way; she stands on her balcony, watching for his return.

The deserted heroine is tortured by *viraha* (estrangement). The fickle is equated with the cloud that wanders in the sky. "All my friends sleep with their husbands, but my own husband is a cloud in another land."

The far-off traveler, who, at pains during the storm on his way home, thinks of his wife.

The passionate night-walkers, where lightnings guide their feet to the trysted place.

The embrace:

> By thunder and lightnings affrightened,
> Every woman hugs her husband close,
> Though well his guilt she knows. . . .

The characters and themes are associated with climatic features, that is, the games, the colors and noises, the flora and fauna typical of the season. Examples are given in Table 3.6.

To fully appreciate this symbolism, we should compare the physiognomy of the Rains with that of another season, in particular the Summertime, since the poets have exploited the contrasts between them to convey the feeling of suspense. Jyestha, the hottest month, brings final decay, or final loneliness. The songs that celebrate the monsoon may begin with a few words of prelude on Jyestha, but then shift to storms:

> In Jyestha the world is on fire,
> Blows the *lūh* (burning wind)
> Which makes clouds of dust swirl up . . .
> This Jyestha burns me out,
> My husband is away, I can't repair the thatch,
> *Viraha* falls on me as so many firebrands . . .
> But Jyestha just has gone, and now comes Asadha,
> From the south new rumbling clouds shape up. . . .

By arranging their pictures into series based on the Twelve Months, and the gamut of musical scales, the Rajput and Pahari miniature painters, like the poets, have exploited the distinctions among the seasons. The schools of Hindu miniature flourished in Rajasthan (Rajputs) and the Punjab hills (Paharis) at the end of the eighteenth century. We can mention but a few scenes here.

First from *The Watch from the Balcony*: a lone young woman, standing on a balcony at the first floor of her house, eagerly looks in the distance for the return of her absent lover. Rambling clouds, rain, and lightnings that are like so many white serpents, are outlined against the dark sky. Rows of white cranes are hurrying towards their nests. A peacock stands on a terrace (19); or peacocks and waterbirds are scattered on cultivated lands in the background (20).

A variant presents *Radha and Krishna Watching a Storm*. Krishna in yellow pajamas sits with Radha who wears a mauve bodice and red skirt in the top story of a white pavilion. Below them are two girl musicians. Outside is a mango tree in fruit, and a night sky with lightning playing in the dark purple clouds. The peacock on the cornice, longing for his mate, raises his head and begins shouting. Ripe mangoes, music, lovebirds, and the lovers united: these are all elements of the month of Sravana, July–August (21).

TABLE 3.6. Climatic Features Associated with the Rainy Season

Games:
  Swing
  Dice

Colors and noises:
  Lightnings meet clouds, clouds fall on the earth
  One cannot tell day from night
  Dark clouds infuriated like mad elephants
  White cranes flying across the black sky
  Strong winds blow fiercely
  Croaking frogs and cooing birds

Plants:
  The thorny *arka* (Calotropis gigantea) and *yavāsa* (Alhagi pseudalhagi) shed leaves
  The *kadamba* (Anthocephalus cadamba), *campaka* (Michelia champaka), *ketakī* (Pandanus tectorius), *mālatī* (Jasminum grandiflorum), and *kāśa* (Saccharum spontaneum) are in blossom

Beasts and birds:
  Dancing peacocks and elephants
  Cooing of the *cātaka* (hawk-cuckoo) longing for raindrops from star Svati
  Proliferation of *indragopas* (cochineals) noticeable on account of their radiant scarlet velvet skin

Rajput painters have also depicted more specific events, like *Women Swinging at Tij*, the festival which is celebrated in Sravana. A group of merry women walk out in procession to join their playmates pushing the swing hung from a blossoming tree. Waterbirds and lightning outlined against the cloudy sky indicate the rainy season. The scene is being watched by the royal couple from the terrace of the palace (20). Illustrations of Bhadrapada (August–September) present in the background an exhilarated elephant dancing with a tiger and a lion. White cranes and whirling clouds fill out the sky. "*One should not leave home during Bhadrapada,*" the poet says, giving the miniature its theme. The hero, ready to set out with his horse and his sword, is requested by his wife not to desert her so soon (20). These are examples from the *bārahmāsā* cycles of miniatures.

Another category of miniatures illustrates the musical modes of Hindustani music. The mode Megha, "Clouds," evokes all the monsoon symbols. Dance and music celebrate the setting in of the rains. The central character is Lord Krishna, distinguished by his three-crested crown, dark blue complexion, and yellow garment, blowing the conch, or dancing with a pink lotus in his right hand, amidst a batch of female musicians. Heavy clouds, in front of which a flight of white cranes is passing by, several peacocks fanning their tails and bursting out into joyful songs, sung in the mode Megha, and pink lotus blossoms symbolize the season (22).

4 CONCLUSION

When, at summer solstice, the sun starts receding southward, the earth comes under the moon's rule. Rains burst out that are nourishing and refreshing but also tinged with sourness and fevers. Travelers will settle for a while, farmers will start growing new rice. At the end of the rainy Four Months, after the harvest of the monsoon crop, a new agricultural year may begin. These are the traditional ideas about the rainy season. The ideal domain of the monsoon thus defined is North India, but the same pattern, duly adjusted to fit in with local conditions, is also relevant to other southeast Asian countries. For example, in Thailand the year is theoretically divided into three seasons of equal length: the hot season (March–June), the wet season (July–October) and the cool season (November–February). Actually the rains fall from May through October, so that the symmetrical arrangement of the seasons appears to be oversimplified, but it is modeled after the conventional triad of Hot, Rains, and Cold found in classical texts. Aranuvachapun and Brimblecombe (23) have recently made known one *Song of the Twelve Months* (this is precisely its title in the Thai language), written in the fifteenth century. This song sets the time of rains bursting out in the beginning of May, thus providing the basis for adjusting the conventional pattern to the facts. This Thai piece also displays poetic devices and symbolic moods that are quite the same as in a Hindi *Song of the Twelve Months*:

> . . . noise fills the sky,
> Heavy rain falls like drops of gold,
> Since last month I cry again and again,
> Separated from my only love,
> Longing for her, I pine away . . .

This passage leads us to believe that the ideas presented in this chapter attributed to Hindu culture are not unfamiliar to other countries of Southeast Asia which have been influenced by India, and perhaps we can assume that they are widely held throughout monsoon Asia.

REFERENCES

1. H. Yule, and A. C. Burnell, *Hobson-Jobson*, *A Glossary of Anglo-Indian Colloquial Words and Phrases*, 2nd ed., John Murray, London, 1903, p. 577.
2. C. D. McLean, Ed., *Glossary of the Madras Presidency*, 2nd ed., Asian Educational Services, New Delhi, 1982, p. 436.
3. S. Stevenson, *The Rites of the Twice-Born*, Clarendon Press, Oxford, 1920, pp. 227, 302, 303.
4. F. Zimmermann, *La Jungle et le Fumet des Viandes*, *Un thème écologique dans la médecine hindoue*, Gallimard/Le Seuil, Paris, 1982, p. 196.

5. J. F. Pugh, "Into the almanac: time, meaning and action in North Indian society," *Contributions to Indian Sociology (New Series)*, **17**, 45 (1983).
6. K. L. Merrey, The Hindu Festival Calendar, in G. R. Welbon and G. E. Yocum, Eds., *Religious Festivals in South India and Sri Lanka*, Manohar, New Delhi, 1982, p. 9.
7. D. L. Eck, *Banaras, City of Light*, Routledge & Kegan Paul, London, 1983, pp. 261, 264.
8. M. Gaborieau, "Les fêtes, le temps et l'espace: Structure du calendrier hindou dans sa version indo-népalaise," *L'Homme, Revue Française d'Anthropologie*, **22**, 11–29 (1982).
9. Varāhamihira, *Bṛhatsaṃhitā* (6th century AD); Sanskrit text and English translation by M. R. Bhat, Motilal Banarsidass, Delhi, Volumes 1 and 2, 1981–1983.
10. F. B. J. Kuiper, *Varuṇa and Vidūṣaka, On the Origin of the Sanskrit Drama*, North-Holland, Amsterdam, 1979, p. 134.
11. E. Leach, *Pul Eliya, A Village in Ceylon*, Cambridge University Press, Cambridge, 1971, pp. 15, 254.
12. Cakrapāṇidatta, Commentary in *Carakasaṃhitā*, Vimāna VIII, 125; quoted in C. Vogel, "Die Jahreszeiten im Spiegel der altindischen Literatur," *Zeitschrift der Deutschen Morgenländischen Gesellschaft*, **121**, 304 (1972).
13. *Carakasaṃhitā*, quoted in F. Zimmermann, "*Ṛtu-sātmya*, The seasonal cycle and the principle of appropriateness," *Social Science and Medicine*, **14B**, 99–106 (1980). A reliable English translation is *Carakasaṃhitā*, P. V. Sharma, Ed., Chaukhambha Orientalia, Varanasi, Volumes 1 and 2, 1981–1983.
14. Vayaskara N. S. Mooss, Ed., *Aṣṭāṅgahṛdayasaṃhitā With the Vākyapradīpikā Commentary of Parameśvara*, Part II, Vaidyasarathy Press, Kottayam, 1963, p. 83.
15. *Suśrutasaṃhitā* (beginning of our era), Yadavji Trikamji, Ed., 3rd ed., Nirnaya Sagar, Bombay, 1939.
16. V. Raghavan, *Bhoja's Śṛṅgāra Prakāśa*, Punarvasu, Madras, 1963, p. 655.
17. Kālidāsa, *Ṛtusaṃhāra* (5th century A.D.), II, 6.
18. S. S. Wadley, "The rains of estrangement: Understanding the Hindu yearly cycle," *Contributions to Indian Sociology (New Series)*, **17**, 58 (1983).
19. E. Waldschmidt and R. L. Waldschmidt, *Miniatures of Musical Inspiration*, Otto Harrassowitz, Wiesbaden, 1967, Fig. 77 and §98.
20. V. P. Dwivedi, *Bārahmāsā, The Song of Seasons in Literature and Art*, Agam Kala Prakashan, Delhi, 1980, Plates 32, 59, 72, and 100.
21. M. S. Randhawa, *Basohli Painting*, Government of India, Delhi, 1959, Plate 34; also in W. G. Archer, *Indian Paintings from the Punjab Hills*, Sotheby Parke Bernet, London, 1973, Basohli No. 21.
22. K. Ebeling, *Ragamala Painting*, Ravi Kumar, Basel, 1973, Plates C32 and C33.
23. S. Aranuvachapun and P. Brimblecombe, "Tawatodsamad Klongdun, An old Thai weather poem," *Weather*, **34**, 459–464 (1979).

4

Monsoons in Agricultural Proverbs in Tamilnadu

Brian J. Murton
Department of Geography
University of Hawaii
Honolulu, Hawaii

INTRODUCTION

The power of the monsoons over the life and death of the land in southern India, and the powerlessness of the peasantry to change its course has long beem commented upon by outside observers. The fickleness with which monsoon i regularities disrupt cultivation accentuates this impression. Nonetheless, the peasantry of southern India has been successful in coping with the vagaries of their environment for thousands of years and have built one of the enduring cultures and civilizations of the world. The seeming vulnerability of indigenous agriculture, cowering beneath the monsoon, belies the fact that a wide array of protective devices has evolved to lessen the impacts of bad seasons. These include various ecological and technological strategies: cropping patterns fitted to the specific properties of a given patch of soil, rotation of mixing and fallowing cycles, manuring of the safest and most lucrative crops, dry farming plowing techniques, and minute—but controllable by comparison with the vastness of the monsoon's capriciousness—supplementary sources of water supply from streams, tanks, and wells.

These specific agricultural techniques, practices, and technologies, however, are but the surface manifestations of an underlying system of knowledge concerning environmental events and resources, aspects of which have already been alluded to by Singh in Chapter 2 and Zimmermann in Chapter 3. No systematic accounting of this knowledge has been compiled for southern India, yet relevant material has been described in a number of places. (See, for example, references 1 and 2). Reviews of cosmological, spatial, and temporal concepts that make nature intelligible and coherent to the region's farmers are just beginning to appear (3). However, this theme is an old one and not without its detractors. Equally convincing evidence can be presented for the ecological rationale of traditional methods and even for the reverse argument that cosmological concepts and kinship relationships are best illuminated by the understanding of physical processes and their management.

Nevertheless, indigenous knowledge concerning the environment of agriculture is considerable. In this chapter agricultural proverbs from the modern Indian state of Tamilnadu are examined in order to see what they tell us about Tamil knowledge and understanding of the monsoon (Fig. 4.1). (Tamilnadu is the state where the Tamil language is spoken and the people are known as Tamils.) To do this, first conditions in Tamilnadu will be described. Then an outline of a framework for

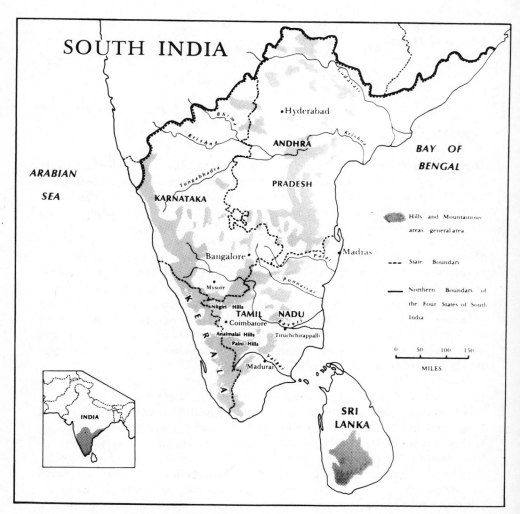

Figure 4.1. Map of South India. The hilly and mountainous areas lying between Kerala and Tamilnadu, and to the north in Karnataka are collectively referred to as the western Ghats. The major gap in the western Ghats lies to the west of Coimbatore.

studying Indian systems of environmental knowledge, into which weather events associated with the monsoon can be placed, will be discussed. Attention is then turned to the agricultural proverbs, and to the relationships of monsoon events and agriculture. The final section is devoted to the conclusions that can be drawn about the *ethnometeorology* of the monsoon.

1 THE MONSOON IN TAMILNADU

Since the length of the growing season in South Asia is limited by low temperatures in only a few areas, precipitation is the principal climate element affecting agriculture and hence the sustenance of human populations. This has led to the truism that almost every aspect of life in South Asia is affected or even dominated by the monsoon. But how far is it true? In the popular western sense the monsoon climate implies a climate with a cool dry season of northerly winds—the northeast monsoon—in December to February giving way to a hot, dry season from March to early June, a hot, wet season of southwesterly winds—the southwest monsoon, *the* monsoon—and a retreat to the dry, cool season around the winter solstice. While this picture is useful, it is important to realize that the subcontinent has several climates rather than a single climate. The popular image of the monsoon climate best fits the tracts lying north and west of a line from Goa to Patna, and even within this area it is upset by the presence of small but significant depression rains of the cooler months in the northwest.

Indeed, the southwest monsoon, which is the lifeblood of agriculture in most parts of India, spends its fury from June to September to the west of the western Ghats in Kerala State. Tamilnadu is a "rain shadow" region during this summer monsoon and the incidence of rainfall in different places depends upon local topography, distance, and location with reference to that mountain range and its gaps (Fig. 4.1). The rainfall from this monsoon is heaviest in the western hills of Tamilnadu (the Nilgiris, Anaimalais, and Palnis), amounting to an average of just over 1000 mm in 50 rainy days, although in places the rainfall exceeds 4000 mm. In those parts of Tamilnadu that lie to the east of the major gap (around Coimbatore) in the western Ghats rainfall is fairly heavy (about 500 mm) spread over 20 to 30 rainy days. Areas to the south of this receive approximately 300 mm in 15 to 20 days, but to the far south, where moisture-laden clouds are completely cut off by the western Ghats, the rainfall from the southwest monsoon is less than 250 mm.

Most of the annual rain over Tamilnadu occurs during the months of October, November, and December. It is associated with the northeast or the winter monsoon season. As the retreat (or withdrawal) of the summer monsoon occurs in September, the monsoon trough commences a southward motion towards the equator. The trough has an east–west orientation and approaches 10°N by around late October, usually when the rainfall over the Tamilnadu commences. The incidence of this rainfall diminishes with the distance from the east coast. Rainfall is heaviest in the coastal areas (750 to 1000 mm) in 20 to 30 rainy days. In other areas in the south, rainfall is somewhat less (400 to 500 mm), spread over 15 to 25 rainy days. The western interior receives 250 to 300 mm in a period of 20 rainy days. The weather from late December to mid-February is cool and dry, while from mid-February to mid-June the weather is hot (in April and May temperatures can reach between 40 and 45°C on the plains). Between 50 and 150 mm of rain from thundershowers normally falls in most places in April and May.

The great spatial and seasonal variation in total rainfall experienced in much of India, described by Shukla in Chapter 14, occurs in Tamilnadu as well. Rainfall is highly variable along the east coast from Madras south to the Kaveri delta. Since the mean annual rainfall of over 1250 mm in this area is more than adequate to meet agricultural needs, even large departures from the mean do not increase crop failures (4). The western interior also has highly variable rainfall, but here rainfall is low as well; thus agriculture can be crucially affected in a bad year. Elsewhere on the plains variability is lower.

As Zimmermann points out in Chapter 3, the traditional Indian conception of monsoon does not involve a relationship between the outburst of the seasonal rains and changes in wind direction. Although there is a vernacular concept of periodical winds, for agriculturalists in South Asia it is the rains, and their beginning and ending, that are important. Certainly, in Tamilnadu variations in agricultural production are directly related to water conditions more than any other variable, physical or social. The traditional farming systems of the area have evolved over very long periods of time, taking water conditions, including periodic disasters, into account. It is little wonder, then, that there exists extensive lore concerning water and agriculture.

2 INDIAN SYSTEMS OF ENVIRONMENTAL AND RESOURCE KNOWLEDGE: A FRAMEWORK

The accumulation and dissemination of knowledge about environmental resources are critical for the development and maintenance of any complex agricultural system. Thought processes and structures underlying agricultural activity can be characterized by a form of "bounded rationality" (5, 6). Two connotations of this concept are important for us. First, environmental knowledge and its articulation as agricultural systems are internally consistent and, to a large extent, self-fulfilling. Premises upon which actions are taken are reinforced when the action is successful: that the crops yield well justifies the system by which they were cultivated. Second, the concept may usefully be viewed as the local empirical validity of traditional knowledge. Traced to ultimate origins, folk knowledge may be rooted in religious beliefs or in cosmographies alien to the tenets of Western science: these ultimate roots belie, however, the validity and efficacy of environmental knowledge.

To understand the place of agricultural proverbs within the indigenous system of knowledge about environment, four questions can be utilized as a framework of interpretation (7):

1. What are the principal categories of resource discrimination and what are the bases upon which they are made?

2. How are limited and unpredictable resources combined in such a manner as to make them less limited and more predictable?

3. How are resource-use activities coordinated in cycles of use and access that regulate the timing of the system?

4. What are the general premises that underpin this body of knowledge and allow the speakers to share similar sets of evaluations and strategies as well as comparable outlooks toward the future?

2.1 General Premises

Before proceeding to detail the system of agricultural knowledge, let us briefly describe some of the general premises that underpin this body of knowledge. Indian folk science is rooted in indigenous religious beliefs and cosmography and is tied closely to the spiritual fabric of traditional society. The general premises include several sets of interlocking principles derived from Indian world view. The basic concepts that allow people to share similar sets of evaluations and strategies, as well as comparable outlooks toward the future are: *samṣāra*, the belief in rebirth; *karma*, the belief that actions in one life determine fortune and status in the next; *dharma*, the defintion of what one's behavior must be in the life status to which one's *karma* has brought one; *moksha*, salvation by release from the wheel of eternal rebirths through the idea of perfect asceticism, of absolute wantlessness; *ahiṃsā*, the belief in the sacredness of all life (8).

The Universe is viewed as consisting of places, each of which may contain gods, men, animals, plants, and things in varying mixtures and proportions (9). In this view there is no separation of Man and Nature and the Supernatural; all have a place, all have a role, all are equally real. However a sharp distinction is often made between the natural and supernatural worlds. The natural world is divided into ground, mountains, forests, hills, oceans, and villages. The supernatural world is divided into separate religious places, such as the loci of Brahma, Siva, and Krishna. Each of the natural and supernatural places, particularly those regarded as on the earth, contains beings, plants, and objects. On the whole this world view appears to be more ecologically oriented than the traditional Western view.

Perhaps a more important general principle is that within each location all phenomena (people, animals, plants, gods, things) are arranged in hierarchical order. This notion of hierarchy is of critical importance in comprehending systems of knowledge about resources because ranking is a feature of many aspects of resource-use: land categories are ranked, soil types are ranked, crops, even varieties, are ranked. In addition, resource-users themselves are ranked, both in terms of socioritual status (*jati* or caste) and in terms of the control and use of land (10).

Indians also segment the continuum of nature: the year is divided into seasons; landscapes are categorized; soils are divided into classes; crop varieties are distinguished. But the Indian mind not only segments phenomena but arranges them in opposite pairs (binary opposites). Many of these opposites are mediated by a third term, which reconciles their bipolar tendencies. For example, in southern India certain castes were considered to be medial between those of the traditional ''left'' and "right" divisions. Furthermore, Indians recognize a few basic substances or elements which underlie the multiplicity of phenomena: earth, water, fire, wind, and sky (11). Each substance is identified with a distinctive quality and each is also a process or the embodiment of a principle for action. Each is identified with a range of other phenomena, so that nature is made intelligible and coherent.

Of equal importance is traditional South Asian understanding of time, which is not thought of in linear progression, but in terms of cycles which repeat themselves endlessly (12). Time moves in cycles within cycles—days within astral periods within months within seasons within years within epochs of approximately 300 million years. Another characteristic of Indian time is that it had no beginning or ending: there is only endless repetition of cycles. Finally, nothing is more indicative of the all-encompassing nature of time than that the gods, no less than man, are part of the eternal cycle. There is no dividing line between the human and the divine; there is a continuum between man and nature and between man and the gods, because all are bound in the movement of time.

2.2 Categories of Resource Discrimination

The principal categories of resource discrimination in India include soil, slope, water, climate, time, space, plants, animals, and fish, all aspects of environmental elements. Such terminological domains are further elaborated in taxonomies. Indians have classifications of landscape morphology, land types, soils, vegetation, crops, macroclimatic and microclimatic resources. These are related to the general principles discussed earlier, and although each can be isolated and reviewed separately, in reality they make up a complex of intersecting, interconnected entities and events.

Agricultural classifications reflect local characteristics of the biophysical environment. Throughout the region, land is initially classified as either wet or dry. This distinction between irrigated and rainfed land is of major importance because distinctive types of management and conservation practices are related to wet and dry land. In the Ganga valley and delta areas, land is further classified in terms of elevation above water level. Soils are most often initially classified according to their color (e.g., red or black), and then in terms of very fine textural discrimination, as well as distance from settlement sites. Crop varieties are classified in terms of obvious seed color, height, seedhead shape, duration, taste, and other visible characteristics.

2.3 Coordination of Agricultural Activities

The coordination of agricultural activities has both temporal and spatial aspects. Traditional systems of dividing time are common to all parts of India. Three methods of computing yearly cycles are used (solar, lunar, *fasli* or revenue year), but in terms of agriculture, the solar year which begins in mid-April is the most important as its beginning more or less coincides with the thundershowers of April which herald the coming monsoon, the beginning of the agricultural season.

As Zimmermann describes in Chapter 3 there are six seasons, each of which is two months in the Indian year. As important as recognized seasons are the sowing periods which are highly localized, reflecting appraisals of the moisture environment in relationship to particular crops. Perhaps the most important devices for regulating the timing of agricultural operations are astral periods: *asterisms* or *lunar mansions* (*kārtti* in Tamil), the *nakṣatras* of the Sanskrit tradition. These are formulated in terms of the relationships of the sun, the signs of the zodiac, and 27 stars. In Tamilnadu there are 27 periods. Agricultural operations and tasks are assigned to each of these astral periods.

Where time is organized into years, months, seasons, and astral periods, space is organized into a number of land–soil–crop associations. Throughout India certain crops or crop associations are related to certain land and soil types, and these interrelationships are models that farmers use to guide them in their agricultural activities.

Of course, Indian agricultural knowledge is not confined to a sum of unconnected parts. Environmental appraisal, as well as agriculture, requires that terms be linked as webs of associations: sorghum intersects with certain land types, soil/water conditions, insect pests; rice with others; wheat with yet others. Different seasons intersect with the availability of certain forage plants, different weather conditions intersect with special diseases, and certain terrains with certain plant types. Thousands of these simple linkages constitute the building blocks of traditional Indian agricultural strategy, the ways in which resources are combined and timed.

2.4 Resource Predictability: The Role of Agricultural Proverbs

When dealing with folk knowledge concerning the monsoon, attention must be addressed to the coordination and timing of resource-use activities and to how limited and unpredictable resources are made less limited and more predictable. At an abstract level, time and space are organized into sequences and segments, linked into webs of association. A wealth of knowledge associated with these serve as a guide to agricultural operations, as well as making the somewhat limited and un-predictable resources of India less limited and more predictable. This accumulation of experience is formalized in the form of folk songs, sayings, and proverbs, already alluded to in the preceding two chapters. The months of the year, the seasons, and especially the astral periods (*kārtti* or *nakṣatras*) all have associated with them a vast array of pithy statements about every aspect of tillage and agricultural opera-tions. In the past, and even today, agricultural operations were guided by such statements, especially by proverbs relating to rainfall and the appropriate timing of cultivation practices.

Proverbs in general are a collective product of many minds in a cultural group. Agricultural proverbs in particular can be regarded as a reflection of a group's accumulated experience in an environment. Traditionally, they were transmitted orally and were normally learned by small boys as they helped their fathers in the fields. In the late nineteenth century the Department of Agriculture of the Madras Presidency collected and published agricultural proverbs in the four major languages of the Presidency (Tamil, Telugu, Kannada, and Malayalam). These first appeared with English translations in the *Bulletin* of the Madras Department of Agriculture in the 1890s, and subsequently were republished a number of times (13). Proverbs had been collected by the British from the middle of the nineteenth century because thoughtful British officials realized that they contained reflective knowledge about the traditional agriculture of the peoples of southern India. Collecting proverbs represented an effort on the part of some officials at the time to come to grips with the realities of peasant cultivators. Similar collections of proverbs exist for most of the languages of India. Singh in Chapter 2 alludes to this and briefly discusses some of those dealing with rainfall from the Punjabi, Mahrati, and Hindi speaking areas.

In total more than 1700 proverbs were collected from different parts of southern India. The largest single collection is 928 Tamil proverbs and it is upon these that the remainder of this chapter is based. The Tamil agricultural proverbs were chosen for the following reasons: first, the large number of Tamil proverbs makes it possible to gain a comprehensive understanding of Tamil environmental knowledge; second, by concentrating on proverbs in one language, problems of intercultural contrasts and differences do not need to be resolved; third, it is easier to interpret the environmental information concerning the monsoon because variations in monsoon regimes are more limited in the Tamil-speaking area than in the entire region or all of India. Even with this limitation to Tamilnadu, difficulties arise because we do not know precisely where certain proverbs were collected. It can be assumed that those pertaining to rice cultivation have their origin in the rice-growing areas of coastal Tamilnadu, including the Kaveri delta, and the valley of the Vaigai River, and that most of those containing knowledge about dry land agriculture and pastoralism came from the interior. But nowhere is this told. This means that any generalizations are for all of Tamilnadu, and that the subtleties of variations in proverbs in relationship to local variations in environmental conditions, which are known to exist, are lost in this analysis.

3 AGRICULTURAL PROVERBS AND THE MONSOON

Just over 20% of the proverbs deal with some aspect of climate and weather. None, of course, uses the term monsoon, but all reflect knowledge of the seasonal cycle, the types of weather expected at different times of the year, problems resulting from deviations from normal weather, and the ways in which an uncertain environment could be made predictable and knowable. Basically, the pertinent proverbs deal with resource predictability. But as there is a temporal element involved, the details given in many proverbs refer to the traditional Tamil systems of dividing time.

3.1 Tamil Time Divisions

In the proverbs the systems of years and months used is derived from the all-Indian solar year system. The solar year begins in mid-April when the sun comes into conjunction with the star Cittirai (the sign of the zodiac, Aries), and consists of 12 months of uneven length, depending on the time the sun takes in passing through the respective signs in the Hindu zodiac (Table 4.1). A month is the specific time the sun takes in going from the first degree in one to the first degree in another. The solar year is closely linked to the monsoon cycle, and thus to agriculture. The beginning of the year in April more or less coincides with the thunderstorms of spring which are of critical importance for agriculture in Tamilnadu. These showers occur after the sun enters the sign of Aries, and they herald the beginning of the season. However, in Tamilnadu, beginning the year on the first day of *cittirai* is a recent practice adopted to correspond with the current national Indian calendar. It is abundantly clear that in traditional Tamil calculations the year began on the first day of the month of *tai* (14). Since *tai* used to begin the calendar year, it is not surprising that a major family feast, a ritual celebration of rebirth and rebeginning, remains associated with the first part of this month. The term *tai* means "green plant" or "young thing" and the *tai* festival is overtly said to be for children. The first day of *tai* originally was at the time of the winter solstice. The traditional calendar does not seem to have taken into account the precession of the earth's axis. The calculations that determined the time of the festival have had the effect of delaying it, so that by now it is celebrated some three weeks late in solar terms. Since one complete solar precession cycle requires about 36,000 years, estimates from the above discrepancy show that the traditional Tamil calendar was developed and put into practice some 1600 years ago. The first day of *tai* is the only day of the year when *poṅkal*, or festival rice, must be symbolically offered to the sun prior to general distribution.

The Tamil calendar has a second major family feast day at the beginning of *āṭi* in mid-July. Thus the traditional Tamil calendar organized the year into two halves. During the first part of the year, from *tai* to *āṭi*, the sun is traveling northward and

TABLE 4.1. The Months in the South Indian Year[a]

Tamil Month[b]	Approximate European Month	Sign of the Zodiac
Cittirai	April–May	Aries
Vaikāci	May–June	Taurus
Āṇi	June–July	Gemini
Āṭi	July–August	Cancer
Āvaṇi	August–September	Leo
Puraṭṭāci	September–October	Virgo
Aippaci	October–November	Libra
Kārttikai	November–December	Scorpio
Mārkaḷi	December–January	Sagittarius
Tai	January–February	Capricorn
Māci	February–March	Aquarius
Paṅkuṇi	March–April	Pisces

[a] From Benson (13).
[b] Zimmermann discusses Sanskrit and Hindi months in Chapter 3.

TABLE 4.2. Tamil Astral Periods[a]

Tamil Name	Approximate European Date
Asvaṇi	April 11–April 25
Paraṇi	April 26–May 9
Kiruttikai	May 10–May 23
Rogiṇī	May 24–June 6
Mirugaṣīram	June 7–June 20
Arudra	June 21–July 4
Puṇarpūcam	July 5–July 18
Pūcam	July 19–August 1
Āyilyam	August 2–August 15
Makam	August 16–August 29
Pūram	August 30–September 11
Uttiram	September 12–September 25
Hastam	September 26–October 8
Cittirai	October 9–October 22
Ṣoṭi	October 23–November 4
Vicākam	November 5–November 17
Anuṣam	November 18–November 30
Keṭṭai	December 1–December 14
Mūlam	December 15–December 27
Pūrāṭam	December 28–January 9
Uttirāṭam	January 10–January 22
Tiruvōṇam	January 23–February 4
Aviṭṭam	February 5–February 17
Catāyam	February 18–March 2
Pūraṭṭāti	March 3–March 15
Uttiraṭṭāti	March 16–March 27
Revatī	March 28–April 10

[a] From Benson (13).

the length of the day is gradually increasing. During the second half, the sun returns on a southward course and the length of the day gradually decreases. Recent research suggests that this stress on a cosmic cycle of increase and decrease, on the sun becoming stronger and weaker, and on the waxing and waning of the moon, can be seen as permeating the entire southern Indian ritual complex. On the philosophical level it corresponds to the Hindu textual tradition of cycles of cosmic flowering versus cosmic decay, outlined by Zimmermann in Chapter 3.

Zimmermann has already discussed the symbolic role of the Sanskrit *nakṣatras*, called *kārṭṭi* in Tamilnadu (Table 4.2), which are the periods of the sun's stay in a particular division of the lunar zodiac, and which were utilized by cultivators to conceptualize the cycle of agricultural operations and related environmental events.

3.2 Reflections upon Rainfall

Of all the weather elements associated with the monsoon, rainfall receives the most attention. A large number of proverbs reflect upon the necessity of rain for agriculture and, indeed, upon the relationships of rain and human existence, a matter dealt with by Singh and Zimmermann in the two preceding chapters. The selection used here (Table 4.3) stresses the role of rain in the rebirth of the land, the dire consequences of the failure of the rains, and the desirability of rain, analogous to the "begetting" of children. At a general level these proverbs iterate the importance of rain for life in Tamilnadu: rain is the life-giver after the dry season. If it fails to come at the appropriate time the land will suffer. This reflects an acute awareness of the possibility that the monsoon rains could be delayed, or could fail at anytime. Indeed, over the past 1000 years there have been catastrophic famines in Tamilnadu, most caused by drought, though some by excessive rainfall, tropical cyclones, or warfare.

3.3 The Weather Cycle

The proverbs contain very complete and accurate descriptions of the weather expected throughout the year. What is emphasized is information that could be used as guidelines to the weather conditions expected throughout the important portion of the agricultural year (Table 4.4). This information is linked to the months of the Hindu year. Most of it pertains to rainfall, but several proverbs allude to heat. The descriptions of rain at different times in these proverbs meshes very well with descriptions of weather found in modern sources (Table 4.5) (15).

3.4 Weather and Agriculture

Information that relates weather elements, especially rain, to agriculture is given in reference to both the system of Hindu months and astral periods (Table 4.6).

TABLE 4.3. Reflections upon Rainfall[a]

1. 'Tis rain works all: it ruin spreads,
   then timely aid supplies;
   As in the happy days before, it bids
   the ruined rise.
2. If heaven grows dry, with feast
   and offering never more,
   Will men on earth the heavenly
   ones adore.
3. Nothing can be done without rain.
4. If the sky fails, the earth will fail.
5. No man has been ruined by begetting
   children or by rainfall.

[a] From Benson (13).

TABLE 4.4. Weather Descriptions from Proverbs[a]

Tamil Month	European Date	Proverb
Cittirai	April–May	After the tenth of *cittirai*, high winds commence and last to the tenth of *aipacci*.
Āṇi	June–July	There will be drizzling in *āṇi* and rain in *āvaṇi*.
Āṭi	July–August	Rain in *āṭi* will be scanty.
Āvaṇi	August–September	*Āvaṇi* drips continually.
Puraṭṭāci	September–October	In the good *puraṭṭāci* month the sun will be so hot as to melt gold and the rain so heavy as to liquefy the earth.
		Heavy showers fall during *puraṭṭāci*.
Aippaci	October–November	Continuous rain in *aippaci*.
		The solar heat in *aippaci* is so great that a skin will dry in the day the animal is flayed.
Kārttikai	November–December	In *kārttikai* the rain will not allow you even to clean your vessels.
		It will rain heavily in *kārttikai*.
		It is one-fourth summer after *kārttikai*.
		Rain will cease after the lamping-lighting festival in *kārttikai*.
Mārkaḷi	December–January	No rain from *mārkaḷi* and no army after the Bharata war.
Tai	January–February	Summer commences in *tai*.
		The soil gets dry at the commencement of *tai*.
Māci	February–March	The dew in *māci* will even soak through a terrace.

[a] From Benson (13).

TABLE 4.5. Modern Weather Descriptions[a]

Month	Description
January	The first dry month after the monsoon rains. Days are hot. Nights are cold with dew and slight showers.
February	Dry month. Fine weather. Dew continues. The days are hot and the nights are cold.
March	Dry month. Fine weather generally. Dew abates. The summer heat is felt about the middle of this month.
April	One of the hottest months. Both the days and nights are hot.
May	The hottest month in the year. Weather improves with the sudden small showers.
June	Hot weather continues and it is mitigated by winds and moderate showers now and then. Southwest monsoon sets in.
July	Southwest monsoon continues.
August	Southwest monsoon continues. Occasional rainfall is heavier than in the preceding months.
September	Close of summer rains. Southwest monsoon lingering with occasional showers. Weather becomes fine.
October	Northeast monsoon is due. Busy season for agriculturalists.
November	Northeast monsoon is vigorous. Cold weather and heavy rains.
December	Northeast monsoon almost closing. Dew weather, chill nights and occasional showers.

[a]From Madras Governmental Printing (15).

TABLE 4.6. Weather and Agriculture[a]

Tamil Month or Astral Period[b]	European Date	Proverb
Asvaṇi	April 11–April 25	If there be rain in *asvaṇi* every crop will be lost.
Paraṇi	April 26–May 9	If it rains in *paraṇi*, the soil will yield well everywhere.
Cittirai	April–May	Rain in *cittirai* renders the earth unfruitful.
Vaikāci	May–June	If the season helps plowing in *vaikāci* the tamarind will blossom abundantly.

(*Table continues on p. 90.*)

TABLE 4.6. (*continued*)

Tamil Month or Astral Period[b]	European Date	Proverb
Āṭi	July–August	Sunshine in *āṭi* is equal to sheep folding.
Āvaṇi	August–September	Freshets in the early part of *āvaṇi* is destructive.
Cittirai	October 9–October 22	If it rains in *cittirai*, the country will be ruined.
Aippaci	October–November	If there be no rain in *aippaci* my field will resemble that of my neighbor.
Kārttikai	November–December	If the rains of *aippaci* and *kārttikai* fail, the elder and younger will be equal.
Mūlam	December 15–December 27	If there is rain in *mūlam*, the grain will suffer.
Mārkaḻi	December–January	If there be rain in the middle of *mārkaḻi*, crops carefully tended will thrive.
Tai–Māci	January–February February–March	There is no grass that does not thrive in *tai*, nor any tree which does not put forth fresh shoots in *māci*.
Tai	January–February	If there be freshets in *tai*, a ryot can feed his mother.
Paṅkuṇi	March–April	If it rains in *paṅkuṇi*, the harvest will be damaged.

[a] From Benson (13).
[b] In this table both Tamil months and astral periods are interspersed. Refer to Tables 4.1 and 4.2 for details.

Many of these proverbs warn of the consequences of rain or a lack of it at certain times of the year. They predict what might happen if the expected or normal seasonal rains did not occur. Interestingly enough one proverb, "sunshine in *āṭi* is equal to sheep folding" emphasizes the significance of fine weather during that month. In parts of Tamilnadu, August is the month during which sorghum and millet sown in late May and June mature. The failure of the northeast monsoon rains also is the focus of attention in proverbs from the months of *aippaca* and *kārttikai*, the benefits of rainfall in *mārkaḻi* and *tai* are noted, and the ill effects of rain in *paṅkuṇi*, the harvest season for rice sown or transplanted during the northeast monsoon, are commented upon.

3.5 Agricultural Tasks and Time Periods

A very large number of proverbs interrelate various agricultural tasks with the months of the Tamil calendar (Table 4.7). Several refer to the benefits, positive or negative, of plowing at certain times. A number tell of the best or worst times to sow. Yet others deal specifically with when to sow or transplant rice, varieties of rice, millet, pulses, oil seeds, and sugarcane.

3.6 Forecasting Rainfall

The largest number of proverbs about rainfall involve forecasting. These include long-term forecasting using natural phenomena, as well as assessments of what would happen in the future if the expected rains did not come. Several proverbs

TABLE 4.7. Agricultural Tasks and Time Periods[a]

Tamil Month or Astral Period[b]	European Date	Proverb
	Plowing	
Cittirai	April–May	Tilth in the month of *cittirai* will be as pure gold.
Kārttikai	November–December	Plowing in *kārttikai* yields very little.
Tai	January–February	Plowing in the *tai* is equal to folding sheep five times. Plowing in *tai* is plowing ghee.
	Sowing	
Cittirai	April–May	If seed be properly sown in *cittirai*, best yields will be obtained.

(*Table continues on p. 92.*)

TABLE 4.7. (*Continued*)

Tamil Month or Astral Period[b]	European Date	Proverb
Āṭi	July–August	Look for *āṭi* and sow.
		Seed sown in *āṭi* and a son begotten at the age of 25 will be a help.
		Sowing finished before the fifth of *āṭi*, transplanted before the 15th of *puraṭṭāci* and a son born in the father's 24th year are like treasures buried by the elders.
		If you sow a seed in *āṭi* you will have a fruit in *kārttikai*.
Āvaṇi	August–September	*Āvaṇi* is a deceptive month for sowing.
Paṅkuṇi	March–April	Even a Pariah will not cultivate in *paṅkuṇi*.
	Sowing Rice	
Āṭi	July–August	*Āṭi* for sowing and *āvaṇi* for
Āvaṇi	August–September	transplanting.
Puraṭṭāci	September–October	To beget a son in our prime and to sow sambah about the middle of *puraṭṭāci* are equal.
Aippaci	October–November	Paddy sown in *aippaci* will not yield enough even for wafers.
Kārttikai	November–December	Cultivation in *kārttikai* gives an armful of straw and paddy that will fill a coconut shell.
	Transplanting Rice	
Āvaṇi	August–September	A son begotten at the age of 24 and transplanting before the 15th of *āvaṇi* are both good.
Āṭi	July–August	The first ten days in *āṭi*, the
Āvani	August–September	second ten days in *āvaṇi*, the
Puraṭṭāci	September–October	last ten days of *puraṭṭāci*,
Aippaci	October–November	and the whole of *aippaci* are unfavorable for transplanting.
Puraṭṭāci	September–October	Transplanting by the 15th of *puraṭṭāci* is best.
		Transplanting in *puraṭṭāci* is like ancestral wealth.

Tamil Month or Astral Period[b]	European Date	Proverb
Aippaci	October–November	Put aside the seedlings in *aippaci*.
		Transplanting in *aippaci* will not yield even grain for wafers.
		Seedlings transplanted in *aippaci* are like a son begotten in a man's 50th year.
Kārttikai	November–December	Throw away the plants after the first of *kārttikai*.
Types of Rice		
Āṇi	June–July	No one has seen *kar*[c] paddy sown in *āṇi* withering nor a dead monkey.
Āvaṇi	August–September	*Kar* in *āvaṇi* is like the flower of a pumpkin.
Tai	January–February	*Kuruvai*[d] sown in *tai* will not yield even bran.
Bulrush (Pearl) Millet (Kambu)		
Mirugaṣīram *Rogiṇī*	June 7–June 20 May 24–June 6	If *kambu*[e] be sown in *mirugaṣīram*, the ears will be like drumsticks; if in *rogiṇī* the grain will not be sufficient even for bread.
Vaikāci	May–June	Every handful of *varagu*[f] sown in *Vaikāci* will produce a *kalam*.[h]
Pulses		
Aippaci	October–November	Horsegrain will strike root firmly in *Aippaci*.
Kārtikkai	November–December	Even if you sow one *kalam* of horsegrain in *kārttikai* you cannot get the seed back.
Gingelly (Sesamum)		
Puraṭṭāci *Chittirai*	September–October April–May	Sow the large variety of the gingelly in *puraṭṭāci* and the small in *cittirai*.

(*Table continues on p. 94.*)

TABLE 4.7. (*Continued*)

Tamil Month or Astral Period[b]	European Date	Proverb
Tai	January–Febuary	Gingelly sown in *tai* remains in the soil.
Māci	February–March	Gingelly sown in *māci* is like money in hand.
Vaikāci	May–June	Gingelly raised in *vaikāci* is certain.
	Sugar Cane	
Māci	February–March	Harvest sugarcane in *māci* and
Paṅkuṇi	March–April	*paṅkuṇi*.
Āṇi	June–July	In *āṇi*, sugarcane is like an elephant's tail.

[a] From Benson (13).
[b] Tamil months and astral periods are interspersed. Refer to Tables 4.1 and 4.2 for details.
[c] The *kar* season for growing rice lasts from June to September (southwest monsoon).
[d] In the Kaveri delta area (Tanjore District), the *kar* season is called *kuruvai*.
[e] *Kambu* is *Pennisetum typhoides*, in English, pearl bulrush, or spiked millet.
[f] *Varagu* is *Paspalam scrobiculatum*, in English, kodo millet.
[h] A *kalam* is a volume measure of approximately two pounds weight.

describe how wind direction affects rainfall. Short-term forecasting of rainfall utilizes a range of natural phenomena.

Most of the long-term forecasts that utilize natural phenomena link those phenomena to the months of the Tamil calendar (Table 4.8). For example, rainbows in *āṇi*, *āvaṇi*, and *puraṭṭāci* were warnings of impending disaster, and the dispersal of clouds in *āṇi* signified a failure of rainfall. Thunder and lightning at different times were considered especially significant in relationship to rainfall, as were the occurrence of eclipses and the positions of the planets and the moon.

A further number of proverbs linked the presence or absence of rainfall in a month or astral period to its future occurrence (Table 4.9). For example, no rain in *āṇi* was a sign that the southwest monsoon had failed, and that the next rain could not be expected until *puraṭṭāci* (September–October), the beginning of the northeast monsoon.

The proverbs that refer to wind direction are of great interest because they demonstrate considerable understanding of the significance of wind from different directions at different times of the season (Table 4.10). Southerly or southwesterly winds, during *āṇi*, *āṭi*, *āvaṇi* (the southwest monsoon period), and even *puraṭṭāci*, are viewed as bringing rain. Later in the year northerly or easterly winds (the northeast monsoon) are seen as essential to the well-being of the countryside. The

TABLE 4.8. Long-Term Forecasts: Natural Phenomena[a]

Rainbows

If a rainbow appears in the east in *āni* and *āvani* there will be a famine.
A rainbow is *āni* forebodes drought in the next month.
A rainbow in *purattāci* forebodes a scarcity of food.

Clouds

If the clouds disperse in *āni*, rain decreases.

Thunder and Lightning

Thunder in the hot weather and lightning during the rainy season bring heavy rain.
If there be thunder in the rainy season and lightning in the hot weather, there will be no
   rain.
If it thunders in *asvani*, there will be no rain for six *kārtti*.
If it thunders on the sixth of *āni*, it will not rain for 60 days.
Thunder on the fifth of *āvani* foretells plenty of rain during the year.
Thunder during the first six days of *aippaci* will injure even seeds preserved in the
   undermost pot.
If it thunders in *aippaci*, grass will grow at the bottom of the wells.
If there be thunder in *kārttikai*, the ears of corn will not fill. [b]

Eclipses, Planets, Moon

An eclipse or shooting stars in *paṅkuni*, *āni*, *purattāci*, or *mārkali* are signs of heavy
   rain; do not then sow on the valleys or the crop will be washed away.
It will rain, when Venus after setting to the south of the moon rises again to the north.
Rain is incessant if Venus enters *ṣoṭī*.
If *kettai* begins and the new moon appears together on a Tuesday, the country will be
   poverty stricken.
If the *tai* new moon slopes to the north it will bring prosperity if to the south, adversity.
If the young moon be level in *tai* and *māci*, and rise a little to the south in *paṅkuni* and
   cittirai and a little to the north in the remaining eight months, the rains will be good; if
   not, there will be distress.

---

[a]From Benson (13).
[b]The corn referred to here is probably one of the varieties of sorghum. If heavy rain falls during the
pollination stage of growth or during the time when grain is being set, the earheads do not develop.

TABLE 4.9. Long-Term Forecasts: Time Periods[a]

Tamil Month or Astral Period	European Date	Proverb
Vaikāci	May–June	If it rains on the fourth day in the dark fortnight of *vaikāci* there will be plenty of rain; if not, clouds will not even be visible, the sea will dry up and gingelly can be raised in the tank beds.
Arudra	June 21–July 4	If *arudra* begins in the daytime, there will be no rain for the next six *kārtti*.[b]
Āṇi	June–July	If there be no rain in *āṇi*, then there will be none for 60 days afterwards.
Āṭi	July–August	If the *āṭi* new moon be obscured, rain will hold off 'til *tai*, the sea will dry up, and there will be much distress.
Āṭi	July–August	If it rains on the new moon day in *āṭi*, paddy will be as dear as rice.
Puraṭṭāci	September–October	Rain in *puraṭṭāci* is uncertain.
Cittirai	October 9–October 22	If it rains in *cittirai*, the country will be ruined.
Aippaci	October–November	If there be no rain in *aippaci* my field will resemble that of my neighbor.
		If there be no rain in *aippaci* the yield obtained in the case of good and bad cultivation will be the same.
		If the rains of *aippaci* and *kārttikai* fail, the elder and younger brother will be equal.[c]
Kārttikai	November–December	Even the grass among the rocks will bear seeds with the rains of *kārttikai*.

Tamil Month or Astral Period	European Date	Proverb
Mārkaḻi	December–January	If there be rain in the middle of *mārkaḻi*, crops carefully looked after will thrive.
Tai	January–February	If there be freshets in *tai*, a ryot can feed his mother.
Paṅkuṇi	March–April	Rain in *paṅkuṇi* causes great loss.
		If it rains in *paṅkuṇi*, the harvest will be damaged.

[a]From Benson (13).
[b]The astral periods begin when certain conjunctions occur. The precise time varies from year to year.
[c]Elder brothers have more prestige, power, and often wealth. Younger brothers live under the shadow of their elders. However, both can be equally impoverished, or starve in the absence of timely rains.

TABLE 4.10. Forecasting Rainfall: Wind Direction[a]

South winds and north winds make crops excellent.[b]
If the north wind blows it will surely rain.[b]
If the north wind blows it will rain immediately; if the south wind, the rain clouds will disperse.[b]
The westerly wind is accompanied by rain.[b]
If an east wind blows in *āṭi* and a west wind blows in *aippaci*, there will no rain that year nor in the next.
If there be an east wind in *āṭi* and a west wind in *aippaci*, well-grown and ill-grown crops alike will fail.
With east winds in *āvaṇi* and west winds in *aippaci* it is useless even to dream of rain.

[a]From Benson (13).
[b]No dates are alluded to in these proverbs, but south or west winds bring rain during the southwest monsoon, and north or east winds during the northeast monsoon. If there is a south wind during the northeast season a dry period results.

dire consequences of wind from the wrong direction at the wrong time also are warned against.

Natural phenomena are employed in the proverbs as devices to predict the immediate occurrences of rain: halos around the sun, rainbows, sky color, clouds, thunder, insect behavior, and animal and bird behavior (Table 4.11). Many of the proverbs indicate either the direction in which rain is falling or the direction from

TABLE 4.11. Short-Term Forecasts: Natural Phenomena[a]

Halo around the Sun

A large halo around the sun foretells rain during the day time.

Rainbows

If a rainbow appears in the east in the evenings or in the west, in the morning, it will rain.
A rainbow in the east foretells rain enough to breach the tank.

Sky Color

A red sky in the morning foretells heavy rain; in the evening, absence of rain.
A red sky in the morning (even in the rainy season) foretells failure of rain.
Red clouds at sunset foretell instant rain.

Clouds

If the sky darkens with clouds in the east, the desired rain will fall.
If it be cloudy in the northeast, there will be breaches in the tank bunds.
Dark clouds gathering in the northeast foretell early rain.

Thunder

Thunder in the morning, a hot sun at noon, and clouds in the evening are forerunners of rain.
A cloudy morning, a hot sun at noon, and thunder in the evening forebode no rain.

Insects

A flight of beetles augurs rain at a distance.
Rain ceases when winged white ants appear.[b]
Should winged white ants come out in the morning, the heavy rain will cease.
Excessive rain follows if white ants take wing in the evening.
If ants crawl up (to higher places) there will be a storm.
The dragonfly is the forerunner of rain.
If mosquitoes be active in the evening, there will be rain.

Animals, Birds

If frogs croak, rain will follow.
If cattle look bewildered toward the sky, it will rain.
If the cranes seek high places, it will rain.

[a]From Benson (13).
[b]The white ants are probably termites.

which it will come. Some of the proverbs are contradictory. This is the case for those concerning sky color. These proverbs probably come from different areas of Tamilnadu and reflect local conditions.

3.7 Summary: Forecasting, Proper Agricultural Operations, and Uncertainty

As we have seen, sets of proverbs are associated with each month of the Tamil calendar. The proverbs represent empirical generalizations of experience in the south Indian environment, and can be regarded as statements of what cultivators expect to occur. For example, rain in the month of *cittirai* (mid-April to mid-May) is viewed as undesirable as it is too early for cultivation to begin. At this time throughout Tamilnadu, rainfall increases, and then usually declines between mid-May and mid-June (*vaikāci*). Throughout most of Tamilnadu the southwest monsoon rains do not begin on a large scale until mid-June to mid-July (*āni*), and the importance of adequate rain in this period is reflected in a number of proverbs. Furthermore, it is believed that if *āni* is dry, *āṭi* and *āvaṇi* (mid-July to mid-September), also are likely to be dry. Other proverbs point out the uncertainty of rain in *āṭi*. The peaking of the rains associated with the northeast monsoon are reflected in numerous proverbs referring to *puraṭṭāci* (mid-September to mid-October) and *aippaci* (mid-October to mid-November). However, the uncertainty of rain, even at this time of year, also is acknowledged.

After mid-November, the decrease in rainfall is reflected in the proverbs. The variable weather of *mārkaḷi* (mid-December to mid-January) is noted as is the beginning of the dry season at the start of *tai* (mid-January to mid-February).

Agricultural tasks are tied to months and weather conditions in many proverbs. The importance of plowing in certain months reflects an accurate appraisal of the weather. Seed time is also related to months; inappropriate times to sow are noted.

Furthermore, the proverbs acknowledge the uncertainty of rainfall. Tamil peasants have made assessments of the chances of rain at different times, exemplified by the use of the word "if" in the proverbs. It is also possible to distinguish three attitude toward uncertainty (Table 4.12). The first is to make it determinate and knowable. In this approach the essential randomness that characterizes an uncertain pattern is replaced by a determinate order in which a sequence of events can be predicted. Another method of eliminating uncertainty is the art of "wishing it away" by denigrating the quality of a natural event to the level of the commonplace, or conversely, of elevating it to a unique position and ascribing its occurrence to a freak combination of circumstances incapable of repetition. A final method of accepting uncertainty is to deny completely the knowability of what is uncertain.

There appears to be no attempt in the proverbs to deny that uncertainty exists. But many proverbs try to make rainfall uncertainty determinate and knowable. Several devices are used to do this. All are concerned with predicting future occurrences of rainfall, or the lack of it, and the consequences of one or the other state. Furthermore, all are connected with natural phenomena or with astrological observations (Table 4.13). The largest number of proverbs are those concerned with rain in months and astral periods.

TABLE 4.12. Attitudes toward Uncertainty[a]

To Make Uncertainty Determinate

A halo around the sun forebodes famine.
A rainbow in the east indicates floods in the west.

To Make Occurrences Commonplace

It will rain heavily in *kārttikai*.

To Make Occurrences Unique

Land watered from wells need never fear famine.

To Deny Completely the Knowability of What Is Uncertain

Even Mahadeva does not know when it will rain or when childbirth will take place.
Nobody knows when it will rain and nobody can say where cattle will graze.

[a] From Benson (13).

The next largest set of proverbs that introduce determinacy deal with forecasts based on the occurrence of thunder and lightning. Lightning is generally held to foretell rain; thunder is deceptive. Thunder during the hot season may indicate rain, but later in the season heralds rain. The position of the planets, sun, moon, and stars are also widely used as predictive devices. Wind direction, often in association with a month or astral period, is also widely noted to bring an element of certainty into the uncertain world. Clouds, halos around the sun, rainbows, and red sky are also cited to reduce uncertainty,

TABLE 4.13. Devices to Introduce Determinacy into Perception or Rainfall Uncertainty[a]

Device to Introduce Determinacy N = 928	Number of Tamil Proverbs
Rain in months and astral periods	78
Thunder and lightning	25
Position of planets, sun, moon, and stars	17
Wind	16
Clouds	13
Rainbows	11
Insects	11
Red sky	5
Larger animals	4
Number of days specified types of rain in a month	3
Halo around the sun	2
Rain on specified days	2
Total	187

[a] From Benson (13).

and the appearance of certain insects are taken as an indication of rain. For example, beetles, butterflies, white ants, dragonflies, and mosquitoes are forerunners of rain. Animals and trees are mentioned in a similar fashion.

4 PROVERBS, ETHNOMETEOROLOGY, AND METEOROLOGY

Tamil knowledge of environment and resources has been transmitted over time in proverbs. Indeed, a catalogue of these agricultural proverbs can be compared to the Sanskrit literary genre, *sūtra*, as an outline of reflective and orally transmitted knowledge. Certainly, proverbs were the primary source of information about the monsoon for Tamil farmers before the coming of modern weather forecasting; however, linkages (which must necessarily exist) between folk knowledge of the monsoon and processes and folk practices of water and crop management have yet to be assessed systematically. Folk perception of the monsoon, folk theories and explanation of the monsoon, and folk perceptions of the relationship between rainfall and crops do emerge from a systematic analysis of Tamil proverbs dealing with climate, weather, and agriculture. However, much needs to be done before a true ethnometeorological understanding of the monsoon emerges. To make such linkages, field research in specific places is required, and even the results of such work could not be generalized for all of the Tamil country.

At a general level, the proverbs contain descriptions of weather events and their timing that accord reasonably well with modern weather descriptions. However, it must be emphasized that the views of the monsoon expressed in proverbs are tied to the fundamental paradigms of knowledge in Indian society. These paradigms ultimately cross the boundary between science and religion, even if day-to-day forces in the environment do not. In fact, to dichotomize between science and religion undoubtedly is unthinkable for many Tamil farmers: a simple pragmatic solution to a problem may well be an appropriate ceremony or prayer.

Today, major changes have begun to manifest themselves with regard to Indian farmers' knowledge and understanding of their environment. More than 100 years of contact with Western scientific ideas, and the acceptance of many of the basic premises of Western science, by educated elites at least, has led to apparent widespread changes in beliefs about the natural world. India has an elaborate and sophisticated scientific and technological community, which is involved in the education process and basic research. These are the people and institutions that are attempting to change existing agricultural technologies and practices into something new, and that new something is a rational, or at least rationalistic, scientific system of agricultural management.

Thus India is the theater of an indescribably complex interaction between the forces of modern science and technology and of an age-old metaphysical tradition. Despite numerous advances, in the village of India, custom is a king not yet and not easily dethroned (though increasingly tottering), and his laws are enshrined in the multitude of pithy proverbs that form a remarkable folk system of knowledge. That this system has survived is fortunate because it is increasingly recognized that indigenous knowledge has something to contribute to efforts to develop agricultural systems. To date, in its arrogance and ethnocentricity, Western science has only too often pushed aside the wealth of folk knowledge and experience that is embodied in traditional agricultural systems. But in an age of potential shortages of energy to fuel the new agrotechnologies, it has become doubly important to have some understanding of systems of traditional agricultural knowledge. In particular, it is being recognized that agricultural systems that have existed for as long as those of India represent at least partially successful solutions to specific sets of man–environment situations, and that the systems of knowledge associated with these complexities may contain valuable information, even if at the moment it appears somewhat anachronistic to those trained in Western science. Certainly, there is a wealth of knowledge about the weather and agriculture contained in Tamil proverbs that should not be forgotten, even in an era of space-age technology and computerized weather forecasting.

REFERENCES

1. C. Balasubramanian and R. Gopalkrishnan, *Madras Ag. J.*, **40**, 55–59 (1953).
2. M. S. Randhawa, M. S. Sivaraman, and I. J. Naidu, *Farmers of India, Volume 2*, Indian Council of Agricultural Research, New Delhi, 1961, pp. 97–105.
3. B. J. Murton, "Traditional Systems of Agricultural Knowledge and Contemporary Agricultural Development in India," in N. Mohammed, Ed., *Perspectives in Agricultural Geography, Volume 5*, Concept Publishing Company, New Delhi, 1981, pp. 401–414.
4. *Economic Atlas of the Madras State 1962*, National Council of Applied Economic Research, New Delhi, India, 1965, p. 21.
5. H. A. Simon, *Psycho. Rev.*, **63**, 129–138 (1953).
6. H. A. Simon, *Amer. Econ. Rev.*, **49**, 253–283 (1959).
7. K. J. Herring, "New Techniques for the Constitution of Ethnosciences," paper presented at Annual Meetings of the Association of American Geographers, Milwaukee, WI, 1975.
8. M. D. Morris, *J. Econ. Hist.*, **27**, 588–607 (1967).
9. A. Beals, *Village Life in South India: Cultural Design and Environmental Variation*, Aldine Publishing, Chicago, 1964, p. 30.
10. L. Dumont, *Homo Hierarchus: The Caste System and Its Implications*, The University of Chicago Press, Chicago, 1970.
11. A. K. Ramanujan, *The Interior Landscape*, Indiana University Press, Bloomington, 1967.
12. A. T. Embree, "Tradition and Modernization in India: Synthesis or Encapsulation," in W. Morehouse, Ed., *Science and the Human Condition in India and Pakistan*, Rockefeller University Press, New York, 1968, pp. 29–38.
13. Those in Tamil appeared in 1896, 1908, and 1933. See C. A. Benson, "Tamil sayings and proverbs on agriculture," *Bulletin*, Department of Agriculture, Madras, India, No. 29, New Series, 1933, pp. 1–75 (Reprint).
14. B. E. F. Beck, *Peasant Society in Konku: A Study of Right and Left Subcastes in South India*, University of British Columbia Press, Vancouver, 1972, pp. 52–56.
15. Madras Government, *A Statistical Atlas of the Chingleput District*, Director of Stationery and Printing, Madras, India, 1965, pp. 9–10.

PART III

IMPACTS AND GOVERNMENT RESPONSE

Abnormal monsoons have enormous economic consequences. Whether responsible for too much or too little rain, monsoons play a central role in the economic, social, and political stability of more than half of the world's population. Primarily monsoons affect agriculture and, as M. S. Swaminathan notes, "if agriculture goes wrong, practically nothing else will go right in most Third World countries."

The authors in Part III discuss the economic implications of weather and climate in monsoon regions. Where Raaj Kumar Sah, economist and academician, is interested in broad questions of how weather and climate information can be utilized in this part of the world, Swaminathan, scientist, administrator, and policymaker, focuses on strategies for coping with abnormal monsoons.

Sah contends that monsoon-region countries do not know how to utilize meteorological data. He states, "The lack of awareness of the potential benefits from weather and climate information has probably been the single most important obstacle to its application to economic problems." While the problem is not unique to tropical and monsoon-region countries, the problem is more acute because their economies are strongly agricultural and the climates are "hostile." Sah argues that proper utilization of weather and climate data would help to solve the economic problems of abnormal monsoons, but to do so will require substantial investments in research by developing countries.

Swaminathan discusses the more immediate problems of developing strategies for coping with adverse monsoons. While he recognizes the need to improve monsoon forecasts by making the long-term investments, he contends that effective management practices that employ available scientific and technological knowledge can solve many of the immediate problems. He concludes that "it may become possible to achieve relatively more stable agricultural production in a world where the abnormal monsoon is 'normal.'"

5

Tropical Economies and Weather Information

Raaj Kumar Sah
Department of Economics, Economic Growth Center,
and School of Organization and Management
Yale University
New Haven, Connecticut

INTRODUCTION

The thesis of this chapter is that investment in certain weather and climate-related research and applications presents an opportunity to many developing countries for making large economic gains. Most less developed countries (LDCs) fall within the tropical and the monsoon regions of the world [see (1) and also Chapter 1 of this volume for a definition of these regions]. Food and agricultural production in these countries is critically influenced by weather and, since these countries are predominantly agricultural, their overall performance is highly sensitive to weather uncertainties. Moreover, many parts of the developing world are climatically "hostile" areas (such as flood-prone, drought-prone regions and deserts) or are vulnerable to weather disasters such as cyclones and typhoons.

Some examples are valuable. During the period 1961–1970, 22 countries in the Far East and the Pacific region sustained damages of $9.9 billion from floods and typhoons, an amount that was almost as large as the World Bank loans to these countries during that period (2). The 1970 cyclone in Bangladesh killed at least a half million people and displaced twice as many. In India, extensive investments and efforts have gone into achieving 2–3% annual growth in agricultural production, whereas one bad growing season brought on by an abnormal monsoon can lower the season's agricultural output by as much as 15% [see (3) and the earlier issues of the same publication]. Because of the economic structure of these societies, weather has a deep impact on the prices and availability of goods, employment and growth prospects, and the economic well-being of literally billions of people.

It is, of course, true that a complete scientific understanding of weather and climate does not exist at present; in fact it might even be unrealistic to expect that such an understanding will be achieved in the foreseeable future. What is not realized generally, however, is that a perfect understanding of weather is neither necessary nor economically desirable. Indeed, beyond some point, the costs of obtaining the information would exceed the gains. The social gain from producing and distributing certain kinds of weather information, on the other hand, can be substantial if the cost of such information and its potential gains are kept in mind.

Many advances have taken place in the last few decades in understanding climate and weather. Some of these advances are directly applicable in developing countries with immediate and large payoffs, while for others it is yet to be demonstrated how they can be applied in a practical sense and more research is needed. In the past, policymakers have not been sufficiently aware of the ways in which the available scientific knowledge in this area can be applied in improving the functioning of economic activities. This chapter is an attempt to provide the needed perspective. Some parts of it draw from earlier works by the author (4, 5).

First, features of developing economies that are important to understanding the economic impact of weather are examined, as are certain aspects of the economics of weather and climate information. This is followed by a discussion of the state of the art of and the economic potential offered by different climate- and weather-related activities, highlighting those specific research activities and applications that might be the most promising candidates for investment and public support.

1 CLIMATE AND DEVELOPING ECONOMIES

The Tropic of Cancer and the Tropic of Capricorn broadly separate the developing and developed countries. Most developed countries lie outside these two tropics in the *temperate* region. Developing countries, on the other hand, are located between the two tropics, that is, in the *tropical* region. Within this region are located most of those countries in which monsoons are a distinctive feature of the annual weather cycle, such as Bangladesh, Burma, India, Indonesia, Malaysia, Malagasy Republic, Tanzania, and Thailand.

Many observations have been made regarding the possible climatic differences between the tropical and the temperate regions. Some of these are:

1. The year-to-year variability in the tropical rainfall is higher than that in the temperate region resulting, at times, in markedly different tropical weather and climate in successive years.
2. The tropical rainfall has greater seasonality, that is, intense rainfall in the tropics occurs only during certain short periods within a year.
3. The tropics have great extremes of climate present side by side, such as arid deserts in the neighborhood of wet and flood-prone plains.
4. In terms of human and animal suffering and loss, weather disasters have a deeper impact in the tropics.

More recently, however, it has been recognized that caution must be exercised when making such generalizations (6). The tropics contain a wide range of climate, and so do the temperate regions. More importantly, the influence of climate on human activities is far more complex than what can be captured through climatic variables alone. A host of mutually interrelated factors are involved, such as topography, soil and vegetation types, land-use patterns, and the structure of the economy.

For this chapter the most important view is that the nature of economic activities in tropical countries is highly sensitive to the variability of weather. The most obvious feature of tropical economies is their dependence on agriculture, which is critically sensitive to weather. In contrast, a larger proportion of economic activities in temperate countries consists of manufacturing and services, which are relatively weather insensitive. Also, the capacity of the temperate societies to absorb climatic shocks is higher. For example, a drought in the United States, a hard winter in the United Kingdom, or an extensive frost in the Soviet Union do have obvious economic ramifications, but these societies are at such levels of economic output that they can generally absorb the damage quickly. In contrast, much of the Sahelian population lives near subsistence and at the same time is dependent on meager and uncertain rainfall. Abnormal monsoons and droughts in the last two decades have caused starvation for more than 10 million of the Sahelian people, wiped out 30–80% of its livestock, and almost all of its other forms of savings as well as exports. This disruption of the Sahelian economy has been a tormenting experience from which recovery can only be slow and prolonged.

An obvious response to the high sensitivity of tropical economies to weather fluctuations has been to change the composition of economic activities toward greater weather insensitivity; for example, to have a greater dependence on irrigation rather than on rainfall for agriculture. This response is clearly desirable and, as we shall see later, it requires substantial information concerning weather and climate. Furthermore, we need to recognize that the scope of this response is limited by natural conditions as well as by resource limitations and time considerations. In India, for example, it is projected that by the year 2000 only 42% of the agricultural land can be irrigated (7). Such projections not only reflect the resource constraints over time, but also the fact that in many regions irrigation is technically not feasible and is prohibitively costly. Also, the dependence of a sizable part of the population on climatically marginal areas (areas in which agricultural productivity is most sensitive to weather fluctuations) is likely to continue, given the lack of alternative employment opportunities.

Under such circumstances, it would be prudent to recognize the potential gain from developing the use of climate and weather information, and consider generation and dissemination of useful information on weather and climate as an important form of productive public investment in tropical countries.

2 ECONOMICS OF WEATHER INFORMATION

Economists have made significant progress in the last two decades in understanding the nature of information and the role it plays in economic behavior. Some of this understanding is directly relevant to how one should think about weather information. [Gilbert and Stiglitz (8) and Stiglitz (9) provide a comprehensive discussion.]

First, information has an economic value only when it changes individuals' actions. If people take one action when they possess certain information, and another action when they do not, then the value of information is the expected gain from the former action *minus* the expected gain from the latter action. Naturally, information is valueless if people take the same action whether they do or do not possess the information.

This economic perspective is often not the one through which physical scientists view information. Information on past climate patterns collected on computer tapes, for example, might be considered highly valuable by atmospheric scientists. From an economic viewpoint, however, value arises only if the data lead to a determination of decision rules that users (say, farmers) adopt (in their own self-interest) while selecting among alternative courses of action. Therefore, the units of measurement of information that physicists and communication engineers typically employ, such as *entropy* or *bits*, are of little or no consequence from an economic viewpoint; what matters is the money value of information *in use*. Further, the same piece of information may have different values for different individuals, depending on its use. One would expect, for example, a large farmer to gain more, in an absolute sense, from agriculture-related climate information than a smaller farmer. If one abstracts from the refinements of social cost–benefit analysis, then the society's gain is the aggregate of all individuals' gains. The net social benefit (or the rate of return) is obtained by deducting the total cost from the aggregate gain.

This approach to measuring the value of information has some consequences in setting up research priorities. Clearly it is better in the short term to focus scientific efforts on generating a type of information that is likely to have higher value to the users. This, however, is not always the kind of research that is most appealing in scientific terms.

Two other points need to be mentioned. First, weather information differs in some important ways from certain other kinds of information. In a stock market, for example, one person's gain from having inside information concerning a corporation depends on how many other people have the same information. In the agricultural use of weather information, any one farmer's output is not affected by the information of others.

Second, it has sometimes been suggested that the users of better weather information might not gain because of what economists call the general equilibrium effects (10). Specifically, it is argued that if all farmers use better information, then their output would be higher which in turn, might reduce the prices of farm output leading to a social loss. This is quite a narrow view; it ignores not only the gains to consumers from lower prices but also a number of other aspects such as the possibilities of international trade and the movement of productive factors across different economic activities within the economy.

A final important feature of weather information in developing countries is that, at least in the near term, it is the governments and not the private sector that will have to supply it. Until tropical weather data sources and forecasting capabilities have reached an acceptable minimum, as they have in developed countries, private firms will not have sufficient incentive to invest in weather and climate research and applications.

3 CLIMATE- AND WEATHER-RELATED ACTIVITIES

The three main categories of climate- and weather-related activities are: (1) climatology, which entails collection and analysis of past weather data, and provides a more rational basis for making choices of activities in various sectors such as irrigation, plant husbandry, housing design, and land use planning; (2) weather and climate prediction, that is, the short-term and medium-term prediction of weather variables such as rainfall, temperature, hail and cyclones, and prediction of average weather conditions on monthly, seasonal, and interannual time scales; and (3) study of climatic trends, on still longer time scales, due to both natural causes and human activities. Though these categories are related scientifically, they differ in how far the available knowledge has advanced, the likely economic potential, the lead time after which the resulting information can be applied operationally, and the costs and benefits in extending the technology beyond what exists at present. A fourth category, not discussed here, consists of weather modification, such as rain formation and hail suppression. See Sah (4) for a discussion.

3.1 Climatology

It is easier to understand climatology if it is distinguished from weather and weather prediction. "Climate" is the statistics of the "weather." Climatology involves collection and analysis of past weather data on rainfall, temperature, sunshine, and wind patterns to obtain reliable inferences on the overall pattern and the range of weather that is likely to occur in the future. These inferences are then employed along with a variety of other data to identify those activities that are likely to be more compatible with the local climate than other activities. Thus, climatological information can lead to a better choice in areas such as irrigation design, groundwater extraction, crop type and cropping pattern, construction and transportation design, and plans for mitigating natural disasters. In contrast, prediction of weather—say, one day to a couple of weeks in advance—is best used in selecting the timing of operations, such as scheduling of sowing, irrigation, harvesting, construction, and recreation. This set of decisions is more operational in nature.

Some examples can present a picture of what climatology can do (11). If the past climatic data at a place shows large year-to-year variation in the number of rainy days in a cropping season, then subject to soil and hydrological characteristics of the region, it might be preferable to use those seed varieties that have shorter growth periods. These varieties may have lower yield in any one year but, over a number of years, the average yield may be better than varieties requiring longer growing periods and hence having higher overall susceptibility to weather fluctuations. These concerns have led to some interest in studying certain crop varieties and farming techniques that might better accommodate the peculiarities of monsoon behavior. Swaminathan discusses this in Chapter 6.

Climatological information has also been found useful in defining favorable periods for sowing and harvesting, estimating the irrigation needs of a region, determining optimal irrigation schemes (to avoid under- and over-irrigation), suggesting useful timing and methods for frost prevention and protecting plants and animals. Climatic variables, for example, have a pronounced impact on the growth and the intensity of attacks by pests, insects, locusts, and a variety of animal diseases. Depending on current conditions and past experience, warning can be quite useful to cultivators on the likely areas and the nature of an impending attack, so that the resulting losses could be reduced.

A landmark example of what can happen when climatological information is not employed is the East African Groundnut Scheme of Great Britian (12). In 1947, the British government implemented a scheme for mechanically cultivating groundnuts on 3.2 million acres of monsoon areas in East and Central Africa. The British government reported the capital cost to be about £24 million at that time, and the operating cost to be about £7 million per year. The project was expected to bring a net annual benefit of £10 million to the British government. Groundnut was the crop choice because the average annual rainfall appeared to be suitable for it. However, no analysis was done of the weekly pattern of rainfall within a year and the year-to-year variability in the timing and the amount of rainfall. It turned out that the variation of rainfall within a year and between years was such that groundnut was a doomed crop. Not surprisingly, the project had to be abandoned within five years with large losses. This financial disaster occurred because climatological factors were not taken into account.

A number of areas exist, other than agricultural, in which climatological information can be substantially beneficial. In land-use planning, such information can lead to a better allocation of land for activities such as agriculture, dwelling, pasture, industry, and recreation. In the past, many settlements have attempted to defy their local climate with unfortunate consequences. The imperial city of Fatehpur Sikri in India, for example, was begun in 1570 by Akbar and abandoned only 16 years later, primarily because of a severe lack of water.

In more recent times, we do not abandon, we just live in misery. The accumulation of industrial pollutants in many newly growing Asian cities is an important example of what can happen when industrial units are located in complete indifference to the prevailing wind direction.

Furthermore, a number of large-scale developmental schemes are being undertaken in developing countries for which no precedent and very little experience exist. Some of these activities include deforestation of large areas for agriculture and habitation, reforestation schemes, introduction of new animal and seed varieties, development of alternative sources of energy, large housing and town planning schemes, and opening up of areas previously uninhabited because of harsh environmental conditions. Not only are many of these projects crucial for the development of the areas concerned but also the success of these schemes is highly dependent on an adequate knowledge of the climate. In fact, an incomplete understanding of the local climate and its interaction with other factors could lead to costly mistakes. In all new agricultural projects it is necessary to insure that the agricultural practices adopted do not erode the soil—an experience common in many parts of the world. Such mistakes may be irreversible. It takes thousands of years for soil to acquire productive properties that it could lose almost instantaneously.

Applications of climatological information are highly desirable for several reasons. First, the state of the art for the collection and analysis of climatological data for many applications is well developed and can be employed immediately. Variations on the applications will obviously be required in different countries since there are climatic differences not only across countries but also within a country; but new research is not required to make use of climatic information. The crucial needs are trained personnel, reliable data collection, and proper analysis of the predominant local activities and the needs. Once these have been achieved, the applications must be tested and the results of the experiments effectively communicated. Second, the cost of climatological information is quite small compared to the possible benefits. For example, the cost of climatological information as a percentage of the total cost of a project has been estimated at about 0.01% for large hydrological works, 0.02–0.03% for housing construction and about 0.5% for town planning. The resulting benefit–cost ratios have been found to range from 50:1 to 2000:1 depending on the application (11).

3.2 Weather and Climate Prediction

A complex and dynamic interaction takes place in the earth's atmosphere which involves solar radiation, water vapor, oceans, icecaps, plants, and animal organisms. These interactions are the key to weather and climate prediction. The interrelationships, however, are only partially understood at present as the reader will discover later in this volume.

The three principal methods of predicting weather, with increasing sophistication, are: synoptic forecasting, statistical methods, and numerical (dynamical) modeling. These methods, or some combination of them, may, in principle, be used for climate prediction. However, the feasibility of using dynamical models for operational climate prediction has only begun to be studied (see Chapter 16).

In synoptic forecasting, the recent trends in weather or climate are examined and compared to similar situations experienced in the past. The forecast is based on the extrapolation of current trends arising out of such an analysis. Forecaster experience plays an important role in this method.

In statistical approaches, a large number of atmospheric and oceanic data at different places in the world are analyzed to obtain the best statistical correlations with the weather or climate at the prediction location. The summer monsoon in India, for example, has been found to have some statistical correlation on monthly and seasonal time scales with parameters such as the rain in South America and the snow accumulation over the Himalayas (see Chapter 16). In the recent past, satellite observations have enhanced the potential of statistical methods by providing worldwide observations (13).

The most sophisticated technique for weather prediction is numerical modeling. Here, the physical laws governing the atmosphere are used in a large computer model of the atmosphere to forecast its future state from current weather data. This methodology has undergone extensive development in and for the temperate regions and is now being used by many national weather services for operational forecasting. Some experimentation has been done with numerical modeling for the monsoon weather system, particularly for the Indian monsoon system. (See Chapters 15 and 16.)

In recent years, a combination of these three methods has been used in many Western countries to make operational weather predictions for the Northern Hemisphere mid-latitudes. The hourly forecasts of rainfall and temperature during winter are reliable up to 36–48 hours in advance and experiments have demonstrated useful forecasting skill up to one week in advance (14). Heavy snow can be predicted with skill 24 hours in advance. The prediction that a thunderstorm or tornado will hit somewhere within a prescribed area can be made several hours in advance. Average daily temperature and total precipitation forecasts are routinely available 3–5 days in advance. Most of these developments have occurred during the last two decades or so, with the advent of computers and satellites, which have facilitated the tasks of data collection, communication, storage, and processing. More recently, empirical or synoptic/statistical methods have been used by a few temperate zone countries for climate prediction on monthly and seasonal time scales. In the United States for example, for the past several years both operational and experimental forecasts have been available (15, 16).

In contrast, weather and climate prediction capabilities in tropical countries vary from nonexistent to qualitative predictions made 10–12 hours in advance, though exceptions exist such as in India (17). (See Chapter 17.) Typical forecasts have extremely limited usefulness for economic activities. Also, the advances that have taken place in the prediction of temperate weather are not always transferable to the tropics. The data requirement in the tropics often is different, and the present understanding of the tropical atmosphere is somewhat limited compared to that of the temperate atmosphere.

In the last decade, special efforts have been made to collect data on the tropical atmosphere. These efforts have been directed and financed largely by countries in the temperate region. A major reason is that some of the data needed for prediction beyond several days in advance in the temperate zones is the same as that needed for short-term and medium-term prediction in the tropical regions. This situation has resulted in bilateral and multilateral basic research efforts, the most ambitious of which was the Global Atmospheric Research Program (GARP), organized jointly by the World Meteorological Organization and the International Council of Scientific Unions (18–20).

Such collaborative programs can be quite useful to many developing countries since their resource limitations may inhibit mounting similar programs on their own. Apart from obtaining global data, a country can acquire access to modern equipment and a well-trained body of international scientists at a relatively small cost. However, these collaborative efforts should be viewed as complementary since the physical processes to be modeled in the tropics are often different from those in the temperate regions. The models constructed for medium-range prediction of temperate zone weather would probably not incorporate some of the aspects of the tropical atmosphere and may not be useful for tropical day-to-day forecasting. It would be necessary, therefore, to construct regional models with greater details of local weather in the tropical countries. This is being done by some atmospheric scientists but their research is in its early stages (see Chapter 16). The operational forecasting centers in developed countries continue to focus on their regional weather and climate requirements and this is not likely to change. Tropical countries must therefore invest in the development of their own forecasting capability.

The most important point here concerns the kind of weather prediction that developing countries should attempt. Alternatively stated, what aspects of weather and climate prediction are likely to be most profitable? The main aspects of a prediction are its lead time (one day in advance versus one month in advance), the time resolution of prediction (hourly forecast versus daily forecast), the spatial resolution (single prediction for 100 km^2 versus 10,000 km^2), and the conditionality of forecast (prediction of rain on the seventh day from today, if there is rain on the third day from today).

From an earlier discussion on the value of information, it should be obvious that finer information (or more detailed information) has a higher value, since more detailed information would never be worse. But at the same time, the technology for obtaining finer information has a higher direct cost; also it takes more time to develop, and the research venture is typically riskier. All of these add to the real cost of the project. Furthermore an examination of economic activities in the tropics indicates, as we shall see below, that the kinds of weather prediction which would bring large economic benefit does not necessarily involve great detail.

In South Asia, about 65% of the food-grain output is produced during the summer monsoon (June to September), the sowing for which is done immediately after the first major shower around June. The success of the sowing depends crucially on the onset of the monsoon, that is, whether the first major shower is followed by days of extensive rainfall. Up to 60% of the seedlings can die because of severe moisture stress if the seven days after sowing are completely dry. This occurs every year in at least some regions in India with varying levels of intensity. The losses from futile sowing can be substantially reduced if a prediction is available that today's rain will not be followed by a week of completely dry weather. This prediction is obviously far less detailed than, say, an hourly prediction of the exact amount of rainfall seven days in advance.

Similarly, South Asian farmers gamble against nature during harvest time which roughly coincides with the withdrawal of the summer monsoon. Heavy rains, which usually accompany the withdrawal, can destroy up to 30% of the ripe crop if the crop is still standing in the field. On the other hand, if farmers opt for an early harvest, then the yield is lower because the crop has yet to mature. A prediction that indicates whether there will be no rain, some rain, or heavy rain in the next week would improve the farmers' prospects substantially. The main point of these examples is that it is important to examine the nature of economic activities to determine the actual requirements and from there decide what is the most desirable direction of research; the most sophisticated research on weather and climate may not be advantageous.

In this context, one should be aware of the time involved in developing weather and climate prediction capabilities. Past experience has shown that the time lag between completing research and routine application is anywhere between 5 and 15 years. Thus while basic research is under way, less sophisticated but immediately available methodologies can be developed and used. Predicting weather more than a few days in advance, while routine in developed countries, is very much at the research stage for the tropics. Thus if sustained efforts are made now by tropical countries, it could take 10–15 years to make reliable weather prediction 1 week to 10 days in advance.

An issue related to weather prediction is disaster amelioration, which is critically important in many tropical countries. The past emphasis has been to alleviate the misery after a disaster has occurred. The best response requires a mixture of long-term planning for disaster-prone areas (based on climatological information) and a reliable prediction and communication system. (Swaminathan discusses strategies to accomplish this in Chapters 6 and 19.) For weather disasters such as cyclones and typhoons, predicting the timing and path of disaster (once the storm has been identified) has been facilitated by the use of satellite, coastal radar, and instrumented aircraft. Beyond that, the prediction of disasters is still in the early stages of research and further work is needed to improve reliability and increase the lead time for which predictions can be made.

It is worth emphasizing here, however, that even with the very best weather prediction, it is nearly impossible to save most of the economic assets in disasters such as major floods and cyclones. Also, these disasters usually occur in the same general locations, that is, they are part of the climatology of a region. Although weather prediction can play an important role in disaster amelioration, it is no substitute for more permanent remedies such as land-use planning based on climatological information.

The cost of creating a major center for weather prediction in a tropical country is estimated to be in the range of $15–18 million. (These and other estimates of investments mentioned later in this chapter are updated from references 4 and 21.) The benefits can be much larger. A modest advancement in the weather prediction capability in India, for example, could save an average of $200–$300 million every year in food production alone. This saving, representing an increase of less than 1–1.5% in the food production, is easy to visualize given that there are widespread crop losses from futile sowing and damaged harvests. In the years in which monsoon abnormalities are more widespread, the country's food production is drastically reduced. For example, the food production in 1979–1980 was 110 million tons compared to 132 million tons in 1978–1979—a drop of 16% in a year, much of which is attributable to weather and/or climatic irregularities.

The gain, in fact, could be even larger since the benefits of weather prediction accrue to the nonagricultural sectors of the economy as well. A notion of the economic desirability of weather prediction can also be obtained from some calculations made for industrialized countries. In some west European countries, for example, the benefit–cost ratio of the investment in 4–10 day weather forecasts was estimated at 25:1 (22).

Though formal social cost–benefit analysis has not been done for LDCs since the detailed data required for such an analysis are not available, it is expected that the gains to these countries from investment in weather forecasting are likely to be quite high. Moreover, the cost of starting centers for weather prediction are primarily fixed costs (e.g., creating scientific institutions); the recurring costs are relatively small. The benefits, on the other hand, will accrue year after year. Furthermore, even if scientific advances are modest, in the sense that any one individual makes only a small gain, the total gain to the society would be quite large because weather affects the economic well-being of such a large proportion of the population in most developing countries.

However, to be of economic value, improvements in forecasts must be accompanied by effective communication of those forecasts and efforts to educate the farmers and other users. This point cannot be stressed too strongly; improved forecasts are of no practical importance unless the people have ready access to them and understand how to use them.

3.3 Study of Climatic Trends

Since the beginning of this century, experts have attempted to understand changes in the earth's climate on different time scales (e.g., changes occurring within a decade versus changes over hundreds of years). A host of factors interact in a way which is intimidatingly complex, and this interaction produces changes in the earth's climate. Some of the factors are the passage of the solar system through the galaxy, variability of solar radiation, varying orbit and wobble of the earth in relation to the sun, continental drift, volcanic activity, atmospheric composition, and the dynamics of the oceans and polar ice caps. A precarious balance exists among these factors, and small changes can have big effects. (Kutzbach and Shukla discuss these interactions in Chapters 10 and 16.)

It is also recognized that the variability of climate is the rule rather than an exception. Thus major climatic changes in the future would be as normal as those that occurred in the past. Numerous climatic changes are recorded within living memory. The Tigris–Euphrates valley and Indus valley once supported large agricultural societies. North Africa was the fertile granary of the Roman Empire. The Little Ice Age occurred in Europe in the sixteenth to nineteenth centuries when carnivals were held on the frozen Thames.

Other climatic changes occur over much shorter time durations. The productive fishery in Peru, Ecuador, and Chile, for example, depends on the cold nutrient water brought up every year to the ocean surface by a combination of winds and ocean currents. Periodically, the absence of this upwelling, known as El Niño, disrupts fishing with serious consequences on the economy of this region (see reference 23 on the recent El Niño). Similarly, the occurrence of drought once in every 5–10 years in Asia and Africa represents the normal variability of climate rather than an abnormality.

Another dimension has been added by the collective human activities which inadvertently could influence the earth's climate. The changes in the spatial and ecological nature of human activities, such as deforestation and urbanization, since the beginning of this century have been rapid compared to those in the preceding history. The possibility that these changes can affect climate has been raised with many persuasive examples. According to one study, 5.7% of the earth's surface, an area larger than Brazil, is man-made desert created in the last two thousand years (24). Another study has indicated that the Sahelian desert is expanding at the rate of about 19 inches an hour (25). The warming of the atmosphere due to increased carbon dioxide emissions from rapidly rising fossil fuel consumption has become a major issue of discussion and concern (26, 27).

More recently, some researchers have asserted that the climate during the last half century has been unduly kind to human beings and that this kindness may not last long. However, current understanding of the complex interactions which determine the earth's climate is so incomplete that the available projections of the long-term trends represent opinions rather than rigorous and widely acceptable scientific conclusions. In fact, final conclusions may not be available for some time to come. Nevertheless, concern over these issues has led several organizations in the West and in the USSR to put considerable effort into investigating long-term trends. The results of this research, when available, could be useful to other countries as well.

4 PRIORITIES FOR THE FUTURE

Weather and climate services have existed in many developing countries since the beginning of this century and, over the years, have grown in size. The incentive for this growth in most less developed countries has come primarily from the meteorological needs of civil and military aviation and navigation; for example, from the needs of international airports that must satisfy the requirements of weather facilities prescribed by international regulations. As a result, most weather scientists in these countries remain specialists in aviation weather; policymakers, on the other hand, have continued to fund ongoing activities without considering the new opportunities offered by weather-related research. As a consequence, many useful applications of weather information in agriculture and industry continue to be neglected. Also, the vital link between weather scientists and other professionals, such as plant scientists, irrigation engineers, hydrologists, and livestock experts, have not been nurtured at all.

In the last two decades, a few institutions in the tropics have attempted to work on weather problems. In India, for example, weather bulletins are issued daily and longer-range forecasts for the northeast and southwest monsoons are issued two months in advance. In Tanzania and Nigeria, agrometeorological bulletins are issued routinely. An agricultural and hydrological institution has been organized in Niger with international support. Also, one of the activities of the International Rice Research Institute in the Philippines is the development of rice and other crop varieties and techniques better suited to monsoon conditions. These organizations, however, are too few and they typically have not received adequate support.

The lack of awareness of the potential benefits from weather and climate information has probably been the single most important obstacle to its application to economic problems. In the past there was little initiative by weather and climate scientists to change this situation; they had made little effort to demonstrate the economic usefulness of their subject. This is changing, particularly in the developed countries (see, for example, references 28–35). Economists, too, have typically disregarded the possible contribution that weather and climate information could make as an input in production and other decision making. Finally, the unavailability of resources and the associated lack of data, processing equipment, and trained manpower have retarded the development of weather and climate information technology in the tropics. Here too, the future looks more promising than the past. In 1979, the World Meteorological Organization established a World Climate Program designed to address these shortcomings (36). This program is a first step towards the establishment of climate data access and research and applications within the LDCs.

Among the categories of weather- and climate-related activities earlier discussed, climatology (especially applied climatology) deserves the highest overall priority in developing countries because of the very high potential benefit (compared to the cost) it offers. The time required to establish organizations that use climatology is relatively short and the technical requirements to create such organizations are not prohibitive. These organizations would require equipment for data collection, a small computer, an interdisciplinary team of applied climatologists, and an effective network for communicating useful results to users.

The collection of local climatological data can be facilitated to some degree by involving users. Similarly, the communication of results to users can be enhanced by giving greater emphasis to the immediate and pressing needs of users, and less emphasis to the questions of longer-term scientific interest. More than one such organization would be required in large countries with widely differing climates in their various regions. However, these centers would be relatively inexpensive, costing initially about $1.5 million each. The lead time in developing them would not be long. Many tropical countries already have some manpower in this field; others can acquire it in 3–4 years.

Another priority should be to create weather prediction capabilities. Setting up one or more major meteorological research and prediction centers would be required for this purpose, and joint efforts by neighboring small countries could be advantageous. The primary goal of these regional institutions should be to provide operationally oriented forecasts for different activities. This should include region-specific routine guidance for agricultural operations, water management and related activities, and with somewhat less emphasis, general purpose weather bulletins to be used by other sectors.

Typically these institutions would cost in the range of less than $15 to $18 million each including the cost of physical facilities, a large computer, and the training of scientists. The time before significant results could be expected could be at least 10 to 15 years. Despite that, the benefits from weather prediction would be more than commensurate, as discussed earlier. These institutions can reduce some of their cost and lead time if they make full use of the global and regional networks for data collection and processing. Also the institutions could draw upon the expertise and information that has been accumulated for more than a century by the World Meteorological Organization and more recently by centers such as the European Center for Medium-Range Weather Forecasts.

The prospects of economic gains from weather and climate information are in some ways similar to those from the Green Revolution. The initial impetus for the Green Revolution was provided by concentrated efforts, amid a great deal of skepticism, leading to certain scientific breakthroughs. These in turn, have brought an unprecedented growth in the yield of several crops in developing countries in the last three decades. What this chapter has argued is that for the world's developing nations, many of which are profoundly affected by the vagaries of the monsoon, a strong case exists for creating capabilities in weather and climate research and applications. Such an effort is inherently risky since scientific efforts of this nature need not always produce the desired outcomes, but it is worth undertaking since the potential gains are exceptionally large compared to the costs involved.

REFERENCES

1. C. S. Ramage, *Monsoon Meteorology*, Academic, New York, 1971.
2. World Meteorological Organization, *The Role of Meteorological Services in the Economic Development of Asia and South-West Pacific*, WMO No. 422, Geneva, 1975.
3. Government of India, *Economic Survey 1982–83*, The Controller of Publications, New Delhi, 1983.
4. R. K. Sah, *World Development*, **7**, 337–347 (1979).
5. R. K. Sah, in D. F. Cusack, Ed., *Agroclimate Information for Development*, Westview Press, Boulder, CO, 1983, pp. 295–300.
6. I. J. Jackson, *Climate, Water and Agriculture in the Tropics*, Longman, London, 1977.
7. Government of India, *Report of the National Commission on Agriculture, Vol. IV, Climate and Agriculture*, Ministry of Agriculture and Irrigation, New Delhi, 1976, p. 139.
8. R. J. Gilbert and J. E. Stiglitz, *Effects of Risk on Prices and Quantities of Energy Supplies*, Vol. 2, Electric Power Research Institute, Palo Alto, CA, 1978.
9. J. E. Stiglitz, *Information and Economic Analysis*, Oxford University Press, Oxford, in press.
10. L. B. Lave, *Econometrica*, **31**, 151–164 (1963).
11. R. Berggren, *Economic Benefits of Climatological Services*, World Meteorological Organization, Technical Note No. 145, Geneva, 1975.
12. A. Wood, *Groundnut Affair*, Bodley Head, London, 1950.
13. H. E. Landsberg, *The Value and Challenge of Climatic Predictions*, IX Meteorological World Congress, World Meteorological Organization, Geneva, May 19, 1983.
14. L. Bengtsson et al., *Bull. Amer. Meteor. Soc.*, **63**, 292 (1982).

15. D. Gilman, Long range forecasting: The present and the future, *Bull. Amer. Meteor. Soc.*, **66**(2), 159–164 (1985).
16. J. Namias, Remarks on the potential for long-range forecasting, *Bull. Amer. Meteor. Soc.*, **66**(2), 165–173 (1985).
17. R. A. Bryson and W. H. Campbell, *Environmental Conservation*, **9**, 51–56 (1982).
18. World Meteorological Organization, *An Introduction to GARP*, Vol. 1, ICSU/WMO GARP Publication Series, Geneva, 1969.
19. Global Atmospheric Research Programme, *The Monsoon Experiment*, GARP Publications Series 18, World Meteorological Organization, Geneva, 1976.
20. A. L. Hammond, *Science*, **188**, 1195–1198 (1975).
21. R. P. Pearce, *Research Priorities and Institutional Requirements in Tropical Meteorology Related to Agriculture and Hydrology*, mimeo, Consultant's Report to the Science and Technology Adviser's Office, The World Bank, Washington, D.C., March 1976.
22. Commission of the European Communities, Prospective Benefits from the Creation of an European Meteorological Computing Centre, Report by the Study Group on Benefit Analysis, No. 4, June 1972.
23. W. Brood, *The New York Times*, August 2, 1983.
24. C. Tickell, *The Climatic Dimension*, mimeo, Center for International Affairs, Harvard University, Cambridge, MA, May 1976.
25. N. Wade, *Science*, **185**, 2234–7 (1974).
26. W. W. Kellogg and R. Schware, *Climate Change and Society: Consequences of Increasing Atmospheric Carbon Dioxide*, Westview Press, Boulder, CO, 1981.
27. Carbon Dioxide Assessment Committee, *Changing Climate*, National Academy Press, Washington D.C., 1983.
28. Economic Commission for Africa and World Meteorological Organization, *The Role of Meteorological Services in Economic Development in Africa*, The World Meteorological Organization, Geneva, 1969.
29. W. J. Maunder, *The Value of the Weather*, Methuen, London, 1970.
30. R. Schneider, et al., *Applications of Meteorology to Economic and Social Development*, World Meteorological Organization, Technical Note No. 132, Geneva, 1974.
31. H. E. Landsberg, *Weather, Climate and Human Settlements*, World Meteorological Organization, Special Environmental Report No. 7, Geneva, 1976.
32. N. Nicholls, Long-range weather forecasting: Value, status and prospects, *Rev. of Geophy. and Space Phy.*, **18**(4), 771–788 (1980).
33. R. L. Winkler, A. H. Murphy, and R. W. Katz, The value of climate information: A decision-analytic approach, *J. Climatology*, **3**, 187–197 (1983).
34. B. G. Brown, R. W. Katz, and A. H. Murphy, On the economic value of seasonal precipitation forcasts: the fallowing/planting problem, *Bull. Amer. Meteor. Soc.*, in press.
35. A. H. Murphy, R. W. Katz, R. L. Winkler, and W. Hsu, Repetitive decision making and the value of forecasts in the cost-loss ratio situation: A dynamic model, *Mon. Wea. Rev.*, **113**(5), 801–813 (1985).
36. World Meteorological Organization, *Procedings of the World Climate Conference*, WMO No. 537, Geneva, 1979.

6

Abnormal Monsoons and Economic Consequences: The Indian Experience

M. S. Swaminathan
The International Rice Research Institute
Manila, Philippines

INTRODUCTION

Throughout history the monsoon-related calamities of droughts and floods have determined the pattern of life in Asia, Africa, and Latin America. Droughts and floods have been the most important natural causes of famine, although rural communities have learned to adjust their life styles and farming systems to excess water.

Millions of people all over the world depend directly on rainfall as their only accessible source of water. In the absence of monsoons that bring adequate rain, life styles can be seriously and sometimes irreversibly disrupted. The first effect of insufficient rains is crop yield reduction or crop death. In addition, recurrent drought may cause a serious shortage of drinking water for humans and animals. Water reservoir levels drop, reducing the generation of hydroelectric power, and the resulting energy shortages limit agricultural and industrial production.

This chapter, discusses the economic consequences of abnormal monsoons—both too dry and too wet—using the experiences on the Indian subcontinent as a focus. Farmers have experienced abnormal monsoons for thousands of years: this text will highlight some current practices that have been adopted to cope with the extreme monsoons. Finally, some promising strategies developed during the last few decades to assist rural communities in alleviating the consequences of abnormal monsoons will be discussed.

1 IMPACT OF ABNORMAL MONSOONS

Famines caused by abnormal monsoons, either too wet or too dry, have occurred frequently on the Indian subcontinent. Indian scientists have studied monsoon behavior for more than a century and, until recently, have found no persistence, cyclicity,

TABLE 6.1. Food Grain Production in India, 1960–1961 to 1981–1982 (millions of tons)[a]

Year	Adjusted Actual Production	Peak Point	Trough Point
1960–61	82.3		82.3
1961–62	82.4	82.4	
1962–63	80.3		80.3
1963–64	80.7		
1964–65	89.4	89.4	
1965–66	72.3		72.3
1966–67	74.2		
1967–68	95.1	95.1	
1968–69	94.0		94.0
1969–70	99.5		
1970–71	108.4	108.4	
1971–72	105.2		
1972–73	97.0		97.0
1973–74	104.7	104.7	
1974–75	99.8		99.8
1975–76	121.0	121.0	
1976–77	111.2		111.2
1977–78	126.4		
1978–79	131.9	131.9	
1979–80	109.0		109.0
1981–82	133.3		

[a]From Swaminathan (1).

trend, or pattern of abnormal occurrence. Recent studies do suggest that monsoon variability is related to other global features. The research on this important subject is described in Chapters 11 and 16.

In recent decades several severe monsoon droughts have occurred on the Indian subcontinent. In 1972 and 1979, deficient rainfall (about 25% below normal) was recorded in one-half to two-thirds of India's plains. In 1965–1966, rainfall was about 20% below normal in 14 of 31 meteorological subdivisions of India. The effects of these droughts on total food grain and rice production in India are indicated in Table 6.1 and Figure 6.1. World rice production shows a fairly regular upward trend but rice production in India, although rising overall, shows quite pronounced fluctuations. In contrast, the production of wheat, which is a winter crop irrigated to a considerable extent with groundwater, is relatively stable and shows a near-steady upward trend (Table 6.2).

Monsoons are also considered abnormal when they result in excess rainfall. Floods are often caused by extreme tropical storms called cyclones in the Indian Ocean. Although cyclones are not part of the monsoon per se, they generally develop during the local monsoon season and are related to the monsoon structure. The floods accompanying cyclones, as well as those resulting from other monsoon-

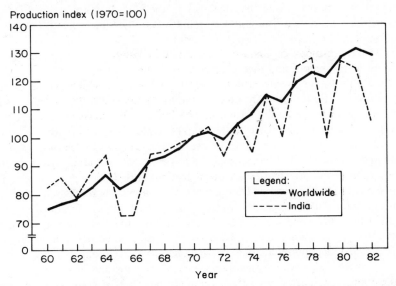

Figure 6.1. Trends in rice production worldwide and in India from 1960 to 1982 (from references 2 and 3).

related weather disturbances, cause much human misery and suffering. However, rural communities suffer less from them than from drought because good crops can be grown after the water recedes. Floods usually deposit silt, thereby adding organic matter and nutrients to the soil. They also recharge the aquifer, thereby improving groundwater availability. Extensive data on aquifer recharge have been collected by the Central Ground Water Board in India. The rate and speed of recharge depend upon soil physical and topographic features. Where there are hard pans just below the surface of the soil, water may not be able to go down to deeper layers. Consequently, there is a heavy surface runoff whenever the rainfall is heavy. Often flash floods

TABLE 6.2. Decline in Production of Food Grains in India in Recent Years of Adverse Monsoons (millions of tons)[a]

Year	Production in Year Preceding Adverse Monsoon		Year	Production in Drought Year	
	Food Grains[b]	Wheat		Food Grains[b]	Wheat
1970–71	108.40	23.83	1971–72	105.17	26.41
1971–72	105.17	26.41	1972–73	97.00	24.73
1978–79	131.90	35.50	1979–80	108.90	31.83
1981–82	133.29	37.45	1982–83	128.35	42.50

[a]Government of India Report (4).
[b]Food grains include wheat.

are caused by the poor water permeability of the soil. In chronically flood-prone areas, patterns of human living and crop cultivation become adjusted to monsoon conditions.

The impact of monsoon abnormalities on human and animal populations varies according to the nature and the severity of the calamity. Most problems relate to the availability of food, safe drinking water, and shelter, which can be destroyed by cyclones and floods. In addition crop failure causes not only a famine of food, but more importantly a famine of work. The Indian experience reveals that if there is work, food can be purchased in the market. Unless there is continuous drought or crop failure in several successive years, there is usually surplus food grains with traders and, in the case of India, with the government. In India, government-controlled grain stocks are distributed through a network of fair-price shops. In the 1960s, the prices were maintained through generous supplies of food grains, particularly wheat under the Public Law 480 (excess foreign currency) program of the United States. In the 1970s and the first half of the 1980s, the government of India had adopted a policy of maintaining substantial grain reserves, both by purchasing within the country and by commercial imports when necessary. Although the prices charged in the fair-price shops are reasonable because of the subsidies involved, people can buy the food only if they have the money. In the case of landless labor families that derive their entire income from daily wages, crop failures result in the loss of the only avenue available for them for earning their daily bread. Thus, lack of purchasing power is an important consequence of weather-caused damage to crops, livestock, and aquaculture.

Three recent examples to indicate the magnitude of the problem caused by abnormal monsoons were cited by a committee set up by the Planning Commission of India to examine disaster preparedness and management: (1) the 1977 typhoon in Andhra Pradesh claimed nearly 10,000 lives; (2) floods in Uttar Pradesh, Bihar, and West Bengal during 1978–1979 damaged 18 million hectares of cropped land, destroyed nearly 4 million hutments, and took a toll of 2,800 human lives and about 200,000 cattle; and (3) the 1979–1980 drought in large areas of northern and eastern India affected more than 38 million hectares of cropped area and endangered the lives of 130 million cattle and more than 200 million people. Food production dropped from 131.4 million tons in 1978–1979 to 109 million tons in 1979–1980 as a result of monsoon failure (5).

2 ACHIEVING STABLE AGRICULTURAL PRODUCTION IN MONSOON ASIA

Stable crop yields are an important objective of all agricultural research and development programs. Can monsoon predictions and early warning of natural calamities be accompanied by timely action to limit the adverse effects of abnormal monsoons? Let us consider how early predictions can be used to prevent or lessen the effect of adverse weather conditions. Rice, since its production is greatly influenced by monsoon behavior, is a good example.

2.1 Rice Production—An Example

Because rice is cultivated under such a great variety of conditions, it is impossible to define a "typical" rice climate. However, for optimum growth and yield the rice crop needs a combination of several conditions: a large volume of water evenly distributed throughout most of the growing season with a relatively dry ripening period, high temperatures throughout the growing season with somewhat cooler night temperatures during ripening, and adequate solar radiation.

According to Huke (6) the Shan upland of Burma and northern Thailand, which were some of the early areas of rice cultivation, probably had a 5- to 6-month growing season. The least amount of rain for any month during the first five months was 180 mm, and rainfall declined markedly during the sixth month. Mean monthly temperatures throughout the growing season ranged from 23.3°C to 27.7°C, the highest temperatures occurring early in the growing season and the lowest well after flowering. The amount of solar radiation received is positively correlated with production (7), and even under substantial cloud cover, the level of radiation is never low enough to hamper crop growth. Temperature and rainfall, however, have a strong effect on rice growth.

Where seasonal temperature is highly variable, low temperature determines the onset of the rice-growing season because it is critical to germination, seedling emergence, and stand establishment. Seedbeds in nonprotected rice fields can be prepared when the minimum temperature is about 18°C. The crop is normally harvested before the temperature drops below 13°C. The duration of the rice crop until flowering is primarily determined by temperature, day length, and the sensitivity of a variety to these two environmental factors. A daily mean temperature of less than 20°C may induce sterility. At flowering, rice is very sensitive to high temperature. Temperatures higher than 35°C may cause sterility. In India, the northern states often experience these critical temperatures during the year while the southern states have a smaller range of seasonal temperature (Table 6.3).

Despite the impact of unfavorable temperatures, rainfall is by far the most distinct climatic constraint to rice production mainly because of its extreme variability. Assuming average water losses from seepage and percolation for most rice soils to be between 1 and 2 mm/day and mean evapotranspiration losses between 4 mm/day in the wet season and 6 mm/day in the dry season, a minimum rainfall of 180 mm/mo is needed in the wet season and 240 mm/mo in the dry. In India however, a long wet season with at least four consecutive months of 200 mm rainfall each, is found only near mountain ranges in Assam.

Farmers traditionally have bunded their rice fields to impound water so that short dry spells are overcome and sufficient water is available throughout the critical parts of the rice-growing season. Submergence of the rice fields stabilizes the water supply of rice plants but may have some deleterious effects because it restricts gaseous exchange between the soil and the atmosphere. It also leads to the accumulation of the products of anaerobic decomposition such as methane, organic acids, and hydrogen sulfide. The accumulation of such compounds impairs the physiologic efficiency of the plant and can lead to yellowing of leaves and poor yields. On the other hand, submergence helps control weeds that compete with the rice plant for

TABLE 6.3. Mean Monthly Maximum and Minimum Temperatures at Some Locations in North and South India[a]

	North India (Chandigarh, 30°44′N, 76°53′E)		South India (Madras, 13°0′N, 80°11′E)	
	Maximum	Minimum	Maximum	Minimum
January	20.8	6.7	27.7	21.8
February	23.4	9.7	29.8	22.0
March	28.6	14.4	32.0	24.2
April	34.4	19.9	34.0	26.6
May	39.3	24.0	36.1	28.5
June	39.8	27.1	35.3	28.1
July	34.1	25.1	33.5	27.1
August	32.9	24.4	32.7	26.3
September	32.9	22.7	32.4	26.1
October	31.8	17.7	30.3	25.1
November	26.7	11.5	27.9	23.5
December	21.8	7.3	27.3	22.3

[a]From International Rice Research Institute (8).

nutrients. The high photosynthetic capacity of local weeds under high temperature and light intensity makes them very efficient competitors. The submerged conditions resulting from monsoon rains or from irrigation protect rice plants from severe competition with such weeds (9). Thus moisture excess has both beneficial and unfavorable consequences.

In some cases, drought and floods may occur in succession so that both must be faced in one season. During unfavorable seasons, a combination of direct seeding and transplanting from a community nursery may help maintain a satisfactory plant population. Under conditions where transplanting must be delayed because of an uncertain rainfall pattern, the rice variety used should be capable of producing stable yields even when old seedlings are transplanted. In general, photoperiod-sensitive varieties do reasonably well when transplanted late, but photoperiod-insensitive varieties respond to delayed transplanting with sharply reduced yield. Therefore it may be best to directly seed photoperiod-insensitive varieties and transplant photoperiod-sensitive varieties from a community nursery. Agronomists should undertake risk distribution research of this kind for every crop grown in a drought-prone area.

In semiarid areas lacking irrigation facilities, moisture conservation and storage techniques must be developed. If surplus water can be stored by developing suitable watershed management, a crop lifesaving irrigation may be possible. Areas having sufficient moisture to sustain a crop during drought years should be identified. In such areas additional extension and production efforts are warranted. These efforts include several improved management techniques such as ensuring adequate plant population, economic use of water, minimal or no tillage, improved weed control, supplying adequate nutrients, promoting better plant protection methods, and introducing improved postharvest technology. In the United States, chemical tillage (i.e., control of weeds using herbicides along with minimum or no plowing) has often been resorted to under conditions of moisture stress. This practice helps to conserve the available moisture.

2.2 Contingency Planning for Abnormal Monsoons

Given a forecast of rainfall, contingency planning demands a knowledge of alternative crops that can be planted and harvested within the time that sufficient moisture is likely to be available. Ideally, if the expected rainfall is not enough for growing sorghum, a short-duration crop such as millet, grain legume, or sunflower would be planted. Such alternative cropping strategies require that seed reserves for the various crops be maintained. In the National Seed Project of India supported by the World Bank, provision has been made to maintain appropriate seed reserves. Fertilizer reserves are also essential for successful midseason corrections in land use.

In most developing countries, major investments in irrigation are being made. Where the irrigated areas are large, people are better insulated from the impact of abnormal monsoons. If groundwater is the major source of irrigation, an adequate supply of energy for pumps is essential for stable production. For example, in the Punjab region of India, energy management is the key to stable crop yields during the southwest monsoon season. Data from 1972 and 1982 show that Punjab farmers can produce high rice yields even during severe drought if there is electricity or diesel fuel for irrigation pumps. Thus rice production in the Punjab was 3.4 million tons in the severe drought year of 1979–1980 as compared to 3.9 million tons harvested during the normal monsoon year of 1978–1979 (10). During the drought years, availability of hydropower goes down and there is a considerable energy shortage. During the past 10 years, the central and state governments of India have accorded high priority to the farm sector in power allocation, by shutting down industries, if necessary.

The efficient use of irrigated areas is important in a drought-management strategy. If necessary, governments can supply inputs such as seeds of high yielding varieties and fertilizers at low prices to enable farmers to increase crop yields. Small farmers seldom apply the recommended input levels because of problems such as inadequate credit availability and unfavorable grain-input price ratios. But in compensatory programs in irrigated areas during unfavorable seasons, a deliberate government decision to make inputs available at low cost can be taken. This is an option which governments should consider along with the option of obtaining grains under either commercial or concessional terms.

Response of this nature demands a "drought code" that clearly defines the necessary steps to maximize production where there is neither irrigation nor adequate soil moisture to grow a crop. The drought code should describe a two-pronged action plan. One set of measures should relieve and rehabilitate the affected human population; the other should seek to increase food production in favorable areas.

A point which is often overlooked is that chronically drought-prone areas need what might be called a "good weather code," that is, during a normal monsoon, measures must be taken to strengthen the ecological infrastructure to favor moisture retention and soil conservation. Unfortunately, funds are made available for relief work only during drought years when programs such as tree planting and watershed improvement cannot be undertaken. Only in years with normal rainfall can extensive afforestation, reforestation, and effective soil conservation measures be promoted. A good weather code assists people in managing subsequent droughts better. It is also important because it can help maximize production during good seasons, thereby contributing to a desirable level of grain reserve and more effective food storage policy. We should recognize that preventive action, such as that which would result from a good weather code, is *always* less costly than remedial action.

2.3 Other Considerations

In chronically flood-prone areas, the most important crop season should be after the floods recede. In many flood-prone areas, the major crop season is when floods are most likely to occur. Fortunately, these areas have a rich aquifer and if energy is available for pumping irrigation water, a good crop can be grown in the flood-free season. In parts of India, even bamboo tube wells are being used by farmers to pump groundwater from riverine areas. When solar photovoltaic systems become economically competitive, new vistas of production can be opened in many of the flood-prone areas. In the past, the photo-sensitive nature of most of the crop varieties cultivated by farmers limited the scope for large alterations in sowing and harvesting dates. The photo-insensitive and early maturing varieties now available in major crop plants provide considerable scope for a radical restructuring of cropping systems in order to mitigate the adverse effects of floods.

Drinking water shortages as well as contaminated and polluted water cause problems for humans and animals during drought and floods. A disaster management strategy must include plans to maintain drinking water security and provide adequate water for farm animals.

Groundwater sanctuaries in drought-prone areas might also be established and water infiltration and aquifer recharge can be improved by planting suitable trees and building percolation tanks. Water from groundwater sanctuaries should not be used except to alleviate an acute drinking water shortage.

Adverse weather, including severe storms, can seriously damage crops but it may be possible to distribute the risks by adopting farming techniques such as the rice garden technique developed at the International Rice Research Institute. Planting and harvesting are done weekly and can be carried out with family labor. Because plantings and harvests are spread throughout the year, risks are also distributed.

An important goal, as mentioned previously, is to increase overall production to build up grain reserves. There are two ways to do this—extending the area under cultivation and increasing the yields per acre. As shown in Tables 6.4 and 6.5, the only pathway open to many countries is productivity improvement because most of the cultivable land is already being used. In India, the area under forest has gone down considerably since the beginning of the century and is now estimated to be

TABLE 6.4. Cultivable and Cultivated Area[a]

Region or Country	Cultivated (ha/capita)	Cultivable (ha/capita)
South Asia	0.27	0.27
East and Southeast Asia	0.22	0.36
China	0.15	0.15

[a] From Colombo et al. (11).

less than 20% of the total land available. Over 45% of the land area is cultivated. Even marginal lands have been brought under cultivation. On the other hand, the population growth is expected to decline only by the end of the century.

Population projections show that 50 years from now India may have a population from 1,231 to 1,375 million depending upon the model used. If the programs now planned for improving female literacy, economic emancipation of women, and widespread voluntary adoption of the small family norm succeed, it should be possible to contain the population size at 1,231 million in AD 2030–2031. India will then need to produce 277 million tons of food grains to meet a per capita requirement of 225 kg per year (5).

Mehra (13) has studied the extent and causes of instability in cereal production in India. She found that after new technologies were introduced, instability in production increased. Hazell (14) compared the sources of increased instability in cereal production in India and the United States. In both India and the United States, the annual growth rate in cereal production during the last three decades has been about 2.7%. In the United States, the increase has come principally from increased yields per acre. In contrast, expansion in the gross cropped area was almost as important as greater mean yields in increasing cereal production in India. The stability achieved in areas such as northwest India, where high-yielding varieties of rice and wheat are grown with the help of groundwater irrigation, helps to offset the instability in overall cereal production in India caused by the uncertainty of monsoon rainfall.

The techniques available for raising crops successfully in semiarid areas fall under the following major groups: (1) land management, erosion control, and

TABLE 6.5. Pathways of Production Increase during 1975–2000[a]

Region	Contribution to Output Growth (%)		
	Arable Land Growth	Cropping Intensity	Yield
90 countries	26	14	60
Africa	27	22	51
Far East	10	14	76
Latin America	55	14	31
Near East	6	25	69

[a] From Saouma (12).

minimum tillage; (2) water conservation and management including the use of mulching practices and economical delivery systems such as drip irrigation; (3) control of weeds, pests, and pathogens; and (4) integrated nutrient supply involving the use of organic and biofertilizers together with small quantities of mineral fertilizers.

2.4 Using Weather Information

Several states of India have set up state-level crop weather watch groups. Such groups monitor the behavior of the monsoon and plan future courses of action in crop management both under rainfed and irrigated situations. They also suggest any mid-season corrections required for standing crops and provide surveillance of pest activities resulting from abnormal monsoons.

While the crop weather watch group can suggest crop planning or mid-season correction, depending on the monsoon forecast made available, it is also necessary that accurate climatological information based on available rainfall data be available and utilized for advance planning, particularly for planning alternative crop strategies.

The 1979 drought was widespread and scientists found that adopting improved practices with moderate levels of fertilizers and using high-yielding varieties insulated considerably almost all the crops against moisture stress. In fact, both in research farms and in the farmers' fields, crops were good in some areas in spite of the drought. Short-duration crops and varieties performed far better than long-duration crops and varieties. The fertilized crops tapped the moisture from lower depths because of deeper root systems and therefore withstood drought effects more than the unfertilized crops.

While the researchers were successful in saving their crops through the timely adoption of the best available management knowhow, farmers could not do likewise. Consequently, the production declined steeply to 109 million tons in contrast to 131.4 million tons harvested in 1978–1979. Rice, which is the most important monsoon season food crop, suffered a severe setback. Monsoon failure during 1979–1980 resulted in a drop of rice production to 42.2 million tons as compared to a production of nearly 54 million tons in the previous year. There are indications that early maturing rice varieties like IR50 could give yields of 2 to 3 tons per hectare in scanty rainfall years, in contrast with traditional long-duration varieties that fail almost completely. If monsoon behavior could be predicted with a reasonable degree of assurance 6 to 8 weeks before the scheduled commencement of monsoon rains, farmers could be advised with greater confidence and precision on the choice of varieties and management practices. Although useful predictions on this time scale will not be achieved overnight, there is growing evidence that it is possible, at least for broad regions (see Shukla's discussions in Chapter 16).

3 LINKS TO THE REST OF THE ECONOMY

In most developing countries, a vast proportion of people (60 to 80%) live in rural areas; in India that proportion is more than 70% (5). To these people, agriculture, including crop husbandry, animal husbandry, fisheries, and forestry, constitutes the most important avenue for food, income, and employment. Hence agricultural progress and stability hold the key to rural and agrarian prosperity.

But more importantly, in the economies of most developing and several industrialized countries, there are major linkages between agriculture and industry (15). Agriculture supplies the raw materials for employment-intensive industries. It stimulates and sustains industrial output through rural household demands for consumer goods and services. It influences industry through government savings and public investment. Its growth requires fertilizer, machinery, tools, chemicals, energy, and labor.

The International Food Policy Research Institute has attempted to quantify such linkages. Using a macroeconomic model, it has been estimated that from 1961 to 1972, a 1% increase in agricultural production in India generated a 0.5% increase in industrial growth. Agriculture increased national income by 0.7% per year (16).

Apart from its impact on national income, instability in agricultural production will affect most the poorer sections of the population. Poor people spend most of any additional income on food. In India, studies show that the poor spend 60% of increments to income on grain and almost 80% of the total on food products. Food and employment are intimately related through the multiplier effects of agricultural growth and development. Hence if agriculture goes wrong, practically nothing else will go right in most Third World countries.

4 CONCLUSION

In developing countries, where agricultural success is closely linked with the behavior of the monsoons, abnormal monsoons can inflict calamity on an enormous scale which affects all segments of the economy of those countries. While this chapter has dealt with only the Indian experience, such information is available for several countries in Asia (17). As the experience summarized in this chapter indicates, it may become possible to achieve relatively more stable agricultural production in a world where the abnormal monsoon is "normal." This requires enlightened agricultural practices and advances in our ability to predict the behavior of the monsoon on virtually every time scale ranging from a few days in advance to a few seasons in advance.

This raises an interesting question. How often are monsoons normal? What are the quantitative and qualitative dimensions of abnormality? Which abnormality has an agricultural consequence? Can we draw for each area a monsoon balance sheet listing the favorable and unfavorable consequences of normal and abnormal monsoons? For example, if drought is the result of an abnormal monsoon, the incidence of pests will be much less. Deep-rooted crops such as cotton, which are prone to pest damage during a normal monsoon, may give a higher yield during years with less than normal rainfall provided there is the minimum essential moisture for crop growth. We need a detailed classification of this type, crop by crop and ecological region by region. The implications of heavy deviation from what is accepted in an agroclimatic region as normal monsoon behavior need to be studied with regard to its potential negative and positive consequences.

With the help of science and technology, human ingenuity and endeavor can convert calamities into opportunities for progress. Through concurrent advances in weather forecasting, prediction of climatic trends, and abnormal monsoon management, humankind can be insulated from hardships caused by droughts and floods to a much greater extent than has been considered possible until now. This is the main message from the experience of India during the last 20 years and it has considerable significance in the context of the African food crisis caused by a succession of droughts in countries of the Sahel.

REFERENCES

1. M. S. Swaminathan, *Science and the Conquest of Hunger*, Concept Publishing Company, New Delhi, 1983, 508 pp.
2. A. C. Palacpac, *World Rice Statistics*, International Rice Research Institute, Los Baños, Philippines, 1982, 152 pp.
3. A. C. Palacpac and R. W. Herdt, *World Rice Facts and Trends*, International Rice Research Institute, Los Baños, Philippines, 1983, 41 pp.
4. Government of India, Planning Commission, "Report of the Working Group on Agricultural Production for the Formulation of Seventh Five Year Plan," New Delhi, India, 1983, 149 pp.
5. Government of India, Planning Commission, "Sixth Five Year Plan, 1980–1985," New Delhi, 1981, 463 pp.
6. R. E. Huke, "Geography and Climate of Rice," in *Proceedings of the Symposium on Climate and Rice*, International Rice Research Institute, Los Baños, Philippines, September 24–27, 1974, pp. 31–50.
7. S. Yoshida, *Fundamentals of Rice Crop Science*, International Rice Research Institute, Los Baños, Philippines, 1981, 269 pp.
8. Agroclimatic Data Bank Climate Unit, International Rice Research Institute, Los Baños, Philippines, unpublished manuscript.
9. I. Tanaka, "Climatic Influence on Photosynthesis and Respiration of Rice," in *Proceedings of the Symposium on Climate and Rice*, International Rice Research Institute, Los Baños, Philippines, September 24–27, 1974, pp. 223–246.
10. M. S. Swaminathan, Plant breeding in preparation for the 21st century, *Proceedings Indian National Science Academy*, B48(1), 1–18 (1982).
11. U. Colombo, D. G. Johnson, and T. Shishido, "Expanding Food Production in Developing Countries: Rice Production in South and Southeast Asia," *Report of the Trilateral Food Task Force*, Trilateral Commission, Bonn, Germany, 1977, 83 pp.
12. E. Saouma, *World Food Report 1983*, Food and Agriculture Organization of the United Nations, Rome, 1984, 64 pp.
13. S. Mehra, "Instability in Indian Agriculture in the Context of the New Technology," Research Report No. 25, International Food Policy Research Institute, Washington, D.C., 1981, 55 pp.
14. P. B. R. Hazell, "Instability in Indian Foodgrain Production," Research Report No. 30, International Food Policy Research Institute, Washington, D.C., 1982, 60 pp.

15. M. S. Swaminathan, "Agricultural Progress—Key to Third World Prosperity," *Third World Quart.* **5**(3), 553–566 (1983).
16. International Food Policy Research Institute, IFPRI Report 1982, Washington, D.C., 1983, 70 pp.
17. M. M. Yoshino, Ed., *Climate and Agricultural Land Use in Monsoon Asia*, University of Tokyo Press, Tokyo, 1984, 398 pp.

PART IV

PAST AND PRESENT CONCEPTS

The earliest knowledge of monsoons came from observations of nature. Whether farmers, mariners, or travelers, people experienced the monsoon and recognized its power and regularity. Attempts to understand the physical causes of monsoons were few. Early "scientific" knowledge consisted principally of descriptive passages by geographers and historians. The common people were content to attribute the mysteries of the monsoon to the gods. Nevertheless, knowledge of the monsoon patterns was employed to advantage by the mariners in their ventures into the oceans and by the farmers in planting and harvesting. Not until the development of modern science would there be an application of physical laws to the study of monsoons.

Bruce Warren begins Part IV with an account of the early mariners who both contributed to and benefited from the knowledge of the reversal of winds and currents. Warren provides tantalizing evidence that as far back as four millenia ago merchants scheduled their ocean voyages according to the monsoon season. By Alexander's time (356–323 B.C.), explorers and historians made specific reference to the monsoon using the terms "annual winds," "periodic winds," and "reversal of winds." Alexander himself was aware of the monsoon as the "untimely" wind changes hindered his military operations in the Indian Ocean.

Most of the advanced cultures of the time knew of the reversal of the winds. Greek and Arab scholars ventured theories about the causes, but as Warren notes, their speculations were based on analogy rather than physical principles.

With the development of modern science in Western Europe in the seventeenth century, the emphasis shifted from the simple recording of observations to a search for physical causes. Beginning with the period of the scientific revolution, Gisela Kutzbach traces the evolution of monsoon meteorology from early concepts based on theories of the general circulation of the atmosphere through scientific inquiry into the monsoon as a regional phenomenon and, finally, to the first attempts to link the monsoon to processes in remote parts of the globe. In following the evolution of the concepts of the monsoon, one is struck by the fact that the early investigators, despite their very limited observational information, more than a century ago deduced the major physical processes at work. To better appreciate the early scientists' insights, the reader may well want to reread Chapter 1 after finishing Kutzbach's chapter.

In the final chapter of this part, John Young presents the modern view of monsoons. In addition to reviewing the main processes that are responsible for monsoons, Young provides the general reader with an introduction to the important terms and principles that are used to describe monsoon physics. Authors in later chapters will assume the reader is familiar with these terms.

7

Ancient and Medieval Records of the Monsoon Winds and Currents of the Indian Ocean

Bruce A. Warren
Department of Physical Oceanography
Woods Hole Oceanographic Institution
Woods Hole, Massachusetts

INTRODUCTION

Merchant seamen have been sailing the Indian Ocean for 5000 years. Inevitably they observed winds and currents, and developed some lore about the great monsoon wind and current systems, knowledge that they could use to advantage in their trade. Until the relatively recent European presence in the Indian Ocean, however, not much of this lore seems to have been written down, and certainly much of that has not survived. In fact, what pre-European records have come down to us are quite dissatisfying, because they are skimpy, unsystematic, and uninformed by any understanding of the phenomena reported. Nevertheless, they are of interest as affording a few glimpses of what people separated from us by one or two millenia knew and thought about these dramatic natural phenomena. The purpose of this chapter is to review that literature up until the late Middle Ages (twelfth century), after which Western Europeans penetrated the Indian Ocean. Not surprisingly, most of this material comes from literary peoples with some curiosity about the world around them, specifically the Greeks and Romans and the early Moslems.

The earliest surviving descriptions of the monsoon winds come from the Hellenistic and Roman periods; they are quoted and discussed in Section 1, accompanied by some introductory commentary on the maritime trade in the preceding millenia. None of these ancient documents implies any knowledge of the seasonal reversal in surface currents, however: that discovery awaited the peoples of the Islamic empire. The pertinent Arabic literature is quoted and discussed in Section 2, which concludes with some brief exploration of medieval Arabic theorizing about winds and currents.

Chapter 13 of this book includes an excellent collection of modern surface-wind and current charts, to which reference may be helpful. The later, European development

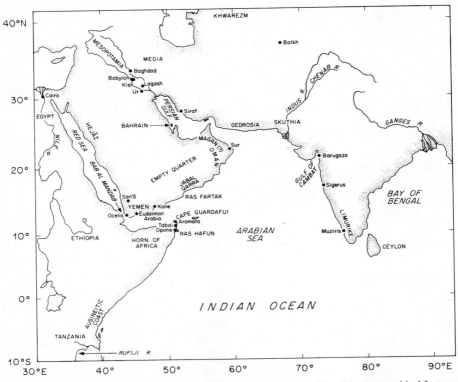

Figure 7.1. Map of the northwestern Indian Ocean and adjacent lands, identifying geographical features and place names cited in the text.

of information and ideas about the monsoon winds is recounted in Chapter 8. Readers interested in the broad historical context of Indian Ocean politics, trade, and seafaring through the ages might wish to consult Toussaint's general outline (1).

Where terms in the quoted material are unfamiliar or unclear, explanations are enclosed in square brackets. French and German translations of source material have been translated in turn into English. Geographical features and place names cited are identified in Figure 7.1.

1 ANCIENT DESCRIPTIONS OF THE MONSOONS

Civilized peoples populated and traveled the littoral of the North Indian Ocean in antiquity, and, especially in the Arabian Sea, also undertook sea voyages, albeit usually along coasts and not far from land. They therefore had direct opportunity to observe, within limited geographical regions, the monsoon winds and rains.

1.1 Northern Rim of the Arabian Sea

Man's oldest known shipping records for the Indian Ocean come from the ancient city-states of lower Mesopotamia. On an inscription dated about 2300 B.C., Sargon of Agade (actual site not yet discovered) boasted that "he made the ships from Meluhha, the ships from Magan, the ships from Dilmun tie up along side the quay of Agade" (2). Dilmun was almost certainly Bahrain, Magan was most likely an as yet undiscovered site in northwest Oman, and Meluhha was probably in the Indus valley, seat of the flourishing Mohenjo–Daro civilization (2, 3). Similarly, Gudea of Lagash claimed two centuries later that "Magan, Meluhha, Gubi [?], and Dilmun brought tribute; their ships came to Lagash with timber" (4). Even as early as the twenty-sixth century B.C.—the age of pyramid-building in Egypt—Ur-Nanshe of Lagash had been able to state about himself, "the ships of Dilmun, from the foreign lands, brought him wood as a tribute" (2). Sailors engaged in this long-standing trade across the Persian Gulf and northern Arabian Sea must surely have noticed the seasonal change in winds there, even though it is not so pronounced as farther south (5), and perhaps they scheduled their voyages according to the monsoons as well.

No mention of those winds seems to have survived from this period, but a few merchant records hint of a seasonality in this maritime traffic that may have been based on the monsoons. Among many Babylonian commercial documents translated by Leemans (3), eight that speak of preparations for, or activities following, individual voyages to Magan or Dilmun also specify the month of the year. Five of these deal with preparations; the earliest, from Lagash in the year 2031 B.C., states (3):

> 70 *gur* [8.5m³] of barley, load for Magan, has from the ensi [governor] of Girsu [city within the city-state of Lagash] B/Pudu received. Seal of Ur-gepar, son of Suna. (From) the storage field Nizina. In the month *še-KIN-kud* of the year Šu-sîn [a king from the Third Dynasty of Ur] 8.

The Sumerian month *še-KIN-kud* came during February–March, which is the time of the most favorable winds for sailing southeastwards in the Persian Gulf (5). Two more records, from Ur in the twenty-first century B.C., are also bills of lading for voyages to Magan, and are also dated *še-KIN-kud*, as is a contract for a Dilmun expedition in 1795 B.C. The fifth record, however, pertaining to a Dilmun voyage in 2029 B.C., is a bill of lading dated July–August, when the prevailing winds are least propitious for such a passage, although even then, favorable winds blow some of the time (5).

The three records bearing on specific return voyages report tithes paid during the twentieth century B.C. to the temple of Ningal (wife of the moon-god) in Ur after trading ventures to Dilmun. An example, dated 1922 B.C., is (3):

> 1 . . . stone from Meluhha, 8 pieces of . . . stone from an expedition to Dilmun, tithe for the goddess Ningal from what has been delivered by Milku-dannum, as shares from

single persons, delivered to the temple of Ningal. In the month *ne-ne-gar* of the year the army of Kish was defeated.

The month *ne-ne-gar* fell in July–August, and the winds at that time of year are the most suited for sailing northwestward from Bahrain (Dilmun) (5). Another record is dated June–July, also a time with generally favorable winds, but the third is dated February–March, when the dominant winds are least convenient for that passage. Even apart from the fact that these records do not state exactly when the cited voyages were actually made, they are far too few to demonstrate an overall seasonality in Babylonian seafaring, let alone to prove that it was governed by the monsoons. But that six of the eight records do suggest sailing during times of predominantly favorable winds is tantalizingly consistent with a recognition and use of the monsoons 4000 years ago.

Contacts between Dilmun and lands to the east appear to have ceased in the first half of the second millenium B.C., perhaps partly as a consequence of the destruction of the Indus valley cities by the Aryan invaders about 1500 B.C. Indo-Mesopotamian commerce resumed in the seventh century B.C. towards the end of the Assyrian supremacy (6), but it died out again during the fifth century B.C. (7).

The conquests of Alexander (356–323 B.C.) opened the Indian Ocean to the Greeks. Striving for ever greater glory, he dispatched his boyhood companion, the Cretan Nearchus (died 312 B.C.), on his famous naval expedition from the Indus delta to traverse and explore the then forgotten maritime route between India and Mesopotamia. Nearchus's own account of the voyage is now lost, but it was still available to Arrian (89?–180?), a Greek historian in the military and civil service of the Emperor Hadrian (76–138), as the basis for his description of the expedition in his *Indica* (mid-second century A.D.). Alexander's fleet was troubled by the southwest monsoon soon after its departure from the head of the delta in September 325 B.C., and Nearchus's discussion of the difficulties, as reported secondhand by Arrian (8, *Indica*, Chapters 21 and 22), seems to be the earliest surviving mention of the monsoon, even though it is only of that local, coastal manifestation:

> As soon as the annual winds were lulled to rest, they [the fleet] started on the twentieth day of the month Boëdromion [the third month of the Athenian luni–solar calendar, probably coinciding approximately with September in 325 B.C.]. These annual winds continue to blow from the sea to the land the whole season of summer, and thereby render navigation impossible.

However, upon reaching the coast about five days later, they found:

> Here great and continuous winds blew from the sea; and Nearchus fearing that some of the barbarians might band together and turn to plunder his camp, fortified the place with a stone wall. The stay here was twenty-four days. . . . As soon as the wind ceased they put to sea.

Evidently the southwest monsoon abated inland that year a month before doing so at the coast, and Nearchus had been deceived into leaving early.

When Arrian described the preparations for Nearchus's voyage in his narrative of Alexander's career and campaigns, *Anabasis Alexandri*, he explicitly mentioned the *reversal* of winds near the mouth of the Indus in autumn (8, Book 6, Chapter 21):

> The season of the year was then unfit for voyaging; for the periodical winds prevailed, which at that season do not blow there from the north, as with us,* but from the Great Sea, in the direction of the south wind. Moreover it was reported that there the sea was fit for navigation after the beginning of winter, that is, from the setting of the Pleiades [a star cluster which set at sunset on about 4 November at the latitude of the Indus delta in Alexander's time] until the winter solstice; for at that season mild breezes usually blow from the land, drenched as it has been with great rains, and these winds are convenient on a coasting voyage both for oars and sails.

Nowadays, at least, the winds continue offshore near the Indus delta from early November through January (5).

Arrian's assertion of "great rains" is puzzling, not so much because he incorrectly attributed the northeast monsoon to them, as because at the present time there is only scant rainfall on the Indus plain and neighboring mountains at all seasons of the year. Moreover, in reporting how a disastrous flash flood had hit one of Alexander's camps during his coastal march to Babylon, Arrian explained (8, Book 6, Chapter 25):

> the country of the Gedrosians [see Figure 7.1] is supplied with rain by the periodical winds, just as that of the Indians is; not the plains of Gedrosia [Makran], but only the mountains where the clouds are carried by the wind and are dissolved into rain without passing beyond the summits of the mountains.

Again, the present-day southwest monsoon does not bring rain far enough to the northwest to alleviate the aridity of Makran at all. During the second millenium B.C., the area at the head of the Arabian Sea was much wetter than now, and Arrian's remarks suggest increased rainfall again during Alexander's time (perhaps connected with known heavier rains over western Europe and northeastern Africa in the middle of the first millenium B.C.), but corroborative geological evidence is lacking (10).

Arrian's perception, however dim, of a causal connection between the rainfall and the winds and orography is striking, because awareness of that connection was not common around the Indian Ocean in medieval times (see Section 2). Arrian probably owed the idea, at least in part, to the principal sources for his description of India, which were the memoirs of Nearchus and of Aristobulus, an engineering

---

* The reference is to the "etesian wind" over the eastern Mediterranean and adjacent lands, which is a northerly wind occurring in summer, and is part of the general airflow toward and around the summertime low-pressure centers over southwestern Asia and the western Sahara. It is so steady that, for example, the wind at Cairo blows from the northwest, north, or northeast 98% of the time in July (9).

officer in Alexander's army, the *Indica* of Megasthenes (ca. 350–ca. 290 B.C.), and the *Geographica* of Eratosthenes (ca. 276–ca. 195 B.C.). None of these works is now available except in fragments quoted or paraphrased by later writers. Megasthenes obtained his information about the geography and people of India while conducting several embassies there around 300 B.C. for Seleucus I (ca. 356–281 B.C.), the general who had won the rule in the eastern part of Alexander's empire after his death. Although Eratosthenes, head of the library in Alexandria, is remembered chiefly as a geodesist and mathematician, he knew enough geography, as reported by the Greek geographer Strabo (64/63 B.C–ca. 23), to attribute the summer rain in India to some extent to the seasonal wind (11, Book 15, Part 1). He may well have learned that from Megasthenes, upon whose treatise he himself relied heavily in describing India, but the linkage is not mentioned in the surviving fragments of his *Indica*. Aristobulus had at least stated that the heavy rains happened concurrently with the onshore summer winds (11, Book 15, Part 1).

From the occurrence of the rainy season in India, Arrian also made a penetrating speculation about the cause of the annual innundation of the Nile (8, *Indica*, Chapter 6):

> India is visited by rain in the summer, especially the mountains, . . . and from these the rivers flow swollen and muddy. In the summer also the plains of India are visited by rain, so that a great part of them are covered with pools; and Alexander's army had to avoid the river Acesines [Chenab River, a tributary of the Indus] in the middle of the summer, because the water overflowed the plains. Wherefore from this it is possible to conjecture the cause of the similar condition of the Nile, because it is probable that the mountains of Ethiopia are visited by rain in the summer, and the Nile being filled from them overflows its banks into the Egyptian country.

Both Aristobulus and Nearchus (11, Book 15, Part 1) appear to have made this analogical argument earlier, and Arrian may have derived it from them.

Arrian did not go so far as to relate the summer rains in India and those correctly hypothesized in Ethiopia to the same large-scale wind system over the Arabian Sea, however, evidently because his awareness of the seasonality of the winds, like that of his sources, was confined strictly to the north coast of the Arabian Sea. Instead he supposed that similar conditions of rainfall in the two countries might be expected in parallel with the similar fauna and physical appearances of the peoples (8, *Indica*, Chapter 6).

1.2 Greater Reaches of the Indian Ocean

A commerce much more poorly documented than the Indo–Mesopotamian one had also developed in antiquity along the East African coast. It was highlighted in the second and third millenia B.C. by the sporadic Egyptian state expeditions to "Punt," which was probably located on the Horn of Africa or southwestern coast of the Red Sea. Apparently the earliest explicit mention of shipping to Punt is contained in a tomb inscription from the reign of King Pepi II in the twenty-third century

B.C., but neither this inscription nor later ones tell us anything about Egyptian sailing practices or winds encountered (12).

In the first millenium B.C. the East African commerce was handled mainly by South Arabians. The great age and geographical extent of this trade are demonstrated, for example, by the name "Ausineitic Coast" (Fig. 7.1) applied by sailors of the second century A.D. to a stretch of East African coastline south of the equator (13). The reference was to Ausan, which was a state that had flourished in Yemen during the seventh century B.C. and had evidently exercised some influence in East Africa. South Arabians traded with India as well during this era, and there were regular voyages around Arabia between Egypt and Persia under the reign of Darius I (521–486 B.C.); but, again, no facts concerning schedules or sailing conditions are available.

Graeco-Egyptian merchants participated in the East African trade during the Ptolemaic and Roman periods, voyaging as far south as present-day Tanzania (13). At least by the first century A.D., and probably for a century or two before, they were also crossing over to India, and Pliny the Elder (23–79) described something of their navigational experience in his *Natural History* (77). An official of the Roman Empire, Pliny had a passionate, if uncritical, curiosity about the natural world, to the extent of venturing so close to Vesuvius to watch its famous eruption as to be killed by poisonous fumes from it. His lengthy encyclopedia is a mix of fact and fancy, but in it Pliny provided the first known report of the large-scale seasonally reversing wind system over the full expanse of the Arabian Sea. After telling of Nearchus's voyage, he stated (14, Book 6, Chapter 26):

In later times it has been considered a well-ascertained fact that the voyage [to the Indus delta] from Syagrus [Ras Fartak], the Promontory of Arabia, reckoned at thirteen hundred and thirty-five miles [the mileage suggests that the coastal route is meant here], can be performed most advantageously with the aid of a westerly wind, which is there known by the name of Hippalus [the southwest monsoon; see below].

The age that followed pointed out a shorter route, and a safer one, to those who might happen to sail from the same promontory for Sigerus [modern Jaigarh], a port of India; and for a long time this route was followed, until at last a still shorter cut was discovered by a merchant, and the thirst for gain brought India still nearer to us. At the present day voyages are made every year: and companies of archers are carried on board the vessels, as those seas are greatly infested with pirates.

Later in the same chapter he gave more detail on the itinerary for the voyage to India, which started on the Egyptian coast of the Red Sea:

Passengers generally set sail at midsummer before the rising of the Dog Star [July 19 at the latitude of Cairo in Pliny's time; the rising of Sirius (Sothis) at sunrise, coinciding roughly with the annual innundation of the Nile, marked the Egyptian astronomical New Year], or else immediately after, and in about thirty days arrive at Ocelis [at Bab al Mandab, the straits at the southern end of the Red Sea] in Arabia, or else at Cane [Kane] in the region which bears frankincense. . . . To those who are bound for India, Ocelis is the best place for embarkation. If the wind, called Hippalus, happens to be blowing, it is possible to arrive in forty days at the nearest mart of India, Muziris [present-day Cranganore] by name. . . . Travellers set sail from India on their return to Europe [sic], at the beginning of the Egyptian month Tybis, which is our December [in Pliny's time], or at all events before the sixth day of the Egyptian month Mechir, the same as our ides of January [January 13]. They set sail from India with a south-west wind, and upon entering the Red Sea, catch the south-east or south [wind].

This is a description strictly of navigational practice, and not at all a scientific or even systematic discussion of meteorology. Nevertheless, it lays out roughly the seasonal reversal in winds over the northern Arabian Sea (although, if the manuscript is not corrupted, Pliny had the direction of the northeast monsoon near India wrong by 90°), and it demonstrates that the Graeco-Egyptian merchants of the first century A.D. were routinely conducting their trade to India in accord with it. A direct passage of 40 days from Ocelis to Muziris, however, if the transcription is correct, implies an average speed of only about 2 knots, which seems remarkably poor with a strong following wind and, probably, favorable currents as well; perhaps these mariners called at South Arabian ports to the east before striking out across the Arabian Sea.

Shortly after Pliny's death, the famous *Periplus of the Erythraean Sea* (13) was written, sometime during the years 95–130, by an unknown trader living in Egypt, probably a Greek. A "periplus" (meaning "circumnavigation") was a pilot book, and this guide gives information about ports, markets, and their imports and exports along the coasts of the Erythraean Sea, which was the name applied by the Greeks to the whole Indian Ocean, including the Red Sea and the Persian Gulf. Not surprisingly, the author of the *Periplus* had much more abundant knowledge than Pliny of Indian Ocean commerce; the trade routes he described began in Egypt on the Red Sea and extended both southward along the coast of East Africa to the Rufiji delta in modern Tanzania, and eastward to the west and east coasts of India, even on to Burma.

Like Pliny, though in more detail, the author of the *Periplus* described the exploitation of the southwest monsoon in the Indian trade, and he explained that the direct route across the Arabian Sea (as distinct from the earlier, roundabout coastal passage), utilizing the monsoon, was pioneered (for the Graeco-Egyptians, at least) by an otherwise unknown mariner called Hippalos (13):

The whole of the circumnavigation described from Kane and Eudaimon Arabia [Aden] was formerly made in small ships by sailing round the bays; but Hippalos was the first navigator who, by observing the positions of the marts and character of the sea, discovered a route across the ocean. Since then, when the winds blow locally from the ocean according to season, as with us [the reference is to the "etesian wind," which is an onshore wind in Egypt—see footnote, page 141], when the monsoon in the Indian Ocean appears to be south-west, it is called Hippalos from the name of the man who discovered the passage across. From then till now, some sail direct from Kane, others from Aromata [near Cape Guardafui], those sailing to Limurike turning the bows of the ship against the wind, and those going to Barugaza [modern Broach, in the Gulf of Cambay] or to Skuthia hold out to the contrary for not more than three days, and for the rest of the voyage keep their own courses clear of the land, sailing past the bays which have been mentioned.

He also stated:

Those who sail with the Indian [winds] put to sea [that is, from Egypt] about the month of July, which is Epiphi. The voyage is risky, but with these [winds] it is the most direct and shortest.

Given the ferocity of the southwest monsoon, one can readily understand that the voyage was considered ''risky.'' In several other instances he mentioned best months for the summer sailing, but, oddly, he never spoke of the northeast monsoon, nor of the times of the year for return trips from India (beyond a glancing remark about some wintertime voyages from India to South Arabia), though he must have been well acquainted with these matters from his own experience.

Apart from tides in harbors, there is only a single reference to currents, and it is a curious one (13):

From Tabai [near Cape Guardafui], after sailing along the peninsula for 400 stades [65 km], drawn by the current, they come to another mart, Opone [at the promontory now called Ras Hafun; both names derivatives of "Punt"] . . . The voyage from Egypt to all these marts [comprising the two above plus those on the southern side of the Gulf of Aden] beyond the straits [Bab al Mandab] is made in the month of July . . .

This is clearly misinformation, because only in winter is there southward flow along the Horn of Africa and these mariners could not possibly have been sailing southward against the full force of the southwest monsoon anyway. Unless successive scribes had merely botched the tenth-century manuscript copy from which the translation was made, it is tempting to think the author of the *Periplus* really did know about wintertime southward flow along East Africa, and carelessly included Opone with the marts in the Gulf of Aden to which one would sail in July.

Other than remarks about winds, the author only once mentioned weather; he spoke of the frankincense country between Kane and Ras Fartak on the South Arabian coast as ''mountainous and inaccessible, with a dense and cloudy atmosphere on account of the trees which bear frankincense'' (13). One does not really expect trees to cause clouds, and Puff (15) and Krümmel (16) considered the reported cloudiness to be fog formed over the cold water that upwells along the eastern part of the South Arabian coast during the southwest monsoon. This seems unlikely, however, because while haze is not uncommon over the cold upwelling regions in the Arabian Sea (17), fog itself seems to be extremely rare (18). Moreover, while reconnoitering South Arabia in preparation for his first traverse of the Empty Quarter, Wilfred Thesiger made a crossing in September 1945 of the mountain range Jabal Qarra, located near the coast some hundred miles east of Ras Fartak; and his remarks support and explain the observation by the author of the *Periplus* (19):

Some peculiarity in the shape of these mountains draws the monsoon clouds, so that the rain concentrates upon the southern slopes of Jabal Qarra, which are in consequence covered with mist and rain throughout the summer and were now dark with jungles in full leaf after the monsoon. All the way along the south Arabian coast for 1400 miles from Perim [island in Bab al Mandab] to Sur, only these twenty miles get a regular rainfall.

Undoubtedly Thesiger was right in attributing the rain and cloudiness to orographic uplift of the monsoon winds, rather than to cold surface water, or to frankincense trees.

Traders from the Mediterranean do not appear to have penetrated very much into the eastern Indian Ocean, where the commerce at that time was handled mainly by Indians and Malays. These people do not seem to have left surviving reports of sailing conditions (7), but several Chinese Buddhists who had traveled to India wrote accounts of their journeys, which contain a few brief allusions to the local seafaring. One of these, Fa Hsien, left China in 399 to obtain Buddhist scriptures in India; in the course of his travels he took passage on a ship sailing from the mouth of the Ganges to Ceylon, and he stated about himself (20):

After this he embarked in a large merchant-vessel, and went floating over the sea to the south-west. It was the beginning of winter, and the wind was favourable; and, after fourteen days, sailing day and night, they came to the country of Singhala [Ceylon].

The matter-of-fact way in which Fa Hsien associated the beginning of winter with the favorable wind suggests that use of the northeast monsoon was commonplace among these traders. On this particular voyage, though, if the text is rendered correctly, they certainly did not make any great speed—about 3 knots; even Pliny (14, Book 6, Chapter 24) had stated, without mentioning winds, that seven days was the usual time of passage from the Ganges to Ceylon.

Thus mariners in classical times were well acquainted with the broad features of the large-scale monsoon system over the Arabian Sea, and perhaps with that over the Bay of Bengal as well. But if anyone inquired into the nature and cause of the phenomenon, he left no mark of it in the surviving reports, which are much more collections of information than systematic treatises. It seems unlikely though, that any theorizing would have been productive then, because contemporary Mediterranean thinking about weather (21) was dominated by Aristotle's nonmechanistic, unverifiable, and ultimately sterile concepts, which even denied that the wind was air in motion, and supposed it instead to be a substance "exhaled" from the earth (see Section 2).

Except for the enigmatic remark by the author of the *Periplus* about southward flow along the Horn of Africa, none of the available writers spoke of ocean currents, let alone of the semiannual reversal in surface currents in the North Indian Ocean. The omission seems significant in the case of the author of the *Periplus*, because he was clearly a knowledgeable mariner, and he wrote his guide to assist other sailors. Had he known much about currents, he would surely have mentioned them;

that he does not seems strong evidence that little was known of the Indian Ocean circulation at that time, even to sailors. Such ignorance is not surprising, of course, because currents could only have been detected as displacement imparted to ships (distinct from that due to wind), and position-fixing was very primitive.

2 MEDIEVAL ARABIC LITERATURE

With the economic decline of the Roman Empire in the third century, Mediterranean traders disappeared almost entirely from the Indian Ocean, and reports of winds and currents are very meager in the next few centuries. In the seventh century, however, the Arabs made their prodigious conquests, and the political unification around the northwestern Indian Ocean set the stage for a great development of maritime commerce. The expansionary enthusiasm extended from conquest and commerce over into exploration and geographical inquiry, and scholars from among the peoples of the Islamic empire expounded their enlarged world view in travel books, geographies, and philosophical discourses. Some of these reported and discussed monsoon phenomena in the Indian Ocean and provided different information from that in the Graeco-Roman records. Indeed the very word "monsoon" comes from the Arabic word "*mawsim*," meaning "season"—as generally used to specify times of year for sailing between different ports, rather than periods of rain or certain prevailing winds (22).

The literature itself is usually called "Arabic" because it was written in the Arabic language, but its authors were not predominantly Arabs. In fact, this wide-ranging intellectual activity, stimulated especially by the Caliph al-Maʾmūn (786–833), flourished most vigorously during the Abbasid Caliphate (750–1258), which was actually rooted in the political and cultural traditions of Sassanid Persia (224–652). The multinational character of the Arabic literature will be evident in the extracts that follow.

2.1 Geographical Descriptions

To people who felt the direct effects of the Indian monsoon on land, its most spectacular feature was certainly the summer rain. While mention of it was common in Sanskrit proverbs and religious and medical texts (see Chapter 3), reports of it by the ancient writers are not extensive (e.g., see Arrian's remarks, as quoted in Section 1), and are surprisingly few even in the medieval geographies. One description, of interest for both its rarity and its detail, was given around 916 by Abū Zayd Hasan in some commentary appended to an earlier account of the trade route from the Persian Gulf to China. Abū Zayd was a scholar living in Siraf, which was the major port on the eastern coast of the Persian Gulf in the ninth and tenth centuries, and he collected information from voyagers there. He stated (23):

The regime of the *bashāra* in India—*bashāra* . . . signifying "rain"—is the following: In summer, the season of the rains lasts through three months, without interruption, night and day, during which the rain never ceases to fall. Before the season of the rains, the Indians prepare provisions. As soon as the *bashāra* begins, they settle into their houses, which are built of wood; their roofs are thick and overlain with thatch. During the season of the rains no one shows himself outside this own house except for important business. It is during this period of forced seclusion that the artisans and workmen devote themselves to their crafts. At this time, the humidity is such that it sometimes rots the soles of the feet. The *bashāra* gives life to the people of the land, for the rain makes it fertile; because, if it didn't rain, they would die of hunger. As it is, they plant rice; they know no other agriculture and have no other food than rice. During the season of rains, the rice is in the *hārāmat*—an Indian word that signifies "field of rice"; it is thrown down into the ground, and it is necessary neither to water it nor to look after it. By the time that the sky clears up and is no longer hidden by rainclouds, the rice has attained its maximum growth in height and bulk. During winter, it doesn't rain.

The "forced seclusion" reported by Abū Zayd may be contrasted with Singh's description in Chapter 2 of happy outdoor activities greeting the onset of the monsoon rains. Also, he was probably describing life in some part of northern India, because there is actually abundant rain over southern India in late fall and early winter during the northeast monsoon (see Chapter 4).

Abū Zayd is a most appealing writer because, while the medieval geographical literature abounds with tedious, nonsensical sea-stories, he expressed a heartening skepticism and restraint (23):

I have abstained from reproducing the lying tales that sailors tell and that they don't even believe themselves. To limit oneself to authentic information, even if in short supply, is preferable.

It is striking that Abū Zayd did not mention any connection between the summer rains and the winds of the southwest monsoon. Perhaps this fundamental physical relationship had not been generally perceived at the time (cf. Chapter 3), despite the ancient Greek glimpses of one. The astronomer al-Bīrūnī (973–1048) from Khwarezm, in his book on the religious institutes of various peoples and sects, *The Chronology of Ancient Nations* (ca. 1000), did cite an Arab verse associating south and west winds with rain, and north and east winds with dust storms (24); but neither the verse nor al-Bīrūnī referred explicitly to the monsoons, and it is impossible to infer from the material any sure knowledge of the monsoon relationship.

In the previous century the first of the great Arabic geographers, Ibn Khurradādhbih (ca. 820–ca. 912), had spoken briefly of the summer rains in southwestern Arabia in his *Book of Routes and Kingdoms*, which was written about 846–847, was revised not later than 885–886, and has survived only in abridged form (25):

In the Hijāz and the Yemen, it rains all summer, but never in winter. At Sanᶜā and around that city, the rain falls in June, July, August, and a part of September, from noon until sunset.

But, like Abū Zayd, he did not relate the seasonality of the rains to that of the winds, nor did he discuss the monsoon wind system at all.

Ibn Khurradādhbih was a Persian employed by the Abbasid government as chief of posts and intelligence in Media; one of his duties was to report to the vizier all information that touched on state security and public order. Presumably it was through collecting such data that he amassed the material for his geography book, which, in the available abridgement, is largely a description of the itineraries and road systems in the known world. His importance in the history of oceanography is that, in an isolated paragraph, he gave the earliest known report of the semi-annually reversing surface currents of the North Indian Ocean (25):

Abd al-Ghaffar the mariner, native of Syria, being questioned about ebb and flood, gave the following explanation of it. This phenomenon appears in the sea of Persia [Persian Gulf] at the rising of the moon; in the great sea, it is divided into two seasons: the summer flood is toward the direction eastnortheast, for six months; at this time, the sea rises in the eastern areas like China, and it subsides in the western areas: the winter flood is toward the direction westsouthwest, during the six other months; the sea rises then in the western regions.*

The medieval Arabic geographers never had any concept of horizontal circulatory gyres in the ocean, and they always spoke of currents as water periodically emptying out of one place and piling up in another. Probably they owed this way of thinking to Aristotle, whose works were being translated into Arabic by the ninth century, and who had claimed that Water could "flow" only in its "natural direction," which, like that of Earth, was downward (the natural direction for his other two primary elements, Air and Fire, was upward): thus the ocean, occupying the lowest places, could only "swing to and fro" (27, Book 2, Section 1). Hence the monsoon currents are conceived here essentially as an annual tide. Nevertheless, the description is roughly accurate for the North Indian Ocean with respect to both direction of flow and season.

Of course, Ibn Khurradādhbih simply related a sailor's report, and he did not speculate at all about the cause of the current reversal. His account does suggest that this knowledge of surface currents was common among the sailors, and it is therefore difficult to judge when (and how) they first learned of the seasonal reversal.

In the next century the remarkable historian al-Mas῾ūdī (ca. 915–956/7), brought up in Baghdad and aroused by a curiosity about the great world, traveled by land and sea, on the Indian Ocean, in India, the Near and Middle East, and East Africa, before settling in Cairo late in life. He was the first Arab to combine history and geography in large-scale treatises, and while his massive encyclopedic works have not survived, a condensed version that he issued in 947 under the title *Meadows of Gold and Mines of Gems* is extant. The style of the book is eclectic rather than systematic, because al-Mas῾ūdī simply merged his own observations, what he had read, and what he had heard from other people, without much critical comment or effort to form a coherent whole. He was sufficiently interested in the oceans and their movements to include a chapter titled "The different opinions on ebb and

---

* An English translation of this paragraph from a different Arabic manuscript copy of the abridged work has been given by Aleem (26); there are no relevant differences between the resulting texts.

flow and all that has been said on this subject,'' and there he too discussed the seasonal reversal of surface currents, first stating (28):

> Many of the Nawàjidah (this is the name for the sailors of Siràf and 'Oman, who are constantly on this sea, and visit various nations in the islands and on the coast) say that the ebb and flow takes place only twice a year in the greatest part of this sea, once in the summer months, then the ebb is six months north-east, during which the sea of China and of other countries of that quarter of the globe is high, for the water flows then from the west; and once in the winter months, then the ebb is six months south-west, for in winter the sea is fuller in the west, whilst the sea of China ebbs.

This description is so close in both substance and form to that of Ibn Khurradādhbih (whose book al-Mas῾ūdī greatly admired) that al-Mas῾ūdī must just have paraphrased that account here, without incorporating anything he might have learned from his own experiences in the Indian Ocean.

Immediately continuing, al-Mas῾ūdī went badly astray (28):

> The motions of the sea cohere with the course of the winds, for when the sun is in the northern hemisphere, the air moves to the south, hence the sea is during the summer higher in the south, for the northern winds are high and force the water there. In the same way when the sun is in the southern hemisphere; the course of the air, and with it the current of water, is from the south to north, and hence there is less water in the south. The shifting of the water in these two directions, from south to north and from north to south is called the hyemal ebb and flow; the ebb of the north is flow, in the south *vice versa*, and if the moon happens to meet with another planet [the translator suggests that the Arabic text is corrupted here, and should read: ''and if the sun happens to meet with the moon or another planet''] in one of these two directions, the warmth is increased by their joint action, and hence the current of the air is strongest towards the hemisphere which is opposite to that where the sun is.
>
> El-Mas᾽údí [translator's spellings here and following] says, this is the hypothesis of el-Kindi and Ahmed Ben et-Taíb es-Sarakhsi, and what we have said is borrowed from them; namely that the motion of the sea coincides with the course of the winds.

Ya῾qūb Ibn Ishāq as-Sabah al-Kindī (ca. 805–ca. 870), born in Yemen and educated and employed in Baghdad, was the first outstanding Islamic philosopher, a polymath with interests ranging from Aristotelian philosophy to music to manufacture of swords to cooking; al-Sarakhsī (ca. 835–899) was his student. Breadth of interest notwithstanding, it is unclear from al-Mas῾ūdī's remarks whether al-Kindī and al-Sarakhsī themselves actually addressed the specific monsoon phenomena of the Indian Ocean in their theorizing; nor can one be sure exactly what they said about winds and currents, because none of al-Sarakhsī's work has survived, and much of al-Kindī's existing work has never been edited and published, or translated from the Arabic. Of al-Kindī's known works (29), the single one that seems pertinent to al-Mas῾ūdī's statement is his treatise, ''On the efficient cause of flood-tide and ebb-tide,'' only parts of which have been translated from Arabic (30, 31). Al-Kindī's theory (discussed more fully below) attempted to identify the flow of air and water with thermal expansion caused by the supposed friction of the planets (sun, moon, Mercury, Venus, Mars, Jupiter, Saturn) moving over the atmosphere. While al-Kindī's recognition of a correspondence between winds and currents is encouraging, his discussion in the translated portion of his treatise is nonspecific, with no application whatsoever to the monsoon. Probably al-Masʿūdī, being acquainted with al-Kindī's general theory of seasonal winds and currents, just added it, in the eclectic, uncritical spirit of his chapter title, to his paraphrase of Ibn Khurradādhbih's account, without noticing or caring that the two descriptions of the current reversal were contradictory, or that the theoretical version had the seasonality of the winds wrong.

Al-Masʿūdī later described sailing conditions in the western Indian Ocean (28):

The difference of the currents and height of water is to be attributed to the direction of the winds, the season when they rise, and other causes. The Persian sea [Persian Gulf] is most stormy, and most dangerous for navigation, at the time when the Indian sea is quiet; and, again, the Persian sea is quiet when the sea of India is boisterous, stormy, dark, and rough. The sea of Fàris [Persian Gulf again] begins to be stormy when the sun enters into the sign Virgo [August 27 to September 23], about the time of the autumnal equinox; it continues so, and storms increase every day, until the sun comes into the sign Pisces [February 20 to March 20]: it is roughest at the end of autumn, when the sun is in the sign Sagittarius [November 23 to December 21]: then it becomes more quiet until the sun enters into Virgo, and it is most quiet at the end of spring, when the sun is in Gemini [May 22 to June 21]. The Indian sea is stormy till the sun enters into the sign Virgo: then begins the navigation on it; for it is easiest when the sun is in Sagittarius.

In support of this description, he then cited statements by the astrologer Jaʿfar Ibn Muhammed Abū Maʿshar (787–886) from Balkh (present-day Wazirabad, in northern Afghanistan) a century earlier in his *Great Introduction to the Science of Astrology* (849–850), which, one suspects, might actually have been al-Masʿūdī's source for the information. According to al-Masʿūdī (28), Abū Maʿshar confirmed

what we have said, that the stormy and quiet seasons on these seas begin when the sun is in the above-mentioned signs of the zodiac; and he relates further, that it is impossible to sail from ʿOmán on the sea of India in the Tírmáh (June), except with first-rate vessels and light cargoes.

The seasonality of the maritime traffic reported here was different from that in the ancient period, and al-Masʿūdī and Abū Maʿshar seem to have been implying that it was determined in their day not so much through seeking favorable wind directions as from avoiding the violent winds and heavy seas of the southwest monsoon. Villiers (32) said much the same thing about the Arab trade along the East African coast a thousand years later when he pointed out that the lateen-rigged dhows did not need following winds but could progress with winds from the side of the ship; that these mariners sailed the East African coast during the northeast monsoon mainly because of the good weather at that season; and that they tried to get home before the southwest monsoon set in at full strength. Although the lateen sail was not entirely unknown during classical times in the Mediterranean (33), the earliest actual evidence for its use in the Indian Ocean is from the ninth century (34); the Arabs must have been using it for some centuries before, but presumably the sailors of whom Pliny and the author of the *Periplus* spoke—as well as their predecessors—were still using square sails, which required that they sail before the wind, and therefore with the southwest monsoon.

In line with al-Mascūdī's remarks, however, Pliny (14, Book 6, Chapter 24) had stated of the local sailing of the Ceylonese:

> They devote only four months in the year to the pursuits of navigation, and are particularly careful not to trust themselves on the sea during the next hundred days after our summer solstice, for in those seas it is at that time the middle of winter [that is, the stormy season].

Al-Mascūdī himself commented (28):

> I have frequently been at sea; as in the Chinese sea, in the sea of er-Rum [Mediterranean Sea], in that of the Khazar [Caspian Sea], of el-Kolzom [Red Sea], and in the sea of el-Yemen: I have encountered many perils, but I found the sea of the Zanj [waters off East Africa] . . . the most dangerous of all.

This is surely another reference to the southwest monsoon.

One of the most famous of the medieval geographies was that by al-Idrīsī, published in 1154. Al-Idrīsī (1100–1165) was born and died at Ceuta in Morocco, but worked in Palermo from 1138 to 1154 for the Norman King Roger II (1093–1154) of Sicily, who sought an up-to-date map and geography of the world. Al-Idrīsī's book was based both on earlier literature and on new information collected from voyages dispatched by his patron for that purpose. He noted the periodicity of the monsoon wind offshore of Southeast Asia (though not mentioning it over the Indian Ocean proper), and he described the semiannual reversal in surface currents in the northwestern Indian Ocean; speaking of ebb and flood, he said (35):

> It is reported that, in the seas of Oman [northern part of the Arabian Sea] and Fars, this phenomenon takes place twice in the year, such that the flood is felt during the six months in the eastern sea, while the opposite takes place in the western sea; then the ebb goes back to the west during the six other months.

Inasmuch as Ibn Khurradādhbih was one of al-Idrīsī's sources, it is not surprising that the language here, even after three centuries, is still reminiscent of Ibn Khurradādhbih. Nevertheless, despite the personal experience and fresh information that al-Mascūdī and al-Idrīsī obtained, one cannot help but wonder in reading their books whether the sailors' knowledge of the seasonally reversing currents would have ever found its way into the medieval literature if Ibn Khurradādhbih had not been in the business of compiling intelligence for the Abbasid government.

Thus, although the Arabic geographers were aware of the seasonal current reversal, their knowledge of it was disappointingly meager. Their reports are little more than repetitions of the scanty information obtained by Ibn Khurradādhbih, and none of these writers seemed sufficiently intrigued by the phenomenon to learn more about it, either from personal observation or from additional inquiry. Not one of them, for example, alluded to the powerful Somali Current, even though Moslems had been on the East African coast at least since the eighth century (1). The sailors may have known a good deal more, of course, but if they did their knowledge was not recorded in any form that has lasted to the present day.

2.2 Arabic Physics

The understanding of winds and currents was even less satisfactory, because the physics of the time was a muddle of Aristotelian natural philosophy, Islamic religious sentiment, astrology, and folklore. Even at its best the theorizing did not lie in deduction from general physical principles, but in qualitative arguments from analogies, often irrelevant, with matters of everyday experience.

The most sophisticated of these theories of fluid motion (albeit an utter failure) was al-Kindī's discussion, in an Aristotelian framework, of flood and ebb. He first described "accidental" flood, the increase of volume of a system resulting from fluid added to the system, and then considered the "essential" or "natural" flood, the possible increase of volume inherent in the nature of the fluid, that is, thermal expansion (31):

> We have therewith treated the flood that arises from addition of matter. The natural flood, that depending on the essence of matter, is generated not through added matter; that is, it is the increase of volume of water in a natural way. . . . This occurs through a heating of the matter, but not otherwise. Every body that is heated requires a greater space, and every body that is cooled requires a smaller space. This can be shown easily with an instrument. . . . One places an inverted flask or something similar of glass in a glass vessel with water. . . . When the air [in the flask] is warmer, a bubbling occurs in the water on account of the air that comes out of the flask; that happens when the air changes into a higher degree of heat than that which it had originally, at the immersion of the instrument. For that reason matter assumes a greater volume. But then the air needs a greater space. Hence it presses on the water in the vessel, and penetrates into it; thereby arise little bubbles, in proportion to the modification of the air to the greater heat. On the other hand, if the air becomes colder . . . then it contracts and must occupy a smaller space. Its volume then becomes smaller and must draw the water along with it. . . . Therefore one sees with his eyes the water rising into the neck of the flask, in which it ascends above the outer vessel so that no vacuum exists. In nature, of course, there can be no vacuum.

Having identified "natural" flood and ebb with thermal expansion and contraction of fluid, al-Kindī then followed Aristotle (27, Book 1, Section 3; 36, Book 2, Section 7) almost exactly in trying to explain the source of heating and cooling in the natural environment (31):

> We therefore show: the warming of the earth, the waters, and the air occur on account of the motion of the bodies found in the heights [that is, the planets and their spheres],

which pass over them in circular motion. We take as true that all objects, when they move over another object, heat the latter, because in this way fire is generated. We find that when wood is moved rapidly on wood, fire comes forth. We observe the same with stones on iron and other softer bodies . . .

We observe further that rapidly moving objects, and especially those in circular motions [Aristotle regarded circular motion to be particularly efficacious, as the only motion that was continuous], take up a heat perceptible to the senses. The part of the air that lies next to them is thereby heated.

At the end of his treatise, al-Kindī (30) went so far as to attribute the tides of the northwestern Indian Ocean being much greater than those of the Mediterranean to the former sea lying more nearly under the ecliptic than the latter. The effect of the motion of the planetary bodies on heating the underlying water—and thereby expanding it—would therefore be greater on the former. [In its Aristotelian basis, and in its general structure of planetary causality, al-Kindī's physics strikes close to the astrology propounded by his rival Abū Maʿshar (37).]

It is easy, then, to see how this line of thinking could lead to the theory of semiannual wind and current reversals that al-Masʿūdī credited to al-Kindī. In its annual progression around the ecliptic the sun passes over the atmosphere of the Northern Hemisphere in northern summer. It would heat the air by its motion over it, and the air would therefore expand toward the south, this expansion being the northerly wind. The underlying water would move southward too, either in consequence of an analogous thermal expansion due also to the motion of the sun over it, possibly reinforced, as al-Masʿūdī indicated, by conjunction with the motion of the moon or another planet, or perhaps, as al-Masʿūdī suggested, in consequence of some sort of drag of the wind on the water.

Of course, not only is the thermal expansion of air and water woefully insufficient to account quantitatively for the effect ascribed to it here, but the predicted seasonality is opposite to that of the monsoon winds (and currents). Indeed, the latter discordance suggests that al-Kindī, in ignorance of the monsoon phenomena, may have been trying simply to improve on Aristotle's theory of seasonal winds (27, Book 2, Sections 4 and 5). Aristotle was concerned with the "etesian winds," which do blow from the north in summer over the eastern Mediterranean, and he conjectured that there were corresponding southerly winds in winter, occurring too far south to be routinely observed; he mistakenly claimed that this seasonality was characteristic of the earth. In his explanation of the alleged phenomenon, he supposed two kinds of "evaporation," or "exhalation" from the earth: a "moist," which produced rain, and a "dry," which, generally carried around the earth by the motion of the spheres, was the wind. Both evaporations, according to Aristotle, were stimulated by heating due to the motion of the sun over the atmosphere, so that when the sun was in the Northern Hemisphere, the substance of the wind was drawn up from the earth to accumulate in the atmosphere and then flow southward in analogy with river flow; with the sun in the Southern Hemisphere, the wind direction was northward. Al-Kindī's theory was slightly more respectable than Aristotle's, in that it was based on a real physical effect, but it was no less unsuccessful. Both schemes groped toward a convective theory of atmospheric circulation, but neither had any awareness of buoyancy forces or pressure gradients—and neither was tuned to the facts of the monsoon.

Other explanations of winds and currents were even flimsier, more speculative, and more whimsical. Al-Mas꜀ūdī seemed to recognize that his own understanding of winds was unsatisfactory (28):

> In some seas the wind comes from the bottom of the sea, stirring up the water; waves rise therefrom as in a boiling kettle, where the particles of the fire come from underneath. In others winds and storms come partly from the bottom of the sea, partly from the air, and in some seas the wind arises wholly from an agitation of the air without any wind coming from the bottom of the sea. Those winds which, as we have stated, come from the bottom of the sea, arise from the winds which blow from the land and penetrate into the sea, from whence they rise to the surface of the water. God knows best how this comes.

[Perhaps the idea of winds passing *through* the ocean was derived from Aristotle's "exhalations," and was related to his notion that earthquakes were produced by wind that had entered the earth and then burst through into the air (27, Book 2, Section 8).]

Al-Idrīsī (35) sought to explain the semidiurnal tide through a sort of sea-breeze driving perhaps adapted from al-Kindī's thermal-expansion theory: that the rising motion of the sun during morning hours generated a wind that piled up the water, and that, when the sun descended in the afternoon, the wind fell, the system relaxed, and the water flowed back. Without stating its cause, he asserted a similar wind at night which produced a high tide then as well. Al-Mas꜀ūdī (28), on the other hand, suggested an astrological lunar driving for tides: the moon, "being congenial with water, makes it warm and expands it."

Al-Mas꜀ūdī also related an unusual hypothesis of air–sea interaction (28):

> It is assumed that the air which is in contact with the water of the sea produces a constant decomposition of it: the consequence of which is that the waters of the sea are expanded and rise, and this is the flow; but in the mean time, the water spreads and produces a decomposition of the air which makes the water return into its former place, and this is the ebb. These actions are constant, and follow each other without interruption, for the water decomposes the air, and the air decomposes the water.

This somewhat opaque notion of mutual decompositions may be based on Aristotle's doctrine of reciprocal transformations among the four primary elements—that is Water, characterized by the qualities "cold" and "moist," would be changed into Air, characterized by the qualities "hot" and "moist," if the "cold" were overcome (somehow) by the "hot"—and vice versa (38, Book 2, Section 2).

Much more successful accounts of the monsoon winds and currents, built upon the careful, systematic observations and disciplined, quantitative reasoning from physical principles that are so conspicuously absent from the ancient and medieval discussions, are provided in other chapters of this volume. One last fancy from al-Mas^cūdī, about the monsoon currents, may therefore serve as a graceful conclusion to this review of old Indian Ocean literature (28):

> Another story of this sort is, that the angel to whose care the seas are confided immerges the heel of his foot into the sea at the extremity of China, and, as the sea is swelled, the flow takes place. Then he raises his foot from the sea, and the water returns into its former place, and this is the ebb. They demonstrate this by an example: If a vessel is only half full of water, and you put your hand or foot into it, the water will fill the whole vessel, and, when you take out the hand, the water will be as before. Some think that the angel puts only the great toe of his right foot into the water, and that this is the cause of the tide.

ACKNOWLEDGMENTS

I am grateful to C. Hurter for tracking down obscure reference material for me, and to M. A. Morrison for explaining Babylonian month-names to me. Preparation of this chapter was supported by the U.S. Office of Naval Research under Contract N00014-82-C-0019 NR 083-004.

REFERENCES

1. A. Toussaint, *History of the Indian Ocean*, translated from the French into English by J. Guicharnaud, University of Chicago Press, Chicago, 1966.
2. G. Bibby, *Looking for Dilmun*, Knopf, New York, 1969; reprint by New American Library, New York, 1974.
3. W. F. Leemans, *Foreign Trade in the Old Babylonian Period*, Brill, Leiden, 1960.
4. H. W. F. Saggs, *The Greatness that was Babylon*, Hawthorn Books, New York, 1962.
5. S. G. Gorshkov (Ed.), *Atlas Okeanov: Atlanticheskii i Indiskii Okeani*, Ministerstvo Oboroni SSSR, Voenno-morskoi Flot, 1977.
6. A. L. Oppenheim, The seafaring merchants of Ur, *Journal American Oriental Society*, **74**, 6–17 (1954).
7. R. K. Mookerji, *Indian Shipping: A History of the Sea-Borne Trade and Maritime Activity of the Indians from the Earliest Times*, 2nd ed., Orient Longmans, Bombay, 1957.
8. Arrian (Flavius Arrianus Xenophon), *Arrian's Anabasis of Alexander and Indica*, translated from the Greek into English by E. J. Chinnock, George Bell and Sons, London, 1893.
9. "Etesian wind," in *The New Encyclopedia Brittanica*, 15th ed., Encyclopedia Brittanica, Inc., Chicago, 1977.
10. H. H. Lamb, *Climate: Present, Past, and Future. Vol. 2: Climatic History and the Future*, Methuen, London, 1977.
11. Strabo, *The Geography of Strabo*, translated from the Greek into English by H. L. Jones, Harvard University Press, Cambridge, MA., and William Heinemann, London, 1961.
12. J. H. Breasted, *Ancient Records of Egypt: Historical Documents, Vol. 1, The First to the Seventeenth Dynasties, and Vol. 2, The Eighteenth Dynasty*, University of Chicago Press, Chicago, 1906.

13. G. W. B. Huntingford, *The Periplus of the Erythraean Sea*, translated from the Greek into English, The Hakluyt Society, London, 1980.

14. Pliny the Elder (Gaius Plinius Secundus), *The Natural History of Pliny*, translated from the Latin into English by J. Bostock and H. T. Riley, Henry G. Bohn, London, 1855.

15. A. Puff, ''Das Kalte Auftriebwasser an der Ostseite des Nordatlantischen und der Westseite des Nordindischen Ozeans,'' unpublished doctoral dissertation, Universität Marburg, Marburg, 1890.

16. O. Krümmel, *Handbuch der Ozeanographie*. 2. *Die Bewegungs-formen des Meeres*, J. Engelhorns Nachf., Stuttgart, Germany 1911.

17. G. Schott, *Geographie des Indischen und Stillen Ozeans*, C. Boysen, Hamburg, Germany, 1935.

18. C. Ramage, Problems of a monsoon ocean, *Weather*, **23**, 28–37 (1968).

19. W. Thesiger, *Arabian Sands*, Longmans, Green; London, 1959. Paperback reprint by Penguin Books, Harmondsworth, Middlesex, England, 1964.

20. Fa Hsien, *A Record of Buddhistic Kingdoms: Being an Account by the Chinese Monk Fâ-Hien of his Travels in India and Ceylon (A.D. 399–414) in Search of the Buddhist Books of Discipline*, translated from the Chinese into English by J. Legge, Oxford University Press, Clarendon, 1886. Reprinted by Paragon Book Reprint and Dover Publications, New York, 1965.

21. H. H. Frisinger, Aristotle and his "Meteorologica," *Bull. Amer. Meteor. Soc.*, **53**, 634–638 (1972).

22. G. R. Tibbetts, *Arab Navigation in the Indian Ocean before the Coming of the Portuguese*, The Royal Asiatic Society of Great Britain and Ireland, Oriental Translation Fund, New Series, **42**, London, 1971.

23. G. Ferrand, *Voyage du Marchand Arabe Solyman en Inde et en Chine, Rédigé en 851, Suivi de Remarques par Abû Zayd Hasan (vers 916)*, translated from the Arabic into French, Les Classiques de l'Orient, Éditions Bossard, Paris, 1922.

24. Abū ar-Rayhān Muhammed Ibn Ahmad al-Bīrūnī, *The Chronology of Ancient Nations, or Vestiges of the Past*, edited and translated from the Arabic into English by C. E. Sachau, Oriental Translation Fund, William H. Allen and Co., London, 1879.

25. Abuꞌl-Qāsim ꞌUbayd Allah ꞌAbd Allāh Ibn Khurradādhbih, *Le Livre des Routes et des Provinces, par Ibn-Khordadbeh*, Arabic text edited and translated into French by C. Barbier de Meynard, *Journal Asiatique*, Série 6, **5**, 5–127, 227–296, 446–532 (1865).

26. A. A. Aleem, Concepts of currents, tides and winds among medieval Arab geographers in the Indian Ocean, *Deep-Sea Res.*, **14**, 459–463 (1967).

27. Aristotle, *Meteorologica*, translated from the Greek into English by E. W. Webster, in W. D. Ross, Ed., *The Works of Aristotle*, Vol. 3, Oxford University Press, Clarendon, 1931.

28. Abū al-Hasan ꞌAlī Ibn al-Husayn al-Masꞌūdī, *El-Masꞌúdí's Meadows of Gold and Mines of Gems*, Vol. 1 (only), translated from the Arabic into English by A. Sprenger, Oriental Translation Fund, London, 1841.

29. N. Rescher, *Al-Kindī. An Annotated Bibliography*, University of Pittsburgh Press, Pittsburgh, 1964.

30. E. Wiedemann, Beiträge zur Geschichte der Naturwissenschaften und der Technick. XXVII, 3: Über die Grösse der Meere nach al-Kindî, *Sitzungsberichte der physikalisch-medizinischen Sozietät in Erlangen*, **44**, 35–37 (1912).

31. E. Wiedemann, Über al-Kindî's Schrift über Ebbe und Flut, *Annalen der Physik* **67**, 374–387 (1922).

32. A. Villiers, *Sons of Sindbad*, Scribner, New York, 1969.

33. B. Landström, *Sailing Ships*, Doubleday, Garden City, NY, 1969.

34. G. F. Hourani, *Arab Seafaring in the Indian Ocean in Ancient and Early Medieval Times*, Princeton Oriental Studies, Vol. 13, Princeton University Press, Princeton, NJ, 1951.

35. Abū ʿAbd Allāh Muhammed al-Idrīsī, *Géographie d'Édrisi*, translated from the Arabic into French by P. A. Jaubert, Recueil de Voyages et de Mémoires, La Société de Géographie, Chez Arthus Bertrand, Paris, 1836 (Vol. 1) and 1840 (Vol. 2).

36. Aristotle, *De Caelo*, translated from the Greek into English by J. L. Stocks, in W. D. Ross, Ed., *The Works of Aristotle*, Vol. 2, Oxford University Press, Clarendon, 1930.

37. R. Lemay, *Abu Maʿshar and Latin Aristotelianism in the Twelfth Century: The Recovery of Aristotle's Natural Philosophy through Arabic Astrology*, American University of Beirut, Publication of the Faculty of Arts and Sciences, Oriental Series No. 38, Beirut, 1962.

38. Aristotle, *De Generatione et Corruptione*, translated from the Greek into English by H. H. Joachim, in W. D. Ross, Ed., *The Works of Aristotle*, Vol. 2, Oxford University Press, Clarendon, 1930.

8

Concepts of Monsoon Physics in Historical Perspective: The Indian Monsoon (Seventeenth to Early Twentieth Century)

Gisela Kutzbach
Department of General Engineering
University of Wisconsin–Madison
Madison, Wisconsin

INTRODUCTION

Over the centuries, the monsoon of southern Asia has shaped the lifestyles and culture of the people within its reach, and, as today, its regularity has been vital for the livelihood of millions of people. Throughout history people have described the striking characteristics of the monsoon in literature. Attempts to predict the abundance or the failure of the monsoon rainfall date back more than two millenia. Methods described in Sanskrit texts ranged from proverbs (Chapter 4) to linkages between the monsoon and the motions of the sun and moon (Chapter 3). However, interest in ascertaining arrival dates of the monsoon season existed primarily in connection with preparing calendars and determining proper times for religious observances, for classical Indian science sought to preserve traditions rather than to search for new insights.

Western contacts with India, reaching back to classical times, established a wealth of practical knowledge on the currents and winds in the Indian seas (Chapter 7). As regular trade routes between Asia and Europe became established in the fifteenth and sixteenth centuries, seamen found an advantage in referring to some advance knowledge of what to expect on their journeys. The observations and sailing directions they accumulated in ship logs often became closely guarded secrets of captains and shipping companies. Even so, this information was fragmentary, and the few attempts of explanation remained speculative—without the benefit of experimental evidence and mathematics.

In the seventeenth century a new perspective of the monsoons emerged with the development of modern science in Western Europe. The new science changed the nature of questions asked about the monsoons; emphasis shifted from collecting simple observations of natural phenomena associated with the monsoon to asking for systematic, experimental evidence, and turned from speculative philosophy to the search for physical causes. The new approach to science proclaimed that the first step toward understanding natural phenomena was experimental study, especially by accurate description, observation, survey, and classification. The second step in the new scientific method was the theoretical interpretation of these phenomena, applying mathematics and physical laws. Prediction was but a third step in which the investigator was to design further experiments or to predict future states that would confirm explanations. Application of these steps to the monsoon problem almost immediately produced highly visible results.

Atmospheric phenomena lent themselves particularly to the initial step in the systematic procedure of scientific inquiry and received early attention. In addition, during this age of rapidly expanding commerce, industry, and trade, people who pursued scientific investigations often had close ties with navigators and merchants. Thus these new scientists were aware of the accelerated need of navigation for accurate knowledge of atmospheric conditions. Because of its striking characteristics and its importance to life and commerce, the monsoon was among the first large-scale atmospheric circulation systems that was investigated using the new methods.

This analysis begins when modern science had already made its first marks on the study of atmospheric processes, and when the basic meteorological instruments were first being used in systematic observations. It ends with the early years of the twentieth century. More recent developments are discussed in the chapters that follow. In Section 1 of this chapter, some of the early studies of land–sea monsoonal circulations are reviewed, including early observational results and the first applications of basic physical laws to explain the monsoon. In Section 2, the focus is on the Indian monsoon because it has been the most intensely studied monsoon circulation for the past three centuries, and because a rich source of historical material has accumulated. In particular the late nineteenth century Indian monsoon studies are used to show the impact of fundamental advances in meteorological theory after mid century, the extraordinary increase in observational knowledge, and, perhaps most important, the emergence of the India Meteorological Department as a world leader in monsoon investigations. Finally, in Section 3, some early attempts at predicting monsoon rainfall are discussed.

1 THE PIONEERS

During the eighteenth century different approaches existed for marine and land studies. At sea the requirements of navigation encouraged the study of the winds. For mariners, knowledge of the large-scale wind patterns over the oceans and the seasonal shifts of these wind patterns was of paramount importance. Seafaring men concentrated on observations that could be made without the aid of instruments. They observed the direction and often the strength of the winds, the state of the sky, clouds, the sea surface, and sometimes ocean currents. They accumulated their observations in countless log books. Beginning with the nineteenth century, some of this data was organized into directions for sailing routes and eventually condensed into sailing charts and sailing handbooks. Some of the findings on the Indian monsoon that resulted from marine meteorology are discussed in Section 1.1.

Section 1.2 focuses on early schemes of the monsoon circulation and how observational data derived from land and ocean regions, as well as theoretical advances, affected the development of these schemes.

While the meteorology of the Indian Ocean had a longstanding tradition, Indian monsoon studies on land developed only gradually and relatively late, with the influx of European merchants and intellectuals. Many of the early observers were associated with the East India Company, which held a monopoly over Indian trade for more than two centuries. The disciplined organization of the company was conducive to keeping regular meteorological records. Investigators on land produced time series of observations of rainfall, temperature, barometric pressure, and other meteorological parameters at single stations and carried out statistical studies of seasonal variations. In general, they paid little attention to the large-scale features of atmospheric motions. This development is discussed in Section 1.3.

1.1 The Charting of the Global Winds

During their travels across the oceans during the seventeenth and eighteenth centuries, mariners could rely not only on a wealth of common knowledge accumulated by explorers and traders, but also on detailed travel accounts and early wind charts. These accounts culminated in the popular travel books by William Dampier (1), the famous English buccaneer and seafarer. Dampier's *Voyages and Descriptions* of 1699 was based on his travel diaries and included exciting mixtures of the events in his adventurous life, nature descriptions, and weather and wind observations. On his wind chart (Fig. 8.1 on p. 162) he noted the monsoon region of the Indian seas as an ''area of shifting trade winds'' and clearly indicated the seasonal reversals of the winds in the Arabian Sea and the Bay of Bengal. Later the first useful descriptions of navigation in the Indian Ocean and of the monsoon winds were prepared by captains of the East India Company. Captain Joseph Huddart's *Oriental Navigator* of 1735 and Captain James Capper's *Observation on the Winds and Monsoon* of 1801 were widely distributed. The authoritative *India Directory* of 1809 by James Horsburgh, hydrographer of the East India Company, also focused on directions for navigation in the monsoon regions (2).

During the late seventeenth century, the men of modern science began to share the interest of mariners and merchants in observing the winds over the oceans. For them the oceans held many unsolved physical problems and presented rich hunting grounds for discoveries. Unfortunately few investigators had an opportunity to study the sea first hand. For this reason, the Royal Society of London, whose members stood at the core of the new science, prepared in 1666 the *Directions for Sea-Men, Bound for Far Voyages* which contained detailed instructions for collecting data. The subject was considered so important that the *Directions* was published in the first volume of the Royal Society's Transactions. While many of the instructions concerned ocean soundings, water sampling, temperature measurements, the tides,

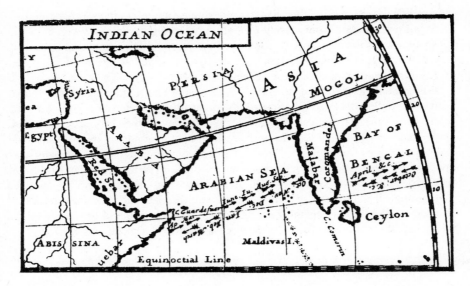

Figure 8.1. Section of wind chart* by W. Dampier (1) showing the seasonal reversals of the winds in the Arabian Sea and the Bay of Bengal and the general southeast trades over the southern Indian Ocean. Arrows in the void spaces show the direction of the "shifting trade-winds" (monsoons), and abbreviations, Aug., Sept., etc., denote the months of the year when these winds blow.

and surveying, two of the instructions related to the observation of atmospheric phenomena at sea (3):

> To keep a register of all changes of wind and weather at all houres, by night and by day, shewing the point the wind blows from, whether strong or weak: The Rains, Hail, Snow and the like . . . especially *Hurricanes* . . . but above all to take exact care to observe the *Trade-Wines* [sic], about what degrees of *Latitude* and *Longitude* they first begin, where and when they cease, or change, or grow stronger or weaker, and how much. . . .

We can assume that Edmund Halley (1656–1742) used the observations supplied by the seamen to the Royal Society in the construction of his famous chart of the winds (Fig. 8.2), found in his paper of 1686 on the causes of the trades and monsoons (4). Halley, sponsor of Newton's *Principia* and secretary of the Royal Society, spent his life investigating the tides, magnetic fields, winds, currents, air pressure, and many other subjects of geophysics. Confronted with the problem of how to represent three-dimensional quantities on paper, he introduced isogenic lines to connect points of equal magnetic variation and constructed the first meteorological flow chart to encompass the globe. The chart depicted the surface winds of the

---

* This and several of the following figures are the best possible reproductions from the originals. Even though their quality is poor, the reader should be able to extract from them the information relevant to the discussions in the text.

trade and monsoon circulations and their boundaries, but in the tradition of marine meteorology, it provided information over the oceans only. Halley introduced it, he wrote, because

> I believed it necessary to adjoyn a Scheme, shewing at one view all the various Tracts and Courses of these Winds. . . . I could think of no better way to design the course of the Winds . . . than by drawing of stroaks—

Each dash had a thick front and pointed tail to indicate direction. Significantly, in the region of the Asian monsoon he represented by "rows of stroaks," alternating in direction, the seasonal variations of the winds, that is, both southwest and northeast monsoon. The chart did not show wind intensities, but Halley observed, the southwest monsoon "blows with rather more force" than the northeast monsoon.

Furthermore, east winds alternating with southwest winds along the east coast of Africa indicated a monsoon circulation over Africa. Interpreting the flow patterns

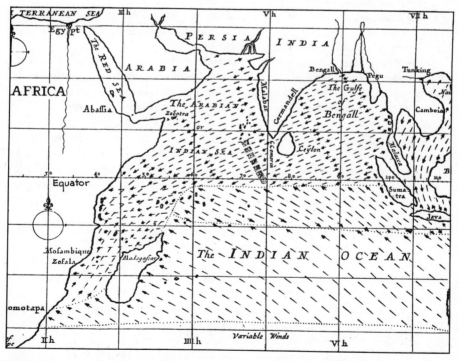

Figure 8.2. Section of wind chart by E. Halley (4), showing seasonal variations of the surface winds of the Asian monsoon. Each dash has a thick front and pointed tail to indicate wind direction, the tail being the direction from which the wind blows. The southwest and northeast monsoons in the Arabian Sea and the Bay of Bengal are both represented by "rows of stroaks" alternating in direction. Wind arrow heads are added for clarity.

over the Indian Ocean, Halley concluded that there was a continuous flow of air across the equator. Halley commented,

'Tis . . . remarkable, that the S.W. [southwest] Winds in [the Indian Sea] are generally more Southerly on the African side, more Westerly on the Indian . . . [Along the African Coast, from Madagascar] and from thence Northwards as far as the Line [equator], . . . from April to October there is found a constant fresh S.S.W. [southsouthwest] Wind, which as you go more Northerly, becomes still more and more Westerly, so as to fall in with the W.S.W. [westsouthwest] Winds.

Halley's account stood at the forefront of knowledge at his time. Reflecting the halting, fragmentary nature of eighteenth century research in meteorology, studies of that period added but little to Halley's work. By the early decades of the nineteenth century, though, Halley's wind chart had become outdated; it only covered the tropics, and, except for the monsoon regions, it did not indicate seasonal variations. Alternative sources of information supplied by the mariners, such as current descriptions of voyages, log books of sailing ships, and sailing directions and tables were not only cumbersome to use but also not always accessible. It was therefore the beginning of a new era in marine meteorology and commerce when Captain Matthew Fontaine Maury (1806–1873) in the United States began summarizing the available data (numbering hundreds of thousands of observations) into wind and sailing charts.

In 1842, after he had resigned from active duty because of a personal accident, Maury was appointed head of the U.S. Navy Depot of Charts and Instruments. He became aware that a vast number of manuscript ship logs had accumulated in the depot. With the practical goal of promoting maritime commerce in mind, he used the data of winds and currents in these logs for charting the general circulation of the atmosphere and oceans. He began publishing his famous winds and current charts in 1847 and issued them free to mariners in exchange for their ship logs. After 1850 he added sailing directions to the charts (5). Through these charts and his book, *The Physical Geography of the Sea* (6), which went through five editions within one year, Maury established a number of important facts about the Indian monsoon.

Maury, a leader in systematizing observations at sea, was instrumental in moving the American and British governments to sponsor the first maritime conference in Brussels in 1853. This conference not only marked the beginning of international cooperation, it also stimulated the formation of national weather services. The international forms for observations agreed upon at the conference subsequently were to be distributed to ship captains and collected and summarized in charts under the direction of individual governments. Figure 8.3 shows one of Maury's section diagrams derived from the information in the forms, for the square 5–10°N and 105–110°E. Each mark indicates an individual observation as reported by a vessel that had passed through the section. The clustering of observations clearly reflects the annual cycle of the northeast and southwest monsoon in Southeast Asia. Pilot charts and annual windroses summarized the sectional diagrams.

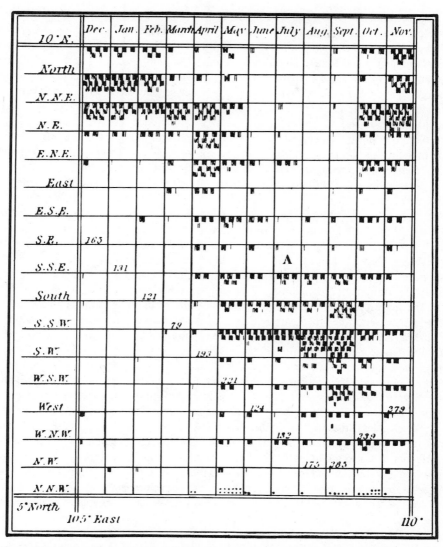

Figure 8.3. Section diagram by M. F. Maury (5), showing wind observations in a section, 5 degrees by 5 degrees square, off the coast of Southeast Asia. The section is subdivided into a system of squares, the months of the year being the ordinates, and the points of the compass being the abscissae. The wind reported by a vessel passing through any part of the section is assumed to have been at that time all over the section. Note the clustering of northeast winds during the winter monsoon and southwest winds during the summer monsoon.

In response to the 1853 conference, the British government established the Meteorological Office in 1855, headed by the experienced hydrographer Captain Robert Fitzroy. Following Maury's advice, Fitzroy immediately set out to contact captains who would make ship observations to be summarized in the widely used pilot charts of the British Board of Trades (7). Other seafaring nations made similar efforts. Thus the establishment of national weather services in mid-nineteenth century made possible the necessary cooperation for obtaining global observations of wind and weather at sea.

Commenting on Maury's system of charts and estimating the commercial value of reducing the length of ocean voyages, the president of the British Association of Science recommended in 1853 to extend this system to the Indian Ocean. He stated approvingly: "If the voyages of those [ships] to and from India were shortened by no more than a *tenths*, [savings] . . . would be about £2,000,000 annually" (8). Subsequently, the merchants and shipmasters of India raised a subscription for establishing a set of wind and current charts for the Indian Ocean. This project was to be coordinated by the Geographical Society of Bombay under the direction of its president, George Buist. Unable to complete the project himself, because of his obligations as supervisor in the East India Company, Buist later sent Maury the first six "skeletal charts" containing observations from more than a hundred ships (5).

The tradition of sailing charts, directories, and sailing handbooks culminated in the last quarter of the nineteenth century. An excellent example of a sailing handbook is the *Segelhandbuch für den Indischen Ozean*, published by the German Hydrographic Office (9). It incorporated the charts of Maury and the Board of Trade, as well as later data, and provided the navigator with an introduction to meteorology and the Indian monsoon. The emphasis was on sailing routes. Instructions often were designed for certain users, for example, the rice-carrying ships returning to Europe in March–April. The general objective was to make the passage as safe and as short as possible.

Nevertheless, the need for sailing instructions soon lost some of its urgency, for the 1880s not only marked the time when sailing ships reached their highest technical perfection but also the time of their final decline in commercial shipping. Steamships, providing a certain independence from winds and currents, began to dominate overseas trade; new shipbuilding techniques, motor development, and the establishment of coaling stations along important routes made long voyages possible at an economic rate. The 1869 opening of the Suez Canal had foreshadowed this outcome. Because the Canal and the Red Sea were unsuitable for sailing vessels, they had to continue the long voyage around the Cape of Good Hope on their route from Europe to India. By the end of the century, the Indian trade had converted almost completely to steamships. Reflecting this newly won independence from the winds, the second edition of the German sailing directory was reduced in size from 600 to 100 pages and its title was changed to sea directory (10).

1.2 The Monsoon and Schemes of the Global Circulation

During the eighteenth and nineteenth centuries, investigators not only worked on backing up experience with systematic observations and on gathering knowledge of large-scale wind patterns over the oceans in tables and charts, but they also took the additional step of venturing theories about the phenomena observed. Early explanations of the monsoon circulation were closely tied to those of the trade wind circulation. In contrast to early observational work, which stressed conditions over the oceans, theories from the beginning considered conditions over both land and sea.

1.2.1 The Indian Monsoon as Part of the Trade Wind Circulation. In 1686, Halley first advanced the idea that global winds followed a pattern and that there was a general circulation of the air over the earth. He coupled the presentation of observations of the trade winds and the monsoons with a qualitative explanation of these circulations based on physical laws. Halley rejected the simple idea, popular in his time, that the westward direction of the trades was due to the lag of the air behind the rotating earth. Instead, he proposed a novel cause (4),

agreable to the known properties of the Elements of Air and Water, and the laws of the Motion of fluid bodies. Such an one is, I conceive, the Action of the Sun Beams upon the Air and Water . . . and the Scituation [sic] of the adjoyning Continents.

Here he identified the role of differential heating of land and sea in producing the general motions of the atmosphere.

I say therefore . . . the Air which is less rarified or expanded by heat, and consequently more ponderous, must have a Motion towards those parts thereof, which are more rarified, and less ponderous, to bring it to an Equilibrium.

Because of the westward movement of the sun the "greatest Meridian Heat" also shifts westward, "and consequently the tendency of the whole Body of the lower Air is that way." Along this meridian, Halley continued, the air is warmer

near the Line [Equator] . . . than at a greater distance from it . . . it follows, that from both sides [the air] ought to tend towards the Line: This Motion compounded with the formerly Easterly Wind answers all the Phaenomena of the general Trade Winds.

Halley elaborated that for reasons of equilibrium "by a kind of Circulation, the North-East Trade Wind below, will be attended with a South Westerly above, and the South Easterly with a North West Wind above." In 1861 John Herschel (11) termed this upper return current the "anti-trade." Alexander von Humboldt and Leopold von Buch (12) referred to the high cirrus clouds from westerly directions above the northeast current as evidence of the antitrades, and they provided observations of west winds above the trades at the Peak of Tenerife in the Canary Islands, located in the trade wind region of the North Atlantic.

Going one step further, Halley explained that the Indian monsoons also result from differential heating effects and constitute a regional modification of the trade wind circulation. In summer, he noted, the reversed temperature conditions of the equatorial regions and the Indian continent result in a reversal of lower and upper currents (see Fig. 8.4): the relatively cool northeast trade is transported into upper layers whereas the southwest current descends to lower layers. "In *April*," he observed, "when the sun begins to warm those Countries to the North, the S.W. *Monsoon* begins, and blows during the Heats till *October*." This observation, he argued, agrees with the principle of differential heating where air moves from cold to warm regions. In winter, however, the normal trade wind circulation—with northeast wind below and southwest wind above—sets in because the northeast wind in the Indian Seas is colder than the equatorial regions: "Because of a ridge of Mountains at some distance within the Land, said to be frequently in Winter covered with Snow [the Himalayas], . . . the Air, as it passes, must needs be much chilled." Accordingly, the trade wind circulation—although modified—was the dominant circulation of the monsoon in Halley's scheme. His paper profoundly influenced later thinking on the general circulation of the atmosphere and the monsoon circulation (see Section 2.1).

Subsequently, in 1735, George Hadley (1685–1758), a London lawyer and fellow of the Royal Society, proposed to improve on Halley's explanation of the trades (13). "The perpetual motion of the Air towards the West," he wrote, "cannot be derived merely from the Action of the Sun upon it." In a short paper of five pages he considered the effect of the rotation of the earth on the north–south currents. For his purpose, Hadley did not need to consider specific data or geographic areas, and he did not mention the monsoon. In contrast to Halley's informative account, Hadley's theoretical essay was of little use to the sailor. The paper did not become well known during the eighteenth century, and the implications of the deflecting force of the earth's rotation for all atmospheric motions were not further evaluated for some time. (See Chapters 1 and 9 for a physical/mathematical description of the deflective force, known as the Coriolis force.)

Both Halley's and Hadley's papers suggested a physical relation between atmospheric pressure, temperature, and wind; they proposed that thermal expansion reduces pressure and that air flows toward areas of reduced pressure. Neither author, however, explored quantitatively the fundamental question of how atmospheric pressure and pressure variations affect the wind circulation. In 1746, to find an

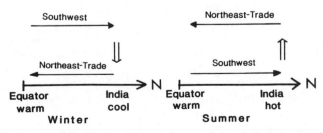

Figure 8.4. Sketch illustrating Halley's circulation scheme for the Indian monsoon in summer and winter. Note that the surface current is directed from relatively cool to warm in both seasons. The northeast trade is a surface current in winter and an upper current in summer.

answer to this question, the Berlin Academy of Science offered a prize for the best work on the laws governing air in motion (14). During the eighteenth century, the principal stimuli for defining research areas and initiating concerted research efforts came from the national academies. The 1746 competition resulted in first attempts to formulate the general equations of motions for fluids. It was not until mid-nineteenth century, however, that these equations were applied to the general circulation of air on a rotating earth (see Section 2.1).

1.2.2 Effect of the Land–Sea Distribution on the Indian Monsoon.

During the early decades of the nineteenth century, Alexander von Humboldt (1769–1859) developed a different approach to the study of atmospheric conditions on a global scale. His starting point was the global patterns of heat distribution rather than of the winds. Humboldt was a devoted humanitarian and a promoter of a worldwide science that would unify the fragmented approach to environmental sciences. His work profoundly affected studies of the monsoon. While investigating the distribution of plants on the earth, Humboldt began to distinguish between the solar climate, which would prevail in the case of a motionless atmosphere, and the real climate, which he said was produced by the winds. In Humboldt's time, studies of the temperature distribution on the globe had already dealt with the effects of seasonal variations in solar radiation at different latitudes. However, looking at observed temperatures at different localities, specifically in Europe and in North America, Humboldt found significant temperature differences along the *same* latitude circle. Contrary to solar radiation calculations and to expectations from Hadley's scheme, the actual mean temperatures did not decrease regularly from equator to poles (15). Because of these findings, Humboldt viewed the land–sea distribution not merely as a modifying influence on the general circulation but as a principal aspect of that circulation.

Perhaps even more significant than this observation was Humboldt's methodology for systematizing data. In 1817, he introduced the representation of the global temperature distribution by isotherms (Fig. 8.5), adapting from studies of the earth's magnetic field the use of isolines for depicting magnetic declination on charts (15). Unfortunately, most of the journal edition of Humboldt's 1817 paper was destroyed by fire, so that his climatic chart became widely known only after republication in 1832.

The chart clearly showed that annual mean isotherms were not parallel to the equator. Humboldt concluded that besides the systematic latitudinal effect, "the foremost effect on the climate of a place stems from the configuration of the continents surrounding it. These general causes are modified by mountains, state of the surface, etc., . . . which are merely local causes."

Following Humboldt's example of collecting and systematizing data for the entire globe, Ludwig Kämtz (1801–1867), a professor of meteorology in Germany and later in Russia, constructed a global temperature chart in 1830. Whereas in the tropics Humboldt had drawn isotherms parallel to the equator for lack of data, Kämtz had sufficient observations to construct more representative isotherms in this region. His chart clearly demonstrated the stronger heating over the continents

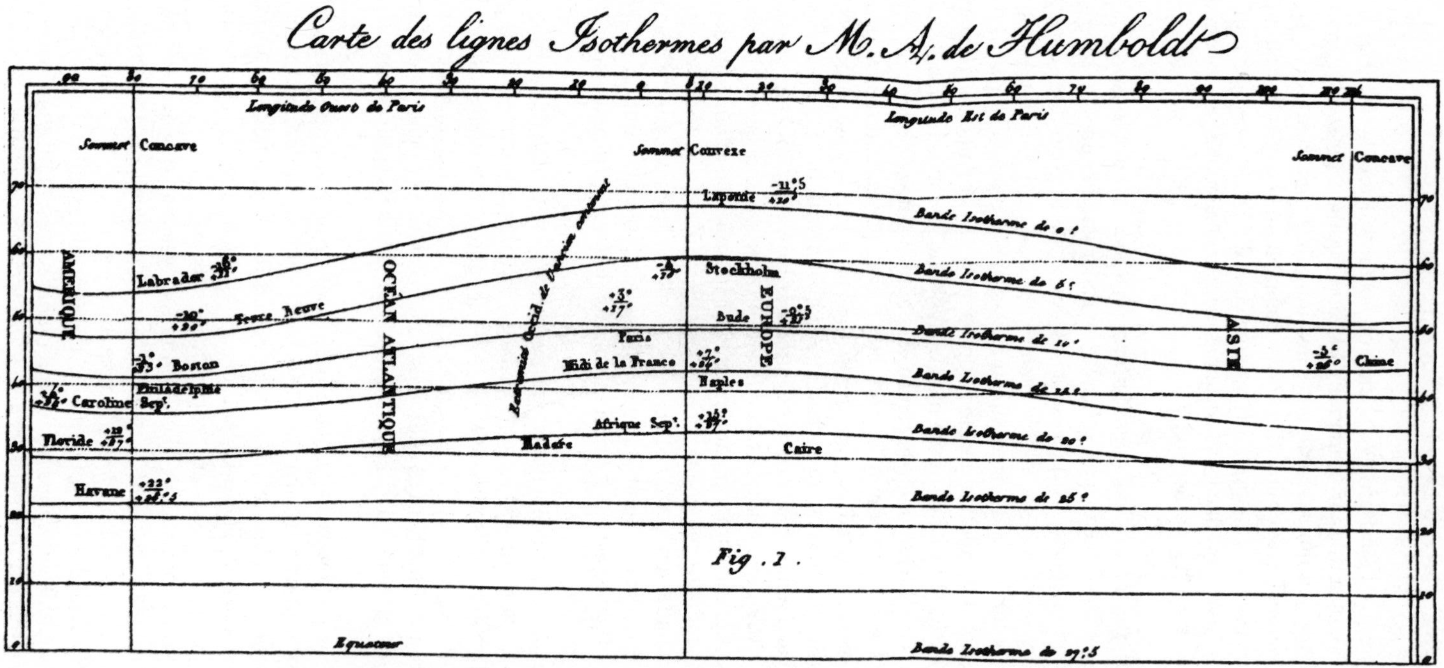

Figure 8.5. Global chart of annual isotherms by A. von Humboldt (1817), showing differential heating along the same latitude circle; outlines of continents are omitted (16). Isotherms, in °C, are drawn for every 5 degrees.

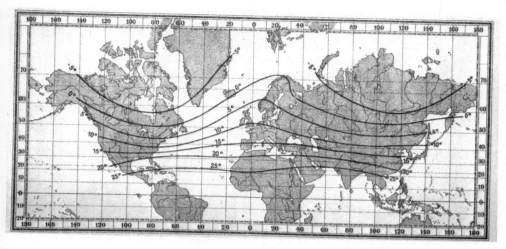

Figure 8.6. Global chart of annual isotherms by L. F. Kämtz (16), showing the stronger heating of continents as compared to the oceans in the tropics. Isotherms, in °C, are drawn for every 5 degrees.

compared to over the sea (Fig. 8.6). In his widely used textbook he gave particular attention to the effects of the land–sea distribution and the monsoon (17).

Kämtz explained the development of atmospheric wind systems as follows (see Fig. 8.7):

> If two neighbouring regions are unequally heated, . . . then the column of [heated] air will expand and its upper limit will be more elevated [than the cool surrounding air]. Air passes in all directions and produces winds, which, as the upper arrow indicates, blow from the hot towards the colder countries. Whilst these phenomena are going on in the higher regions of the atmosphere, the equilibrium is also destroyed at the level of the ground.

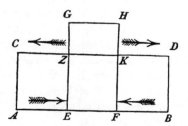

Figure 8.7. Circulation produced by unequal heating of neighboring regions by L. F. Kämtz (1847). Heating takes place in the region above EF and causes expansion and lifting to GH. To restore equilibrium, outflow takes place from the region ZKHG, from warm to cold. Barometric pressure decreases over EF and increases over AE and FB, resulting in low level inflow from cold to warm (18).

As the pressure at the surface of the neighboring regions increases because of inflow of air above, a current develops at the surface "from the colder towards the hotter regions, in the direction of the lower arrows."

Building on Halley's and Hadley's work, Kämtz proposed that the monsoon was controlled by two factors, the differential heating and cooling of land and sea and, equally important, the effect of the deflecting force of the earth's rotation. However, as did Hadley, Kämtz still considered only the east–west component of this force (18):

> In winter southern Asia is colder than the Indian Ocean, and therefore a northwind will blow in the lower layers of the atmosphere and a southwind above. In summer, however, Asia is warmer than the sea to the south, and for this reason the wind must come from the south at the surface. However, that these two winds are not from N[orth] and S[outh], but from N[orth]E[ast] and S[outh]W[est], that is due to the rotation of the earth.

One of the most important followers of the Humboldt tradition was Heinrich Wilhelm Dove (1803–1879), a professor of experimental physics in Berlin and founder and director of the Prussian Meteorological Institute (1848). Dove extended Kämtz's ideas (19):

> The southeast winds, when they cross the line [equator], turn from south to southwest, due to the decrease in velocity of rotation of the earth with distance from the equator; however, northeast winds, when they come from the northern hemisphere, must change into north and northwest when entering the southern hemisphere.

Dove's global perspective of the Indian monsoon is evident in his ideas on the geographic origins of the northeast and southwest monsoon. He suggested that the northeast monsoon was a "polar" current originating over central Asia and thus an integral part of his scheme of the general circulation, the principal features of which were polar and equatorial currents. Like Maury, Dove stressed that in summer the trades of the Southern Hemisphere crossed the equator and intensified the southwest monsoon. Dove explained that as the equatorial calm belt moves north with the sun towards the Indian continent, "the air masses of the torrid zone are incorporated" into the monsoon circulation. Maury confirmed this statement, using 421 ship observations selected at random. These showed that cross-equatorial flow took place "without any intervening calm. This in summer is the rule" (5).

In general, writers of the first half of the nineteenth century continued to agree on the close relation between trade winds and monsoons. Thus, they accepted Halley's early statement that in summer the normal northeast trade current was blowing above the southwest monsoon. Their physical explanations of this relation, however, varied and were at times overly simplistic. In the region of the Indian Ocean, for example, Maury elaborated (5),

> a force is exerted upon the northeast trade-winds of that sea by the disturbance which the heat of summer creates in the atmosphere over the interior plains of Asia, which is more than sufficient to neutralize the forces which cause those winds to blow as trade-

winds; it arrests them and turns them back; . . . for the wind blows toward that place where there is least atmospheric pressure.

Similarly, John Herschel, in his widely used textbook of 1861, still maintained that the lower-level northeast monsoon flow is "in fact no other than the undisturbed north east trade wind" (11). (See Chapters 1 and 9 for current views.)

1.2.3 Effect of Moisture on Monsoon Intensity.

Whereas many early writers had concentrated on the analysis of the temperature differences and the resulting wind distribution during the monsoon, Maury pointed out still another source of energy effective in the monsoon circulation: the release of latent heat in convective currents. Only to a certain extent, he stated, are monsoons like sea breezes in which "the heat of the sun by day and the radiation of caloric by night are alone concerned." In monsoon circulations, he explained, the latent heat of water vapor that is set free in the moist southwest current over land is also a "powerful agent" (5). Maury here applied the 1841 work (20) of James P. Espy (1785–1860), who had helped organize a meteorological network in the United States and became the first American professor of meteorology. Espy pioneered by demonstrating with experiments and observations that the release of latent heat enhances the strength of convective currents and therefore cloud growth. The torrential rains of the monsoon season obviously came from vast convective currents of moist air that intensified with the release of latent heat. The process appeared to be especially important in the development of tropical storms in the Bay of Bengal (see Section 1.3.2).

Also, water vapor content of the air, taken as an indicator of the southwest monsoon flow, was used to estimate the vertical extent of the current. Dove found the southwest monsoon was of rather shallow vertical extent, quoting as evidence water vapor measurements at Poona (1700 feet) which in April averaged only half the value of those in Bombay (at sea level).

1.2.4 Transition Periods between Winter and Summer Monsoon.

Perhaps in part to gain a better understanding of the relation between the trades and monsoons, investigators produced many detailed accounts of the transition from the winter to the summer monsoon, that is, the transition from northeast winds, identified as the trades, to southwest winds. Kämtz, who did not distinguish between the progression of the southwest winds and the onset of rains, noted that the southwest monsoon winds reach the tip of the Indian subcontinent in March and then gradually progress northward (17). Maury's descriptions reflected the improved observations as compared to those of the time of Kämtz 20 years earlier, but they were unclear about the processes of the transition period. Thus, using 11,697 observations along the Indian coast of the Bay of Bengal, Maury concluded that "the southwest monsoons commence at the north, and 'back down,' or work their way toward the south" with an average speed of 15 to 20 miles a day (5). According to Maury, it takes six to eight weeks for the southwest winds to advance from Calcutta south to the equator in May. He added that, while southwest winds begin to predominate in Calcutta already in February, "it is not till the southwest monsoons have been extended far out to the

sea that they commence to blow strongly, or that the rainy season begins in India.''
His description of the transition from the northeast to the southwest monsoon did
not clearly distinguish between winds and monsoon rains and conflicted with later
observations; yet it was quoted in the meteorological literature until the end of the
century.

Both Kämtz and Dove investigated the relation between rainfall and temperature
during the transition period in the tropics. Analyzing the data series that already
existed for Madras and Calcutta, Kämtz found that temperatures were highest during
the dry seasons, before and after the rains, when the sun was still high in the zenith,
and that the rainy season of the monsoon was associated with a relative temperature
minimum. Using data from a number of stations between 7 and 23 degrees latitude,
Dove traced the north to south movement of the initial heating during the late stage
of the northeast monsoon and the gradual build up of the heat low over the continent,
followed by a reduction of temperature once the rains set in, and a second temperature
maximum in the lower latitudes after the rains ceased (19).

The interest of western investigators in the monsoon lessened during the 1860s
and 1870s when they became preoccupied with the development of synoptic charts
and the study of mid-latitude cyclones (21). As we shall see later, new incentives
for monsoon studies arose during the 1880s, largely with the establishment of the
India Meteorological Department in 1875 and the subsequent dramatic increase in
monsoon investigations by India-based meteorologists. First, we shall look at early
nineteenth century observational studies of the Indian monsoon and monsoon disturbances by observers in India.

1.3 The Meteorologists in India

British officials of the East India Company carried out most of the early land
observations in India. Chartered in 1600 during Queen Elizabeth's reign and holding
sovereignty over the British colonies in India after 1702, the East India Company
maintained three separately administered provinces or presidencies, with headquarters
in Bombay, Madras, and Calcutta. For the company officials of the early 1800s,
knowledge of the meteorological and climatological characteristics of India was not
only of economic importance, but in a sense it also added an intellectual dimension
to the political control of the country. They pursued science in a gentlemanly way,
as dedicated amateurs, like others back in England at this time. Professional recognition
of their efforts generally did not come until the 1860s, when the British government
took control of India.

1.3.1 Early Observations in India. For the most part, doing science alongside
their regular duties was not an easy matter for the busy officials of the East India
Company. Depending on the location and persistence of the observers, most observational series of the first half of the nineteenth century covered only short
periods. Also the data were difficult to compare because of nonstandarized instruments
and observational procedures. In 1835 Colonel William Henry Sykes (22) remarked
on the obstacles he faced in making quality observations:

I was confined to two of Thomas Jones's barometers: they required to be filled when employed, and were destitute of an adjustment for the change of level of mercury in their cisterns, unless the position of the cistern had been altered at each observation. . . .

Generally, instruments were made available to medical officers and revenue collectors. Instruments were not standard and were sent without instructions. The hospital assistants, clerks, and other personnel entrusted with taking observations "in most cases [were] untrained or very imperfectly trained, and with other peculiar views of the requirements for accuracy or punctuality of observation." Still another problem was the publication of observations from these sources. No attempt was made to coordinate and evaluate them, and they remained "practically inaccessible for the purposes of scientific inquiry" (23).

Fortunately, the East India Company also established in all British provinces meteorological observatories, which produced high quality, continuous observational records. The Madras Observatory, the first observatory in Asia, was established in 1792. The Colaba Observatory at Bombay followed in 1826, and the high elevation observatory on the Dodabetta Peak in the Nilgherry Hills opened in 1845. Other observatories were built at Trivandrum, Simla, and Calcutta. These observatories regularly issued publications of meteorological observations and discussions.

In addition, company officials organized a number of learned societies and journals. The Asiatic Society of Bengal, founded in 1784, served as a forum for intellectual debate. The Society began publishing a journal in 1832, which was owned by Society secretary, James Prinsep, the master of the mint in Benares and an avid meteorological observer. In 1838, the Geographical Society of Bombay was formed and also published a journal. Its first president, George Buist, director of a school of industry in Bombay, had a special interest in the cyclones of the Indian seas. Subsequently some of the meteorological data collected by higher company officials were published in occasional meteorological articles in these journals or even in the publications of the Royal Society of London.

Around 1850 the East India Company ordered that observations be made at all principal stations of British troops. Soldiers were trained for this task; nevertheless observations remained poor and were completely suspended during the Indian mutiny of 1857–1860. After the British government took control of India from the East India Company in 1860, interest in the study of meteorological records surged. The British government contracted with the German company Schlagenweit to carry out a scientific mission in India (1861–1863). Schlagenweit representatives collected many of the existing local observational series of rainfall for analysis and eventually deposited them in Munich. In 1878, Henry Blanford (1834–1893), first director of the India Meteorological Department, examined these and other rainfall data from the period before 1865. Unfortunately he found them too unreliable to be useful for meteorological investigation (24), but they did shed some light on the meteorological conditions preceding the severe droughts and famines of the first half of the century.

East India Company officials also studied in great detail the cyclonic storms of the Bay of Bengal. In fact, since the destructive forces of these storms threatened the Company's economic interests both at land and at sea, it supported the investigation of these storms more than any other activity in meteorology.

1.3.2 The First Systematic Observations of the Bay of Bengal Cyclones. The

intense storms of the West Indies, the Indian Ocean and the China Sea had always been perceived as posing grave "difficulties" for seafarers. Thus, once marine and land meteorologists were familiar with the broad scale wind patterns over the oceans, a fair number began investigating the nature and movement of tropical cyclones.

In 1838, Colonel William Reid (1791–1858), of the Royal Engineers and governor of Bermuda, published a comprehensive treatise on the law of tropical storms, including charts, storm tracks, sailing directions, and traditional descriptions of famous storms (25). He had been inspired by the work of William Redfield (1789–1857), first president of the American Association for the Advancement of Science and an authority on the tropical cyclones of the West Indies. Because Reid thought it important to extend the study of tropical cyclones to all regions wherever they occurred, he proposed to the director of the East India Company that the Company initiate a program to observe such storms in the Bay of Bengal and the Arabian Sea.

In 1839 in response to Colonel Reid's proposal, the Company distributed a circular to its officers that provided an organizational backing for the systematic collection of meteorological data and singled out the observation of tropical cyclones (26):

> The public officers of the different settlements and stations of India, are accordingly invited and requested, upon the occurrence of any hurricane, gale, or other storm of more violence than usual, to note accurately the time of its commencement, the direction from which the wind first blows . . . also to note, with as much accuracy as possible, the changes of direction in the winds. . . . The variations of the thermometer and barometer . . . will also be of importance, if the means are forthcoming of making such observations.

Captain Henry Piddington (1797–1858), who had retired from active duty with the Company and was the curator at the Calcutta Museum of Oeconomic Geology, was appointed the coordinator of these observations. He analyzed every piece of information on storms transmitted to him, and between 1839 and 1855 he published 40 articles on storms in the *Journal of the Asiatic Society of Bengal*, discussing extracts from ship log books and summarizing this information in charts. The following quote from the log book of the *Belle Alliance* describes vividly how the vessel struggled through the center of a cyclone on its way from Madras to Calcutta. Figure 8.8, indicating positions of vessels enroute and of the storm center for a period of three days, shows how the *Belle Alliance* was carried far to the south by the winds associated with the storm before it was able to resume its course to the north:

Figure 8.8. Section of chart showing the track of the Bay of Bengal cyclone of April 27th to May 1st, 1840, and positions of ships within its reach, by H. Piddington. Extracts from the log book of the *Belle Alliance* are quoted in the text. The vessel, on its way from Madras to Calcutta, was diverted from its course by the strong northeast winds, reached the center of the storm, and finally resumed its course northward when southwest winds set in (27).

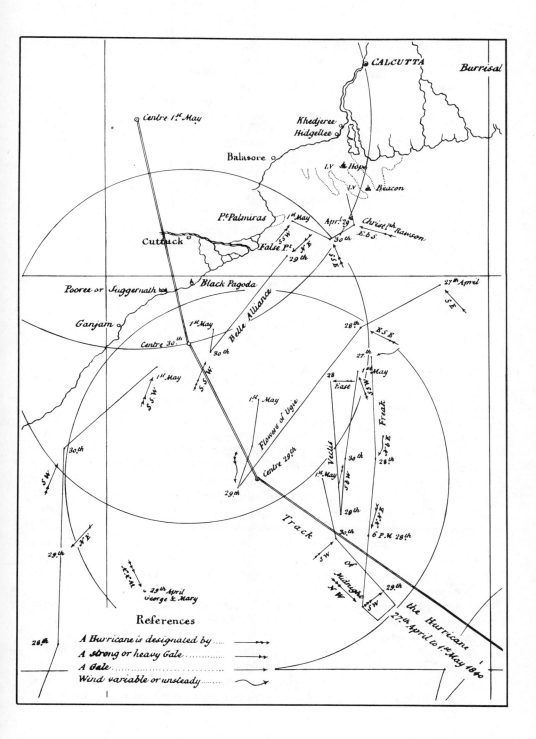

CALCUTTA   Burrisal

Khedjeree
Hidgellee

Balasore

I.V   Hope
              I.V   Beacon

P.t Palmiras   1.t May   Apr.l 29   Christoph. Rawson
                    S.S.W   E.b.S.
Cuttuck   False P.t   N E   30 th
                    29 th   S.S.E
                                            27th April
Pooree or Juggernath   Black Pagoda   S.E
                                    28th
Ganjam   1.t May   Belle Alliance   E.S.E
        Centre 30th   27th
                    30th   1.t May
            1.t May   S.S.W   Freak
    S.S.W   S.S.W
            1.t May   S.E
        Flowers of Ugie   1.t May   30th   28th
                    28   S.S.W
                    East   Vellis   29th
    30th   1.t May
                            29th   N.E
    S.W   6 P.M. 28th
                    30th
        N.E   Track   S.W   29th
    29th   of   Midnight
                            N.W   S.W
                    29th April   the Hurricane
    28.th   George & Mary   27th April to 1.t May 1840

        References
        A Hurricane is designated by .... ⟶
        A strong or heavy Gale .... ⟶
        A Gale .... ⟶
        Wind variable or unsteady .... ⟿

[29th April, 1840].—At daylight, an increasing breeze from NE [northeast] and squally. At Noon an increasing gale with hard squalls, distance on the log 71 miles. . . .

30th April.—1 A.M. to 5 A.M. The gale increasing, with violent gusts of wind and heavy rain; furled topsails; ship laying to under storm main staysail. At 6 A.M. ship plunging deep, with a heavy confused sea; carried away the flying jibboom; cut away the wreck. At 8 A.M. trying to strike topgallant masts; ship laying over and plunging deep could not, and obliged to cut away fore and main topgallant masts; in so doing the head of the foretopmast broke above the rigging; a heavy confused sea. At 9 A.M. *a sudden calm*, struck main topgallant mast; ship labouring much, from the heavy sea. At 10[A.M.], a violent gale from SSW. [south-southwest] with most awful gusts of wind and heavy rain. . . . Moderating after midnight, with thick hazy weather.

The barometric pressure observations of the *Belle Alliance* during the final phase of the storms show a pressure drop of 0.9 inches of mercury in a period of two hours (27):

30th April,	10 p.m.	Bar.	29.40 [inches]
	12 "	"	29.20
1st May,	3 a.m.	"	29.00
	5 "	"	28.10
	8 "	"	28.30
	4 p.m.	"	28.90
	9 "	"	29.10

Redfield, Reid, and Piddington recognized many of the essential characteristics of tropical storms and summarized their results graphically on synoptic charts and charts of storm tracks. Studying the relationship between the wind and pressure distribution in tropical cyclones, Redfield established rotation as the salient feature of these storms. He also noted a "slight vortical movement" and in one case showed an "average convergence, or inward inclination, of about six degrees" (28). Subsequently Reid formulated a number of rules for the seafarers for navigating around these revolving storms. Piddington coined the term "cyclone" to describe the wind circulation associated with tropical storms. He defined cyclones as storms where the wind blows around a common center, using the expressions breeze, gale, storm, and hurricane to denote the strength of the wind rather than the nature of the storm (29).

Although Redfield, Reid, Piddington and others were aware that the wind trajectories in tropical storms had a component toward the storm center, they treated this characteristic as nonessential; conveniently, but unfortunately for the seamen, their schematic representations of tropical storms showed circular rather than inward spiraling winds. Redfield's theory of cyclones, for example, which became known as centrifugal theory, explained that the circular motion resulted because centrifugal forces balanced pressure gradient forces. Piddington's famous transparent hornplates, which he designed for navigation and which were included in his book, had diagrams of the winds in tropical storms in both hemispheres drawn on them that also showed tangential motion. Based on this, seamen were advised that the center of a tropical cyclone is located along a line going perpendicular and to the left of the observed wind, and this misinformation caused many mishaps at sea.

In the early nineteenth century, the systematic investigation of tropical cyclones was still in its infancy. Piddington, like other scientists, was primarily interested in collecting facts. In his early accounts of cyclones, numbering hundreds of pages, he almost never considered the causes of these storms, although he did later, through his own work as well as his interpretation of others. He rejected Espy's theory that ascending currents of warm moist air and the release of latent heat were essential for the formation and maintenance of tropical storms because Espy insisted on unrealistic (as viewed by Piddington) centripetal flow into storms. Piddington also rejected the suggestion that the rotary motion in cyclones was maintained by the inflow of two currents of opposite direction, a condition that presumably existed along the boundary of the monsoon and the trades. This theory was held by Alexander Thom (30), an Army surgeon stationed in Mauritius and self-appointed meteorologist, and later by Charles Meldrum (1821–1901), who in 1848 was appointed professor of mathematics at the Royal College in Mauritius and became an expert on tropical cyclones in the southern Indian Ocean.

Late in his career Piddington ventured to present preliminary views on the development of tropical cyclones because he thought it was important for determining the paths of these storms. Knowing the average tracks of the cyclones, Piddington noted, was of course not only of scientific interest but desirable from a nautical point of view: "It is upon the track of the Storm (as upon that of a pirate or enemy) that [the seaman's] manoeuvers must depend" (31). He summarized data from 1800 to 1846 in a chart of the Bay of Bengal cyclone tracks. Examining these tracks and extending them backward to their probable sites of origin to the southeast, he became convinced that the development and the track of tropical storms were related to places of volcanic activity in Southeast Asia (31):

If we produce by eye or by ruler the various tracks of the Storms backwards to the Eastward on the same line we shall find them all *tending* as it were, to some focus of volcanic activity. Beginning from the South, the first set appear to come from some of the numerous Sumatran Volcanoes . . . The next set . . . will mostly be found to arise from the Barren Islands . . . while a third, the Dacca and Kuook Phyoo hurricanes seem traceable from the volcanic centers of Cheduba.

The early studies of the tropical storms in the Bay of Bengal, concentrated on gathering descriptions and on charting the wind field and tracks of storms. The systematic investigation of the relation between their occurrence and frequency and the monsoon cycle became possible in the second half of the nineteenth century (see Section 2.4), with the advances of modern meteorology and the organization of the India Meteorological Department.

2 THE EVOLUTION OF MONSOON METEOROLOGY

The last decades of the nineteenth century were a period of fundamental change in meteorology. The coincidence of a large number of factors brought about the emergence of meteorology as a discipline and initiated its organization on a national and international scale (21). Let us briefly consider these general developments in relation to monsoon meteorology.

2.1 The Development of Modern Meteorology

By 1870, many countries, including most European countries and the United States, had created national weather services. The primary activity of the new weather services was establishing a station network and systematically collecting and publishing observations in the form of daily records, averages, and synoptic charts. The need for international cooperation in constructing daily weather charts had become apparent. In 1873, at the First Meteorological Congress in Vienna, steps were taken toward uniform observation procedures, standardization of data, and systematic communication and publication.

Also, during this time meteorology was recognized as a profession. The national weather services required a trained scientific staff. The first professorships in meteorology were established during the 1870s and 1880s. Regular communication of scientific results and discussions was accomplished through meteorological societies and journals.

With the organization of meteorology, empirical knowledge of the atmosphere grew on an unprecedented scale. Land and marine meteorologists began to interact effectively, that is, both groups used the same parameters, such as barometric pressure data reduced to sea level, and statistical and synoptic charts that extended over land and sea. Furthermore meteorologists began to explore systematically the three-dimensional structure of the atmosphere by means of cloud observations, kites, and balloonsondes.

Throughout this period, scientists strengthened the theoretical basis of meteorology as they applied the basic laws of hydrodynamics and thermodynamics to atmospheric processes. The fundamental papers of William Ferrel (1859–1860) and Cato Guldberg and Henrik Mohn (1876 and 1880) on the equations of motion on a rotating earth contained theoretical models that featured the so-called Hadley cell in the tropics and inward and upward spiraling motions in cyclones. Papers of Theodor Reye (1864) and Henri Peslin (1868) clearly tied Espy's early experiments on the release of latent heat in convective currents to the first law of thermodynamics (32).

As part of this general development, British meteorologists in India began to organize and advance Indian meteorology. During the late 1850s, the Asiatic Society of Bengal called for the establishment of a unified, government-controlled meteorological service for India. Additional pressure came from the business and shipping community of Calcutta, calling for an efficient storm warning system. Furthermore the military urged that meteorological activities be coordinated because of interest in examining the relationships between climate and weather and disease in India.

As a first result, five independent provincial systems were formed in 1865, covering about one third of the total area of India. Unfortunately, there was little coordination among them. As Eliot (23) remarked 40 years later, "the parochial system of dealing with the meteorology of a well-defined meteorological area like the Indian monsoon region was self-condemned from the first." Soon, "the defects of the provincial system, and the failure of these isolated systems of observation to throw light on the meteorology of India" led to renewed efforts to create a general meteorological system. In 1875, upon the final recommendation of the Commission of Enquiry into the Orissa and Bengal Famine of 1866, the government of India authorized the organization of the present centralized India Meteorological Department (24). Publication of the *Indian Meteorological Memoirs* began in 1884.

Partly because of the numerous and high quality publications of meteorologists in India, beginning in the 1880s, Western meteorologists showed renewed interest in the study of the monsoon during the last decades of the nineteenth century. In addition, renewed emphasis on the general circulation stimulated the expansion of the meteorological station network, and the first systematic upper-air observations further encouraged the study of the monsoon as a part of the global circulation.

2.2 Changing Concepts of the Monsoon Circulation

It was fortuitous during the early phase of modern meteorology in India that its political boundaries coincided to a large degree with the extent of the southwest monsoon. India, like the United States, could introduce uniform procedures for observations and their evaluation over an extended area that encompassed the large-scale features of the atmospheric circulation. In contrast, European meteorologists had to communicate with as many as 15 governments and deal with diverse observing systems and standards in order to obtain the pertinent information for a single storm.

In addition to accumulating systematic observations of local phenomena associated with the Indian monsoon, investigators sought to reveal the physical relations of these phenomena and to determine the role of the monsoon as part of the global atmospheric circulation.

2.2.1 Effect of the Land Configuration on the Indian Monsoon Circulation. In 1864, Henry F. Blanford of the Indian Geological Survey was appointed honorary secretary of the Meteorological Committee that the Asiatic Society of Bengal had formed in 1857. Soon after he became the first meteorological reporter to the government of Bengal and, in response to a disastrous cyclone that earlier had struck Calcutta, he organized a storm warning system for that port. By far the most recognized meteorologist in India at the time, Blanford was appointed first director of the India Meteorological Department in 1875. Subsequently, he established the best meteorological observational network in Asia. The data from this network formed the basis for numerous studies. Blanford's work culminated in his classic monograph *The Rainfall of Northern India* (33).

Blanford's book of 1877, the *Indian Meteorologist's Vade-Mecum*, became the most widely used textbook on tropical meteorology for the rest of the century (34).

Well-read in meteorological literature, Blanford applied to the monsoon the concepts and results of what was then called the "new meteorology," such as Ferrel's and Guldberg and Mohn's applications of the equations of hydrodynamics to horizontal atmospheric motions and Reye's and Peslin's thermodynamic treatment of convective processes.

Center of the Monsoon Circulation. A starting point in Blanford's monsoon studies was the determination of the center of the monsoon circulation. Detailed factual knowledge of the conditions in the center presumably offered some hope for predicting the development of the monsoon. Contrary to the widely held opinion that the monsoon was one large circulation centered over Asia, he viewed the Indian monsoon as a subdivision of the Asian monsoon, effectively shielded from the larger continental circulation by the high mountain chains in the north and the Arabian and Bengal seas to the west and east (34).

> The Asiatic monsoon . . . consists . . . not of one current flowing alternately to and from Central Asia, but of several currents, each having its own land centre. . . . The centre of the Indian Monsoon lies south of the Himalayan chain.

The center is located in the upper Punjab, he noted, where the surface pressure is lowest and temperature highest in summer and vice versa in winter (35).

> The north-east monsoon [originates] in the plains of the Punjab, Upper and Central India and Assam; probably also on the southern slopes of the Himalaya. . . . The south-west monsoon is produced by the heating of the land surface of the peninsula and a superincumbent air to a temperature much above that of the sea to the southward. . . . But with the advance of the sea-winds and upward diffusion of condensation of water vapour, the heat is also diffused to higher levels. . . . By this diffusion of heat and increasing temperature of the ground surface and the lower strata of the air under nearly vertical sun, the pressure falls steadily and the sea-winds are drawn from a greater distance south . . . and the southeast trades or perhaps only a portion of it crosses the line [equator] and brings the monsoon rains to Bengal and the west coast of India.

The centers of low and high pressure over the Punjab, Blanford observed, generate patterns of cyclonic inflow in summer and anticyclonic outflow in winter that deviate considerably from the earlier ideas of the simple southwest and northeast monsoon. Accordingly, Blanford proposed to use the terms "summer" and "winter" monsoon instead of southwest and northeast monsoon.

The Russian meteorologist Alexander Woikoff (1842–1916) supported Blanford's views (36), citing pressure differences as evidence that the Indian monsoon circulation was centered over northern India rather than over the Tibetan Plateau. He noted that the surface pressure difference between the Punjab and Ceylon is twice as large in summer as in winter and that wind velocities vary correspondingly. Moreover, the pressure difference between February and July in the Ganges valley was more than three times that at Simla at more than 6000-ft elevation, indicating a greater intensity of the monsoon effects in the surface layers. The fact that the monsoon effect diminishes with altitude over northern India, Woikoff noted, supported the view that the center of its circulation was located in northern India.

John Eliot (1839–1908), who in 1889 succeeded Blanford as director of the India Meteorological Department, also assumed that the monsoon was centered over northern India. His analysis (37) of two years of daily weather charts covering the Indian monsoon area left him with no doubt that

> the lower air-current of the North-east Monsoon has its origin as a horizontal movement [outflow] in Upper India. It is undoubtedly fed, to some extent, by air-drift down the passes. . . . It is not, however, the continuation of a horizontal current advancing from Afghanistan or Beluchistan, or across the Himalayas from Tibet.

Nevertheless, the question of the monsoon's circulation center remained unresolved, partly because no data were available from the Tibetian plateau. Instead, investigators examined the factors that might enhance or restrict the effect of land–sea contrasts on the formation of the monsoon.

Effect of Latitude. Julius Hann (1839–1921), director of the Austrian weather service and an avid student of world climates, stressed the effect of latitude on the global distribution of the monsoons (38). Monsoons, he noted, depend on the summer–winter temperature differences between land and sea, which increase with latitude, but are practically nonexistent in the equatorial belt and only small in the tropics. Thus, ''an entirely tropical continent located between about 15°N and [15°]S, without extended landmasses beyond the tropics, could have only strong land and sea breezes, but no monsoon winds to speak of'' because, he suggested, of small seasonal temperature contrasts, and, equally important, because the deflective force of the earth's rotation is too small in this zone to produce large-scale cyclonic or anticyclonic circulations. However, he went on, the equatorial and tropical zones could become part of strong monsoon circulations centered along or beyond the boundaries of the tropics. The most favorable conditions for the development of monsoons, he stressed, occur in latitudes near and beyond the tropics. The Indian monsoon, which includes the tropical and equatorial zone, develops because of the strong heating of the extratropical landmasses surrounding the Indian Ocean and because much of India lies north of 15 degrees latitude where the deflective (Coriolis) force becomes important.

Hann further noted that in mid- and high latitudes the monsoon–producing forces are weakened by the general strong north–south temperature and pressure gradients between pole and equator (38):

> For this reason they do not reach that development in higher latitudes which they could attain on account of existing temperature differences. The monsoon character of the winds is often confused, and becomes clearly recognized only when the general mean direction of the wind is subtracted from the individual wind directions.

Effects of Elevation. The American meteorologist William Ferrel (1817–1891), who had merged theory and facts in the first widely acclaimed mathematical theory of the general circulation of the atmosphere and oceans, added an important restriction to the differential heating argument (39). He pointed out that all great monsoons occur near high mountains and not in the Sahara or Arabia, despite the large temperature gradients between land and sea there. Ferrel argued that, in fact, not a vast column, but "only a comparatively thin stratum of air next to the earth's surface . . . is very much heated above that of the ocean." Consequently, he suggested, even large temperature differences between flat country and adjacent ocean would lead to only small horizontal motion, and unstable vertical temperature gradients would result in only "small local eruptions through the strata above." However, he continued, if the "continent . . . has highlands with long slopes," the intensity of the summer monsoon "is very much increased" because in this case the stratum of heated surface air is inclined upward, encouraging rising motion along the slope. Barometric pressure decreases (39)

> in the warm rarified stratum where it rests upon an inclined surface[;] the greater pressure of the heavier air everywhere else at the same level tends to drive the lighter surface air up the slopes . . . The tendency of the warmer stratum of air to flow up an inclined surface increases with increasing differences of temperature, and increasing length and steepness of slope.

Ferrel noted further that ascending motion is intensified in the case of high plateaus, such as the Tibetian Plateau, which often heats up as much as the plains and thus is much warmer than the surrounding atmosphere at the plateau level.

Although Ferrel acknowledged that latent heat release adds to the strength of the summer monsoon, it appears that he did not fully appreciate the essential role of convection of moist air and cloud formation in the Indian monsoon. The heating, Blanford clarified, is not confined to a shallow surface layer; rather, water vapor significantly increases the mean temperature of the air column above the surface layer because moist air is cooling less rapidly than dry air as it rises (due to the added heat of condensation). In May, Blanford observed, "the rise of temperatures over India at 7000 and 8000 feet proceeds as rapidly as on the plains." Lack of water vapor—and not so much lack of mountains—Blanford suggested, explains the puzzle that no monsoons were found in the Sahara (34). Subsequent attempts to determine more precisely the conditions for cloud formation and precipitation shed new light on the vertical structure of the monsoon.

2.2.2 Vertical Structure of the Monsoon.

Studies of the vertical structure of the monsoon were part of the late nineteenth century development to explore and treat quantitatively the three-dimensional nature of atmospheric processes. Once investigators had accumulated a basic knowledge of flow patterns and atmospheric conditions near the surface of the earth, they realized that in order to better understand atmospheric circulations they needed to use observations at different altitudes and to consider processes in the vertical as well as the horizontal. The first mountain observatories were established at this time. Hugo Hildebrand Hildebrandsson (1838–1925), representing the International Meteorological Organization, organized a cloud observation network encompassing the entire globe; the first International Cloud Year (1896–1897) was organized; the French meteorologist Teisserence deBort (1855–1913), who later discovered the stratosphere, constructed the first global pressure charts for 4000 m altitude; and weather services and private organizations began to operate kite stations and experimented with the first balloonsondes (48).

As in Europe, investigators in India showed great interest in these new meteorological observations in the upper layers of the atmosphere and the new observational techniques, although, during the nineteenth century, Indian upper air data remained restricted to mountain surface station and cloud observations.

The vertical depth of the lower monsoon currents attracted early attention. In particular, meteorologists attempted to determine the altitude of the monsoon's neutral pressure plane, separating the lower-and upper-monsoon currents. In the vertical, this plane would be the point of wind reversal, that is southwest currents over the northeast monsoon or northeast currents over the southwest monsoon. They also hoped to learn about the vertical temperature structure of the monsoon currents. This information would be important in investigating the causes of monsoon rainfall.

Wind Reversal. Blanford was the first to attempt to determine the plane of neutral pressure. It was fortunate that India had stations at both high and low altitudes. Using pairs of hill and plain stations in Ceylon in the south and Sikkim and Assam in the north, Blanford (35) calculated the mean elevation of the neutral pressure plane between Ceylon and Sikkim for winter and summer. In winter this plane varied little in altitude from north to south, but in summer the southwest monsoon increased dramatically in vertical extent from south to north. This finding agreed with Blanford's and others' concept that the southwest (or summer) monsoon low-pressure center was located over northern India, fed by inward and upward spiraling currents of moist air from the southwest. Blanford's methods were also used by Samuel Alexander Hill (41) and E. Douglas Archibald (42), both scientific assistants at the India Meteorological Department.

Because of the moderate height of the northeast monsoon current it was comparatively easy to find evidence for the upper return current of the winter monsoon. Blanford (35) showed that in winter above 5000 to 6000 feet an "anti-monsoon," that is the upper return current, could be observed, similar to the antitrade. These upper southwest winds in winter, still relatively moist, Blanford explained, brought the snows observed in January and February to the slopes of the Himalayas.

Searching for evidence of the postulated upper return current above the summer monsoon was less successful. Blanford cited cloud observations from the Dodabetta Peak in southern India which indicated northwest flow at 8600 feet. On the other hand, re-evaluating the data available, Woikoff concluded in 1887 that no upper return current above the summer monsoon had yet been demonstrated, principally because of the great elevation of this current. The antimonsoon had to occur at much higher elevations than the antitrades, he reasoned, because the thermal heating of the air over India in summer was much greater and reached higher than the heating of the air over the oceans in the calm belt between the trades (36).

Hann (38) pointed out a difficulty with the assumption of a simple upper return current. He noted that the monsoon currents reach altitudes where they are affected by the general circulation of the atmosphere. For example, the strength of the upper level westerlies over Asia in winter dominates the flow in that region and essentially masks the upper branch of the monsoon circulation. As Hann said,

> the great Asian winter anticyclone is fed at upper levels probably almost exclusively by the generally dominant upper west winds in these latitudes and not by inflow from all sides, so that a regular cyclonic motion cannot develop above the lower anticyclone.

Cloud observations collected by several investigators during the 1880s (43, 44) supported Hann's idea that the upper return currents of the monsoon circulation are not strict reversals of the lower currents but instead are complicated and, in general, dominated by the general circulation. However, the concept of the neutral pressure plane found further application in some interesting studies on cloud formation in the lower monsoon current.

Vertical Temperature Structure and Cloud Formation. In 1841 Espy had recognized convection as the active process in the formation of cumulus clouds. He had shown in particular that large vertical temperature gradients favored the formation of ascending currents and clouds and hence, precipitation. He emphasized that the maintenance of these ascending currents depended largely upon the absorption of heat at the earth's surface and the release of latent heat in the rising, moist air. Using station pairs at low and high altitudes, Blanford (35) determined the vertical temperature gradient in the center of the Indian monsoon circulation to be three times as high in the summer as in winter. Eliot reported (45) that in the case of India the summer monsoon circulation persists into July–August because

> it is maintained for a considerable period not by the continuation of the hot weather conditions in upper air but mainly by its self contained energy liberated by the rainfall during its existence over India.

Subsequently, meteorologists in India attempted to get a rough estimate of rainfall amounts by estimating the thickness of cloud layers in the southwest monsoon current. It was necessary to calculate levels of condensation above which cloud formation could theoretically take place. Using observed vertical temperature gradients and moisture data for the period 1870–1880, Archibald (42) calculated condensation levels for different regions in India during the summer monsoon. He tried in particular to detect a relation between the thickness of the cloud layers—the stratum of the monsoon current between the level of condensation and the neutral pressure level—and the amount of rainfall (see Fig. 8.9). His results, however, remained inconclusive.

All in all, the late nineteenth century monsoon studies illustrate not only the great interest and insights into the three-dimensional structure of the monsoon but also the need for more and better observations: analysis of the upper air data then available allowed no more than tentative conclusions.

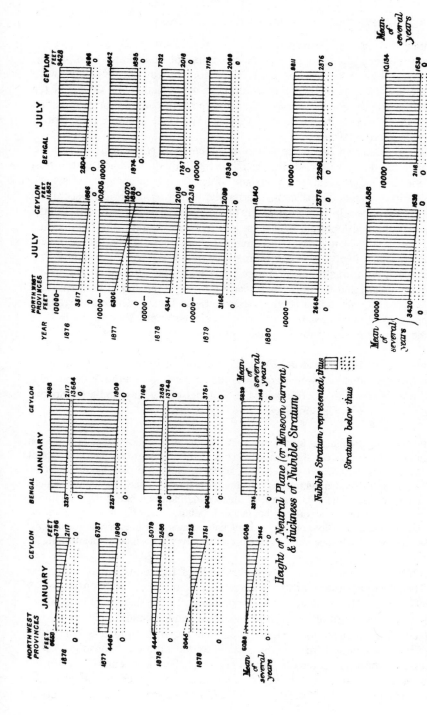

Figure 8.9. Height of the neutral plane separating lower and upper monsoon current and thickness of the nubible (nubilous) stratum, by D. Archibald. Cross sections are from northwestern India to Ceylon and from Bengal to Ceylon, in January (left) and July (right) for the years 1876 to 1880. The nubible stratum is the layer between the level of condensation and the neutral pressure plane in which cloud formation is theoretically possible. Note the dry northwest in January and the great depth of the nubible stratum in July (42).

2.3 Onset and Establishment of the Southwest Monsoon

The beginning of the rainy season in India is a much anticipated event. For the meteorologists of the 1880s, to gain an understanding of this complicated process was an essential step on the way to their ultimate goal—prediction. Knowing what triggered the sudden start of the rains would help in predicting the timing of the monsoon. Knowing the source of the current and the major sources of its moisture would help in predicting the amount of rainfall. We shall examine how British meteorologists dealt with these questions.

In general, the data available were still too fragmentary to allow firm conclusions, and resources for doing any kind of research were limited; burdened with a host of daily duties, including weather reports and forecasts, official meteorologists had little time left for research. The scientific staff at the India Meteorological Department was kept to an absolute minimum, consisting of the director, two scientific assistants, and the meteorological reporters of Bengal, the Western Region, the Punjab, and Madras. John Eliot, when he was director of the weather service, worked with "tireless energy . . . twelve hours or more every day of the week . . . and would sometimes speak with longing of the future years when the Director would have a trained scientific staff and facilities for research" (46). At the end of his term Eliot finally succeeded in obtaining a personal assistant, the Indian meteorologist, Hem Raj. Despite these limitations, meteorologists achieved impressive results in their study of the establishment of the southwest monsoon.

2.3.1 The Burst of the Monsoon.
Colorful descriptions of the onset of the rains, the "burst" of the monsoon, were part of the Indian monsoon literature throughout the nineteenth century. Sir E. Tennant (47), Colonel of the East India Company, reported on this event in Ceylon:

> [After the] burning droughts of March and April . . . the air becomes loaded to saturation with aqueous vapour . . . the sky, instead of its brilliant blue, assumes the sullen tint of lead . . . the days become overcast and hot, banks of cloud rise over the ocean to the west . . . At last the sudden lightnings flash among the hills, and sheet through the clouds that overhang the sea, and with a crash of thunder the monsoon bursts over the thirsty land, not in showers of partial torrents, but in a wide deluge. . . . This violence, however, seldom lasts more than an hour or two, and gradually abates after intermittent paroxysms, and a serenely clear sky supervenes. For some days heavy showers continue to fall.

Nineteenth century scientists viewed the burst of the monsoon as a peculiar puzzle: "This, the most striking weather-change in the whole year," Ralph Abercromby (1842–1897) observed, "is associated by no change in the shape of the isobars" (48). This was contrary to all other experiences of sudden weather changes.

In search of an explanation, Blanford studied the meteorological conditions during the transition period from the hot season to the southwest monsoon. As the hot weather conditions are gradually building up in northern and central India during April and May, he noted, sea breezes develop near the surface and displace the northeast winds upward. In 1874, Blanford (35) illustrated in a vertical cross section, from about 20°N in the Bay of Bengal to the Khasi Hills to the north, that during the hot season the sea breezes are rather shallow and that the northeast monsoon is still present on top (Fig. 8.10). The rainy season, Blanford suggested, sets in when the sea breezes are replaced by the much more powerful, warm and moist southwest current that extends high into the atmosphere.

According to Abercromby, analyses showed that the surface pressure field over India rarely changes. However, when the analyses were extended to include the oceans surrounding the Indian subcontinent, a relationship could be seen between the onset of the monsoon rains and the surface pressure distribution. That is, the onset of the southwest monsoon coincided with the disappearance of high pressure ridges in the Bay of Bengal and in the Arabian Sea. Abercromby himself noted, "the burst is apparently coincident with the disappearance of the belt of high pressure to the south of the Bay of Bengal" (48). Blanford stated more specifically: "In June, the ridge of high pressure over the sea, which has steadily receded southward since February, is obliterated" (35). Anticyclonic circulations in March and April were clearly evident over the Arabian Sea and the Bay of Bengal on surface wind charts (see Fig. 8.11). These high pressure areas, Woikoff remarked, gradually diminish because of heating and outflow, and disappear by the end of May or beginning of June (36). Köppen (1846–1940) in the *Segelhandbuch* (9) noted that the burst takes place when the southwest winds over northern and southern India connect and increase in strength. This occurs when the last remnant of the pressure maximum, around 15°N in April in both seas, disappears and an uninterrupted pressure decrease is established from the southern to the northern tropics.

2.3.2 Cross-Equatorial Flow. Early descriptions of the southwest monsoon current, including those by Halley and Maury, had suggested that the flow originates south of the equator. Investigations during the last quarter of the nineteenth century provided much supportive evidence. Blanford (35) noted in 1874 that "the southeast trade, or perhaps only a portion of it, crossing the line [equator], brings the monsoon rains to Bengal and the west coast of India." John Eliot substantiated this connection. Eliot had come to India in 1869, teaching physics and mathematics at the Rookee Engineering College, the Muir Central College at Allahabad, and at the Calcutta Presidency College. Like Blanford, he began his meteorological career as meteorological reporter to the government of Bengal.

In 1896, Eliot (37) published a case study of the cross-equatorial flow during the summer of 1893. The data he used were derived from one year of daily weather

Figure 8.10. Vertical cross section by H. F. Blanford, illustrating the sea breeze circulation in March between the Khasi Hills and the Bay of Bengal (35).

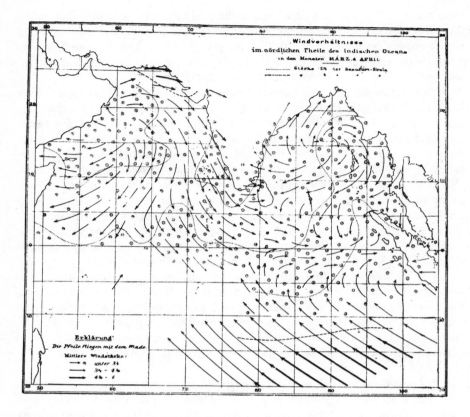

Figure 8.11. Surface winds in the Indian monsoon area in March and April. Note the areas of high pressure with outflowing currents over the Arabian Sea and the Bay of Bengal. Arrows go with the wind; dots indicate wind speeds of less than 3 knots, wind speed increases with length and thickness of arrows (9).

charts covering an extended area, the so-called "Indian Monsoon Area Charts" (see Fig. 8.12). (For two years the government funded the large-scale data collection and analysis project which produced the charts. It might be regarded as the first monsoon experiment.) Examining in detail the wind and pressure changes over the Indian monsoon region during the transition period, Eliot found that in May, before the burst (Fig. 8.13*a*), the isobars correspond to the thermal conditions over the interior, while in June, during the southwest monsoon, they are "directly related to the prevalence of a strong, steady, and massive air current from south to north" (Fig. 8.13*b*). "This distribution of pressure is clearly not the product of the actual thermal conditions over India." Rather, Eliot said, the burst of the monsoon in India represents a "definite large change of weather conditions" in the second half of May associated with the development of "a steady and continuous horizontal current from south to north across the equatorial belt" (37).

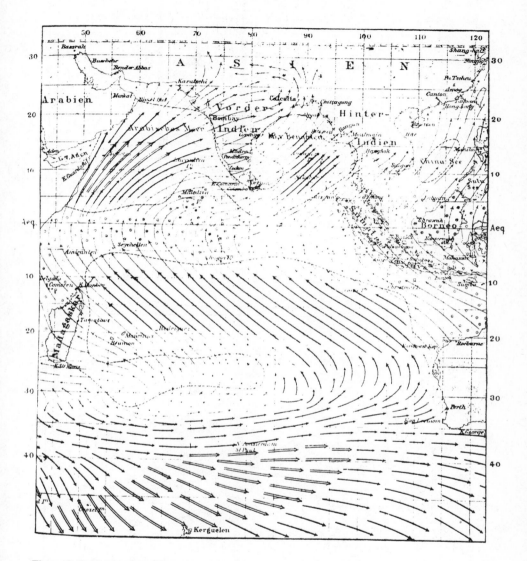

Figure 8.12. Surface winds in the Indian Ocean for July (9). Arrows indicate direction and strength of the wind. Note the strong winds along the eastern coast of Africa and Arabia. Long arrows denote constant winds and short arrows variable winds; dashed lines enclose regions of winds with speeds greater than 6 m/sec, and dotted lines enclose regions of winds with speeds of less than 3 m/sec; → below 6 m/sec; ⟶ 6–9 m/sec; ⟶ 9–12 m/sec; ⟹ over 12 m/sec.

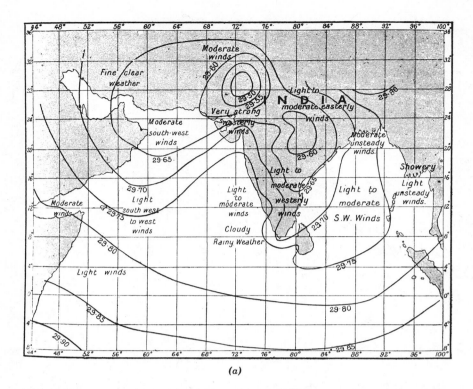

(a)

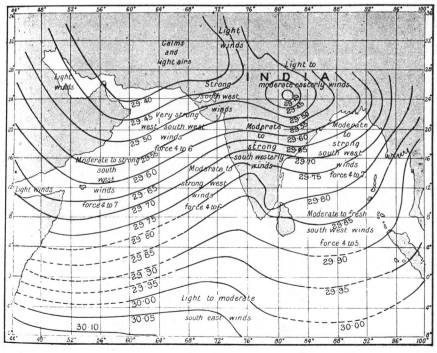

(b)

Eliot considered it particularly important to study the western part of the Indian Ocean because the air currents are stronger and steadier than in the eastern part of the equatorial belt. A wind chart published by the German hydrographic service in 1892 (9) clearly showed the great strength of the current (Fig. 8.12). In the Arabian Sea the southwest monsoon had average velocities that did not occur anywhere else in the tropics, about 6–8 on the Beaufort scale (15 to 21 m/sec). Apparently, even steamships would not readily attempt the return from India to Europe during their reign. The western boundary of the southwest monsoon appeared to form a distinct line from Cape Guardafui and along the southwest coast of Arabia. Along the eastern side of the cross-equatorial current (south of the eastern Arabian Sea) winds were much weaker (under 6 m/sec) and used to advantage by the seamen. "During the height of the southwest monsoon," Blanford had noted (35),

there is a tract lying between the equator and 9 north latitude, and extending nearly from Ceylon to Socotra, in which the winds are light and the sea smooth. This is known to navigators as "the soft place in the monsoons," and is taken advantage of, more especially by steamers.

2.3.3 Sources of Rain for the Southwest Monsoon.

Studying the origin and flow patterns of the southwest monsoon also shed light on the sources of rain. In particular, the large-scale, regional variation in rainfall in the African–Indian monsoon regime appeared to be related to variations in the basic southwest currents. "Most frequently," Blanford wrote (35), the rain

originates in this variable but on the whole westerly and rainy current over the equator in the South-west Monsoon, which is doubtless fed by the South-east Trades [of the Southern Hemisphere] . . . but it represents only a portion of the air poured by the Southern Trade winds into the equatorial region, the remainder ascending convectively. . . . Nor is the Monsoon drawn from this source alone. It is probably recruited . . . from more northern latitudes. . . . [T]he Northern Indian Ocean presents an expanse of evaporating surface, which, supplemented by that of the Bay of Bengal, and that of India itself, is sufficient to supply all the rain that falls annually.

Eliot was convinced that variations in the strength of the cross-equatorial flow near the equator are mirrored in the southwest monsoon current and precipitation. The data of the 1893 summer monsoon show very consistently, Eliot said, that the rainfall in India "was heaviest and most general when the air current across the equatorial belt was strongest, and vice versa" (37).

Eliot suggested that the current across the equator supplies the precipitation for three areas, the White and Blue Nile in Africa, India, and Burma. He hypothesized that variations in precipitation depended on the general strength of the current and that changes in the main current result in similar variations of rainfall in all three

---

Figure 8.13. Charts of the Indian monsoon area, showing surface pressure distribution and dominant winds (37); *(a)* May 2, 1893, *(b)* June 21, 1893. Data were extracted from the log books of ships on passage to Bombay and Calcutta; barometers were checked in the ports.

areas. He pointed out that Sir William Willcocks, the director-general of the Reservoirs of Egypt, also suspected such a relation. When he compared flood levels of the Nile with Indian harvest data, he found that "famine years in India are generally years of low flood in Egypt" (37). However, Eliot's analysis of flood levels of the Nile and approximations of the monsoon rainfall in India from 1875 to 1892 did not produce any conclusive evidence for Willcocks' statement.

Later, Eliot was to reverse his earlier suggestions and propose that, to some extent, there is a compensatory relationship between the three areas of Africa, India, and Burma; that is, prolonged rainfall in one will be associated with decreased rainfall in other. For example, when investigating the air currents over Africa and Burma at the time of the drought of August 1893 in India, he found that "the air current across the equatorial belt . . . was much more Easterly or less Westerly, thus indicating that the current was determined in larger amount to the Upper Nile and Abyssinia precipitation areas." Similarly in 1899, the failure of the Indian monsoon was coupled with an unusual occurrence of rain over Africa (49). Although Eliot was convinced of a relationship between the cross-equatorial flow and the southwest monsoon rainfall, he found that his data base was insufficient to conclusively prove any correlation.

2.4 The Bay of Bengal Cyclones in Relation to the Monsoon Cycle

The amount of rainfall—the most important parameter of the monsoon for those living within its reach—decidedly depends on local conditions and processes. Although investigators showed that the Indian monsoon rainfall was related to events taking place far to the south of India (Section 2.2) or even to variations in the sun's radiation (Section 3.2.1), it was evident already at the time of Piddington that rainfall in northern India was closely related to two factors: orographic features of the Indian subcontinent and the storms of the Bay of Bengal. While orographic features constituted a permanent, nonvarying influence, the rain-bringing cyclones of the Bay varied in frequency and strength from year to year, making the determination of their effects a complex problem.

In the Bay of Bengal, conditions for observing these storms were particularly favorable. Meteorological stations surrounded the Bay on three sides, and the heavy shipping traffic ensured numerous observations at sea. Piddington's work on the tracks and frequency of these storms had been one of the first concerted efforts in Indian meteorology. Both Blanford and Eliot continued these efforts, focusing on the tracks, intensity, and formation of the tropical storms. Utilizing Buist's catalogue of the Bay of Bengal cyclones (going back to 1737), Blanford (34) collected data on 115 cyclones. Eliot (50), who became the recognized expert on Bay of Bengal cyclones, organized a system for collecting ship observations of storms in the Bay of Bengal and the Arabian Sea. He assembled monthly track charts that contained 211 storms observed over a period of 25 years and published them in a two-volume handbook on these storms (Fig. 8.14).

Eliot made a clear distinction between the many small and moderate storms, frequent throughout the rainy season, and the truly dangerous cyclones, prevalent during the seasons of transition—particularly at the end of the summer monsoon. He found that in May and October one out of three cyclones is dangerous. These storms, Eliot noted, often are of considerable extent and accompanied by considerable rainfall in Bengal and northern India. Eliot proposed to reserve the term cyclone for storms with a wind force greater than 10 Beaufort (26 m/sec) and to call storms of lesser force cyclonic storms. Cyclones with long paths across the Bay were generally more intense storms, but they dissolved soon after reaching land. Moderate storms, which appeared to develop at the head of the Bay, traveled considerable distances northwest in the Ganges valley.

These findings by Eliot supported earlier studies on the formation and maintenance of tropical cyclones which had focused on the convective process. In 1872, T. Reye (51) used precipitation data of the great Cuba storm of 1844 to compute the energy available in tropical storms from the release of latent heat and showed that this energy was sufficient to explain the kinetic energy of the winds. Following Reye, Blanford suggested that the ascending convective current associated with condensation of water vapor carried by moist ocean air was the essential feature of the storms of the Bay of Bengal (34).

Blanford believed that specific local conditions favor the formation of tropical cyclones. He hoped that eliminating these local differences from cyclone data would eventually reveal the general laws for their formation. Two conditions characteristic for their formation in the bay, Blanford suggested, are calm or only light and variable surface winds over the bay, that is, small surface pressure differences, and little or no rain on the east and northeast coast of the bay. These conditions, he believed, support the development of a convective current and prevent the products of evaporation and condensation from being carried away. In addition, since intense cyclones never seemed to form at the head of the bay or in the lee of Madras, an inrush of saturated air from the southwest or west-southwest appeared to be important for formation (34).

Eliot, who supported Blanford's conclusions, described the formation of the monsoon depression of early July 1883 (52):

The atmospheric whirl was fed and maintained by a very strong southwesterly air current moving northward up the Bay near the Burma and Arracan coast. It was apparently formed in front of this air current, and was causing winds to draw round over the northwest of the Bay. The indrought from that quarter, however, was feeble and unimportant, except as an indicator of the bad weather to the southeast. . . . Energy given out during the process of aqueous vapour condensation on the large scale is the motive power of cyclones, and the rainfall must be localized and concentrated over a considerable area, for a period of one or more day, in order to produce the continuous and rapid accumulation of energy which characterises a large cyclonic disturbance.

Seen in perspective, during the 1870s and 1880s progress in the understanding of tropical cyclones in the bay, coupled with an extensive observational network and considerable statistical results of the storms' behavior, made possible a functioning storm warning system. Even so, Eliot lamented in the preface to the second edition

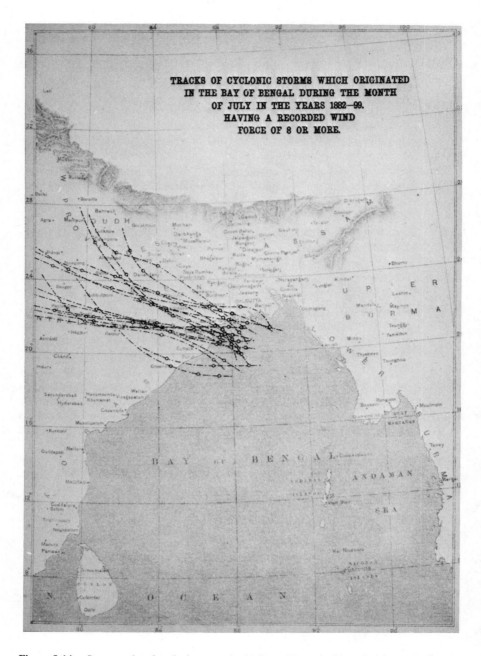

Figure 8.14a. Storm tracks of cyclonic storms in the Bay of Bengal with a wind force of 8 Beaufort (15 m/sec) or more, by J. Eliot during July in the years 1882–1999. Open dots indicate daily position of storm centers, advancing from sea toward land (50, Vol. 2).

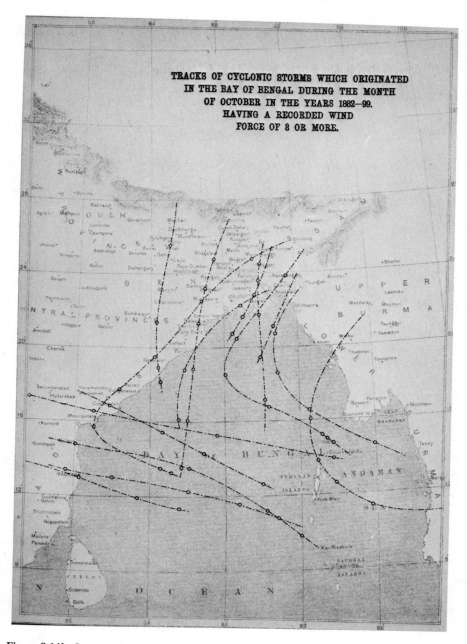

Figure 8.14b. Storm tracks of cyclonic storms in the Bay of Bengal with a wind force of 8 Beaufort (15 m/sec) or more, by J. Eliot during October in the years 1882–1999. Open dots indicate daily position of storm centers, advancing from sea toward land (50, Vol. 2).

of his *Handbook*, ''we have still too much theory and too little accurate information of the details of cyclone phenomena'' (50).

3 THE GOAL: PREDICTION

When Blanford was appointed director of the India Meteorological Department in 1875, he developed an organization which incorporated both research and service. Blanford proclaimed two main objectives of the new meteorological service: (1) the systematic study of climate and weather over India as a whole, and (2) the application of this knowledge to the issue of storm and flood warnings and weather forecasts. The following is a brief discussion of the early development of these goals.

3.1 Short-Term Forecasting

During the 1860s a strong need for reliable monsoon forecasts led to the first steps of organizing a forecasting system. After a devastating cyclone struck Calcutta in October 1864, causing extensive flooding and terrible destruction of life and property, the influential shipping and merchant community of Calcutta and also the Commission of Famine declared the urgency for advance knowledge of storms and floods. Soon after that, Blanford was appointed first meteorological reporter to the government of Bengal. In one of his first actions, Blanford established a storm warning system for the port of Calcutta, using the network of newly installed telegraphic stations along the coast of the Bay of Bengal. By 1880 these storm warnings were extended to the west coast, and by 1886 the storm and flood warning system included all Indian ports.

To aid in the preparation of these forecasts, Blanford established a system to collect local weather reports, via telegrams, from all parts of India and Burma. Daily regional weather reports, at first issued two days after the actual date, began in 1878, preparation of daily weather charts began in 1887, and after 1888 the daily report included forecasts for the next day. The reports also included the daily progression of seasonal rainfall.

More information on this system may be found in *Hundred Years of Weather Service*, published by the India Meteorological Department (24).

3.2 Long-Term and Seasonal Forecasting

Unlike most other weather services around the world, the India Meteorological Department, almost from the beginning, was pressed to develop seasonal forecasts of monsoon rainfall, principally to prepare for possible famines. Thus, after extensive droughts during 1877 and considerable delay of the monsoon rainfalls in the northwest in 1878, the government requested from the Meteorological Department a forecast of the character of the monsoon season. Eliot, who at that time officiated for Blanford, prepared a reasonably successful forecast. After 1878, forecasts of monsoon rainfall were issued regularly, first confidentially and from 1886 published in the Indian *Gazette*.

During the nineteenth century two major approaches evolved in long-term monsoon prediction: (1) correlation with cyclical variations in solar radiation, and (2) correlation with certain meteorological parameters in India and in other parts of the globe.

3.2.1 Correlation of Monsoon Rainfall with the Sunspot Cycle.

After the influence of the 11 year sunspot cycle on variations of terrestrial magnetism had been demonstrated during the 1850s, it was a small step to consider such an influence on meteorological parameters. Serious investigations of this subject began during the 1870s. Meldrum, for example, the expert on tropical cyclones of the South Indian Ocean and director of the observatory at Mauritius, studied the relation of sunspots and cyclone frequency during the 1870s (53). His results apparently were so convincing that, in the words of one of his admirers, "the number of wrecks which came into the harbour . . . and the number of cyclones observed in the Indian Ocean could enable anyone to determine the number of spots that were on the sun about that time" (54). Meteorologists in India were motivated to study the relation between sunspots and monsoon rainfall because drought years and wet years tended to occur in runs, and this observation encouraged the inference that they were manifestations of cyclical physical causes.

Strictly speaking, it was first necessary to establish experimentally that solar radiation indeed varied with sunspot activity. As early as 1801, Sir William Herschel surmised sunspots "to be symptoms of a copious emission of light and heat" of the sun (55). He stated that years of remarkably abundant or deficient sunspots had been noted for their high or low general temperature, and especially for abundant and deficient harvests as reflected in the price of wheat. In the early 1870s, Blanford's spectroscopic observations and measurements with blackened thermometers at numerous stations in Bengal for six years (using only data with clear sky and temperature differences rather than absolutes) appeared to confirm "the original idea of Sir W. Herschell, and that the solar radiation is greatest in years of abundant sun-spots and vice versa" (56).

The first results of sunspot research in India appeared to indicate a clear relation between monsoon characteristics and the sunspot cycle. For example, in 1871 Sir J. N. Lockyer (54) was informed by the editor of the *Ceylon Observer* "that everyone in Ceylon recognized a cycle of about thirteen years or so in the intensity of the monsoon. . . . Afterwards it turned out that the period in Ceylon was really of eleven years." Blanford's analysis of 64 years of rainfall data from six stations in India showed a distinct cyclical variation: "It appears that the average minimum rainfall somewhat anticipates the [sunspot minimum], while the maximum rainfall tends to occur three or four years after the [sunspot maximum]" (56). He also found that the variation was much greater and more distinct at Madras than at any other station. He noted that the cyclone records in the Bay of Bengal were too imperfect to allow a correlation analysis with the sunspot cycle as Meldrum had done for the South Indian Ocean.

In agreement with Blanford, S. A. Hill (41) showed that more than average rainfall occurs in the first half of the sunspot cycle after the maximum (even though the rainfall curve did not follow the sunspot cycle curve), but he also found a reversed correlation for the winter rains of northern India, with the maximum more than one year before sunspot maximum activity and a minimum immediately following the sunspot maximum. Further attempts to interpret these contrasting results remained unsatisfactory.

Nevertheless, being impressed by the initial results of sunspot research, Lockyer remarked in his book *Solar Physics* (54):

> Surely in meteorology, as in astronomy, the thing to hunt down is a cycle, and if that is not to be found in the temperate zone, then go to the frigid zones and look for it, or the torrid zones and look for it, and if found, then above all things, and in whatever manner, lay hold of it, study it, record it, and see what it means.

Many people appear to have followed this advice. However to identify cycles often turned out to be easier than to "see what it means."

For example, it was unexpected when Köppen (57) found in 1873 that, "in the tropics, maximum temperature coincides more nearly with the minimum than with the maximum of sun-spots; preceding the former, however, by one to one and a half years." Blanford (56) attempted to explain this paradox in a qualitative fashion: the temperature of the land stations Köppen dealt with

> must be determined, not by the quantity of heat that falls on the exterior of the planet, but by that which penetrates to the earth's surface. . . . The greater part of the earth's surface being, however, one of water, the principal immediate effect of the increased heat must be to increase the evaporation; and, therefore, . . . the cloud and the rainfall. Now, a cloudy atmosphere intercepts a great part of the solar heat; and the re-evaporation of the fallen rain lowers the temperature of the surface from which it evaporates, and that of the stratum of air in contact with it. The heat liberated by cloud condensation doubtless raises the temperature of the air at the altitude of the cloudy stratum. . . . As a consequence, an increased formation of vapour, and therefore of rain, following on an increase of radiation, might be expected to coincide with a low air-temperature on the surface of the land.

Sir Gilbert Walker (1868–1958), who in 1903 was appointed third director of the India Meteorological Department, continued the investigation of sunspots in relation to meteorological parameters on a global scale. Since no long-term pressure or temperature series was available for many parts of the globe, he thought it appropriate to consider first the effect on rainfall (58). He used data after 1850, covering five to six sunspot cycles. His computation confirmed earlier results of Meldrum and Blanford, and he concluded "that the variations in solar activity affect the monsoon as a whole, but not the irregularities in the geographical distribution of the rainfall of India as a whole."

We shall return later to Walker, who had a profound influence on monsoon meteorology in the early twentieth century.

3.2.2 Correlation of Monsoon Rainfall with Weather in Other Regions.

Preoccupation with synoptic charts, which typically encompassed an area not much larger than that covered by a storm, had led many meteorologists to restrict their attention to the atmospheric processes taking place within a small area. Hann (38) called this practice the "church tower politics" of meteorology, that is, interest not going much beyond the field of vision from a church steeple. Only toward the end of the century was it generally accepted that atmospheric phenomena could not be considered in isolation. Attention began to shift again to the study of the general circulation, and meteorologists sought to explore relations and correlations of atmospheric phenomena over the entire globe. Important examples of this type of investigation during the 1880s and 1890s were studies on the compensating effects between the centers of action in the North Atlantic, namely the Icelandic Low and the Azores High (59). Among the earliest attempts to trace relations between atmospheric characteristics over India and distant regions was an 1880 study by Blanford which demonstrated compensation of mean barometric pressure over India and Russia in winter (60).

Important for long-range forecasting were attempts to relate monsoon rainfall to variations of meteorological parameters in the period preceding the monsoon. Widely discussed was Blanford's suggestion that "unusually heavy and especially late falls of snow on the North-Western Himalaya" are followed by "deficient summer rainfall on the plains" (61). Blanford based this hypothesis on the analysis of 20 years of data from which he noted that during years of below normal summer rainfall, abnormally high pressure occurs over an extensive region, including northern India. In particular, this happened during the droughts of 1876 and 1877. During these years, "dry NW [northwest] winds down the whole of Western India" were associated with high pressure in northwest India and Bombay, and low temperatures in the northwest. He suggested that the dry northwest winds in summer were "fed by the descent of air from an upper stratum, namely, from a current moving at a considerable elevation from west to east." Some sparse cloud observations made at Simla appeared to support this suggestion, indicating upper winds from some westerly quarter during the hot season. Blanford inferred "that some cooling influence more potent than usual was at work . . . condensing the lower strata of the atmosphere and causing an unusual outflow of cooled air." He traced this cooling influence to above average winter rains and unusual snowfall in the northwestern Himalayas, which lowered the snowline 3000 feet below normal by the end of April. Blanford's observations were supported by Archibald and Hill who reported in 1879, "that the winter rains are heaviest when the summer rains are defective, and vice versa" (62).

Blanford's hypothesis was so convincing and attractive that it "seemed to justify . . . provisional adoption [of this hypothesis as a criterion] for forecasting the basic characteristics of the [summer] monsoon rains." In June 1883 Blanford made a forecast in the *Gazette* of India of a prolonged drought based on a previous wet winter. Since the forecast was reasonably successful, he felt encouraged to continue.

In 1894, Eliot began to systematically analyze weather characteristics in coastal regions and islands of the South Indian Ocean. The relation between Indian monsoon rainfall and barometric pressure over the South Indian Ocean appeared most significant. Its explanation seemed to be in the relation between the southeast trades of the Southern Hemisphere and the southwest monsoon of the Northern Hemisphere. Subsequently, Eliot based his forecasts in part on his finding "that the 'burst of the monsoon' is produced by the advance of a humid current from the equatorial belt [i.e., the continuation of the Southern Hemisphere southeast trades] and that the monsoon rains are due to the invasion of India by this current" (37). Although Eliot provided elaborate, detailed justifications for his forecasts, their accuracy improved little with time. Failure to forecast the disastrous drought of 1899 led to suspension of the publication of forecasts in the *Gazette*.

Eliot's successor, Sir Gilbert Walker, faced a challenging situation. Seasonal forecasts of monsoon rainfall—in particular, forecasts of deficient rainfall—were highly desirable and perhaps the most important task for the head of the weather service; yet, as Walker observed, no quantiative theory of the monsoon existed that would support objective forecasts. Unlike Blanford and Eliot, who had both held the position of Bengal meteorological reporter before assuming the directorship of the India Weather Service and could in their forecasting efforts draw on years of personal experience, Walker was a newcomer to meteorology. He had been lecturer in mathematical physics at Cambridge University between 1895 and 1902, with interests in electrodynamics and dynamical problems such as projectiles. Upon his appointment to the foreign service in 1903, therefore, he completed a six-month course as Eliot's assistant. However, the Meteorological Department was still understaffed, and after Eliot's departure, Walker as the new director could consult with few officials. As had his predecessors, Walker had to rely for the most part on his own resources.

Even though public forecasts had been suspended after the 1899 drought, Walker was to deliver confidential forecasts to the government. Following his own axiom that "what is wanted in life is ability to apply principles to the actual cases that arise," he applied statistical methods to discover relationships that were relevant for Indian monsoon rainfall (63). Throughout his 20 years in India and for 30 more years afterwards he worked toward perfecting the application of these methods to long-range forecasting—not only of the Indian monsoon rainfall but of world weather. He brought great organizational and administrative skills to the monumental task he had set.

Walker's studies covered a number of different but related topics and were motivated by extended periods of deficient rainfall. He found that the Indian subcontinent could not be regarded as a homogeneous rainfall area, and so he divided it into four subunits for forecasting purposes. He investigated and then ruled out the possibility that prolonged deficiencies in monsoon rainfall were related to human activities, such as deforestation. In 1910, in a paper examining rainfall data in India and flood levels of the Nile, he concluded that "there is no proof of any permanent climatic change in India" (64). He also noted, "the variations of monsoon rainfall . . . occur on so large a scale," that it can be assumed they are "preceded and followed by abnormal conditions at some distance . . ." (65). Walker concentrated therefore on interrelations of monsoon and general circulation. Also, as discussed in the previous section, Walker explored at length relations of the Indian monsoon rainfall with sunspot activity, and concluded that sunspot activity intensifies existing conditions but is not the dominant control of monsoon rainfall (66).

By 1908, Walker had narrowed his choice of indicators for Indian monsoon rainfall to six forecasting parameters, which he combined into a forecasting formula in the form of a regression equation (for details, see Chapters 16 and 17). He applied this formula for the first time in the forecast of the 1909 rainfall, having developed it with rainfall data from 1865 to 1903. Based on the results, he cautioned, "it will only be possible to make a reliable inference that rainfall will be in excess or deficient when the calculated departure is relatively large" (65).

Walker could draw on a number of statistical correlations that had been worked out by others. For example, the Lockyers (67) had identified a pressure "see-saw" between South America and India, and in 1897 H. H. Hildebrandsson (68) noticed an opposition of barometric pressure between Sydney and Buenos Aires. T. deBort had introduced the concept of centers of action, which Walker named "strategic points of world weather" (69), and Felix Exner (70) had analyzed statistically weather anomalies of the Northern Hemisphere. Walker was able to organize these and his own findings into patterns of atmospheric behavior. In 1923 (71) he identified three major swayings of the atmosphere:

> We can perhaps best sum up the situation by saying that there is a swaying of press. [pressure] on a big scale backwards and forewards between the Pacific Ocean and the Indian Ocean, and there are swayings, on a much smaller scale, between the Azores and Iceland, and between the areas of high and low press. [pressure] in the N. Pacific.

The most important of these swayings, the "Southern Oscillation," appeared to be associated with an "increase" in the general circulation (stronger westerlies). "By the Southern Oscillation," Walker affirmed, "is implied the tendency of pressure at stations in the Pacific . . . , and of rainfall in India and Java (presumably also in Australia and Abyssinia) to increase, while pressure in the region of the Indian Ocean . . . decreases" (72). Walker's final charts of correlation coefficients for the Southern Oscillation for summer and winter (Fig. 8.15) obtained from a large network of stations, clearly show relations of observations of pressure, temperature, and rainfall in the Pacific Ocean with those in the Indian Ocean.

In the Southern Oscillation, Walker stated, in the important winter period "S. America may be described as an 'active' rather than a 'passive centre'" (72), as it appears to have exceptional control of subsequent conditions in other parts of the world. The sea temperature and the icebergs around Antarctica suggest, Walker wrote, that this control depends on a sea current leaving South America in winter (66, 72), but he later concluded "that if some Antarctic factor dominates the Southern Oscillation it has not yet been found" (73). In addition, he stated, the Indian seasonal rainfall, while closely related to pressure conditions six months earlier in South America, was significantly correlated with many subsequent events; thus the Indian monsoon also stood out as an active rather than a passive center. As Sir Charles Normand, director of the India Meteorological Department from

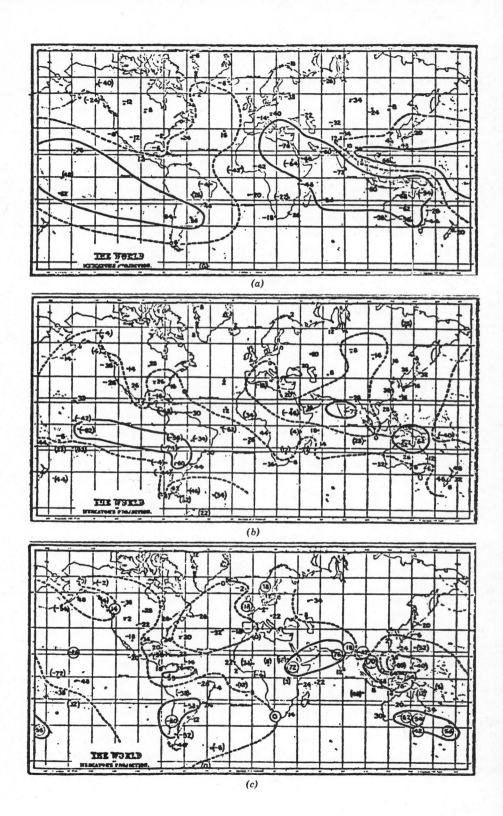

(a)

(b)

(c)

1928 to 1944, observed, "on the whole Walker's worldwide survey ended by offering more promise for prediction of events in other regions" than for prediction of the monsoon rainfall in India, the initial goal (74).

By 1924, Walker retired from his directorship to succeed Sir Napier Shaw as professor of meteorology at the Imperial College in London. Although he had perfected his forecasting formula over the years, he was careful to point out the limitations of his results; for example, the use of the Southern Oscillation failed in the monsoon forecast of 1911—South American pressure conditions were "favorable" for a strong monsoon, but the rains failed. Forecasts using the statistical method, he concluded, would be reasonably successful only in years of strong relationships; but, since "the word 'forecast' is associated with prediction on every occasion," he preferred to use the term 'foreshadowing' as "indicating a vaguer prediction" (75).

Despite his reservation, Walker maintained that the statistical relations expressed in his forecasting formula indicated real physical relations. Even though he was unable to provide explanations for these relations, Walker sincerely hoped that his work might provide a productive starting point for a theory of global teleconnections.

4 CONCLUSION

During the more than 200-year period discussed in this chapter, monsoon meteorology emerged as a major area of meteorological inquiry. Contributions to solving problems associated with the monsoon came from scientists who lived in regions beyond its reach as well as within, but by the end of the period the center of activity in monsoon research had clearly shifted to the Indian region itself. After an initial period of largely observational studies, investigators turned to theoretical considerations as well, paralleling developments in main stream meteorology. Because of the importance of the monsoon for the survival of millions of people, very early on attention focused on understanding and eventually predicting the timing and the amount of monsoon rainfall.

In the course of this chapter we observed three major areas of activities important in monsoon research. First, monsoon investigators drew on an ever expanding observational record that provided the basis and means for verification of theories concerning phenomena of the monsoon. Second, the application of physical principles increased understanding of concepts related to monsoon physics such as the land–sea breeze, formation of monsoon rainfall, and the monsoon as a dynamical circulation system. Third, the establishment of the centralized India Meteorological Department in the 1870s provided the organizational structure for systematic investigations of the monsoon and for the first attempts at seasonal forecasts. Throughout, the diverse backgrounds and personalities of individuals and the unique circumstances of the Indian nation significantly shaped the direction and scope of monsoon investigations.

Figure 8.15. Correlation coefficients between seasonal values of (a) pressure, (b) temperature, and (c) rainfall and the Southern Oscillation in winter (June to August), by Sir G. Walker. Parentheses indicate less than 30 years of data, circles indicate rainfall of regions rather than individual stations (73).

ACKNOWLEDGMENTS

A research grant from the National Science Foundation's History and Philosophy of Science Division (Grant No. SES-82-09159) supported this work. The author thanks the reviewers of this paper for their helpful suggestions.

REFERENCES

1. W. Dampier, *Voyages and Descriptions*, J. Knapton, London, 1699, includes "Discourse of the Trade Winds," with chart; J. Masefield, Ed., *Dampier's Voyages*, Vols. 1 and 2, Dutton, New York, 1906, includes a reprint of Dampier's chart.

2. Captain J. Huddart, *The Oriental Navigator*, Laurie and Whittle, London, 1785; Captain J. Capper, *Observations on the Winds and Monsoons*, Wittingham, London, 1801; Captain J. Horsburgh, *India Directory, or Directions for Sailing to and from the East Indies, China, etc.*, Vols. 1 and 2, 2nd ed., Black, Parbury and Allen, for the East India Company, London, 1817.

3. *Phil. Trans., Roy. Soc. London*, **1**, 142 (1666), reprinted in M. B. Deacon, Ed., *Oceanography: Concepts and History*, Benchmark Papers in Geology (distr. by Academic), Vol. 35, pp. 15–18, Dowden, Hutchinson and Ross, Inc., Stroudsburg, Pa., 1978.

4. E. Halley, Historical account of the trade winds and monsoons, *Phil. Trans., Roy. Soc. London*, **16**, 153–168 (1686).

5. M. F. Maury, *Explanations and Sailing Directions to Accompany the Wind and Current Charts*, 6th ed., E. C. and J. Biddle, Philadelphia, 1854.

6. M. F. Maury, *Physical Geography of the Sea*, Harper, New York, 1855.

7. R. P. W. Lewis, The founding of the meteorological office, *Marine Observer*, **70**, 86 –93 (1982).

8. Earl of Harrowby, Presidential address, *Report 24th Meeting British Assoc. Adv. of Sci.*, **54**, 1xi (1855).

9. Deutsche Seewarte, *Segelhandbuch für den Indischen Ozean*, Friedrichsen, Hamburg, 1892.

10. Deutsche Seewarte, *Seehandbuch für den Indischen Ozean*, Eckardt & Messtorff, Hamburg, 1915.

11. J. Herschel, *Meteorology*, Black, Edinburgh, 1861.

12. L. von Buch, *Physikalische Beschreibung der Canarischen Inseln*, Akademie der Wissenschaften, Berlin, 1825.

13. G. Hadley, Concerning the cause of the general trade winds, *Phil. Trans., Roy. Soc. London*, **39**, 58–62 (1735).

14. J. le Rond d'Alembert, *Réflexions sur la Cause Générale des Vents*, Davis, Paris, 1747.

15. A. von Humboldt, *Fragmente einer Geologie und Klimatologie Asiens. Untersuchungen über die Ursachen der Beugung der Isothermen*, translated from French by J. Loewenberg; List, Berlin, 1832, pp. 178–255; and Récherches sur les causes des inflexions de lignes isotherms, *Mémoirs de Physique d. l. Soc. d'Arcueil* (Paris) **3**, (1817).

16. W. Meinardus, "Die Entwicklung der Karten der Jahres-Isothermen," in *Hundertjährige Wiederkehr von A. v. Humboldt's Reise nach Amerika*, 7th International Congress of Geographers, Gesell. für Erdkunde Berlin, W. H. Kühl, Berlin, 1899.

17. L. W. Kämtz, *Vorlesungen über Meteorologie*, 1840, translated by C. V. Walker, *A Complete Course of Meteorologie*, Baillière, London, 1845.

18. L. W. Kämtz, Über die Windverhältnisse an den Nordküsten des alten Festlandes, *Bull. de l'Acad. St. Petersburg, Class. Sci., Phys.-Math.*, **5**, 294–314 (1847).

19. H. W. Dove, *Meteorologische Untersuchungen*, Sander'sche Buchhandl., Berlin, 1837.

20. J. P. Espy, *The Philosophy of Storms*, Little and Brown, Boston, 1841.

21. G. Kutzbach, *The Thermal Theory of Cyclones: A History of Meteorological Thought in the Nineteenth Century*, Historical Monograph Series, American Meteorological Society, Boston, 1979.

22. W. H. Sykes, On the atmospheric tides and meteorology of Dukhaun (Deccan), East Indies, *Phil. Trans., Roy. Soc. London*, **125**, 161–220 (1835).

23. India Meteorological Department, *Climatological Atlas of India*, J. Eliot, Ed., Bartholomew, Edinburgh, 1906.

24. India Meteorological Department, *Hundred Years of Weather Service (1875–1975)*, Director General of Observatories, Poona, India, 1976.

25. Sir. W. Reid, *The Law of Storms*, J. Weale, London, 1838.

26. H. T. Prinsep (Elder), Notification to Officers of East India Company, Calcutta, September 11, 1839; through Maury (5).

27. H. Piddington, Third memoir with reference to the theory of the law of storms, *J. Asiatic Soc. Bengal*, **9**(2), 1017–1018 (1840).

28. W. Redfield, Remarks on the prevailing storms of the Atlantic coast of North America, *Amer. J. Sci.*, **20**, 17–51 (1831).

29. H. Piddington, *The Sailor's Hornbook for the Law of Storms*, Wiley, New York and London, 1842.

30. A. Thom, *An Inquiry into the Nature and Cause of Storms*, Smith, Elder and Co., London, 1845.

31. H. Piddington, Note to accompany a chart of the Bay of Bengal, with the average courses of its hurricanes from A.D. 1800 to 1846, *J. Asiatic Soc. Bengal*, **16**(2), 848 (1847).

32. W. Ferrel, The Motions of fluids and solids relative to the earth's surface, *Mathematical Monthly*, **1** (1859), **2** (1860); C. Guldberg and H. Mohn, *Études sur les Mouvements de l'Atmosphère*, parts 1 and 2, Brøgger, Christiania, 1876 and 1880; T. Reye, Über vertikale Luftströme in der Atmosphäre, *Zeits. für Mathematik und Physik*, **9** (1864); H. Peslin, Sur les mouvements généraux de l'atmosphère," *Bull. Hebd. l'Ass. Sci. France*, **3** (1868).

33. H. F. Blanford, The rainfall of northern India, *India Meteor. Dept. Memoirs*, **3**(1), 658, 1886.

34. H. F. Blanford, *The Indian Meteorologist's Vade-Mecum*, Part 2, *Meteorology of India*, Government Printing Office, Calcutta, 1877.

35. H. F. Blanford, The winds of northern India, in relation to the temperature and vapour constitutents of the atmosphere, *Phil. Trans., Roy. Soc. London*, **164**, 563–653 (1874).

36. A. Woikoff, *Die Klimate der Erde*, Vol. 2., Costenoble, Jena, 1887.

37. J. Eliot, On the origin of the cold weather storm of the year 1893 in India, *Quart. J. Roy. Meteor. Soc.*, **22**, 1–37 (1896).

38. J. Hann, *Lehrbuch der Meteorologie*, 3rd ed., Tauchnitz, Leipzig, 1915.

39. W. Ferrel, *A Popular Treatise of the Winds*, Wiley, New York, 1889.

40. H. H. Hildebrandsson, The international observations of clouds, *Quart. J. Roy. Meteor. Soc.*, **30**, 317–343 (1904); T. deBort, Étude sur la Circulation Générale de l'Atmosphère, *Ann. du Bureau Centrale Météorologique*, **1**, 34–44 (1885); W. Köppen, *Erforschung der Atmosphäre mit Hilfe der Drachen*, Deutsche Seewarte, Hamburg, 1902.

41. S. A. Hill, The meteorology of the northwest Himalaya, *India Meteor. Dept. Memoirs*, **1**(12), 427, 1881.

42. E. D. Archibald, The height of the neutral plane of pressure and depth of monsoon current in India, *Quart. J. Roy. Meteor. Soc.*, **10**, 123–139 (1884).

43. H. H. Hildebrandsson, Werth der Messungen von Zugrichtung und Höhe von Wolken, *Aus dem Archiv der Deutschen Seewarte*, **14**(5), 1–6 (1891).

44. J. Eliot, *India Meteor. Dept. Memoirs*, **14**(8), 1891, and **15**(1), 1903; reviewed by Sir N. Shaw, in, On the general circulation of the atmosphere, *Proc. Roy. Soc. London*, **74**, 29 (1904); and W. A. Harwood, The free atmosphere in India, *India Meteor. Dept. Memoirs*, **21**(7), 1924.

45. J. Eliot, "Droughts and Famines in India," *Report of the International Meteorological Congress*, Chicago, pp. 444–459, 1893.

46. Sir G. Walker, Recent meteorological work in India, *Meteor. Mag.*, **68**, 15 (1933).

47. R. Abercromby, *Weather, a Popular Exposition of the Nature of Weather Changes from Day to Day*, Kegan Paul, London, 1887; quote of Colonel E. Tennant through Abercromby, pp. 297–298.

48. R. Abercromby, *Weather, a Popular Exposition of the Nature of Weather Changes Day to Day*, Kegan Paul, London, 1887.

49. L. Dallas, A discussion on the failure of the south-west monsoon rains in 1899, *India Meteor. Dept. Memoirs*, **12**(1), 1900.

50. J. Eliot, *Hand-book of Cyclonic Storms in the Bay of Bengal*, Vols. 1 and 2, 2nd ed., Government Printing Office, Calcutta, 1900–1901.

51. T. Reye, *Wirbelstürme, Tornadoes und Wettersäulen*, Rümpler, Hannover, 1872.

52. J. Eliot, Account of the south-west monsoon storm of the 12th to the 17th of May in the Bay of Bengal at Akyab, *India Meteor, Dept. Memoirs*, **4**(1), 1884, and *Report on the Madras Cyclone of May 1877*, Government Printing Office, Calcutta, 1879.

53. C. Meldrum, On cyclone and rainfall periodicities in connexion with the sun-spot periodicity, *Proc. Roy. Soc. London*, **144**, 297–308 (1873).

54. J. N. Lockyer, Simultaneous solar and terrestrial changes, *Nature*, **69**, 351–357 (1904).

55. Sir W. Herschel, Observations tending to investigate the nature of the sun, in order to find the causes or symptoms of its variable emission of light and heat, *Phil. Trans. Roy. Soc. London*, **91**(2), 265–318 (1801).

56. H. H. Blanford, On some recent evidence of the variation of the sun's heat, *J. Asiatic Soc. Bengal*, **44**(2), 21–35 (1875).

57. W. Köppen, Über mehrjährige Perioden der Witterung, *Zeitschr. für Meteor.*, **8**, 241–248 and 257–268 (1873).

58. Sir G. Walker, Sunspots and rainfall, *India Meteor. Dept. Memoirs*, **21**(10), 1915.

59. T. deBort, Étude sur les causes qui déterminant la circulation de l'atmosphère, *Ass. Franc. pur l'Adv. des Sci.*, Congress de Reims, **44**, 1880.

60. H. F. Blanford, On the barometric seesaw between Russia and India, *Nature*, **21** (1880).

61. H. F. Blanford, On the connexion of the Himalaya snowfall with dry winds and seasons of drought in India, *Proc. Roy. Soc. London*, **37**, 3–22 (1884).

62. E. D. Archibald, *Nature*, **16**, 339 (1879); S. A. Hill, *Report on the Rainfall of the N.-W. Provinces and Oudh*, Government Printing Office, Allahabad, 1879; and S. A. Hill, Variations of rainfall in northern India, *India Meteor. Dept. Memoirs*, **1**(3), 1879.

63. Sir. G. Walker, Presidential address, *Proc. 5th Indian Science Congress*, Asiatic Soc. Bengal, Lahore, p. 3, 1918.

64. Sir G. Walker, On meteorological evidence for supposed changes of climate in India, *India Meteor. Dept. Memoirs* **21**(1), 1910.

65. Sir. G. Walker, Correlation in seasonal variations of weather, *India Meteor. Dept. Memoirs*, **21**(2), 1910.

66. Sir G. Walker, Further study of relationships with Indian monsoon rainfall, *India Meteor. Dept. Memoirs*, **21**(8), 1914.

67. Sir N. Lockyer and W. Lockyer, On some phenomena which suggest a short period of solar and meteorological changes, *Proc. Roy. Soc. London*, **70**(2), 500–504 (1902).

68. H. H. Hildebrandsson, Sur la compensation entre les types des saisons simultanes en differentes régions de la tèrre, *Kongl. Svenska Vet. Akad. Handl.*, **45**(2), 1909, and **45**(11), 1910.

69. Sir G. Walker, Correlations in seasonal variations of weather, X, *India Meteor. Dept. Memoirs*, **24**(10), 1924.

70. F. Exner, Über monatliche Witterungsanomalien auf der nördlichen Erdhälfte im Winter, *Sitzber. Akad. Wissenschaften Wien*, **122**(IIa), (1913).

71. Sir G. Walker, Correlations in seasonal variations of weather, VIII, *India Meteor. Dept. Memoirs* **24**(4), 1923.

72. Sir G. Walker, Correlations in seasonal variations of weather, IX, *India Meteor. Dept. Memoirs*, **24**(9), 1924.

73. Sir G. Walker, World weather V, *Memoirs Roy. Meteor. Soc.*, **4**, 68 (1932).

74. Sir C. Normand, Monsoon seasonal forecasting, *Quart. J. Roy. Meteor. Soc.*, **79**, 469 (1953).

75. Sir G. Walker, World weather IV, *Memoirs Roy. Meteor. Soc.*, **3**, 87 (1930).

9

Physics of Monsoons: The Current View

John A. Young
Department of Meteorology
University of Wisconsin–Madison
Madison, Wisconsin

INTRODUCTION

The Asian monsoon is notable in its broad extent and strength, its domination of weather over large regions, and its influence on the fate of their human populations. Physically, the monsoon system is a complex of seemingly disparate parts: two fluids, the mobile air and the slowly changing ocean below; deserts and areas of torrential rain; the dramatic seasonal progression of winter to summer; winter and summer hemispheres linked by strong wind and ocean currents across the equator; mountain complexes that assist or inhibit rising air motion; water evaporation and condensation; clouds that alter most properties, from wind to radiation; cyclones that grow over the ocean and drive deadly coastal surges. The total description of these elements spans the scales of modern observations, from planetary to regional to local rain systems. The net effect of these interacting weather phenomena is to produce the great climate machine of the Asian monsoon.

From a scientist's viewpoint, monsoonal mechanisms are numerous and intriguing. Because of its size and strength, the Asian monsoon generates signals in the pressure and wind fields that move across the planet within days; descriptions of these teleconnections require concepts of waves and currents in both atmosphere and ocean. Monsoons span the equatorial belt so that the influence of the earth's rotation is intricately varied. Both the ocean and air encounter solid earth barriers which intensify ocean currents, atmospheric jet streams, and rain systems. The boundary layer structure of each fluid reflects the strong interaction across the interface of the atmosphere and ocean. The system of forced, free, unstable, and nonlinear fluid modes of circulation invites imaginative inquiry.

While the extensive Asian monsoon is the strongest and most dramatic, monsoons affect other regions, such as Australia and West Africa. An introduction to the basic physics common to monsoons is given in Chapter 1. The combinations of processes which affect wind, temperature, and water, the essences of all monsoons, are considered here. First is an explanation of the basic physical laws that apply to the monsoon elements, then a discussion of the many distinct features of monsoons, and last a description of the complete dynamical systems. Since mathematics is the language of science, relevant elementary equations are included in this chapter. However, the reader may ignore the equations and rely upon the text to illustrate the physical ideas.

1 WHY ARE THERE MONSOONS?

Monsoons are characterized by their seasonality, geographical preference, and strength. Monsoon rain and winds are the end result of heating (Q) patterns produced by the sun and the distribution of land and ocean. The heating may be positive or negative (cooling). Monsoons may be understood by ignoring vertical structure and daily changes in the heating and emphasizing typical horizontal patterns that change slowly with the seasons. It is convenient to think of these patterns separated into three idealized categories:

$$Q(x,y,t) = Q_{GC}(y) + Q_{IH}(y,t) + Q_{EC}(x,t) \qquad (1)$$

Here x and y are distances measured eastward and northward, respectively, and t is time. (Quantities in parentheses indicate functional dependence.) Q_{GC} represents the heating which would be produced on a homogeneous (ocean or land covered) earth on the average, given a sun fixed directly above the equator. A symmetric (with respect to the equator) *general circulation* would result. Q_{IH} represents *interhemispheric* differences brought about by the *annual cycle*. On a homogeneous earth these would occur in response to the oscillation of the overhead sun between hemispheres: the subtropics (the latitude belts roughly 10° to 30°) experience overhead sun at the summer solstice while the equatorial zones (about 10°S to 10°N) experience it at the two equinoxes. The strength of Q_{IH} is increased by the uneven distribution of land on the real earth. Broadly speaking, the Southern Hemisphere is a maritime hemisphere dominated by ocean while the Northern Hemisphere is a more continental one. This distinction is especially pertinent in the Eastern Hemisphere, where the massive Asian continent lies to the north of the expansive Indian Ocean. The heating Q_{EC} is associated with *equatorial continentality*. This zonal asymmetry arises because of the uneven longitudinal distribution of continents in the tropical belt. It may also have an annual cycle, but as we will see later, Q_{EC} is determined by intricate physical processes and is the hardest component to understand.

Monsoon temperature, wind, and rain patterns may also be split into three corresponding parts. Thus, there is a temperature field T_{GC} associated with basic general circulation patterns: the warmest temperatures and lowest surface pressures are, on the whole, found on the equator. The surface winds are caused by air being pulled toward the equator from both hemispheres and deflected westward by the Coriolis force that arises from the earth's rotation about its polar axis. [This deflection can also be understood to be a consequence of the conservation of angular momentum for the total motion (earth's motion and relative wind) seen by an observer fixed in space (1).] Finally, the converging winds near the equator cause upward motion concentrated along the *intertropical convergence zone* (ITCZ). This is the warm rising branch of a thermally direct circulation (of the sort envisioned by Hadley, discussed in Chapters 1 and 8) that produces rain in the general vicinity of the lower, equatorial pressures. The winds higher in the atmosphere return poleward and, deflected by the Coriolis force, flow roughly opposite to the surface winds.

The interhemispheric temperature contrast T_{IH} and wind can be enhanced by the radical differences between the heat storage of dry land versus ocean. This is most dramatic in the Asian sector because of the latitudinal land distribution. As discussed in Section 1.3 the heat storage in dry land is relatively small, partly because land is a solid which conducts heat slowly through relatively thin layers. This is in contrast to the ocean, which can hold large quantities of heat obtained from the sun in deep layers typically 50 m thick. Thus the ocean acts as an extensive *thermal reservoir* with great thermal inertia; it experiences relatively moderate temperature changes with time, even over one season. In contrast, the land rapidly adapts its temperature to the incoming solar radiation and air overhead, thus accounting for the strong seasonality of a continental surface temperature. The net result is the familiar tendency for land masses to be warm relative to oceans in the summer and relatively cold in the winter. For the Northern Hemisphere summer, the Southern Hemisphere oceans are a source region of relatively cold air while the Northern Hemisphere Asian land mass becomes relatively warm. Now, lower surface pressures over the land mass draw air *across* the equator. The air is first deflected westward by the Coriolis force and, subsequently, after crossing the equator, eastward. Upon meeting the Asian land mass, the air rises, forming clouds and rain. Thus the annual heating cycle and distribution of land can explain why the Northern Hemisphere summer monsoon is wet and strong. (In the Indian region the northern winter monsoon draws air from the land over the sea, so it is drier. It is also weaker, as the Tibetan Plateau blocks the draining of coldest air from Asia.)

A different land–ocean orientation can produce local east–west temperature differences T_{EC} which may induce convergent, rising air motion over warm equatorial land areas. Poleward of the equator, Coriolis forces cause a deflected surface wind pattern with counterclockwise* (clockwise) spinning in the Northern (Southern) Hemisphere at low levels. High in the atmosphere the flow tends to be reversed; this clockwise circulation is observed above the Tibetan Plateau during the Northern Hemisphere summer monsoon.

Figure 9.1 summarizes possible flow patterns in terms of simple scenarios. First consider the effect of the earth's rotation. Monsoons on a nonrotating (i.e., no Coriolis force) earth (Fig. 9.1a) would be like relatively weak "sea-breeze" systems with no spin. The effect of the earth's rotation away from the equator is to deflect winds. Figure 9.1b shows the wind pattern in the vicinity of mid-latitude coastlines.

---

* Counterclockwise about a local vertical axis looking down at the earth's surface.

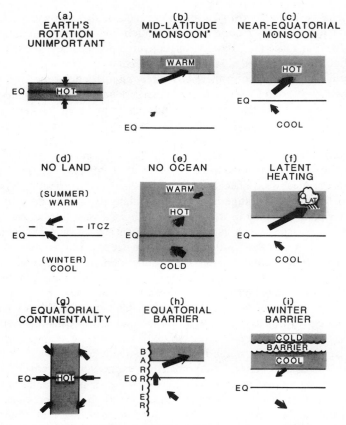

Figure 9.1. Various influences on monsoonal surface winds. Land areas are shaded; winds are shown by arrows, strong winds are bold. Panels (*a*)–(*c*) illustrate the importance of land location on the spinning earth. (*a*) The earth's rotation does not influence a small equatorial continent. (*b*) A mid-latitude continent produces no cross-equatorial flow. (*c*) A near-equatorial continent produces cross-equatorial flow. Panels (*d*) and (*e*) illustrate flows without the land–sea contrast. (*d*) An ocean-covered earth produces a weak convergence zone (ITCZ) in the warmer summer hemisphere. (*e*) A land-covered earth produces strong flow into the summer hemisphere. Panels (*f*) and (*g*) illustrate heating effects. (*f*) Rainfall enhances the heating (Q_{lat}) and the inflow for a near-equatorial monsoon [cf. (*c*)]. (*g*) Heating concentrated on the equator draws air inward along the equator, with cyclonic spin in the subtropics. Panels (*h*) and (*i*) illustrate barrier effects. (*h*) An equatorial mountain barrier helps to deflect the winter hemisphere easterlies into the summer hemisphere westerlies. (*i*) A subtropical mountain barrier can stop the drainage of coldest air and reduce a northern winter monsoon's winds [cf. (*c*)].

When the land is sufficiently close to the equator, air can be drawn across the equator from the winter hemisphere (Fig. 9.1*c*), a fundamental characteristic of major monsoon flows.

The next row of diagrams shows how the distribution of continents influences the interhemispheric system. In the absence of land (Fig. 9.1*d*), the convergence zone with light winds and denoted by the ITCZ would be found somewhat displaced from the equator into the summer hemisphere. For a land-covered earth (Fig. 9.1e) the convergence is found at higher latitudes, near the sun's directly overhead position. Strong winds could result, but the lack of a water source from evaporation would restrict rainfall. The combination of land and ocean hemispheres interacting (Fig. 9.1f) thus allows for the simultaneous development of strong winds and rain. The attendant rain over land would release latent heat of condensation and intensify the circulation.

The last row of diagrams shows the effects of equatorial continentality and orographic barriers. Figure 9.1g shows that equatorial continentality would create only moderate winds of complex patterns. However, in (Fig. 9.1h) a north–south mountain barrier (e.g., the East African highlands) can accentuate the flow across the equator, feeding a strong southwest monsoon in the Northern Hemisphere. In the winter, an equally strong northeast monsoon is prevented by the damming of mid-latitude cold air by a mountain barrier (Fig. 9.1i), such as the vast Himalayas.

Of course, these descriptions oversimplify the role of land in real monsoons. Moistening of the land surface allows evaporation to occur. Evaporation lessens the warming of the land by the sun, causing the thermal contrast with surrounding oceans to decrease. Nevertheless, the total heating experienced by an air column would be increased, for the latent heat of condensation released in the rain is a powerful heat source for the monsoonal system. The net result is to concentrate atmospheric heat sources over land. This interdependence of monsoon heat sources, winds, and rain introduces a very intricate *feedback process* which affects monsoon variability and is not fully understood.

2 MONSOON MOTIONS

The wind is an integral part of the entire water cycle of the monsoon: it scours water from the ocean surfaces, transports the resulting vapor as a humid air mass over land, and finally rises to produce clouds and rain. This last phase strongly heats the air and reminds us that the thermodynamic description of the monsoon, together with the hydrodynamic one, is essential. This section considers how these concepts can be united to explain monsoon motions.

2.1 Ways of the Wind

On a weather map, the flow pattern can be represented by *streamlines* showing direction, and *isotachs* enclosing areas of fast or slow wind speed. In a steady flow (such as a persistent monsoon circulation) the streamlines would show the trajectories of different moving air masses. Persistent regions of fast speeds are called *jet streaks*; examples are the low-level Somali jet and the high-level tropical easterly jet of the northern summer monsoon.

The horizontal wind field has two fundamental kinematic properties: its (relative) vorticity ζ and its horizontal divergence δ. The *vorticity* measures the instantaneous rate of spin of a fluid parcel about a local vertical axis. In the Northern Hemisphere, cyclonic vorticity has counterclockwise turning; in the Southern Hemisphere, it is clockwise. Relative vorticity can be evaluated from two terms: the curvature and the shear. The curvature contribution is calculated from the wind speed divided by the radius of curvature of the flow. Cyclonic curvature is important in monsoon depressions, while anticyclonic curvature marks the circulation around the high pressure cell over the Tibetan Plateau in summer. The shear vorticity is proportional to the wind-speed changes in a direction perpendicular to the flow (or streamline). Shear vorticity is especially prominent on the flanks of jet streaks. Figure 9.2a and b illustrate these two categories for positive vorticity; in each case parcels spin in a counterclockwise sense as they move.

Analogous to the relative vorticity ζ, the earth's vorticity is defined as the local vertical spin of an area fixed on the rotating earth, and is denoted by f, the Coriolis parameter. Numerically the Coriolis parameter is $f = 2\Omega \sin \phi$, where Ω is the angular velocity of the spinning earth and ϕ is latitude. An important property of f is that it is positive in the Northern Hemisphere, negative in the Southern Hemisphere, and vanishes at the equator. The Coriolis parameter helps determine the deflecting effect of the rotating earth, to be discussed shortly.

The horizontal *divergence* δ has the same units as vorticity ($\sec^{-1}$) but represents the horizontal expansion of an air mass. It is numerically equal to the fractional rate of change of area for a given parcel of fluid. Divergence is generally smaller than vorticity for circulations wider than about 100 km. The divergence can be described as difluence and speed changes. The difluence term is proportional to the speed and the horizontal spreading of the flow streamlines; its opposite is confluence, which occurs when streamlines converge downwind. The speed divergence term is

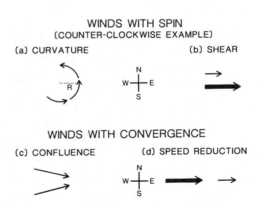

Figure 9.2. Wind patterns illustrating spin (vorticity) and convergence. Top: Winds with counterclockwise spin (positive vorticity) that are cyclonic in the Northern Hemisphere. (*a*) Spin due to curved streamline with radius of curvature R. (*b*) Spin due to wind shear. Examples of winds with negative vorticity would correspond to reversing the indicated wind direction in (*a*) and (*b*). Bottom: Winds with horizontal convergence. (*c*) Convergence due to confluence of neighboring streamlines. (*d*) Convergence due to slowing of speed along a streamline. Examples of winds with positive divergence correspond to reversing the indicated wind directions in (*c*) and (*d*).

proportional to the changes in wind speed with distance downwind. The lower part of Figure 9.2 (*c* and *d*) illustrates the two kinds of convergence ($\delta < 0$).

Because wind has both speed and direction, it is represented as a vector. The total wind vector **V** can be separated into a nondivergent component $\mathbf{V}_\psi$ and a divergent component $\mathbf{V}_\chi$. The nondivergent part $\mathbf{V}_\psi$ is generally the larger component, and contains all of the vorticity information and none of the divergence information. Conversely, $\mathbf{V}_\chi$ is smaller but contains all of the divergence information and represents the horizontal branch of a *secondary circulation*. Secondary circulations are gentle but important overturning processes in the vertical plane; two examples are the planetary *Hadley* and *Walker* circulations, which will be discussed in detail later. The vertical component of the secondary circulation is the vertical velocity *w*. Although *w* is very small when averaged over large monsoonal areas, it is fundamentally associated with monsoonal rain. The two branches of the secondary circulation are linked diagnostically by an approximate conservation of mass statement:

$$\delta + \frac{1}{\rho} \frac{\partial}{\partial z} (\rho w) = 0 \tag{2}$$

where ρ is density and *z* is altitude. This equation links the horizontal divergence δ to changes of vertical motion with height in a fluid column. For example, horizontal convergence ($\delta < 0$) over a flat portion of the earth's surface requires rising motion at higher levels above the surface which in turn is accompanied by divergence at still higher levels in the upper troposphere.

In general, the vertical motion depends both upon divergence and its value w_s at the earth's surface:

$$w_s = \mathbf{V}_s \cdot \nabla h^* \tag{3}$$

where ∇h^* is the gradient of the earth's topography h^* and $\mathbf{V}_s$ is the surface velocity. This expression shows the mechanical effect of mountains as barriers to atmospheric flow: air in contact with the sloping ground must flow parallel to it. Thus *upslope winds* may occur when the terrain rises along the wind. Alternatively in some situations, no vertical motion occurs. The wind, blocked by the topographic barrier, must then blow around the obstacle. Both effects are present in the monsoon flow over and around the Tibetan Plateau. Upslope flow is also found over the Ghat mountains of western India, while blocking is illustrated by flow parallel to the East African highlands.

2.2 The Temperature Connection

Vertical motions must be consistent with Newton's second law of motion, which states that vertical acceleration is caused by the net imbalance of the upward pressure gradient force and downward gravity (*g*) force (weight). For a resting state (i.e., no vertical motion) these two forces balance each other and the vertical pressure gradient $\partial p / \partial z$ is negative, that is, the pressure p decreases with height z. Even when the air is not at rest, this *hydrostatic balance* is a very good approximation for large monsoonal systems because their vertical motions are so small. The hydrostatic statement

$$\frac{\partial p}{\partial z} = -\rho g \tag{4}$$

implies the following. (1) The surface pressure p_s is proportional to the mass of air overhead, hence changes of pressure horizontally or in time reflect mass redistributions by winds at various heights. (2) The steady decrease of pressure with height allows its use as a vertical coordinate in atmospheric models: 1000 mb (1 mb = 10^3 dynes/cm^2) is near the earth's surface, 500 mb is near the center of gravity of an atmospheric column and 100 mb is near the top of the troposphere, the upper limit of the monsoonal rain systems. (3) The hydrostatic balance reflects a fundamental coupling in the vertical which has important implications for large-scale dynamics. It states that the rate at which pressure decreases with height is proportional to the air density. Density is linked to pressure and temperature by the equation of state for air $p = \rho R T_v$, where T_v is virtual temperature (absolute temperature increased slightly by the presence of water vapor) and R is the gas constant. It follows that pressure decreases more rapidly with height in cold air columns than in warm air columns. This is equivalent to saying that the *thickness* of an air column bounded by two fixed pressure surfaces is proportional to the absolute virtual temperature of that column. Thus an isolated warm column of air may be accompanied by some combination of anomalously low pressure at low altitude (or high pressure at high altitude). Such a condition is found in monsoon depressions and tropical cyclones. A low-level example is the *heat low* which is typically found over desert regions (e.g., Arabia or over pre-monsoonal India). Such a pressure anomaly is proportional not only to the temperature anomaly but also to the thickness of the layer that the temperature anomaly occupies; strong pressure anomalies require significant temperature anomalies over deep layers. Thus a low-level temperature anomaly confined to a relatively thin (2 km) layer of the atmosphere might yield a pressure deviation of only 2 mb, which would not create a significant flow change except possibly near the equator. This concept is also pertinent to understanding the possible influence of sea surface temperature on the atmosphere (discussed in Chapters 11 and 16).

2.3 Heating and Cooling

Pressure differences create a force that causes winds, and the hydrostatic relation links these to the temperature field. The logical next step is to identify the processes that can create temperature differences. The most straightforward answers are provided by the *first law of thermodynamics*:

$$\frac{d\theta}{dt} = \frac{\theta}{c_p T} Q \tag{5}$$

where c_p is the specific heat of the air and θ is the potential temperature (temperature T of an air parcel brought down, and compressed, to a pressure of 1000 mb without adding heat.) This equation states that the potential temperature of a moving air parcel can change in time at a rate proportional to the external heat source influence Q, called the *diabatic* heating rate. [When $Q = 0$, the processes are *adiabatic*, and Eq. (5) implies that $d\theta/dt = 0$.]

The total diabatic heating within the atmosphere can be divided into types

$$Q = Q_{rad} + Q_{sens} + Q_{lat} \tag{6}$$

where Q_{rad} is the net *radiational* heating or cooling, Q_{sens} is the *sensible* heating, and Q_{lat} is the *latent heating*. These heating components contribute to the differential heating in Eq. (1) creating the monsoon in different ways.

Q_{rad} includes (1) the positive heating effect of the sun's (shortwave) radiation absorbed by air, and (2) the net cooling by infrared (longwave) radiation. This cooling is due to strong energy emission by the air to space and to the ground below. The energy loss is partly balanced by absorption of longwave energy from below. The radiative transfer depends upon the distribution of radiatively active constituents (e.g., water vapor, clouds, and dust). Q_{rad} affects the broad planetary scale environment: strong cooling (a few degrees Celsius per day) in the winter hemisphere and weak cooling or warming in the summer hemisphere create *temperature gradients* over broad regions.

Q_{sens} is due to turbulent diffusion of heat in the air away from (toward) the warmer (colder) earth's surface. As explained in Section 1, variations in Q_{sens} between air over solar-heated land surfaces and air over the ocean can produce strong temperature differences within a given hemisphere, resulting in strong sea breezes of both small-scale (near coastlines) and continental-scale monsoonal circulations. Q_{sens} is normally concentrated in the lowest 2 km above ground. The extent of its influence is affected by the height of the earth's surface. For example, the Tibetan Plateau acts as an important elevated (4 km) heat source relative to the surrounding atmosphere.

Finally, the latent heating Q_{lat} is released during *condensation* of water vapor when clouds and rain are produced. Q_{lat} creates regions of strong heating which depend upon the flow in a very complex way. This complexity has been a major obstacle in understanding and predicting monsoons and other precipitating atmospheric circulations.

Heating gradients are ultimately responsible for creating horizontal variations in temperature, concentrated in areas of contrast called *baroclinic zones*. Some of the most prominent baroclinic zones in the tropics are found in association with the monsoon. These regions contribute to the *available potential energy* of the tropics, which is closely proportional to the horizontal variance of θ. Conversions of available potential energy will be discussed in Section 2.6.

The role of atmospheric motions in altering temperature distributions can be explored by considering for the moment the adiabatic limit, where $d\theta/dt = 0$; a parcel of air moving three dimensionally conserves its potential temperature. The parcel perspective (a Lagrangian one) is unfortunately not practical since most observations are made and expressed in Eulerian coordinates fixed to the earth. From calculus the local (Eulerian) tendency for the adiabatic case is written as

$$\frac{\partial \theta}{\partial t} = - \mathbf{V} \cdot \nabla \theta - w \frac{\partial \theta}{\partial z} \tag{7}$$

$$\underbrace{\phantom{- \mathbf{V} \cdot \nabla \theta}}_{\substack{\text{horizontal} \\ \text{advection}}} \quad \underbrace{\phantom{w \frac{\partial \theta}{\partial z}}}_{\substack{\text{adiabatic} \\ \text{change}}}$$

Changes in local temperature are brought about by the two terms on the right-hand side. The first term is the *horizontal advection* of potential temperature θ; it quantitatively represents the transport of air with different potential temperature from the upwind direction. It simply reflects the horizontal rearrangement of air masses of different temperature; no new centers of warm or cold air are created by the process. The second term is analogously the vertical advection of θ which is called the *adiabatic change* term. Since over much of the atmosphere, θ increases with height ($\partial \theta / \partial z > 0$), this process results in a negative local tendency ($\partial \theta / \partial t < 0$) for rising motion ($w > 0$) and the reverse for sinking. Physically, this term arises from temperature changes experienced by expansion of rising air parcels and compression of sinking ones.

Persistent aspects of monsoon dynamics can be understood for *steady state* conditions defined by the local change rate $\partial \theta / \partial t = 0$. Even then, moving air parcels may still experience major temperature changes ($d\theta / dt \neq 0$) due to *diabatic heating Q*. If we return to the case where the diabatic heating is not equal to zero, Eqs. (5) and (7) indicate that it may produce a local temperature tendency or bring about a balance by horizontal advection and/or vertical motion influence. Many tropical regions, especially those experiencing concentrated latent heating, Q_{lat}, reach an approximate balance of the adiabatic and diabatic terms; heated regions tend to rise and cool adiabatically. When these regions are warmer than their surroundings a *direct* secondary circulation, a generalized form of convection, occurs. (In an *indirect* circulation the rising branch is relatively cooler.) Latent heating both depends upon and also affects the vertical motion so that a closed *feedback loop* between vertical motion and latent heating exists. This feedback is normally positive, that is, increased rising motion and latent heating accompany each other. It follows that latent heat release enhances the *unpredictability* of monsoonal weather systems on all scales, from individual cumulus clouds to rain areas on the planetary scale.

2.4 Forces

Newton's second law of motion states that the horizontal acceleration of a parcel of air or water, $\mathbf{A} = d\mathbf{V}/dt$, is caused by the sum of the horizontal forces influencing it:

$$d\mathbf{V}/dt = \mathbf{P} + \mathbf{C} + \mathbf{F} \tag{8}$$

P is called the *pressure gradient force*, **F** is the friction force, and **C** is the *Coriolis force*, an apparent force due to air motion relative to the spinning earth. The force **P** is unique to fluid motion. It can be expressed as $\mathbf{P} = -(1/\rho)\nabla p$; the force is directed down-gradient away from high pressure toward low pressure centers. On weather maps, the pressure gradient force at all points is perpendicular to the isobars (lines of constant pressure). The force arises because any volume of fluid is subject to inward pressure forces by surrounding fluid; the net force is due to spatial variations of pressure. Because pressure is related to temperature, the pressure field links the hydrodynamic statement, Eq. (4), and the thermodynamic processes, Eqs. (5) and (6).

If we neglect the only other real physical force **F** (friction) for a moment, the remaining term in Eq. (8) is the Coriolis force ($\mathbf{C} = -f\mathbf{k} \times \mathbf{V}$). Since **k** is the vertical unit vector, the direction of **C** is perpendicular to the horizontal fluid motion. It is a deflecting force which acts to change the fluid direction, but not its speed. Although it does no direct work on a fluid parcel, its indirect effects strongly influence monsoonal motions. This force is proportional to the Coriolis parameter f (introduced earlier) which changes sign between hemispheres and is zero at the equator. Thus this deflecting force does not operate at the equator, but increasingly influences motion away from the equator. The rate of change of f with distance from the equator along a meridian is given by the symbol β, and is the source of the *beta effect* discussed in Section 2.5.

The Coriolis effect is evident in monsoonal winds crossing the equator into the summer hemisphere. The air initially moves poleward, but it is subsequently deflected toward the east, following a curved path typical during solstice seasons. The strong variations of Coriolis influence are also responsible for an entire class of large-scale *equatorial wave* motions which are fundamental to monsoons and discussed in Section 4.2. For the moment it suffices to recognize that weather patterns can be described as a sum of many sinusoidal-like waves, with each member growing, moving, interacting, and dying in a distinct way.

In the absence of friction **F**, an important concept follows from Eq. (8). When the horizontal acceleration is small (i.e., $d\mathbf{V}/dt \approx 0$) the pressure gradient and Coriolis forces **P** and **C** are then in a state of balance. Reordering the terms, we obtain the *geostrophic wind*

$$\mathbf{V}_g = \left(\frac{1}{f\rho}\right) \mathbf{k} \times \nabla p \qquad (9)$$

This hypothetical frictionless wind is often used as a first approximation to the real wind. On a weather map, this wind blows parallel to the isobars at a speed inversely proportional to their spacing. This wind is essentially nondivergent so that it gives no direct information on the secondary circulation. The geostrophic wind $\mathbf{V}_g$ is oriented with low pressure on its left and high pressure on its right in the Northern Hemisphere; the reverse is the case in the Southern Hemisphere. (The low-level wind in Fig. 9.3 illustrates these properties.) Thus cyclonic circulations are asso-

WINDS IN GEOSTROPHIC BALANCE

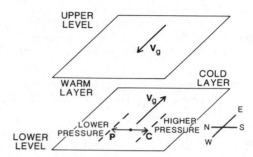

Figure 9.3. Geostrophic motion in three-dimensional perspective in the Northern Hemisphere, typical of the summer monsoon over the Arabian Sea. Lower level: the geostrophic wind $\mathbf{V}_g$ flows parallel to isobars (dashed lines) separating lower and higher pressures. The balance of forces $\mathbf{P}$ (perpendicular to isobars) and $\mathbf{C}$ (perpendicular to wind) is indicated. Upper level: the geostrophic wind has changed with height according to thermal wind relationship for the horizontal temperature distribution shown in the intermediate layer. In this example, the temperature contrast is so large that the geostrophic wind has reversed direction, with a level of no wind within the intermediate layer.

ciated with low-pressure centers in both hemispheres, although their spins are reversed. The geostrophic wind possesses relative vorticity ζ_g, which can be calculated from the pressure pattern alone. The definition [Eq. (9)] shows that, for a given geostrophic speed, the pressure gradient magnitude must decrease as the Coriolis parameter f decreases from middle to low latitudes. Thus pressure differences at low latitudes can be important, but they may be small and difficult to measure.

The vertical change of geostrophic winds obeys the revealing *thermal wind* formula, which can be written approximately as

$$\frac{\partial \mathbf{V}_g}{\partial z} = \left(\frac{g}{fT_v}\right) \mathbf{k} \times \nabla T_v \tag{10}$$

The geostrophic wind shear $\partial \mathbf{V}_g/\partial z$ reflects change of the wind direction and/or speed with height. It is oriented along isotherms (lines of constant temperature) with cold air on the left and warm air on the right in the Northern Hemisphere (the reverse in the Southern Hemisphere). The formula indicates that small temperature gradients are capable of sustaining significant wind shear at low latitudes (where f is small). It is especially useful for describing the zonal (u component, directed eastward) wind component. For example, the common decrease of temperature toward the poles is associated with increase of the westerly (eastward) wind component with increasing height, thus accounting for the upper tropospheric westerly jet streams of middle latitudes. In the heart of the monsoon this shear may be reversed. Figure 9.3 illustrates the summer circulation over the western Indian Ocean where warmer air is found on the poleward side of the lower and upper wind streams; hence the lower stream (*southwesterly monsoon*) decreases in intensity with height and ultimately yields the oppositely directed *tropical easterly jet* in the upper troposphere.

The thermal wind relation is a clear link between the baroclinity (∇T_v) and the circulation structure. If baroclinity is absent (no horizontal thermal contrast) the atmosphere is barotropic, and according to Eq. (10) the geostrophic wind will not change with height. This partly accounts for the strong influence of mountains on high-level jet streams. The low-level flow splits to move around mountains, and the geostrophic wind far above tends to split as well. The influence of the Tibetan Plateau upon jet streams operates partly in this way. (See Chapter 12, particularly the discussion on Taylor columns in Section 2.1.)

While these geostrophic properties are important, the complete physics of monsoonal circulations requires knowledge of the ageostrophic component of the wind, $V_{ag} = V - V_g$. If we continue to ignore friction, we may write the equation of motion (8) as

$$A = -f \mathbf{k} \times V_{ag} \qquad (11)$$

Small accelerations are associated with small ageostrophic winds perpendicular to them. When A is sufficiently small compared with C the *quasi-geostrophic* approximation is made:

$$A = \frac{dV_g}{dt} \qquad (12)$$

Its powerful applications are discussed later in Section 4.1.

Three primary kinds of acceleration and frictionless ageostrophic winds are shown in Figure 9.4. In Figure 9.4a the ageostrophic *isallobaric wind* blows toward centers of falling pressure. It is important in transient monsoon circulations, especially propagating subtropical jet features and drifting monsoon depressions. The remaining two kinds of ageostrophic wind can exist in steady conditions such as cross-equatorial flow (Fig. 9.4b) or persistent jet streaks (Fig. 9.4c).

Figure 9.4b shows strong curved *flow across the equator*, characteristic of the summer monsoon over the western Indian Ocean. At the equator the centripetal acceleration A is directed eastward and the Coriolis force $C = 0$, so the pressure gradient force P must be oriented eastward as well. Higher pressures are found to the west on the equator and the cross-equatorial flow is in local *cyclostrophic* balance ($A = P$). Away from the equator, Eq. (11) shows that the ageostrophic component is directed downwind in the Northern Hemisphere and upwind in the Southern Hemisphere; the former may account in part for the intense Somali jet core of the southwesterly monsoon.

Figure 9.4c shows the simplest representation of air flowing into a zone of concentrated speed called a *jet streak*. In this case the air parcels speed up and later slow down as they move along the streak. Thus the ageostrophic wind blows across the main axis, toward low pressure on the upwind (entrance) side and toward high

WINDS OUT OF GEOSTROPHIC BALANCE
(FRICTIONLESS FLOW)

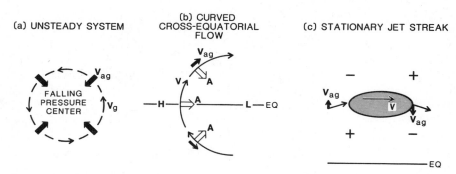

Figure 9.4. Frictionless winds with an ageostrophic component V_{ag} (short, bold arrows) caused by accelerations. (*a*) Transient systems produce a convergent V_{ag} (isallobaric wind) into centers of falling pressure. A developing geostrophic circulation is shown along dashed curve. (*b*) The trajectory of flow across the equator, **V**. Centripetal acceleration **A** to the right. At the equator, **A** is produced by **P** which is directed along the equator from high (H) to low (L) pressure and the flow is cyclostrophic. Away from the equator the flow is enhanced (retarded) by V_{ag} in the Northern (Southern) Hemisphere. The relative vorticity due to curvature is everywhere negative in this example. (*c*) Flow **V** (arrows) through a jet streak (wind speed maximum shaded) first experiences acceleration and then deceleration, producing lateral V_{ag} and centers (labeled + or −) of divergence.

pressure on the downwind (exit) side. A good example of a jet streak entrance zone is found in the upper troposphere over southeast Asia during the winter monsoon.

Finally, we must remember that *friction* **F** must play an ultimate role in dissipating kinetic energy of motions. Indeed some monsoonal elements, such as strong low-level flows, are directly altered in obvious ways by friction. This is because the effects of friction are normally concentrated in the *planetary boundary layer* or the lower 1.5 km of the atmosphere. The origin and representation of the friction force in terms of the overall flow may be quite complex. In fact, in the ocean the force may cause ocean currents whereas in the atmosphere it frequently tends to oppose the wind. This will be discussed in more detail later.

If we consider the atmospheric case first, then Eq. (8) shows that the presence of friction **F** can disrupt the geostrophic balance between the pressure gradient and Coriolis forces, **P** and **C**. Realistic representations of friction show that **F** is usually strongest near the surface and decreases upward through the planetary boundary layer. It follows immediately that the Coriolis force and its associated wind must change with height. These properties are demonstrated by the famous solution to Eq. (8) which, when accelerations are negligible, is called the *Ekman spiral*. The Ekman spiral describes winds that increase in speed and turn clockwise (in the Northern Hemisphere) with height until the geostrophic wind is reached at the top of the boundary layer where the flow is essentially frictionless. Thus friction causes both vertical shear and ageostrophic wind (see Fig. 9.5). We see convergence into centers of low pressure producing rising motion above such centers. This frictional Ekman pumping can be a critical mechanism for initiating low clouds and precipitation in monsoon trough zones and depressions. Thus the earth's surface influences the secondary circulations at higher altitudes through the intervening friction layer.

A dramatic connection to the secondary circulation of the ocean also exists. Wind over the ocean is accompanied by a stress on the ocean surface that can drive a deflected Ekman circulation in the upper layer of the ocean. (See Fig. 9.5.) It can be shown that the vertical mass transport (ρw) is the same for the ocean and the atmosphere; thus, frictionally induced ascent over low-pressure centers in the atmosphere is accompanied by *upwelling* of colder ocean water from deeper layers. The corollary holds: the summer air circulation over the western Arabian Sea is anticyclonic, so frictionally induced subsidence helps to suppress rain and bring about downwelling of the upper layers of the Indian Ocean away from the coast. Near the African coast the oceanic Ekman flow drives water away from shore, which in turn produces strong upwelling of cold water near the Horn of Africa. (A similar upwelling effect is found on the equator in the Pacific Ocean.)

FRICTIONAL FLOW

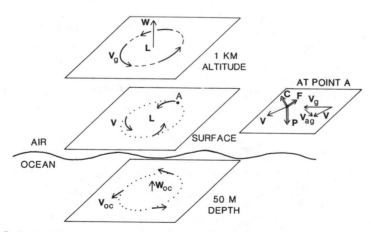

Figure 9.5. Frictional flows in three-dimensional perspective in the Northern Hemisphere. Low pressure area (enclosed by dotted lines) is marked by L at each level. Top level—above the friction layer, the wind is strong, geostrophic, and circular, with no convergence. Gentle vertical motion (w) is the result of air rising from below. Mid-level—in the friction layer, the wind is less strong and converges inward, producing rising (w) aloft. Comparison of wind direction with that for top level shows clockwise turning with height. Bottom level—in the oceanic friction layer, the current has an outward component that causes upwelling (w_{oc}). Right inset: The Ekman balance of forces (**C**, **P**, **F**) relative to the wind for point A is shown. The relation among the actual wind, **V**, the geostrophic wind, **V**$_g$, and the ageostrophic wind, **V**$_{ag}$, is also shown.

2.5 Spin Sources

One good way to understand the great spinning of monsoon circulations is to use the prediction equation for vorticity introduced by the pioneer dynamicist, C. G. Rossby. The predictive *vorticity equation* is approximately

$$\frac{d}{dt}(\zeta + f) = -(\zeta + f)\delta + \text{tilting} + \text{curl } \mathbf{F} \tag{13}$$

which states that the *absolute vorticity* $(\zeta + f)$ of a moving fluid parcel can be changed by three mechanisms. The first (called the divergence term) is usually the most important in geophysical flows. It produces an increase of the absolute vorticity given by the rate of convergence of the air. This mechanism is closely associated with the conservation of angular momentum for converging (horizontally contracting) sheets of air. The second term accounts for the tilting of the horizontal vorticity component into the vertical direction. It requires strong vertical wind shear and horizontal gradients of vertical motion and is typically important only in intense localized circulation systems, such as those near mountains or monsoon depressions. The last term is the production or destruction of vorticity by the curl (spinning influence) of the friction force distribution. It is generally important in the boundary layers of the lower atmosphere or upper ocean, or in convective zones (such as monsoon depressions) where mixing by cumulus clouds is strong.

Some important understanding may be gained by first assuming all terms on the right side of Eq. (13) are zero; the result is the statement of conservation of absolute vorticity typical of simple *barotropic dynamics*. In such a case, fluid moving northward (in the Northern Hemisphere) increases its value of f, and hence its relative vorticity must decrease at the same rate. It can be shown that initial north–south motion induces circulation changes that restore fluid parcels toward their original latitude and shift the original patterns westward. A train of propagating *Rossby waves* may thus arise from the periodic exchange of vorticity between the wave and the variable earth's vorticity. These simple waves are stable, so that their overall circulations do not intensify.

In contrast, an analogous vorticity exchange between a wave perturbation and a mean horizontal shear flow is responsible for *barotropic instability*, and growing Rossby waves. For example, if an east–west (zonal) flow has strong north–south shear, then it may fold up at an increasing rate into an amplifying wave pattern. The necessary criterion for this to occur is that the absolute vorticity of the mean flow $(\bar{\zeta} + f)$ be an extremum. The rate of growth of the waves has an order of magnitude given by $\bar{\zeta}$, corresponding to amplitude doubling in a day or two. The size of the waves or *eddies* is comparable to the width of the mean shear zone. This process is a generalization of the common shear instabilities that produce eddy motions throughout the fluid environment. It has the effect of redistributing vorticity into more complex patterns and reducing the regions of strongest initial flow. Thus the mechanism of barotropic instability can create eddy patterns in air or ocean and it can slow down jets, such as the tropical easterly jet or Somali current. This instability may also help initiate the amplification of monsoon depressions over the Bay of Bengal. These depressions will be discussed later in Part V of this volume.

Descriptions of *wind driven circulations* in the ocean require the friction term in Eq. (13). For example, away from the African coast in the Indian Ocean, vorticity changes are brought about by the friction curl term which is negative due to the clockwise spinning of the air over the sea surface. Some of the influence is reflected in development of currents with negative relative vorticity, but away from the coast ocean masses drift southward (producing the so-called *Sverdrup* transport) toward regions of smaller earth's vorticity f. Near the coast, a very strong northward current may exist where the increase of earth's vorticity is balanced by a strong friction effect opposing the associated circulation. This simple idea may account for the presence of the very strong *Somali current* off the eastern African coast during the summer monsoon (see Chapter 13, Section 2.2).

Now let us turn to the effect of the divergence term that is critical in many applications. It is sometimes referred to as the *stretching* term because the magnitude of absolute vorticity is increased when vertical columns are stretched vertically, corresponding to horizontal convergence. For adiabatic motions it is possible to show that the *potential vorticity*, given approximately by

$$q = (\zeta + f)/\Delta h$$

is conserved in the absence of friction. [Here ζ is evaluated on a constant potential temperature (θ) surface; in the tropics this surface is nearly horizontal.] This equation states that, with q constant, changes in the absolute vorticity of a layer between two θ surfaces must be in the same sense as changes in the thickness Δh of the layer. Equivalently, the absolute vorticity must change opposite to the static stability $\partial\theta/\partial z$. In the absence of forcing, both the thickness and absolute vorticity of large-scale tropical motions tend to remain nearly constant in time. However, air motions forced to ascend or descend mountain ranges will experience layer thinning or thickening, respectively, and hence reductions or increases in absolute vorticity will result. The increase in vorticity for a descending layer in the lee of a mountain is often responsible for *lee-cyclogenesis*. Cyclones developing in the lee of the Tibetan Plateau may subsequently trigger cold surges during the winter monsoon of Southeast Asia (see Chapters 12 and 18). For tropical regions away from mountains, strong diabatic heating may produce secondary circulations and, with them, stretching. The production of low-level cylonic motions within heated areas of both deserts and monsoon depressions are examples. In fact, the maintenance of monsoon depressions has been thought of as a kind of forced stretching; a related instability process (known by the acronym CISK) will be discussed later.

Finally, the stretching mechanism can account for development of new vorticity centers through the mechanism of *baroclinic instability*. This phenomenon occurs when a baroclinic mean flow (one with sufficiently strong horizontal temperature contrast) supports growing wave perturbations (eddies) that rearrange the horizontal temperature field in much the way that barotropic instability rearranges the vorticity field. A critical result of this process is that vertical stretching and shrinking create new centers of cyclonic and anticyclonic vorticity. These developing eddies transport heat down-gradient so as to reduce large temperature differences between warm and cold air regions. While the stretching mechanism is important for intense baroclinic zones on the fringes of monsoon regions, theory indicates that the small Coriolis influence near the equator renders the mechanism ineffective. Thus baroclinic instability is thought to be a major explanation for the waves of middle latitudes, but seldom for the low latitudes.

2.6 Energy

Since the thermodynamic laws discussed earlier are really statements about thermal energy processes, it makes sense to examine the energy implications of hydrodynamical laws. From Eq. (4), it can be shown that the prediction equation for the kinetic energy per unit mass (K) of a moving air parcel is:

$$\rho \, \frac{dK}{dt} = \underbrace{- \, \mathbf{V} \cdot \nabla p}_{\substack{\text{pressure} \\ \text{gradient} \\ \text{work}}} + \underbrace{\mathbf{V} \cdot \rho \mathbf{F}}_{\substack{\text{friction} \\ \text{work}}} \qquad (14)$$

This energy is changed only by the work performed by actual physical forces. Thus the Coriolis (apparent) force can do no direct work to change the kinetic energy. However, its indirect effect of changing the direction of the flow alters the way that the physical forces **P** and **F** can create or destroy energy. The work done against friction is negative for bounded fluid systems; however, in individual layers of air or water the friction work can vary significantly and can even be positive, as in the case of a wind-driven ocean current. The pressure gradient work term $- \, \mathbf{V} \cdot \nabla p$ is positive for wind flowing across the isobars toward lower pressures; it is zero for geostrophic flow (parallel to the isobars). Thus, ageostrophic motions are required to create kinetic energy by this mechanism. In most monsoon circulations (e.g., monsoon depressions), the divergent component of the secondary circulation accomplishes this work.

The pressure gradient work term implies a powerful link to other processes. This is seen by rearranging the differentiation and using Eq. (2). In the simplest case where ρ is constant in Eq. (2) we obtain

$$- \, \nabla \cdot (p\mathbf{V}) - \frac{\partial}{\partial z} \, (pw) - \rho wg$$

The first term is the convergence of the horizontal *energy flux*, $p\mathbf{V}$. This flux represents the net energy transfer by pressure work between neighboring fluid volumes, and is an essential mechanism for horizontal *propagation of wave energy* of all kinds. The second term is the vertical counterpart of the first: it represents a

ENERGY PROCESSES

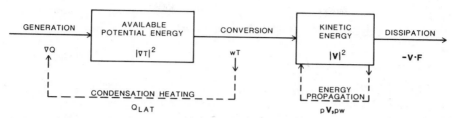

Figure 9.6. Energy of monsoon weather systems (boxes) and associated processes (arrows). Starting at the left, horizontal heating differences may generate temperature differences and available potential energy; available potential energy may be simultaneously converted by thermally direct secondary circulations to kinetic energy of the winds; the rising branch of the secondary circulation may cause rain and net condensational latent heating, which may be a positive feedback on the generation; wind energy may propagate to other locations or altitudes; and wind energy is ultimately lost through dissipation, primarily in the frictional boundary layers.

convergence of vertical energy flux. This flux pw links the kinetic energy of different layers of the atmosphere. For example, the rising motion above surface low-pressure centers extracts kinetic energy from the frictionless layers overhead, causing a spindown of those circulations. On the other hand, this flux can also communicate energy from active source regions to other layers: monsoonal tropospheric motions can transfer their energy into the stratosphere by this mechanism.

These flux mechanisms are very important for understanding the spreading of pre-existing energy through space. The last term ($-\rho wg$) is essential for creating the kinetic energy. For eddies, this *conversion* term is proportional to the product wT_v and represents a conversion from *available potential energy* (A) into kinetic energy. (Available potential energy is that portion of total potential plus internal atmospheric energy that can be converted to kinetic energy.) This conversion is positive when, on average, warm air rises and cold air sinks. This is known as a *thermally direct circulation* and corresponds to the lowering of the center of gravity of a fluid column, net upward heat transport, and an increase of static stability. When we combine this picture with the thermodynamic energetics discussed earlier, we see (Fig. 9.6) that monsoonal circulations are ultimately driven by differential heating which creates available potential energy. Secondary circulations arise that convert this energy to kinetic energy of horizontal wind systems. The winds may propagate energy into other regions, where it is eventually dissipated by frictional processes.

3 VERTICAL MIXING

Because the atmosphere is relatively thin, it is easily influenced by physical processes that cause air from one layer to travel vertically to a neighboring one. These motions are gentle for the broad monsoon circulations discussed in the last section. However, small-scale circulations may create strong vertical motions which cause *mixing* between layers. Common examples range from dispersion of smokestack effluent at low levels to tropical rain clouds extending through the depth of the troposphere. These important processes are explained in this section.

3.1 Small Eddies

Weather maps show synoptic-scale (roughly 1000 km) weather systems which change from day to day. The maps represent observations or numerical model data (see Chapter 15) based on information at regularly spaced gridpoints (often 100 km or more horizontally, 1 km in height, and several hours in time). Thus the patterns really represent volume and time averages (denoted below by overbars). Predictions are made from these variables. Much smaller scale features such as local wind gusts, sea breeze systems, cumulus clouds, and undulations caused by hills are not explicitly represented in these average fields and often are referred to as *turbulent eddies* (which should not be confused with the very different synoptic-scale eddies discussed in the preceding section). Eddies are mathematically defined as deviations from the averages: $(\ )' = (\ ) - (\overline{\ \ })$. Most turbulent eddies may appear chaotic, but they are not truly random. Instead they have a systematic impact in transporting sensible heat, moisture, and momentum. The vertical transport rates are called *turbulent fluxes*. Quantitatively, they represent the amount of a quantity passing across a unit horizontal area per unit time. The most important ones for our purposes are

$$\tau = -\rho \overline{w'\mathbf{V}'}$$

$$H = \rho\, c_p \overline{w'T'} \tag{15}$$

$$M = \rho \overline{w'q'}$$

where q' is specific humidity of water vapor and τ, H, and M are, respectively, the turbulent momentum, heat, and moisture fluxes. The vertical change of a flux represents the net effect of small-scale eddies on larger volume averages and hence is an effective source or sink term in the prediction equations discussed earlier. Thus the friction force is

$$\mathbf{F} = \frac{1}{\rho}\frac{\partial \tau}{\partial z} \tag{16}$$

where τ is known as the Reynolds stress. Similarly sensible heating $Q_{\text{sens}} = -\partial H/\partial z$ and the net moistening is $-\partial M/\partial z$. An example of a process producing these fluxes is cumulus convection: each cloud carries warm and moist air upward $(H, M > 0)$. The favored directions for wind gusts under the clouds (e.g., westerly gusts in the winter subtropics) indicates that these motions may also transport horizontal momentum from higher levels down to the ground.

A fundamental problem of fluid mechanics, and especially atmospheric dynamics, is the *parameterization*, or implicit representation of these small-scale eddy processes in terms of large-scale (averaged) fields alone. This, of course, is necessary if the effects of the eddies represented are much smaller than the grid boxes of weather prediction models. The simplest approach is to use the mixing length theory, which has modest success if the eddies are very small and obey simple physics. According to this theory the eddy transports have effects somewhat similar to the mixing by random molecular motions. Each flux is written in terms of equivalent *diffusion* of heat, moisture, and momentum and is proportional to an eddy transfer coefficient. These eddy diffusive fluxes transport quantities down-gradient much faster than molecular processes, thereby destroying extremes in the vertical profiles and making the patterns smoother and more nearly homogeneous. For example, this kind of mixing is capable of slowing down the air in the Somali jet core and increasing the speed of adjacent fluid layers, including those in the ocean. In the atmosphere, strong mixing tends to dissolve differences across the interface between warm dry continental air (north and top side of the jet) and cool moist maritime air (south and bottom side), while the wind stress transfers momentum to the Somali current of the upper Indian Ocean.

Refinements of such mixing models have yielded very useful formulae developed from Eq. (15) for the fluxes over the ocean surface, denoted by the subscript $_s$:

$$\tau_s = \rho\, C_\tau\, |\mathbf{V}|\, \mathbf{V}$$

$$H_s = \rho\, C_H\, |\mathbf{V}|\, (T_{sea} - T_{air}) \qquad (17)$$

$$M_s = \rho\, C_M\, |\mathbf{V}|\, (q_{sea} - q_{air})$$

The transfer coefficients (C's) are typically of order 2×10^{-3} over monsoonal oceans. These forms emphasize that the exchange between atmosphere and ocean is proportional to the wind speed and the difference in properties between the ocean and air. The surface stress τ depends most strongly upon the wind speed. The sensible heat flux represents a direct influence of sea surface temperature T_{sea} on the atmosphere; its *seasonal* variability over the Arabian Sea is great: pre-monsoonal warm waters undergo a rapid cooling after the onset of the monsoon. Over the eastern Pacific Ocean, large *interannual* variations of T_{sea} (a few degrees Celsius) may be important in climate variations. Finally, the evaporation M_s is also sensitive to sea surface temperatures. This is because q_{sea} (the specific humidity of air in saturated equilibrium with the sea surface) increases strongly with temperature T_{sea}. (The Clausius–Clapeyron equation describes this effect and is explained in reference 3.) This sensitivity of evaporation to sea temperature undoubtedly influences the seasonal cycle of water vapor fuel to the monsoon.

Over land, relations similar to Eq. (17) become more complex: the coefficients C are larger, and the quick response times of the land surface processes are significantly different. For example, over dry land the solar insolation absorbed during the day may be given directly back to the atmosphere as H_{sens} with little internal storage. This is the case over desert regions on the fringe of the monsoon areas, particularly Arabia. The presence of vegetation and soil moisture greatly changes the fluxes. The flux of sensible heat into the atmosphere over land is reduced with moist soil because the solar energy that causes heating is also used to evaporate water, contributing to the moisture flux M. Naturally, representation of M under these conditions is more complex, particularly when the effects of vegetation are taken into account.

When the turbulent processes farther above the ground in the planetary boundary layer (Fig. 9.7) are considered, the concept of local mixing by small-scale eddies becomes less useful. This is especially true in highly convective situations, such as strong daytime surface heating over deserts or air flowing over warm oceans. Under such conditions convection manifests itself as deep thermal plumes which may or may not be topped by cumulus clouds. The simplest model of a strongly convective boundary layer is the *mixed layer* where potential temperature θ is nearly

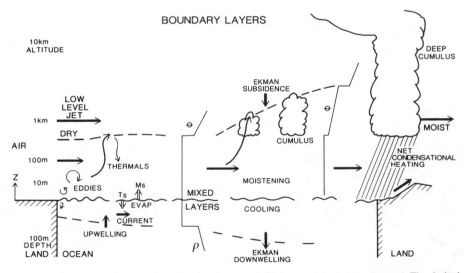

Figure 9.7. Downwind evolution of monsoon boundary layers in atmosphere and ocean. The dashed lines delineate the stable edges of thickening layers. Beginning at the left, strong winds over the ocean often create turbulent eddies and buoyant thermals which transport momentum (τ), moisture (M), and heat (H) vertically in such a way as to create homogeneous mixed layers. The mixed layer profiles of θ and ρ are shown for atmosphere and ocean, respectively. In the ocean, the air–sea interaction produces ocean currents, coastal upwelling, mid-ocean downwelling, and cooling. In the atmosphere the boundary layer air moving toward the right moistens and deepens until clouds are formed and a more complex marine boundary layer structure evolves, as seen in the second θ profile. Ultimately the moist, destabilized air causes the eruption of clouds out of the boundary layer into deep cumulus rainstorms over the ocean, mountain slopes, or land. The organized rain systems help to sustain the energy of the overall monsoon flow.

constant throughout most of the depth (implying thoroughly mixed air). At the top of the layer the potential temperature increases rapidly in an inversion. The heat flux H_{sens} ideally decreases linearly with height. This model has also been used with less success to describe the profiles of moisture and wind.

The boundary layer in the *upper ocean* can be very active in monsoons because the mixing is sustained by the strong surface wind stress and oceanic convection. This convection can be driven by evaporation alone, since evaporation promotes a top-heavy density structure due to surface cooling and an increase in salinity. Simple mixed-layer models are thus useful for the ocean as well as the air.

Over large parts of the ocean, convective air parcels form small cumulus clouds. Their circulation and mixing are altered by the release of latent heat of condensation producing a new kind of *marine boundary layer*, which is exemplified by the trade wind boundary layer. This layer is broadly divided into two sections, a well-mixed subcloud layer and the cloud layer in which θ, q, and $\mathbf{V}$ vary linearly with height.

There are some situations for which the fluxes cannot be described by simple profiles through relatively thin (2 km) boundary layers. For example, air flow over mountains is known to produce a drag force which depends upon the flow structures and spectrum of hills in complicated ways. The clear air turbulence experienced by jet aircraft is evidence of concentrated high-level momentum stresses τ which can be caused by upper-level jets or mountains far below.

3.2 Deep Cumulus Rain Clouds

Away from mountains, the greatest difficulty in parameterization may be tropical convection. Tropical rain systems exist on a variety of scales, some too large to characterize as small-scale eddies. These rain shower systems reflect complex physics which makes it difficult to characterize their effects by a simple parameterization. A precise parameterization of tropical convection may not be possible, and this may limit the inherent *range of predictability* of large-scale monsoonal motions. On the other hand, improvements in this parameterization could improve predictions of global and monsoonal circulations up to a week in advance.

Since modern turbulence theories are ultimately based upon the physics of turbulent elements, an appreciation of cumulus parameterization requires an understanding of fundamental cumulus processes which occur in most monsoonal rain systems. The *vertical equation of motion* for small cumulus scales can be written as

$$\frac{dw'}{dt} = -\underbrace{\frac{\rho'}{\rho}g}_{\text{buoyancy}} - \underbrace{\frac{\rho_l}{\rho}g}_{\substack{\text{water} \\ \text{drag}}} - \underbrace{\frac{1}{\rho}\frac{\partial p'}{\partial z}}_{\substack{\text{pressure} \\ \text{gradient}}} + \underbrace{F_z}_{\text{friction}} \qquad (18)$$

The primes represent deviations of cloud properties from the surrounding environment (represented by the overbar) and ρ_l is the density of liquid water contained in cloud and rain drops. The critical term for production of upward motion is the *buoyancy* term, which is approximately equal to $+ (T'_v/\overline{T_v})g$; it is upward for warm cloud air and downward for cool cloud air (relative to the environment). Although it is numerically a small fraction of gravity (typically on the order of 1%) it may create substantial vertical motions if it acts over several minutes of a cloud's lifetime. The liquid water drag term is somewhat analogous to negative buoyancy because it represents a downward force numerically proportional to the weight of liquid water in an air volume. (The term arises because the droplets reach a terminal fall velocity in which their weight is balanced by the drag force as they fall through the air.) It is normally smaller than the buoyancy term but can help to create downdrafts in the cloud. The third (pressure gradient) term is often neglected in simple theories because it depends upon the full three-dimensional fluid dynamics. It tends to oppose the net effect of buoyancy and drag and is similar to the resistance that a buoyant cloud "bubble" would have in trying to rise through a motionless environment. Modern theories and numerical models take this force into account. Finally, the friction force F_z is undoubtedly significant, as the "bumpy" visible edges of cumulus clouds suggest strong small-scale mixing.

The development of cumulus clouds requires positive buoyancy and we consult the first law of thermodynamics, Eq. (5), to identify its source. (See reference 3 for details of the following.) We first consider unsaturated parcel ascent prior to cloud formation. The rate of temperature decrease is given by

$$\frac{dT}{dt} = -\Gamma_d w' \tag{19}$$

where $\Gamma_d = g/c_p$ ($\sim 10°C/km$) is the *dry adiabatic lapse rate* of temperature. Once saturation is reached and the cloud forms, the heat source Q_{lat} is activated. Q_{lat} is proportional to the rate of ascent w'. With the addition of latent heat, a rising saturated cloud parcel cools at a rate less than the unsaturated parcel, or:

$$\frac{dT}{dt} = -\Gamma_m w' \tag{20}$$

Here Γ_m is the *moist* adiabatic lapse rate ($\sim 5°C/km$ near cloud base). The reduced cooling allows the parcel to maintain its buoyancy longer because it remains warmer than the environmental air. The other factor contributing to a parcel's buoyancy is the distribution of temperature with height outside the cloud. For an observer ascending with the parcel speed w', the rate of change of environmental temperature $\overline{T}$ is

$$\frac{d\overline{T}}{dt} = -\Gamma w' \tag{21a}$$

where $\Gamma = -\partial \overline{T}/\partial z$ is the actual environmental temperature lapse rate. The buoyancy change in a parcel can be obtained from Eqs. (20) and (21a) as

$$\frac{dT'}{dt} = (\Gamma - \Gamma_m)w' \tag{21b}$$

Thus when the temperature lapse rate outside the cloud is greater then Γ_m, a saturated air parcel displaced upward will become increasingly warmer than its environment and more buoyant. This is the physical basis for cumulus growth, and this process is known as *conditional instability*.

Any process which increases the lapse rate of temperature Γ may create conditions favorable for cumulus convection. Synoptic weather systems typically have large-scale rising motion increasing with height in the lower troposphere which tends to reduce its stability and eventually promote cumulus convection. Once this convection intensifies over a large region it may have a systematic impact on the temperature of that region. Latent heat release is concentrated in the convective area; not all of it is used to drive the cumulus scale circulation. The cumulus activity promotes forced descent around the outside of the clouds and a net downward heat transport ($H < 0$) producing warming in the lower layers. The total *cumulus heating* effect will promote further large-scale rising motion and destabilization. Thus the cumulus convection organized on the larger scale can create its own positive feedback involving *moist processes* which has been called *Conditional Instability of the Second Kind (CISK)*. This feedback is undoubtedly significant in many monsoonal processes, but there remain many unresolved questions at this time. Monsoon depressions seem to be the best example of synoptic disturbances maintained by the assistance of organized cumulus convection. The cumuli moisten the upper layers of the atmosphere and may cause the exchange of momentum between the lower and upper layers by vertical mixing. The momentum mixing influence (known as *cumulus friction*) on monsoon circulations is not presently understood.

Many monsoon rain systems may be so close to a state of saturation over large areas that the large-scale rising motion alone releases the latent heat; if so, cumulus parameterization in models would be neither necessary nor correct. More research is under way to confirm the extent of cumulus activity and its practical parameterization in different monsoon rain systems.

4 MODERN VIEWS: MONSOONS AS ADJUSTING SYSTEMS

An attempt has been made to illustrate how individual processes affect wind, temperature, and pressure, but in reality these fields interact as a system to produce monsoon weather. Since about 1950, fundamental theoretical bases for understanding such systems have been developed; two key ones are *quasi-geostrophic* (QG) theory and *equatorial wave theory*. They provide fundamental links to the monsoon theory and prediction topics discussed throughout much of this book.

4.1 Subtropics: Quasi-Geostrophic Theory

Monsoon circulations extend to subtropical latitudes poleward of 15° latitude where the geostrophic approximation is useful and implies a strong pressure–wind link. Yet it has already been seen that the QG equation of motion allows some ageostrophic flow. The complete QG theory is obtained by closing the system mathematically with the first law of thermodynamics, Eq. (5), where the ageostrophic secondary circulation appears only in the adiabatic cooling term. Two equations result: the QG potential vorticity equation and the omega equation.

The QG *potential vorticity equation* is

$$\frac{\partial q_g}{\partial t} + \mathbf{V}_g \cdot \nabla q_g = \frac{\partial}{\partial z}(aQ) + b\,\mathbf{k} \cdot \nabla \times \mathbf{F} \tag{22}$$

$$\underbrace{\phantom{\frac{\partial q_g}{\partial t} + \mathbf{V}_g \cdot \nabla q_g}}_{\text{advection}} \quad \underbrace{\phantom{\frac{\partial}{\partial z}(aQ)}}_{\substack{\text{diabatic} \\ \text{heating}}} \quad \underbrace{\phantom{b\,\mathbf{k} \cdot \nabla \times \mathbf{F}}}_{\text{friction}}$$

where q_g is the geostrophic approximation to the potential vorticity discussed in Section 2.5 and a and b are positive coefficients. This predictive equation states that q_g is conserved following the horizontal geostrophic flow, apart from the diabatic heating and friction terms on the right side. In the absence of the latter, changes in QG systems involve horizontal advective redistribution. The equation describes mechanisms that are responsible for: (1) stable Rossby waves, which propagate westward due to the β effect; (2) barotropic instability of flows with strong horizontal wind shear, and (3) baroclinic instability of flows with strong horizontal temperature differences. The inclusion of cumulus heating in Q allows CISK to be studied with this equation. Since q_g is related to the three-dimensional Laplacian of pressure, this equation identifies mechanisms causing pressure change. Mechanism 1 is *mechanical*: advection of cyclonic vorticity (or cyclonic vorticity production by $\mathbf{F}$) tends to create falling pressures and cyclonic circulations. Mechanism 2 is *thermal*: vertical increase in temperature advection or diabatic heating tends to force falling pressure centers as well. Rising pressures are predicted by reversing these rules.

This equation is very useful for studying forced monsoonal motions and *energy propagation* through space. First consider a periodic diabatic heating pattern with longitudinal wavelength $2\pi/k$ (k is called the zonal wave number) moving with speed c_Q. A maximum response is expected when this forcing *resonates* with a free wave having the same wave number k and speed of propagation c. For free Rossby waves we have c related to k by

$$c = U - \frac{\beta}{k^2 + m^2} \tag{23}$$

where m is a real constant. Equation (23) shows that Rossby waves retrograde to the west relative to a mean flow U. The resonance condition for stationary forcing ($c_Q = 0$) is that the free Rossby wave is stationary ($c = 0$), which can only exist in a westerly ($U > 0$) flow. The major stationary waves of middle latitudes are thought to be dominated by quasi-resonant mechanisms of this sort.

The north–south as well as vertical spreading of Rossby waves from source regions can be studied by setting Q and $\mathbf{F}$ of Eq. (22) equal to zero. The mathematical theory becomes rather complex but shows there may be conditions where wave energy is *reflected* toward the source region, leading to a concentration of energy there and a relative lack of wave energy on the other side of the reflecting latitude and altitude. In addition there are *critical barriers* along locations where $U = c_Q$;

they also inhibit propagation and may even absorb wave motion. For stationary monsoonal waves these barriers are located at the boundary ($U = 0$) between westerly and easterly winds (see Chapter 11).

The *omega equation* is the popular and powerful QG statement on secondary circulations. It is written as the elliptic equation

$$\nabla^2 w + \frac{\partial^2}{\partial z'^2} w = \frac{\partial \mathcal{M}}{\partial z'} + \nabla^2 \mathcal{T} \tag{24}$$

where z' is $(N/f)z$, a stretched vertical coordinate. Here $N = [(g/\theta)(\partial\theta/\partial z)]^{1/2}$ is the frequency of a simple, stable buoyancy oscillation. The equation states that two kinds of mechanisms produce vertical motion. The first, ($\partial \mathcal{M}/\partial z'$), involves *mechanical forcing* $\mathcal{M}$, which is a spin (vorticity) source. This term shows that cyclonic vorticity advection increasing with height (or an equivalent effect by curl **F**) is associated with rising motion. The second, ($\nabla^2 \mathcal{T}$), depends on the *thermal* source $\mathcal{T}$; warm temperature advection or diabatic heating Q is associated with rising motion. This reminds us that the vertical motion w acts as a safety valve to lessen the impact of the forcing. For example, rising motion would partly balance the effects of diabatic heating by producing adiabatic cooling. In monsoon circulations latent heat release makes this compensation (negative feedback) relatively ineffective and may even lead to CISK (positive feedback). The w' pattern, which obeys Eq. (24), gives direct information on divergence and convergence in neighboring layers. The rules for subsidence are obtained by changing the signs of the forcing terms in Eq. (24).

Both of the QG equations imply that the Rossby *radius of deformation* λ is a natural horizontal length scale. It is defined as

$$\lambda = \frac{N}{f} h \tag{25}$$

where h is the depth of the circulation system (about 10 km for major monsoonal elements). Monsoon circulations of various widths L will behave differently, depending upon their ratio L/λ. This ratio varies with different monsoon applications because there are many scales L and λ varies with latitude. Regime 1 ($L > \lambda$) is called a *large-scale regime*: the QG equations imply that diabatic heating Q is efficient in changing the temperature field and geostrophic circulation. It is less efficient than mechanical forcing in creating secondary circulations. In regime 2, ($L < \lambda$), the *smaller-scale regime*, the results complement those of regime 1. For example, diabatic heating Q produces strong vertical motion (a branch of the secondary circulation), but significant geostrophic circulations are more efficiently created by the mechanical forcing field.

While these quasi-geostrophic results are powerful, they cannot be extended to the deep tropics, where significant monsoon forcing occurs, because the Coriolis force goes to zero. These shortcomings of QG theory are especially true for small-scale weather disturbances (regime 2).

4.2 Deep Tropics: Equatorial Waves

In the deep tropics (equatorward of about 15° latitude) the geostrophic relation between pressure and wind does not hold. Instead, the wind and pressure are continually adjusting to each other and no single relationship can be identified. A major contribution of wave theory is that distinct pressure and wind patterns can be identified for different kinds of wave motions. Three broad categories of waves exist in the atmosphere and ocean: sound waves, gravity waves, and Rossby waves. Sound waves are unimportant for weather patterns and ocean circulations. Rossby waves can produce day-to-day changes in circulations, and gravity waves may be important in some monsoon circulations in the deep tropics. Each wave type also has a clear link between space and time changes: Rossby waves move at speeds on the order of typical wind speeds, whereas gravity waves move faster.

The wave description is useful if the circulation systems are not too localized; broad patterns on weather maps may be thought of as Fourier sums, or the super-position of different wave lengths and shapes. The strength and phases of these waves change in time by linear dynamics and nonlinear advective processes. Most theories concentrate on the former but *spectral models* predict the nonlinear evolution of each wave component.

The pressure–wind adjustment at low latitudes can be qualitatively understood by considering the *linear dynamics* of individual waves. We can appreciate equatorial wave theory by first reviewing the properties of simple gravity waves.

A very simple *gravity wave* that is familar to all readers is the wave on a water surface, such as that caused by a boat moving across a lake. The water oscillates coherently in both space and time in a stable manner; the restoring force is gravity. The actual motion of individual fluid particles is basically a secondary circulation: closed orbits in the vertical plane are reflected by the bobbing up and down and back and forth of a floating object. Looking down on the water, no spin in the horizontal plane is evident; the simplest gravity wave is a horizontally divergent motion. This strongly distinguishes the simple gravity wave from a Rossby wave, which has great spin and small divergence.

The wave appearing on a water surface is called an *external* wave, but *internal* wave motions may occur within the water body. Each internal wave is characterized by reversal of currents with depth in a water column. Observations have shown a clear example in the upper Indian Ocean along the equator during July. An internal wave structure is also an essential part of atmospheric motions; its dynamics are analogous to that for a simple layer of water with *equivalent depth h*. Large-scale gravity waves obey the hydrostatic relationship and move with a speed expressed as

$$c_0 = (gh)^{1/2} \tag{26}$$

For waves confined to internal layers of depth h, Eq. (26) is modified by replacing g by reduced gravity g', which is a measure of buoyancy restoring forces across the layer. For internal tropospheric motions c_0 can be up to 50 m/sec, much faster than average monsoonal air speeds of 10–20 m/sec. This represents the fastest adjustment and reminds us that circulation patterns may propagate through the tropics in a time much shorter than the travel time for individual air parcels. Since c_0 is a constant, simple gravity wave motion is nondispersive: patterns and energy propagate at the same speed for all wavelengths. The equatorial oceans may respond to seasonal winds by such waves propagating from west to east.

Gravity waves can be modified by the earth's rotation if they are sufficiently large scale. An idealized expression for *gravity–inertia* wave frequency ω ($= 2\pi/$ period) is

$$\omega = [f^2 + c_0^2 k^2]^{1/2} \tag{27}$$

The influence of the earth's rotation is to create a minimum frequency equal to f and to cause the waves to be *dispersive*, with energy of wave groups propagating at speeds less than c_0. (Individual wave centers appear to propagate at phase speeds greater than c_0.)

These waves are fundamentally out of geostrophic balance and any weather system with strong nongeostrophic winds contains them. Since they disperse their energy away from the local unbalanced region, leaving a more geostrophic wind there, they are a basic mechanism for the *adjustment* of circulations toward geostrophic balance.

In the near-equatorial zone where f becomes small, its variations with latitude become important and it can be shown that the gravity inertia waves are reflected as they propagate poleward and experience stronger Coriolis forces. Thus the waves are trapped in an *equatorial duct* where they may propagate in the east–west direction. Other wave types may be similarly trapped. This duct extends to about 20° of latitude away from the equator for many monsoonal circulations. Its width is proportional to the Rossby deformation radius at the equator given by

$$\lambda = (c_0/\beta)^{1/2} \tag{28}$$

The maximum period for gravity–inertia waves in the duct is about two days, emphasizing the potentially fast north–south adjustment of tropical weather systems.

Linear theory for the duct shows that there are three other wave types that are important in deep tropical circulations. The unique signatures of each of these freely propagating waves are summarized in Figure 9.8. First note that the *Rossby* wave is not confined to the extratropics, but can exist near the equator; its circulation resembles a geostrophic one with the strongest zonal winds on the equator. The *Kelvin* wave also has strong geostrophic winds that are perfectly zonal and centered on the equator. In contrast, the *mixed Rossby–gravity* wave (a kind of hybrid between Rossby and gravity–inertia waves) shows strong ageostrophic motion across the equator. The pressure and temperature patterns for each wave are distinctly different. Rossby and Kelvin waves are symmetric about the equator: identical Rossby wave centers straddle the equator, while the center of the Kelvin wave lies on it. The mixed wave has pressure doublets straddling the equator; this asymmetry can describe

SOME IMPORTANT EQUATORIAL WAVES

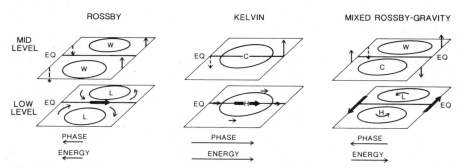

Figure 9.8. Some fundamental traveling equatorial waves in three-dimensional perspective. Lower level: curved lines represent isobars enclosing high (H) or low (L) pressure centers. Arrows depict winds; the strongest speeds are boldest. Mid-level: curved lines enclose warm (W) or cold (C) air centers. Arrows show vertical motion. Strongest phase and energy propagation vectors typical of long wavelengths are indicated. Each wave type shows a distinct pattern of pressure, temperature, and wind. Rossby wave wind is nearly geostrophic with gentle vertical motion. Kelvin wave wind is partly geostrophic with stronger vertical motion. Mixed wave has wind which is mainly nongeostrophic with strong vertical motion; its characteristics resemble those of the gravity-inertia wave.

a systematic monsoonal temperature gradient across the equator. The implied secondary circulations for the three waves are very different: the Rossby wave has small secondary circulations, but those of the Kelvin and mixed wave are strong and distinctive. The Kelvin wave has maximum secondary circulations in the equatorial plane; these are often referred to as *east–west* or *Walker* type circulations. The secondary circulations for the mixed wave connect across the equator and are called *north–south* or *Hadley* type circulations.

The east–west propagation picture is fascinatingly complex, even in the simplest category of long zonal wavelengths. The Rossby and mixed wave propagate circulation features towards the west while the Kelvin wave propagates very rapidly toward the east. The directions of strongest energy propagation are not analogous: the Rossby wave sends energy slowly westward while the mixed and Kelvin waves propagate energy toward the east most rapidly. The fastest of all of these speeds (c_0) corresponds to a travel time around the earth of 10 days. This time scale is comparable to that of some monsoon *break circulations*, so global wave adjustment may be a factor in local monsoon variability. Other monsoon oscillations of longer period (40 days) have also been discovered. They propagate slowly eastward around the entire globe, but they appear to be strongest at Asian longitudes where local northward propagation is also evident. They are probably modified by energy sources and dissipation which are effective over periods in excess of about one week.

Since these wave patterns are building blocks for describing low latitude circulations, they may be used to describe the response to average monsoonal heating. Recall that such monsoon heating fields have two important properties: east–west variations near the equator and north–south variations across the equator. The theoretical results are shown in Figure 9.9. The first example, the east–west equatorial heating Q_{EC} in Eq. (1) gives a lopsided response which is narrower and stronger to the west. This corresponds to a Rossby wave traveling slowly westward while being dissipated; the broader region to the east is Kelvin wave energy traveling faster (and hence further) while being dissipated. The wind is strongest along the equator, converging towards the heating region. Due to subsidence, the warm centers are displaced toward middle latitudes and westward relative to the heating. The secondary circulation consists of very strong rising motion in the heating center with compensating subsidence on the equator to the east, and poleward to the west. The former accounts

(a) FORCED EQUATORIAL CIRCULATIONS

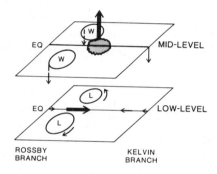

(b) CROSS-EQUATORIAL HEATING/COOLING

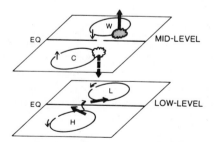

Figure 9.9. Equatorial circulations in three-dimensional perspective forced by heating. Mid-level: diabatic heating (shaded) or cooling (unshaded) centers are enclosed by scalloped lines. Vertical motion is indicated by arrows. Curves enclose warm (W) or cold (C) air centers. Lower level: curved lines are isobars enclosing high (H) or low (L) pressure centers. Winds are shown by arrows, with strongest winds boldest. (a) Heating centered on the equator. At low levels, air converges along the equator under the heating, moves upward there, and eventually descends at other locations. Moderate geostrophic circulations are sustained away from the equator. (b) A cross-equatorial heating/cooling pair tends to create pressure centers which maintain strong winds on their equatorial flanks and promote flow across the equator. Again, vertical motion is strongest in heating/cooling centers.

for the Walker circulation found to the east of Indonesia over the equatorial Pacific Ocean. The latter accounts for weak Hadley overturning. When mean easterly winds ($U < 0$) flow past the stationary heat source, the eastern circulation is intensified and narrowed, since the Kelvin wave is nearer resonance. Analogously, the western side is intensified and narrowed by a mean westerly flow ($U > 0$) since the Rossby mode is brought closer to resonance.

In the second example, *cross-equatorial* heating similar to Q_{IH} in Eq. (1), an eastern (Kelvin) branch is not excited so that the response is confined to the heated region and neighboring western areas, described by Rossby and mixed Rossby–gravity modes. This time the zonal wind reverses across the equator, a characteristic shared by actual summer monsoon winds over the Indian Ocean, namely, the southeast trades/southwest monsoon combination at low levels or their reverse in the upper troposphere (tropical easterly jet and winter Southern Hemisphere westerlies). Some flow across the equator is evident, but this linear theory cannot account for the observed intensity over the western Indian Ocean in summer. The warm and cold air mass centers are poleward of the thermal forcing, the secondary circulation is mainly of Hadley type across the equator, with little overturning in the equatorial plane. The presence of a mean westerly flow localizes and increases the intensity of the circulation due to improved Rossby wave resonance, while easterly flow decreases it.

The response to more realistic heating distributions can be thought of as sums of patterns such as those shown in Figure 9.9 but generalized to include many wave patterns and speeds. Because the monsoon winds can be very strong, linear results are modified by nonlinear advections and so only *numerical prediction models* are capable of showing the complete response to monsoon heating. Nevertheless, these simple wave descriptions give comprehensive insight into deep tropical monsoon flows which are not achieved by other reasoning.

5 SOME FINAL COMMENTS

This chapter has taken us on a journey of ideas extending from the classical physics of two centuries ago to modern views that are based upon recent monsoon observations and theory. Many concepts have been discussed and they will be further explored in the remaining chapters. It is remarkable that these numerous concepts arise from just a few basic physical laws.

The great variety of monsoon phenomena reflects the many modes of fluid motions influenced by gravity and the earth's rotation. Indeed, some aspects of monsoon behavior can be understood solely in the context of *geophysical fluid dynamics*, as is suggested by the last section. Monsoons are also affected by *internal processes* that are particularly intricate: these include the evaporation and precipitation of water which affect both air and sea, and small-scale mixing processes. They add complexity that is partly the source of monsoon variability for time scales of hours, days, and possibly longer. In contrast, other chapters will illustrate the impact on monsoons of *external conditions*, such as the distribution of land, ocean, mountains, and the march of the seasons. These also add complexity on a wide range of longer time scales, including the interannual and beyond. These ideas are explored in the chapters that follow.

REFERENCES

Concepts and phenomena introduced in this chapter are treated in more detail in portions of the following books. The books are ordered according to increasing degree of complexity and/or specialization.

Elementary

1. G. T. Trewartha and L. H. Horn, *An Introduction to Climate*, McGraw-Hill, New York, 1980, 416 pp.
2. S. D. Gedzelman, *The Science and Wonders of the Atmosphere*, Wiley, New York, 1980, 535 pp.
3. J. M. Wallace and P. V. Hobbs, *Atmospheric Science—An Introductory Survey*, Academic, New York, 1977, 467 pp.

Intermediate

4. S. Pond and G. L. Pickard, *Introductory Dynamical Oceanography*, Pergamon Press, Oxford, 1983, 329 pp.
5. H. Riehl, *Climate and Weather of the Tropics*, Academic, London, 1979, 611 pp.
6. A. Wiin-Nielsen, *Compendium of Meteorology, Volume 1, Part 1—Dynamic Meteorology*, World Meteorological Organization No. 364, Geneva, Switzerland, 1981, 334 pp.

Advanced

7. C. S. Ramage, *Monsoon Meteorology*, Academic, New York, 1971, 277 pp.
8. J. R. Holton, *An Introduction to Dynamic Meteorology*, Academic, New York, 1979, 391 pp.
9. M. J. Lighthill, Ed., *Monsoon Dynamics*, University Press, New York, 1981, 735 pp.
10. A. E. Gill, *Atmosphere–Ocean Dynamics*, Academic, New York, 1982, 666 pp.
11. G. J. Haltiner and R. T. Williams, *Numerical Prediction and Dynamic Meteorology*, Wiley, New York, 1980, 476 pp.
12. C. P. Chang and T. N. Krishnamurti, Eds., *Reviews in Monsoon Meteorology*, Oxford University Press, New York. (Expected date of publication 1987).

PART V

INTERACTIONS AND VARIABILITY

In some ways the monsoon is one of the most consistent, predictable events in nature. One often speaks of "the regular pulse of the monsoon." Close examination shows, however, that the character of the monsoon is quite complex and variable on any number of temporal and/or spatial scales. In addition, the monsoon is not an isolated phenomenon. Its interactions with other atmospheric circulation systems, the oceans, and even mountains are many and involve complex feedbacks.

Changes in the timing, extent, and strength of the monsoon go well beyond a simple interannual variation to decades, centuries, even millenia. While historical and rain gauge records can be used for "recent" times, scientists have turned to geologic records for evidence of the monsoon thousands of years ago. What the records reveal is that major changes have occurred in the location and strength of the monsoon. In Chapter 10, John Kutzbach reviews the evidence of change as revealed by many types of paleoclimatic data, from lake pollen to ocean sediments. He also presents hypotheses for how these changes occurred and explains how modern climate simulation models can be and currently are employed to test these hypotheses.

On far shorter time scales, the monsoon is characterized by internal variability and complex interactions, both regional and global. For example, that the mid-latitude upper-level jet streams may be affected by and, in turn, influence monsoons is not immediately obvious. In Chapter 11, Peter Webster presents observational evidence for this and other interactions and then offers hypotheses to explain how they occur.

In Chapter 12, Takio Murakami explores the relationship between the Asian monsoon and the major orographic feature of Asia, the Tibetan Plateau. The massif plays vastly different roles in the monsoon circulations of the summer and winter seasons. It is remarkable how far above the Plateau its influence is felt. Murakami provides convincing evidence that the power of the southwest and northeast monsoons of Asia would be greatly diminished in the absence of the Plateau.

The perspective changes in Chapter 13. Robert Knox is an oceanographer. Rather than consider the effect of the ocean on the monsoon, he looks at the monsoon as a force that drives ocean circulations. Thus it is the monsoon's winds, not its rainfall, which are of interest. Knox discusses how the major currents and gyres and the unusual upwelling patterns along the Somalian coast are related to the cycle of the monsoon winds.

In Chapter 14, J. Shukla uses 81 years of precipitation records from India to demonstrate the interannual variability of the monsoon. Even within one monsoon season there is considerable variation in the amount of rainfall that occurs. As Shukla points out, although "the dates of the onset (May to June) can fluctuate by more than 30 days," the amount of rainfall received in "June as well as for the whole monsoon season is not related to the date of onset." Add to this the spatial variations from one part of India to another and one has an incredibly complicated rainfall pattern, a pattern in which one hopes to find some order, some predictable aspect.

Shukla then introduces his theory about the causes of interannual variability. He shows evidence of how the conditions at the boundary of the atmosphere can influence subsequent monsoon activity. These relationships may serve as a basis for long range forecasting, a subject which Professor Shukla discusses in detail in Chapter 16.

10

The Changing Pulse of the Monsoon

John E. Kutzbach
Center for Climatic Research
Institute for Environmental Studies
University of Wisconsin–Madison
Madison, Wisconsin

INTRODUCTION

The basic seasonal rhythm of the monsoon, cloud and rain alternating with clear and dry, is controlled with clockwork precision by the seasonal cycle of the sun's track across the sky. To be sure, year-to-year differences exist in the date of welcome arrival of the rains. Likewise, the regional distribution of the rain is somewhat different each year. Nevertheless, there is a fundamental regularity and dependability to the monsoon that sets the yearly rhythm of life across much of Africa and South Asia. Aspects of this regularity are molded into the traditional culture and the proverbs of the people as described by Zimmermann in Chapter 3 and recorded by ancient mariners as discussed by Warren in Chapter 7.

Given this reassuringly regular annual cycle, it is perhaps surprising to learn that the character of the monsoon has altered dramatically in the past—but so it has. The following pages describe the monsoon's changes over the centuries and millenia, and suggest possible causes of this variability.

1 MONSOONS OF PAST DECADES AND CENTURIES

Keen observers have long been alert to the comings and goings of wind and rain. As a result, there are records, now of great value, from which we can study variability of the monsoon.

1.1 Rain Gauge Records

Rain gauges, used in India for more than a century, record the unsteadiness of the monsoon from one decade to another. In 1910, Sir Gilbert Walker (1), then Director-

General of Observations of the India Meteorological Department, used gauge records begun in 1840 to describe the variability of Indian rainfall. The drought of 1905 had been particularly severe and, following upon previous drought years, the question of long-term monsoon changes and their consequences for India was no doubt prominent. Walker found two periods of greatest rainfall deficiency, 1843–1860, and 1895–1907.

The analysis of rainfall records for monsoon trends, runs, or cycles has continued to this day. The interval of deficient rains that was noted by Walker persisted until about 1920; it was followed by a remarkably low frequency of drought for the next 30 or more years—until drought became more common again in the 1960s (2). An index of Indian drought for the period 1891–1975 (3) portrays these decade-scale shifts (Fig. 10.1). Of 14 major drought years in the 85-year record, eight occurred in the first 30-year period (1891–1920); there was only one in the second 30-year period (1921–1950). In the last 25 years (1951–1975), five major drought years were recorded. That the drought years (and also flood years) occur in runs, rather than scattered randomly through the years, has been confirmed through statistical analysis of African rainfall records by Kraus (4), who concluded:

> The clumping or persistence of dry and wet anomaly values which are exhibited by the [rainfall] record must be assumed to be a manifestation of a real physical process. The biblical story of the seven fat and the seven lean years in Egypt may well have had a basis in fact.

Monsoon fluctuations are not only persistent through time, producing runs of wet or dry years, but they also tend to occur simultaneously over large areas of Africa and South Asia. In the past century, North Africa and northwest India

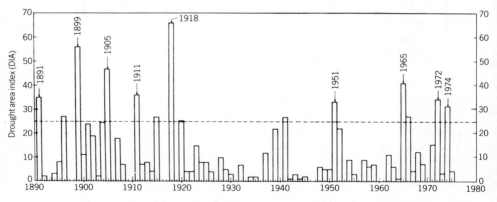

Figure 10.1. A yearly drought area index (DAI) based on the percentage of the total area of India experiencing moderate, severe, or extreme drought. From rain gauge records of the period 1891–1975. Selected years of extreme drought are labelled. Reprinted with permission of the American Meteorological Society from Bhalme and Mooley, *Monthy Weather Review* (reference 3).

experienced common intervals of drought (4). Long rainfall records from East Africa (5) and West Africa (6) show decade-scale fluctuations of the African monsoon which are similar to the changes of rainfall in India. In North Africa, where the weak monsoons of the period 1968–1973 claimed international attention because of the drought and famine conditions of the Sahel, the generally decreased rainfall of the 1960s and 1970s (compared to previous decades) paralleled decreased rains in India. These drier conditions have persisted into the 1980s. Earlier West African drought episodes occurred in the pre-1920 period (as in India) and around 1940.

1.2 Historical Records

Historical records of climatic events that preceded the widespread use of rain gauges show century- or millenium-scale variations of climate upon which decade-scale variability, of the sort found in rain gauge records, is superimposed. Nowhere are there longer historical records than in the monsoon lands. Aristotle (7) said of changes of climate in Greece and Egypt:

> The same parts of the earth are not always moist or dry, but they change according as rivers come into existence and dry up. But we must suppose these changes to follow some order and cycle.

In this particular context, Aristotle was describing possible local climatic changes associated with the process of silting of a river bed.

Historical records of river flow, of both flood levels and low-stage levels, are climatic indicators; so too are reports of the water level of lakes and the occurrences of famine. Some of the earliest historical evidence of the annual rhythm of the monsoon and of its variation over the centuries comes from the Nile. Rainfall in equatorial East Africa (the headwaters of the White Nile) and summer monsoon rains in the highlands of Ethiopia (the headwaters of the Blue Nile) contribute to the Nile flow and produce a strong annual rhythm (8):

> . . . each year the river is lowest in about April and May, rises rapidly in July, and is high through late August and September. This pattern is due to the flood of water in the Blue Nile bringing down the water from the summer monsoon rains.

As an historical aside, the knowledge that the annual Nile flood was an indicator of the annual pulse of monsoon rains far to the south was not so obvious once. Herodotus (9) who described accurately the importance of the annual Nile flood to Egypt, searched for the mechanism that produced it:

> About why the Nile behaves precisely as it does I could get no information from the priests or anyone else. What I particularly wished to know was why the water begins to rise at the summer solstice, continues to do so for a hundred days, and then falls again at the end of that period, so that it remains low throughout the winter until the summer solstice comes around again in the following year.

The question of the cause of the *annual* rhythm of the Nile is now largely settled, but one can still echo Herodotus' frustration concerning the lack of information, "from priests or anyone else," on the precise cause of the long-term variability of the Nile.

Sir Gilbert Walker (1), mentioned earlier, used the Nile records to infer possible variability of the Indian monsoon:

> Of the countries affected by the monsoon the only area for which reliable data extend over a satisfactorily long period is Egypt, where the Nile data extend back as far as 1737. . . . Inasmuch as the Nile flood is determined by the monsoon rainfall of Abyssinia, and as the moist winds which provide this rainfall travel in the earlier portion of their movement side by side with those which ultimately reach the north of the Arabian Sea, there is a tolerably close correspondence between the abundance of the Nile flood and that of the monsoon rains of northwest India. . . .

Walker's analysis showed that the Nile flood was below average for most of the period 1895–1907 (matching the Indian rainfall record) and that comparable earlier Nile flood deficits had occurred from 1781 to 1797 and from 1825 to 1839.

After Walker's early work, a much longer Nile record was published in 1925 by Prince Omar Toussoun (10). Since A.D. 622, the highest and lowest levels of the Nile had been noted each year at a gauge or "nilometer" located on the island of Roda at Cairo. This 1300-year record indicated sizeable departures from the "average" annual cycle of the Nile flow. Brooks, among others, used this long record of the Nile to infer variations in the African monsoon rainfall occurring far to the south of the Cairo nilometer (11). In a recent study, the Nile stage records were used by Riehl and Meitin (12) to chart the cumulative departures of discharge from the long-term mean (Fig. 10.2); on their graph, a positive slope (cumulative discharge departure increasing with time) indicates a period of above average discharge—and, by inference, above average African (and Indian) monsoon rainfall. Several major climatic fluctuations—of duration 50–100 years—were identified in the 1300-year record. There was also a long period with only minor variations of discharge (A.D. 950–1200) when, perhaps coincidentally or perhaps significantly, a period of unusually mild climate existed in much of western Europe.

The most recent sequence of African monsoon changes, as reflected in the Nile records (Fig. 10.2), began with increased discharge about 1850 and was followed by decreased discharge after about 1900. The shift from above to below average discharge around 1900 paralleled the shift from above to below average Indian rainfall at about the turn of the century. To the extent that the relatively good monsoon rains of India for the period 1920–1950, mentioned earlier, are reflected in the Nile record, they are associated with the brief interval of average discharge that interrupts the long interval of decreased discharge during the twentieth century.

As noted at the outset, the record of Nile floods is by no means the only source of information on monsoon variability through the centuries. The water level of Lake Chad in West Africa has been estimated for the past 1000 years (13), and the Lake Chad and Nile fluctuations are well-correlated (14). Lake level records, travel

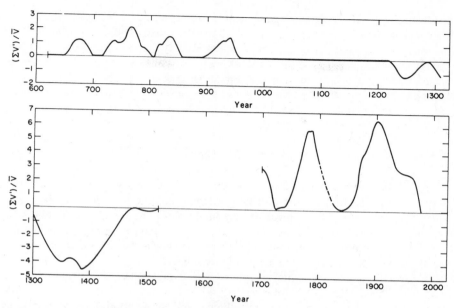

Figure 10.2. Time history of short-period climate changes of the Nile River from 622 to 1976. The dashed part of the curve indicates interpolation. The curve after 1870 is based on measurements at Aswan. The vertical axis is scaled as cumulative sum of discharge departures from the average ($\Sigma V'$) divided by the average discharge ($\overline{V}$). A positive slope to the curve (cumulative discharge increasing with time) indicates that discharge is *above* the long-term average; a negative slope indicates that discharge is below the long-term average. From reference 12, copyright 1979 by the American Association for the Advancement of Science.

reports, and chronicles of famine and drought have been combined to summarize the climatic history of West Africa for the past millenium (6, 15); several previous episodes of Sahelian drought have been identified: 1681–1687, 1738–1756, 1770–1774, 1790–1796, and 1823–1839 (16). At the eastern limit of the monsoon lands, historical records have produced a 500-year chronology of drought and flood in China (17).

Paging back through past millenia, the extant historical records become fewer in number, and more difficult to interpret. In Egypt, high water marks of the Nile flood were scratched on rock cliffs that overlook the river, and comparing these marks for different periods suggests higher flood levels in the First Dynasty, around 5000 B.P. (before present), than in subsequent dynasties (8, 18). These earliest markings may be indicators of the last stage of a remarkable interval of strong monsoons (and high Nile floods) that characterized the period 12,000 to 5000 B.P. After the time of the First Dynasty, the historical records, although very incomplete, indicate low Nile floods and, by inference, weak monsoons. Bell (18) related the change from strong to weak monsoons to a "first Dark Age of Ancient History" that began in Egypt around 4000 B.P., at the end of Dynasty VI, when ". . . a very

stable society, with seeming suddenness fell into anarchy.'' This was also the
approximate time of onset of the decline of the Harappan civilization in India (next
section). A ''second'' Dark Age of Ancient History, around 3200 B.P., saw the
simultaneous decline of powerful states in Greece, Anatolia, Egypt, and Mesopotamia;
these events have also been related to broad patterns of drought (18–20).

In the following section, major shifts of the monsoon that predate the systematic
records of early civilizations will be traced in the geologic record.

2 GEOLOGIC RECORDS OF PAST MILLENIA

It is perhaps paradoxical that some of the most quantitative evidence for large
variations in monsoons of the past comes from geologic records. This fact goes
against our intuition that the mists of time might blur our view of ancient climates.
But the combined efforts of geologists, biologists, archaeologists, chemists, and
physicists have produced numerous techniques for reading and dating the environmental
record accurately—Rosetta stones of another kind.

In particular, geologic evidence exists for a several-millenium period of strong
monsoons during the early Holocene (12,000–5000 B.P.), a period that is quite
unlike our experience of more recent millenia, and for a period of weak monsoons
at the time of the most recent glacial maximum (about 18,000 B.P.). A more
comprehensive review of past climates in tropical lands, spanning hundreds of
thousands of years, is found elsewhere (21).

2.1 The Monsoons of the Early Holocene

Lakes and rivers change size, the flora and fauna evolve, and subtle shifts in physical
or chemical properties occur. Some of these changes, insofar as they are related to
monsoons, are reviewed here. The spade, the drill, the microscope, and the radioactivity
clock are the rain gauges, thermometers, and calendars of the past.

2.1.1 Lakes. Early scientific exploration of Africa revealed the existence of relic
shorelines of huge lakes. An early twentieth-century observer (22) remarked: ''The
existence of these great volumes of water where we now find only small drying
lakes must be due to former moister climate during so called Pluvial epochs.'' As
evidence for former tropical *pluvial* climates accumulated, debate centered upon
whether tropical pluvials and polar *glacials* occurred simultaneously or in opposition,
or whether any temporal correlation existed. Although there were proponents for
all three points of view, perhaps the most common view was that pluvials and
glacials were synchronous. However, the advent of radiocarbon dating led to a
dramatic reversal of thought. Beginning in the 1960s and culminating in the early
1970s, it became clear that the tropics were relatively dry at the time of maximum
glaciation (around 18,000 B.P.) and for several thousand years thereafter. Then
with only brief interruptions, more humid conditions prevailed from about 12,000
to about 5000 B.P. (23, 24)—that is, during the maximum of interglacial ''growing
season warmth'' at high northern latitudes.

The geographic extent of the climatic changes of the early Holocene has only recently become apparent (25, 26). A vast region of the tropics, stretching from Africa across Arabia to northwest India, experienced high lake levels simultaneously (Fig. 10.3). In some cases, the changes were extreme. Lake Chad was 40 m deeper than at present and had a surface area equal to that of the modern Caspian Sea—an areal expansion of tenfold or more. Lakes in tropical East Africa and Ethiopia also expanded (23, 27, 28).

Referring to the high lake levels of the early interglacial, Street-Perrott and Roberts (29) noted that

> there was an unsteady northward migration and broadening of the intertropical runoff peak, suggesting an enhanced monsoonal circulation over the African and Arabian continents, probably accompanied by a northward shift in the mean Intertropical Convergence Zone.

Through these and other studies, a description of the monsoon circulation of the earlier Holocene is now emerging (30–32).

The changed environment was by no means restricted to the physical appearance of lakes. According to Street and Grove (25):

> . . . biological productivity and biomass greatly increased in tropical Africa, and the surface albedo was substantially diminished by the expansion of open water, marshes, and forest. Elephant, giraffe and antelope ranged comparatively unchecked over the Sahara, which must at times have existed only as relict desert areas isolated by corridors

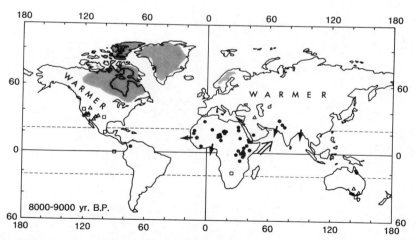

Figure 10.3. Composite chart (90°N–60°S) of selected climatic indicators for the period around 8000-9000 B.P. The figure shows continental outlines, extent of land ice (heavy shading), and lake status (●, high; □, intermediate; △, low). Other items, indicating changes compared to present, are added as follows: stronger southwesterly monsoon currents (⇒); increased fresh water discharge from tropical rivers (→); more mesic vegetation (*); and, warmer growing season conditions in northern mid-latitudes (warmer). Ocean surface temperatures were comparable to present except near certain coastlines where changes in upwelling occurred. In part from reference 26, revised and reprinted with permission of *Quaternary Research.*

of gallery forest and swamp along major wadis. Overflowing lakes established links between previously separate drainage systems. . . . Traces of hippopotamus and crocodile are widespread even in the central Sahara. . . .

Hunting–fishing–collecting communities became established around Saharan lakes by about 9000 B.P. (33), and it is possible that the most ancient Saharan rock art, including engravings of elephant, wild buffalo, and hippopotamus were created at this time (Fig. 10.4).

2.1.2 Rivers. Studies of Nile river deposits in conjunction with analyses of the lake and vegetation history along the Nile valley have provided a synthesis of the

Figure 10.4. Hippopotamus hunt, Tassilli Plateau Region (24°N, 9°E), early Holocene to 5500 B.P. (From reference 34).

geological record that extends far beyond the historical records of Nile floods used by Walker or his followers. Quoting Adamson et al. (35):

> During the intertropical cold dry phase from 20,000 to 12,500 B.P., the aggrading Nile was a braided, highly seasonal river. With a headwaters change to warmer, wetter conditions [after 12,500 B.P.], it became an incised, sinuous, suspended load river. Overflow from Lake Victoria and severe floods in Egypt heralded the change in Nile regime.

2.1.3 Deserts and Semi-arid Lands.

Analyses of the location of fossil sand dunes across the subtropical deserts of Africa and Asia show arid conditions around glacial maximum and indicate the onset of wetter conditions by about 12,000–10,000 B.P. (36). Of particular interest are the monsoon changes that occurred in India. The evidence was found buried in the Rajasthan Desert of northwest India—in pollen and sedimentary sequences of salt lake deposits. The Rajasthan Desert is located northwest of the region of maximum penetration of India's monsoon rains at present (except in rare years), but the material uncovered by Singh (37) shows that the modern period of saline lakes or desiccated lake beds extended back only to about 4000 B.P. During the preceding 6000 years, roughly the period 10,000 to 4000 B.P., the sediments and pollen suggest freshwater lakes and a surrounding landscape that supported significantly more vegetation. Cerealia-type pollen and carbonized materials that might have resulted from scrub burning were present in the sediment cores as early as 9300 B.P., prompting Singh to raise the question whether some sort of primitive cereal agriculture was introduced into the area at that time (37).

The magnitude of the monsoon change in the northwest of India was striking. For much of the period 10,000 to 4000 B.P. the annual precipitation may have been 200–250 mm above the present value of 250 mm, that is, almost a doubling (37–39). The pollen record contained freshwater aquatic species now seen in Rajasthan only several hundred kilometers to the southeast, where annual rainfall currently exceeds 500 mm. Freshwater conditions must have prevailed in lakes that are now desiccated or saline. Tree pollen was more common then than now, and the pollen of some desert shrubs was either absent or less abundant in the early Holocene than now.

Toward the end of this long period of generally strong monsoons, the Harappan civilization flourished. It extended for more than a thousand kilometers along the Indus river valley. Harappa and Moenjodaro were its major cities, with economies built on cereal cultivation and the use of domesticated animals—cattle, water buffalo, horses, camels (40).

The decline of the Harappans came around 4000 B.P. (41, 42). Singh (37) suggests that:

> the onset of aridity in the region around 1800 B.C. probably resulted in the weakening of the Harappan culture in the arid and semi-arid parts of northwest India. . . . The extinction of the Indus culture may have thus been initiated through gradual decline as a result of climatic change, but the process may yet have been completed by successive invasions from the northwest by the Aryans.

2.1.4 Other Lands. Further to the east, the same climatic themes are repeated. The monsoon strengthened at the end of the glacial period and then weakened by mid-Holocene time. Pollen records in central Taiwan (43) show a rapid replacement of cool-temperate species by subtropical and warm-temperate species about 10,000 B.P. In China, the early postglacial climate was warm and moist (44). Teilhard de Chardin and others (45, 46) reported former high water levels of lakes in East Mongolia and in Honan, Shansi, and Shensi; although these studies lacked the time-control of radiocarbon dates, the lake-level evidence was believed to be from the early Holocene. Early Holocene climates were also generally warmer and wetter in Australia and New Zealand (47).

2.1.5 Ocean Sediments. The secrets of past monsoons are also revealed by microscopic fossils found in deep-sea sediment cores. Key findings have come from the Arabian Sea, which is also a focal point for modern monsoon studies. The Arabian Sea southwesterlies that convey moisture to India in summer also produce intense cold-water upwelling along the Somali and Arabian coasts—thereby providing an environment of cool, nutrient-rich waters that is ideal for upwelling-associated planktonic foraminifera such as *Globigerina bulloides*. Important to the monsoon story are changes through time in the number of these microscopic planktonic skeletons that reach the ocean floor and become incorporated in the sediment. Prell (48) retrieved deep-sea cores from this region, and identified and evaluated the planktonic "message." There is a maximum in the abundance of *Globigerina bulloides* around 10,000 B.P., indicating increased upwelling and stronger monsoon southwesterlies at that time. This of course, is the same period when evidence of strong monsoons is found in the land and lake-level records.

A limited but growing number of deep-sea cores at other strategic locations along the African–Asian coastline are furnishing additional evidence of the strong early Holocene monsoon. Analyses of the abundance and isotopic composition of planktonic foraminifera from sediment cores in the Bay of Bengal indicate the presence of low salinity surface waters during this period (49, 50); the changed salinity is attributed to increased fresh water runoff from the Ganges. Pollen from East Africa, carried offshore by the monsoon winds and deposited in Arabian Sea sediments, indicates a more humid climate and stronger southwesterly flow in the early Holocene— relative to the period before and after (51). Increased sedimentation rates, along with changes in oxygen isotopic composition of foraminfera, show that discharge from the Niger into the Gulf of Guinea increased at roughly the same time (52). On the west coast of Africa just north of Dakar, discharge from the Senegal River increased, and the abundance of Mediterranean and Saharan pollen taxa decreased relative to tropical, humid pollen taxa (53). Heavy monsoon rains in Africa, channeled northward by the Nile, could also have produced a low-salinity surface layer in the eastern Mediterranean; a low-salinity surface layer would help explain the presence of black muds (sapropels) in the sediments, because the stable density stratification would have depleted the oxygen content of the bottom waters and helped to preserve this organic matter (35, 54, 55). The most recent sapropel layers are dated around

12,000 to 8000 B.P. (with a brief interruption around 10,000 B.P.) and are therefore consistent with the wealth of other evidence for stronger monsoons during this period.

2.2 The Monsoons of the Late Glacial

A 10,000-year period of relatively weak monsoons preceded the strong monsoons of 12,000 to 5000 B.P. These weak monsoons coincided with the time of maximum glaciation in polar latitudes (culminating around 18,000 B.P.) and with the initial phase of deglaciation.

A global synthesis of glacial-age climate conditions has been produced by the CLIMAP (Climate/Long Range Investigation Mapping and Predictions) project (56). Members of the CLIMAP project used a combination of land- and ocean-based geologic records to chart the sea surface temperature, sea level, sea-ice extent, land vegetation, and ice-sheet topography for the glacial world at 18,000 B.P. Figure 10.5 portrays some of this information, along with lake-level status, following the same format as Figure 10.3. Large ice sheets covered northern North America and Scandinavia, the domain of polar sea ice extended equatorward, the ocean's surface was colder, and tropical lands were more arid. As an additional consequence of the large volume of land ice, the sea level was about 100 m lower and shallow sea floors were exposed, especially between southeast Asia and northern Australia, thereby significantly increasing the land area of the monsoon region.

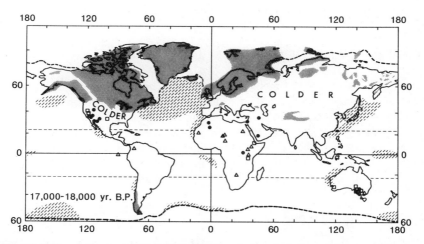

Figure 10.5. Composite chart (90°N–60°S) of selected climatic indicators for the period around 17000–18000 B.P. The figure shows continental outlines, extent of land ice (heavy shading), extent of sea ice in August (dashed line), and lake status (●, high; □, intermediate; △, low). Changes of continental outlines associated with lowered sea level are not shown. Other items, indicating changes compared to present, are added as follows: ocean surface temperatures at least 4°K lower than present (dashed shading from reference 56); and colder growing season conditions in northern mid-latitudes (colder). In part from reference 26, revised and reprinted with permission of *Quaternary Research*.

General circulation models, of the same type that simulate the behavior of the modern climate and the monsoon, used the surface boundary conditions for ice, land, and sea that were estimated by the CLIMAP project to simulate the global and regional patterns of atmospheric circulation of the ice-age. The simulated glacial climate was substantially cooler and drier, compared to present, especially over the continents of the Northern Hemisphere. There was an enhanced anticyclonic circulation (high pressure area) over the North American ice sheet and a generally weakened summer monsoonal circulation (57–59).

The CLIMAP project provided a "snapshot" view of events at 18,000 B.P.; other researchers, many of them associated with the CLIMAP team, have added a broad temporal perspective. The volume of glacial ice on land was at a maximum from about 22,000 B.P. to about 15,000 B.P., and remained substantial until about 12,000 B.P. This long interval of glacial conditions at middle latitudes corresponds very closely to the period of weaker monsoons in the tropics: lake-levels in Africa were low (23, 26), the pollen record from land shows drier conditions in Africa (24) and northwest India (37), and there were more extensive regions of active sand dunes in Africa, Arabia, and northwest India (36). Around glacial maximum, the mountains of the Tibetan Plateau were snow-covered and glaciated (60), as were mountains in tropical Australasia (61). The planktonic record from the Arabian Sea indicates warmer temperatures in coastal waters (reduced upwelling and reduced southwest monsoon flow) for the interval 22,000 to 12,000 B.P. (48), the isotopic record from the Bay of Bengal suggests decreased river discharge for the same period (50), and wind-blown pollen deposited in the Arabian Sea records the existence of subarid steppe conditions on the land nearby (51).

3 POSSIBLE CAUSES OF MONSOON VARIABILITY

There is no "one cause" of monsoon variability. The climate system incorporates phenomena operating on many time scales; it may possess a certain intrinsic level of variability due to so-called "internal" climate mechanisms of the atmosphere, ocean, land surface, and biota. Moreover external factors such as variability of the sun, changes of the earth's orbit, or the action of geologic processes (volcanic activity or, at long time scales, plate tectonics) may influence the climate. Some of these possibilities have been examined specifically with regard to monsoons.

3.1 Scale of Decades and Centuries

One of the causes of monsoon variability over past decades may have been changes in the level of volcanic activity. Bryson (2, 62) showed that in periods of increased volcanic activity the temperature of the Northern Hemisphere was lowered and the monsoon rains were suppressed. For example, the frequency of drought was low from about 1920 to 1950 (Fig. 10.1), a period when volcanic activity was at a minimum relative to preceding and following decades. Relationships between monsoon fluctuations and variability of the sun have also been suggested (3). Still others have studied possible physical mechanisms that could link the monsoon circulation with climatic events far removed, such as possible correlations between decade-length fluctuations in the circulation of both hemispheres (4), or between monsoon rainfall and ocean temperature or snow extent (1) or soil moisture (see Chapter 16). An idea behind these studies of links is that the climate system might retain a "memory" of past conditions that could in turn influence current events, and thereby produce internal fluctuations of climate.

Historically, the lack of quantitative and complete indices of volcanism and of solar activity has made it difficult to assess the possible role of external factors in causing monsoon changes. However, significant progress has been made in developing such indices. Extensive records of radiocarbon-dated volcanic eruptions have been assembled (63). Acidity profiles from well-dated Greenland ice cores provides records of deposition of volcanic material for the past several thousand years (64, 65). Astronomical observations, extending back to the seventeenth century or even earlier, are providing an increasingly accurate history of the variability of the sun for recent centuries (66). A history of solar activity for the past several thousand years is also being developed (67, 68). Records of carbon-14 content of tree-rings (of known calendar date) are used to estimate past levels of atmospheric carbon-14 which in turn, are related to the modulation of the incoming cosmic ray flux by solar storms.

As knowledge of these external conditions becomes more complete, it may be possible to isolate the relative importance of external forcing and internal feedback in producing fluctuations of the monsoon.

3.2 Longer Time Scales

The earth's orbit about the sun changes slowly over the millenia. More than 2000 years ago, Hipparchus, who preceded Ptolemy by some 250 years, discovered the precession of the equinoxes by comparing his observations of the longitudes for certain stars with the results of earlier observations. The speed of precession is now known to be about one degree of arc every 60 years, or one full cycle approximately every 22,000 years. Thus the solstices no longer occur in Cancer and Capricornus, as known to the ancients, but in Gemini and Sagittarius–a displacment of about 30° of arc (about one month's time) in about 2000 years. For the same reason, the classical Indian calendar date that marks the seasonal change from rain to dry has also shifted about one month (see Chapter 3).

Orbital changes can modify the climate as well as the calendar. One effect of precession is to change the season of perihelion and thereby the season of maximum and minimum solar radiation. The tilt of the earth's rotational axis also changes in an approximate 40,000-year cycle and this produces changes in the latitudinal distribution of solar radiation. Variations of the earth's orbital parameters appear to be an important factor in producing the glacial–interglacial fluctuations of the past several hundred thousand years (69–71).

Studies of the so-called "astronomical" or "Milankovitch" theory of climatic change have concentrated on the role of orbitally induced solar radiation changes at high or middle latitudes, an emphasis that was apparent in the pioneering work of Croll in the 1860s and Milankovitch (72) in the 1920s and 1930s, and in subsequent work (73–75); see Imbrie and Imbrie (71) for an historical review of the contributions of Croll, Milankovitch, and others. The possibility that solar radiation changes at low latitudes might be important for understanding climatic changes of the tropics has been examined in detail only recently. In the 1950s and 1960s, Zeuner (76) argued that precession might provide an explanation for tropical pluvials and Bernard (77) developed a theoretical framework for the role of orbital parameter variations in producing climatic changes in the tropics. Much more recently, Prell (48, 78) interpreted Indian Ocean sediment records to be the result of major monsoon changes and suggested that the intensified monsoon circulation and rains of the early Holocene were related to low-latitude solar radiation changes caused by orbital variations.

3.3 General Circulation Model Experiments

Detailed tests of the idea that the earth's orbital parameters (obliquity, precession, eccentricity) may have influenced the monsoon circulations through their effect on the seasonal cycle of solar radiation were made by Kutzbach (79) and Kutzbach and Otto-Bliesner (80). They conducted general circulation model experiments using solar radiation values for 9000 B.P. in place of modern values. As illustrated in Chapter 16, the general circulation model is a powerful tool for testing ideas about the seasonal monsoonal circulations of the present; this same tool has helped test ideas concerning the dramatic monsoon changes of the past.

At 9000 B.P., the obliquity, or tilt of the earth's rotational axis, was 24.23° (the modern value is 23.45°), the date of perihelion was July 30 (the modern value is January 3), and the eccentricity of the earth's orbit was 0.0193 (the modern value is 0.0167); these factors combined to produce increased solar radiation in the Northern Hemisphere in July and decreased radiation in January. The seasonal radiation changes were of magnitude 25–35 W/m^2, or about 7% of modern values, over a broad band of latitudes (79, 81). Although the values of solar radiation used in the general circulation model experiment applied specifically to 9000 B.P., increased July and decreased January radiation (compared to present) occurred for several thousand years before and after this time.

The model's ocean surface temperature distribution for 9000 B.P. was assumed to be the same as today based upon observational evidence and the argument that the large heat capacity of the ocean would effectively dampen the seasonal temperature response of the upper ocean to the altered seasonal radiation cycle. The 9000 B.P. ocean temperatures differed from modern conditions in coastal waters where changes of wind currents caused changes in upwelling, but such areas were probably small compared to the total area of the ocean.

The results of the experiment showed clearly that the increased solar radiation produced a stronger summer monsoon. The land surface temperature for June to August 9000 B.P. was 2–4° K higher than for the modern control experiment over much of Eurasia (Fig. 10.6*a*). In response to the increased solar radiation and the resultant higher temperature of the African–Eurasian land surface (relative to the

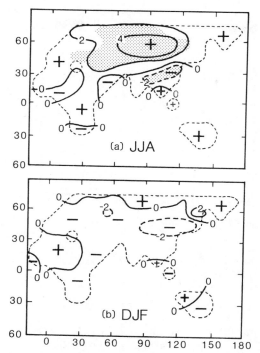

Figure 10.6. Simulated land surface temperature difference for Africa–Eurasia for June to August and December to February (9000 B.P. minus Modern), in °K, for the sector 0–180°E. Grid points where the temperature differences exceed three standard deviations are indicated with shading. Temperature difference at ocean gridpoints is set to zero. Reprinted with permission of the American Meteorological Society from Kutzbach and Otto-Bliesner, *Journal of the Atmospheric Sciences* (reference 80).

surrounding ocean), the summer monsoon low pressure intensified considerably compared to the present (Fig. 10.7). The low-level cyclonic inflow of air strengthened, including the Arabian Sea southwesterlies and the cross-equatorial flow from the Southern to the Northern Hemisphere over Africa and the western Indian Ocean. In the upper troposphere, the tropical easterly jet stream intensified. Precipitation and precipitation-minus-evaporation increased over North Africa, the Middle East, and South Asia as shown in Table 10.1.

In the January simulation, the Eurasian land mass was colder at 9000 B.P. than now (Fig. 10.6b) and the winter monsoon circulation was slightly stronger. The annual average temperature changed less than the seasonal averages because the seasonal extremes compensated each other at 9000 B.P. (Table 10.2).

One surface boundary condition was significantly different from present at 9000 B.P. The North American ice sheet still extended across Canada from west of Hudson Bay to the Atlantic coast, although it was considerably reduced in volume as compared to glacial maximum (82, 83). The Eurasian ice sheet had already melted. A model simulation that incorporated both the North American ice sheet and the

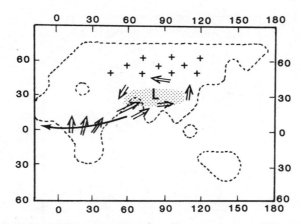

Figure 10.7. Schematic of simulated temperature, pressure, wind, and precipitation difference, 9000 B.P. minus Present, for Africa–Eurasia for July: higher temperature ($+$), intensified monsoon low (L), strong low-level wind ($\Rightarrow$), strong upper-level wind ($\rightarrow$), increased precipitation (shading). Reprinted with permission of the American Meteorological Society from Kutzbach and Otto-Bliesner, *Journal of the Atmospheric Sciences* (reference 80).

solar radiation regime for 9000 B.P. showed a pattern of June to August temperature change over Eurasia that was similar to that in Figure 10.6*a*; likewise, the intensification of the African–Asian summer monsoon circulation (compared to the modern control) was similar to the simulation without the North American ice sheet (Table 10.1).

3.4 Comparison of Model Experiments with Observations.

A detailed comparison between simulated paleoclimate and observed paleoclimate requires that a detailed climatic record be assembled and analyzed and that a high-resolution and accurate climatic model be available for simulation studies. For this reason, the experiments just described are being repeated with a general circulation model of higher spatial resolution and improved physics (84), and parallel efforts aim to provide a more detailed climatic record. The presently available model results agree with many of the paleoclimatic observations mentioned in the previous section and indicated in Figure 10.3 (84, 85). The region of increased precipitation and precipitation-minus-evaporation across North Africa–South Asia agrees broadly with the locations of high lake levels reported by Street-Perrott and others (26, 29). The magnitude of the simulated increase of precipitation, or precipitation-minus-evaporation (Table 10.1), is also consistent with the magnitude of the increases that have been inferred from paleolake and paleovegetation studies (84).

Outside of the monsoon central region, considerable evidence shows that the increased solar radiation produced a warmer growing season climate during the period 10,000–5000 B.P. In Siberia, for example, pollen and macrofossil evidence (reflecting primarily summer half-year conditions) indicate that vegetation zones

TABLE 10.1. Simulated Precipitation and Precipitation-Minus-Evaporation for the Land Region of North Africa (30°N to Equator) and the Middle East and Asia (east of 30°E and south of 40°N)

	9000 B.P.	Modern	Difference	% Change	Difference with N. A. Ice Sheet
JUNE TO AUGUST					
Precipitation	546 mm	434 mm	112 mm	(26)	63 mm
Precipitation-minus-evaporation	28 mm	−55 mm	83 mm		(15%) 72 mm
ANNUAL					
Precipitation	1548 mm	1590 mm	−42 mm	(−2.5)	
Precipitation-minus-evaporation	−154 mm	−241 mm	87 mm		

The last column refers to differences for the experiment with a North American ice sheet. From Kutzbach and Otto-Bliesner (80).

shifted northward 200–400 km around 10,000–9000 B.P., followed by cooling and southward shifts around 5000 B.P. (86).

The period of increased Northern Hemisphere summertime insolation (compared to present) extended from about 16,000–6000 B.P. This interval corresponds closely to the interval of high lake levels and increased monsoon precipitation (Fig. 10.3), except that the Late-Glacial aridity persisted in Africa until about 13,000 B.P. (29). Prior to about 13,000 B.P., the continental ice sheets were still large, sea ice was more extensive than now, and the oceans were cold; these factors, along with others

TABLE 10.2. Simulated Surface Temperature for Northern Hemisphere Land, Southern Hemisphere Land, and the Global Average of Land and Ocean for 9000 B.P. Compared to Modern Values

Space Average	9000 B.P. (°C)	Modern (°C)	Δ (K)	S.L. (%)
JUNE TO AUGUST				
Northern Hemisphere, land	25.0	23.8	1.2	0.1
Southern Hemisphere, land	2.5	1.7	0.8	
Global, land and ocean	17.8	17.5	0.3	1
ANNUAL				
Northern Hemisphere, land	13.3	13.6	−0.3	
Southern Hemisphere, land	7.5	7.2	0.3	
Global, land and ocean	15.9	15.9	0	

The difference between 9000 B.P. and the present is denoted by Δ. The significance level (S.L.) is determined from the ratio of Δ to the model standard deviation. From Kutzbach and Otto-Bliesner (80).

such as the lowered concentrations of carbon dioxide and increased aerosol content of the glacial-age atmosphere, may help explain why the strengthening of the monsoon lagged behind the increase of solar radiation (84).

Even if certain large and long period variations of climate can be attributed to orbital changes, there are many wiggles, bumps, spikes, and valleys in climatic time series that must be explained in other ways. For example, there were short but dramatic lake-level regressions around 10,000 and 7500 B.P. that briefly interrupted the several-thousand year period of high lake levels (29). Thus while radiation changes associated with orbital variations may *contribute* to an explanation of the changes of monsoon climates of the early Holocene, they could not have been the only cause. Shorter period orbital changes, volcanic or solar activity, long-term tidal fluctuations, or internal processes must play a role.

3.5 The Past 100,000 Years.

Further evidence that orbital variations are related to variations in monsoon climates comes from comparing long time series of the orbitally induced radiation changes and the climatic changes. The planktonic record of the large changes of the monsoon, as found in the Arabian Sea sediment cores, covers about 150,000 years. Within this long time series, there are several periods of strong monsoons. A variance spectrum analysis shows a tendency for strong monsoons to occur at about 22,000-year intervals (the approximate period of the precession cycle); moreover, Prell finds a significant correlation between the Arabian Sea planktonic record and the record of solar radiation (48, 78). Sediment cores from the Mediterranean Sea, mentioned earlier, have also recorded a succession of major monsoon changes and related their timing to orbital variations (53–55). The approximate 22,000-year period of climatic fluctuations is apparent in many other locations, and may be one of the fundamental pulse beats of global-scale climate (82).

3.6 And Before That.

The changing pulse of the monsoon can be seen in rain gauge, historic, and geologic records, for tens of thousands of years. But that is only a glimpse at what could be a much longer perspective. Since the start of the Cenozoic, about 65 million years ago, there has been a succession of major geologic events—the glaciation of Antarctica, mountain building in the region of the Rockies, the Alps, and the Himalayas, and just several million years ago, the establishment of sea-ice cover in the Arctic Ocean and the commencement of mid-latitude Northern Hemisphere glacial–interglacial oscillations. All of these changes could have had profound consequences for monsoon climates (21, 32, 87). At still earlier times, such as the Cretaceous, the shifted location of the continents must have produced vastly different climates (88, 89). Nevertheless, if there were land–ocean contrasts and if the earth's rotational axis were inclined to the orbital plane so as to produce seasons, then monsoonal winds would have blown across these ancient lands (whose very geography is still known only approximately). Moreover the sedimentary record (90) suggests that even these early monsoons may have waxed and waned in response to shifts of the earth's orbit about the sun.

4 THE PULSE OF THE MONSOON

As emphasized throughout this book, the timing of the monsoon is closely associated with the seasonal track of the sun across the sky—from tropic to equator to tropic, and back again. This fundamental rhythm was noticed by careful observers of sun, rain, wind, and flood thousands of years ago. The geologic records—the plankton, the pollen, the rivers, and lakes—speak to us of major monsoon changes, thereby seeming to contradict the pattern of seasonal regularity. However, if it can be established that major monsoon changes were produced by changes of the cosmic calendar itself—the season in which the earth reaches perihelion and the degree of its axial tilt—or if other external factors, such as volcanic eruptions, can be shown to have modified the seasonal radiation cycle, then even the changing pulse beat of the monsoon can be viewed as an example of nature's consistency.

ACKNOWLEDGMENTS

Research grants to the University of Wisconsin–Madison from the National Science Foundation's Climate Dynamics Program (NSF Grants ATM79-26039 and ATM81-11455) supported this work. The computations for the 9000 B.P. general circulation model simulation were made by P. Guetter (University of Wisconsin) at the National Center for Atmospheric Research (NCAR), which is sponsored by NSF, with a computing grant from the NCAR Computing Facility. M. Woodworth is thanked for preparing the manuscript.

REFERENCES

1. G. T. Walker, *Indian Meteor. Memoir*, **21**(1), 1–21 (1910).
2. R. A. Bryson, *Environ, Conserv.*, **2**, 163 (1975).
3. H. N. Bhalme and D. A. Mooley, *Mon. Wea. Rev.*, **108**, 1197 (1980).
4. E. B. Kraus, *Mon. Wea. Rev.*, **105**, 1009 (1977).
5. R. Rhode and H. Virji, *Mon. Wea. Rev.*, **104**(3), 307 (1976).
6. S. E. Nicholson, *Mon. Wea. Rev.*, **108**, 473 (1980).
7. N. Shaw, *Manual of Meteorology*, Vol. 1, Cambridge University Press, London, 1926; the quotation from Aristotle is on p. 84.
8. H. H. Lamb, *Climate: Present, Past and Future*, Vol. 2, Climatic History and the Future, Methuen, London, 1977, 835 pp.
9. Herodotus, *The Histories*, Book 2, translated into English by Aubrey de Selincourt, Penguin Books, Harmondsworth, Middlesex, 1954, p. 108.
10. O. Toussoun, "Memoire sur l'histoire du Nil," *Mem. Inst. Egypt*, **9**, (1925).
11. C. E. P. Brooks, *Climate Through the Ages*, 2nd rev. ed. reprinted by Dover, New York, 1970, p. 395.
12. H. Riehl and J. Meitin, *Science*, **206**, 1178 (1979).
13. J. Maley, "Travaux et Documents de l'ORSTOM," Office de la Recherche Scientifique et Technique Outre-Mer, Paris, No. 129 (1981).

14. F. A. Hassan, *Science*, **212**, 1142 (1981).
15. S. E. Nicholson, "Saharan Climates in Historic Times," in M. A. J. Williams and M. Faure, Eds., *The Sahara and the Nile*, A. A. Balkema, Rotterdam, 1981, 400 pp.
16. H. Flohn and S. E. Nicholson, *Palaeoecol. Africa*, **12**, 3 (1980).
17. S. Wang, Z. Zong, and C. Zhen-hua, *GeoJournal*, **5**(2), 117 (1981).
18. B. Bell, *Am. J. of Archaeology*, **75**, 1 (1971); **79**, 223 (1975).
19. R. A. Bryson, H. H. Lamb, and D. Donely, *Antiquity*, **43**, 46 pp. (1974).
20. B. Weiss, *Clim. Change*, **4**, 173 (1982).
21. F. A. Street, *Progress in Physical Geography*, **5**, 157 (1981).
22. E. Nilsson, *Geogr. Annlr.*, **13**, 249 (1931).
23. K. W. Butzer, G. L. Isaac, J. A. Richardson, and C. Washbourn-Kamau, *Science*, **175**, 1069 (1972).
24. D. A. Livingstone, *Ann. Rev. Ecol. Systematics*, **6**, 249 (1975).
25. F. A. Street and A. T. Grove, *Nature* (London), **261**, 335 (1976); see also F. A. Street-Perrott and A. T. Grove, Chapter X, in A. Hecht, Ed., *Paleoclimate: Data Analysis and Modeling*, Wiley, New York, 1985.
26. F. A. Street and A. T. Grove, *Quat. Res.*, **12**, 83 (1979).
27. F. A. Street, *Palaeoecology of Africa*, **11**, 135 (1979).
28. S. Hastenrath and J. E. Kutzbach, *Quat. Res.*, **19**, 141 (1983).
29. F. A. Street-Perrott and N. Roberts, "Fluctuations in Closed Basin Lakes as an Indicator of Past Atmospheric Circulation Patterns," in F. A. Street-Perrott, M. Beran, and R. Ratcliffe, Eds., *Variations of the Global Water Budget*, Reidel, Dordrecht, 1983, pp. 331–345.
30. J. Maley, *Nature* (London), **269**, 573 (1977).
31. P. Rognon and M. A. J. Williams, *Palaeogeogr., Palaeoclim., Palaeoecol.*, **21**, 285 (1977).
32. H. Flohn, "Possible climatic consequences of a man-made global warming," Intern. Inst. for Applied Systems Analysis, Laxenburg, Austria, 1980, 81 pp.
33. S. K. McIntosh and R. J. McIntosh, *American Scientist*, **69**, 602 (1981).
34. H. Lhote, *The Search for the Tassili Frescoes: The Story of the Prehistoric Rock-Paintings of the Sahara*, Hutchinson, London, 1959.
35. D. A. Adamson, F. Gasse, F. A. Street, and M. A. J. Williams, *Nature*, **287**, 50 (1980).
36. M. Sarnthein, *Nature*, **272**, 43 (1978).
37. G. Singh, The Indus valley culture, archaeology and physical anthropology, *Oceania*, **VI**, 177–189 (1971).
38. R. A. Bryson and A. M. Swain, *Quat. Res.*, **16**(2), 135 (1981).
39. A. M. Swain, J. E. Kutzbach, and S. Hastenrath, *Quat. Res.*, **19**, 1 (1983).
40. B. M. Fagan, *People of the Earth, An Introduction to World Prehistory*, Little, Brown, Boston, 1980, 412 pp.
41. R. A. Bryson and D. A. Baerreis, *Bull., Amer. Meteor. Soc.*, **48**, 136 (1967).
42. R. A. Bryson and T. J. Murray, *Climates of Hunger*, University of Wisconsin Press, Madison, 1977, 171 pp.
43. M. Tsukada, *Palaeogeog., Palaeoclim., Palaeoecol.*, **3**, 49 (1967).
44. Kwang-chih Chang, *The Archaeology of Ancient China*, Yale University Press, New Haven, 1968, 483 pp.
45. P. Teilhard de Chardin, *Bull. Geol. Soc. of China*, **16**, 195 (1936–1937).
46. J. G. Andersson, *Bull. of the Museum of Far Eastern Antiquities* (Stockholm), **19**, 1 (1947).

47. A. B. Pittock and M. J. Salinger, *Clim. Change*, **4**, 23 (1982).
48. W. L. Prell, "Glacial/interglacial variability of monsoonal upwelling: Western Arabian Sea," in *Evolution des Atmospheres Planetaires et Climatologie de la Terre*, Centre National d'Etudes Spatiales (France), 1978, pp. 149–156.
49. J. L. Cullen, *Palaeogeogr., Palaeoclim., Palaeoecol.*, **35**, 315 (1981).
50. J. C. Duplessy, *Nature*, **295**, 494 (1982).
51. E. Van Campos, J. C. Duplessy, and M. Rossignol-Strick, *Nature*, **296**, 56 (1982).
52. L. Pastouret, M. Chamley, G. Delibrias, J. C. Duplessy, and J. Thiede, *Oceanologica Acta*, **2**, (1978).
53. M. Rossignol-Strick and D. Duzer, *Meteor.-Forsch-Ergebnisse*, **C1**, 30 (1979).
54. M. Rossignol-Strick, W. Nesteroff, P. Olive, and C. Vergnaud-Grazzini, *Nature*, **295**, 105 (1982).
55. M. Rossignol-Strick, *Nature*, **303**, 46 (1983).
56. CLIMAP Project Members, *Science*, **191**, 1131 (1976); CLIMAP Project Members, *Geological Society of America Map and Chart Series MC-36*, Geological Society of America, Boulder, CO, (1981).
57. W. L. Gates, *Science*, **191**, 1138 (1976).
58. W. L. Gates, *J. Atmos. Sci.*, **33**, 1844 (1976).
59. S. Manabe and D. G. Hahn, *J. Geophys. Res.*, **82**, 3889 (1977).
60. G. Singh and D. P. Agrawal, *Nature*, **260**, 232 (1976).
61. P. J. Webster and N. A. Streten, *Quat. Res.*, **10**, 279 (1978).
62. R. A. Bryson and G. J. Dittberner, "A hemispheric mean surface temperature model applicable to monsoon studies," in K. Takahashi and M. M. Yoshino, Eds., *Climatic Change and Food Production*, University of Tokyo Press, International Symposium, October 4–7, 1976, Tsukuba and Tokyo, 1978, pp. 359–378.
63. R. A. Bryson and B. M. Goodman, *Science*, **207**, 1041 (1980).
64. C. U. Hammer, *Nature*, (London), **270**, 482 (1977).
65. C. U. Hammer, H. B. Clausen, and W. Dansgaard, *Nature* (London), **288**, 230 (1981).
66. J. A. Eddy, *Science*, **192**, 1189 (1976).
67. P. R. Damon, "Solar induced variations of energetic particles at one AU," in O. R. White, Ed., *The Solar Output and its Variations*, Colorado Association University Press, Boulder, 1977, pp. 429–448.
68. M. Stuiver, *Nature* (London), **286**, 868 (1980).
69. J. D. Hays, J. Imbrie, and N. J. Shackleton, *Science*, **194**, 1121 (1976).
70. J. Imbrie and J. Z. Imbrie, *Science*, **207**, 943 (1980).
71. J. Imbrie and K. P. Imbrie, *Ice Ages: Solving the Mystery*, Enslow, Short Hills, NJ, 1979, 224 pp.
72. M. Milankovitch, *K. Serb. Akad. Geogr. Spec. Publ. No. 132*, 1941, 484 pp. (English translation published in 1969 by *Israel Program for Scientific Translations*. Available from the U.S. Department of Commerce, Clearinghouse for Federal Scientific and Technical Information, Washington, D.C.)
73. M. J. Suarez and I. M. Held, *Nature* (London), **263**, 46 (1976).
74. M. J. Suarez and I. M. Held, *J. Geophys. Res.*, **84**, 4825 (1979).
75. S. H. Schneider and S. L. Thompson, *Quat. Res.*, **12**, 188 (1979).
76. F. E. Zeuner, *The Pleistocene Period—Its Climate, Chronology and Faunal Successions*, Hutchinson, London, 1959, 447 pp.
77. E. A. Bernard, "Theorie astronomique des pluviaux et interpluviaux due quaternaire African," *Acad. Roy. Sci. Outre-Mer (Brussels), Classe Sci. Tech., Memo. in 8°*, **12**(1), 1962, 232 pp.

78. W. L. Prell, "Variation of monsoonal upwelling: A response to changing solar radiation," in J. Hansen and T. Takahashi, Eds., *Climate Processes and Climate Sensitivity*, Maurice Ewing Series, 5, American Geophysical Union, Washington, D.C., 1984, 368 pp.; W. L. Prell, "Monsoonal climate of the Arabian Sea during the Late Quaternary: A response to changing solar radiation," in A. Berger, J. Imbrie, J. Hays, G. Kukla, and B. Saltzman, Eds., *Milankovitch and Climate: Understanding the Response to Astronomical Forcing*, Reidel, Dordrecht, 1984, pp. 349–366.

79. J. E. Kutzbach, *Science*, **214**, 59 (1981).

80. J. E. Kutzbach and B. L. Otto-Bliesner, *J. Atmos. Sci.*, **39**(6), 1177 (1982).

81. A. L. Berger, *Quat. Res.*, **9**, 139 (1978).

82. W. F. Ruddiman and A. McIntyre, *Science*, **212**, 617 (1981).

83. G. H. Denton and T. J. Hughes, Eds., *The Last Great Ice Sheets*, Vol. 1, Wiley, New York, 1981, 484 pp.

84. J. E. Kutzbach and P. J. Guetter, "The Sensitivity of Monsoon Climates to Orbital Parameter Changes for 9000 Years B.P.: Experiments with the NCAR GCM," in A. Berger, J. Imbrie, J. Hays, G. Kukla, and B. Saltzman, Eds., *Milankovitch and Climate: Understanding the Response to Astronomical Forcing*, Reidel, Dordrecht, 1984, pp. 801–820; see also J. E. Kutzbach and P. J. Guetter, *Annals of Glaciology*, **5**, 85 (1984), J. E. Kutzbach and F. A. Street-Perrott, *Nature*, **317**, 130 (1985), and J. E. Kutzbach and P. J. Guetter, *J. Atmos. Sci.*, **43**, in press (1986).

85. J. E. Kutzbach, "Monsoon Rains of the Late Pleistocene and Early Holocene: Patterns, Intensity, and Possible Causes of Change," in F. A. Street-Perrott, M. Beran, and R. Ratcliffe, Eds., *Variations of the Global Water Budget*, Reidel, Dordrecht, 1983, pp. 371–389.

86. N. A. Khotinskii, "Transkontinental'naia Korreliatsiia etapov istorii rastitel'nosti i klimata Severnogo Evrazii v. Golotsene," in M. I. Neishtadt, Ed., *Problemy Palinologii*, Nauka, Moscow, 1973, pp. 116–123 (English translation by G. M. Peterson, available from Center for Climatic Research, University of Wisconsin, Madison, Wisconsin).

87. T. J. Crowley, "The geological record of climatic change," *Rev. Geophys. and Space Phys.*, **21**, 828 (1983).

88. E. J. Barron, J. L. Sloan II, and C. G. A. Harrison, *Palaeogeog., Palaeoclim., Palaeoecol.*, **30**, 17 (1980).

89. E. J. Barron, S. L. Thompson, and S. H. Schneider, *Science*, **212**, 501 (1981).

90. A. G. Fischer, "Climatic oscillations in the biosphere," in M. Nitecki, Ed., *Biotic Crises in Ecological and Evolutionary Time*, Academic, New York, 1981, pp. 103–131.

11

The Variable and Interactive Monsoon

Peter J. Webster
Department of Meteorology
Pennsylvania State University
University Park, Pennsylvania

INTRODUCTION

Of all the major weather phenomena on earth, the monsoon systems of Africa, Asia, and Indonesia–Australia are the most vigorous, persistent, and energetic, with circulation features that are readily identifiable over the entire Eastern Hemisphere during all seasons. The strong seasonal cycle of the monsoon together with its phases, the summer and winter monsoons* was described in detail in Chapter 1 [see especially, Figures 1.1*a* and *b*; see also (1)]. Despite variations in the intensity of the circulations from year to year, or in the amount of rainfall at any one location, the annual monsoon cycle is most remarkable for its geographic and temporal consistency.

Given its vigor and scale, it is no surprise that the monsoon system exerts influences well beyond its immediate circulation regime and, indeed, is influenced by other circulation systems. There is evidence that such a two-way interaction occurs on time scales ranging from synoptic (weather) to interannual (climate). This multitude of influences characterizes the interactive nature of the monsoon and is the major concern of this chapter.

To comprehend the complex interaction of the monsoon with other features of the atmosphere, it is useful to identify circulation features of the atmosphere that have temporal and spatial scales comparable to the monsoon. That variations of other circulations occur simultaneously with variations of the monsoon suggests an interaction among them.

---

* It is traditional to use Northern Hemisphere seasonal chronology for the phase of the annual monsoon cycle. Rather than referring to the hemisphere in which the major precipitation is located, the terms *summer* and *winter* monsoons refer merely to the particular phase of the monsoon during the Northern Hemisphere *summer* and *winter*, respectively.

Some relationships are explainable in terms of cause and effect. For example, it will be shown that the latent heat release in the summer hemisphere influences the extratropical circulation by strongly affecting the jet streams of the winter hemisphere. Also some relationship exists between the variation of the sea surface temperature in the eastern Pacific Ocean and the vigor of the monsoon: the strongest correlation is between the sea surface temperature and rainfall in the winter monsoon over Indonesia.

Other aspects of the monsoon system do not have a clear cause. Such is the case for variations on subseasonal time scales, for example, the rather strong 20- to 50-day variability in the monsoon flow. On these time scales considerable modulation in the intensity, timing, and location of the precipitation maxima occurs. Other major features of the monsoon, the *active* and *break* periods for example, are also variable in their form and timing. There is considerable debate as to whether these modulations of the monsoon occur because of external influences or because they are intrinsic properties of the monsoon circulation itself. Either way, they are intriguing and indicate the real complexity of the *variable and interactive monsoon*.

This chapter covers a mix of diagnostic studies of the observed monsoon and simple theoretical and physical interpretations of these observations. The observations will be used to establish the character of the monsoon, its variability, and its interactions and hypotheses will be posed to explain them. Theoretical considerations of the interactions will be used to examine their physical nature and to test the validity of the hypotheses.

1 OBSERVED RELATIONSHIPS AND STRUCTURES

For more than two decades the atmosphere has been surveyed from space. Meteorological satellites have provided both pictures of the evolving cloud structures and measurements of temperatures throughout the depth of the troposphere. For the extratropics, a knowledge of the temperature structure allows an approximation of the wind field because of a strong geostrophic balance (see Chapter 9). However, at low latitudes the geostrophic balance does not hold and it is difficult to make a direct interpretation of the wind field from temperature measurements. Since the late 1970s, the entire tropics have been observed by as many as five geostationary satellites whose orbital characteristics are such that each remains over the same geographic location along the equator. Because they provide continuous surveillance of the tropical atmosphere, they allow the tracking of discrete cloud features and thus, a fairly good estimate of the wind field, at least at cloud level. In addition, both geostationary and polar orbiting satellites measure the infrared temperature of the atmosphere. We will see later that infrared data provide important information about the state of the atmosphere.

In addition to the satellite observations, special experiments have been conducted aimed at obtaining a better set of observations of the monsoon regions. These were the International Indian Ocean Expedition (1965–1966), the Winter Monsoon Experiment (Winter MONEX, 1978–1979) and the Summer Monsoon Experiment

(Summer MONEX, 1979). The last two experiments were subprograms of the Global Weather Experiment (1978–1979).

With more than two decades of satellite data, the special data collected from the monsoon experiments and the meteorological data gathered around the world on a regular basis, a large data archive is available for the study of monsoons. These archives form the basis for the observational studies discussed in the following sections.

1.1 The Principal Circulation Components

To facilitate the investigation of monsoon structure, it is helpful to separate the rather complicated circulation of the tropical regions into distinct components; one which is zonal (i.e., east–west) and one which is primarily meridional (north–south). Although artificial, this dissection allows a considerable conceptual simplification. Furthermore phenomenologically there is a separation that can be seen in the temperature and wind fields that will be discussed later.

Satellite data, especially infrared data, are useful in the tropics because they provide a measure of cloud-top temperatures. Since cloud-top temperatures are close to the environmental temperatures at the same level, the infrared data implicitly give a measure of cloud height if the distribution of temperature with height is known. In the tropics the temperature varies only slightly from the climatological average value. Thus once the temperature of the cloud top is obtained we know fairly accurately the height of the cloud. Cold infrared temperatures sensed by the satellite indicate regions of very deep clouds. Again from climatology, we know that in the tropics regions of high clouds are always regions of deep convection and precipitation. Thus cold infrared areas of the tropics are the zones of convection and precipitation.

Figure 11.1 shows the distribution of the mean seasonal infrared radiating temperatures (or infrared irradiance) measured from satellite for the Northern Hemisphere winter (December, January, and February; DJF) and summer (June, July, and August; JJA). The radiating temperatures (a measure of convection and precipitation, as discussed above) are averaged for each season from the five years of data archived during the period 1974–1978.

The most significant areas of convection are in the Asian–Indonesian and Australasian regions. The greatest convection and precipitation occur in JJA between Indonesia and south and southeast Asia. During DJF, the area of maximum convection slips southward into the Southern Hemisphere although a large area remains straddling the equator in the western Pacific Ocean. It appears that on the average convection and precipitation always occur in the near-equatorial regions of the western Pacific Ocean.

There are smaller, but still very important, areas of cold infrared temperature observed over equatorial Africa and South America. The African and American convective zones lie over the rain forests of central Africa and the Amazon Valley. In many ways, the rain forests are similar to warm ocean regions; that is, tropical oceans and rain forests are both very warm and excellent sources of water vapor.

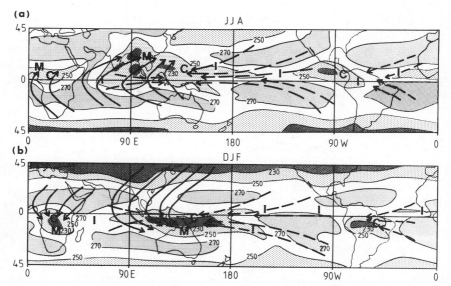

Figure 11.1. Distribution of the five-year mean (1974–1978) infrared radiating temperature (units °K) as sensed by geostationary satellite for June, July, and August (JJA) and December, January, and February (DJF) (from reference 2). Cold temperatures (250°K and less) are stippled and represent regions of persistent and intense convection and precipitation. The shaded areas indicate warm infrared temperatures and thus an absence of convective clouds. Regions with temperatures 270°K and warmer are virtually cloud free. The solid and dashed lines show the low-level monsoon flow and near-equatorial flow, respectively.

The major convective regions, identified above, are marked by the letter C in Figure 11.1.

More or less continuous bands of cloudiness connect the three equatorial convective zones (the Cs). Designated by an I, these bands represent the Intertropical Convergence Zone or, as it is sometimes termed, the near-equatorial trough.

It is no surprise that the Indonesian and western Pacific regions are the most persistent convective regions. These are areas with the warmest sea surface temperatures on earth. As discussed in Chapter 1, this means that the air near the surface in these areas is very moist. Since the ocean provides an infinite supply of water vapor, the warm surface air is very close to saturation and potentially unstable.

Some areas have a dramatic annual variability. Marked M in Figure 11.1, the zones correspond to the *monsoon regions* of Africa and Asia–Australia. A more detailed set of diagrams of the mean surface flow for both DJF and JJA may be seen in Figure 1.1. Chapter 1 also suggests that the physical mechanisms that dominate the monsoon circulation are somewhat different from the warm sea surface or the moist rain forests that control the near-equatorial convection. In the monsoon regions, it is the temperature gradient between the land and the sea produced by solar heating and the associated pressure gradient that drives the monsoons. Since the Asian, African, and Australian continents extend well into the subtropics, the heating of the land is linked strongly to the migration of the sun. The convection that results from the moist maritime air being drawn toward and forced to rise over the heated continent has a strong seasonal signature.

Superimposed on Figure 11.1 are arrows that represent a rather simplified view of the low-level flow in the tropics. The flow is divided into the two distinct circulation forms that are associated with the Cs and the Ms of the figure. For convenience we will refer to these forms, respectively, as the *near-equatorial* flow (dashed lines) and the *monsoonal* flow (solid lines).

By comparing the two panels of Figure 11.1, we see that the near-equatorial flow component is fairly *symmetric* about the equator. That is, the flow is oriented east–west with little or no north–south component. Indeed, the circulations are called "east–west" cells (3-6), or the *Walker Circulation* after the pioneer meteorologist, Sir Gilbert Walker, who first noted their existence. Note that the near-equatorial convection or Walker Circulations appear to have only a small annual variation as they are locked to the sea surface temperature distribution which normally varies little in position.

In sharp contrast, the monsoonal flow exhibits a distinctly seasonal character with a strong cross-equatorial flow that emanates from the winter hemisphere and moves toward the convective regions of the heated continents (i.e., the Ms of Fig. 11.1). At the equator, the monsoon flow is almost north–south in direction.

We can examine the characteristics of the two components of the tropical circulation more quantitatively by looking simultaneously at the infrared satellite and wind data. If we average the zonal wind components around the globe for each season, a predominance of easterlies at low latitudes is found as well as strong westerly winds at higher latitudes. The mean distributions of the zonal wind may be seen in Figure 11.2. For each season, we then subtract these mean zonal winds from the total wind to see how much the zonal wind varies from longitude to longitude. The same procedure is also followed to obtain the variation of the meridional wind component.

Figures 11.3a and b show the residuals of the wind components (the actual winds minus the zonal average) along the equator for DJF and JJA. Here, both the east–west (zonal) component (u) and the north–south (meridional) component (v) are displayed for the lower troposphere (1000 mb), the middle troposphere (500 mb) and the upper troposphere (200 mb). Because the meridional winds are much smaller than the zonal winds, the v scale has been expanded. To facilitate the comparison of winds and satellite data, similar cross sections of the infrared radiating temperature are shown in the lower panels. Here we have inverted the temperature scale so that the cold temperatures (representing regions of convection) appear as relative maxima.

From the residual wind structure (referred to below as simply *winds*) shown in Figure 11.3, we can make the following observations:

1. Strong easterlies and westerlies exist along the equator in the lower and upper troposphere (i.e., 1000 and 200 mb) but not in the middle troposphere (500 mb) where there are relatively weak winds.
2. The equatorial easterlies and westerlies have little seasonal variation.

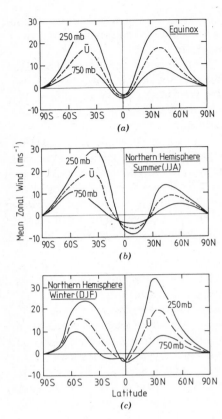

Figure 11.2. Mean seasonal zonally averaged zonal (east–west) wind component (m/sec) for (*a*) the equinox, (*b*) Northern Hemisphere summer (JJA), and (*c*) Northern Hemisphere winter (DJF), based on 20 years of data. Upper (250 mb) and lower (750 mb) tropospheric winds are shown. The dashed line indicates the tropospheric vertical average. Westerly winds are positive, easterly winds are negative.

3. The zonal winds in the lower and upper troposphere are nearly completely out of phase; for example, in both seasons the Eastern Hemisphere is dominated by upper-level easterlies whereas the low-level flow is basically westerly. In the Western Hemisphere the situation is reversed.

4. The meridional winds along the equator have a strong seasonal variation.

5. The meridional winds, like the zonal winds, are out of phase in the vertical. For example, in JJA, (Fig. 11.3*b*, right panel) southerlies lie over northerlies near 90°E. The 500-mb meridional winds are also much weaker than those at 1000 and 200 mb.

6. The *u* and *v* components at all levels, heights, and seasons do not appear to be correlated with each other.

Viewing the residual wind structure simultaneously with the infrared data shown in the lower panels of Figure 11.3, it is apparent that:

1. The zonal wind component appears to be correlated with the convective heating (i.e., cold infrared temperatures) along the equator. On either side

of the convective heating, winds converge (negative slope of u around the heating) in the lower troposphere and diverge (positive slope of u around the heating) in the upper troposphere. Such deep flow probably explains the reason for the weak winds at 500 mb. Inflow and outflow occur in the lower and upper troposphere whereas the middle troposphere is a connecting zone of maximum vertical motion but little horizontal motion. This location is sometimes referred to as the level of nondivergence.

2. The meridional wind component along the equator does not appear to be associated with the near-equatorial convection but rather with the convective heating in the subtropics of the summer hemisphere. For example, in JJA, (Fig. 11.3b) the coldest infrared temperatures, and therefore the strongest convection, exists at 15°N. The low-level cross-equatorial flow is toward the convective areas, or areas of latent heating but away from them and toward the winter hemisphere in the upper troposphere.

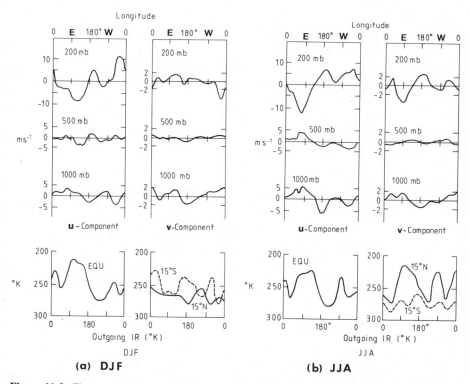

Figure 11.3. The seasonal mean east–west (u) and north–south (v) velocity components in the 5°N–5°S latitude belt along the equator for (a) DJF and (b) JJA. The zonal average wind has been removed. The lower graphs show the infrared radiating temperatures (scale inverted) determined from satellite along the equator and at 15°N and 15°S.

From the observations described above we infer the following regarding the large-scale near-equatorial and the monsoonal flow:

1. The near-equatorial flow may be thought of as a very deep circulation filling the entire troposphere with low-level air flowing toward the convective regions in an almost zonal direction, rising motion within the convective region and upper tropospheric flow, also almost zonal, away from the convection. A schematic view of the near-equatorial flow along the equator is shown in Figure 11.4a. During an El Niño episode (see Section 1.2.1) a major change in the configuration occurs as shown in Figure 11.4b.

2. The monsoon flow may be envisioned as a strong *local* Hadley circulation (see Chapter 1) with low-level flow out of the cool winter hemisphere across the equator and into the monsoon convection area of the summer hemisphere. Upper-level flow returns to the winter hemisphere. (Both the summer and

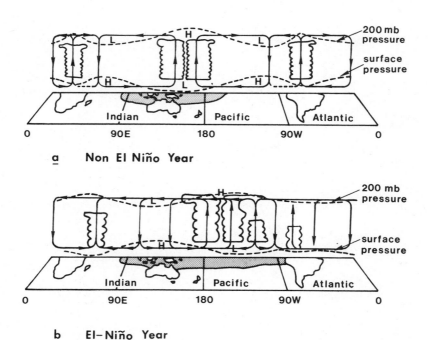

Figure 11.4. Schematic view of the near-equatorial circulation or Walker Circulation described in the text for (*a*) the non-El Niño years when the sea surface temperature (SST) is maximum in the western Pacific and (*b*) for El Niño years when the SST maximum is much further east. Note the correspondence of the position of the convective cell to the SST maximum. Dashed lines represent pressure variations at the surface and in the upper troposphere, and the stippled regions represent areas where the SSTs are greater than 27°C (from J. Zillman, private communication).

winter monsoon circulations are presented schematically in Fig. 11.9 and will be discussed subsequently.)

3. In a fundamental sense, the two components of the low-latitude flow can be thought of as large-scale, deep circulations or overturnings that are roughly at right angles to each other. The near-equatorial flow, however, has little annual variation, whereas the meridional or monsoon flow changes polarity with the annual march of the sun.

4. The monsoon flow is interhemispheric.

1.2 Evidence of Interaction and Variability

In the previous discussion, we have treated the monsoon and near-equatorial flows independently. However, this was for convenience and we now consider how these two very different circulation features of the atmosphere interact.

1.2.1 Relationships between Monsoons and Near-Equatorial Circulations. The earliest search for connections between the monsoons and other global phenomena was made by Sir Gilbert Walker earlier this century. His pioneering work is summarized in a number of articles (7, 8) and treated in its historical context in Chapter 8. Using data sets collected over the entire globe, Walker showed that the behavior of many atmospheric variables over the Indian and Pacific Oceans and beyond, is related. Walker constructed an index (the Southern Oscillation Index or SOI) which represents the degree of association between the two regions. The index was a rather complicated composite of a large number of climatological variables. However, it was shown later (8) that Walker's index could be simplified and represented by the difference in surface pressure of Darwin (12.4°S, 130.9°E) and Tahiti (17.5°S, 149.6°W). When the surface pressure is anomalously high (low) in the monsoon regions of Asia–Australasia, it is low (high) in the eastern Pacific Ocean. This simpler index is sometimes referred to as the *Troup Index* (or TI) after the Australian meteorologist and climatologist Sandy Troup and is now commonly used to represent the rather unwieldy SOI.

Troup's contribution (8) goes far beyond merely simplifying the SOI to a surface pressure difference between two stations. The difference underlies a very basic physical property of oscillation. Troup showed that as the SOI or TI swings back and forth on an interannual time scale, a transfer of mass takes place between the monsoon regions and the Pacific Ocean. In fact, the large-scale circulation that connects these remote geographic regions and is responsible for the aperiodic movement of mass is the near-equatorial flow or the Walker Circulation. Figure 11.4 shows the Walker Circulation for the extreme values of the TI; low pressure over Indonesia, high pressure in the eastern Pacific (Fig. 11.4a) and relatively high pressure over Indonesia and low pressure in the eastern Pacific (Fig. 11.4b). Note that the convective regions follow the low pressure, which in turn, corresponds to the region of maximum sea surface temperature (stippled zones). Times of negative TI correspond to the El Niño periods in the equatorial and eastern Pacific Ocean.

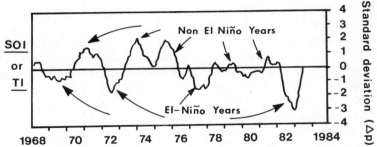

Figure 11.5. The difference between the standardized sea level pressure anomalies at Tahiti and Darwin. This is the simplified Southern Oscillation Index (SOI) or the Troup Index (TI). The surface pressure at both locations has been standardized by removing the monthly mean values (see Table 11.1) and dividing the monthly anomalies by the standard deviation for that particular month's pressure at each location. To interpret the figure, refer to Table 11.1. A 3.5 standard deviation drop in pressure difference in 1982–1983 represents a pressure difference between Tahiti and Darwin of about 4 mb.

A time series of the TI* from 1968 to 1983 is shown in Figure 11.5 (see also Table 11.1). The index progresses through a number of distinct variations with negative extrema (greater than one standard deviation) occurring in 1968–1969, 1972, 1977, and 1982–1983 when the equatorial circulation was similar to that displayed in Figure 11.4*b*. During large positive excursions of the TI, during 1970–1971, 1973–1974, and 1975–1976, the circulation was similar to that shown in Figure 11.4*a*.

A time series by itself, even one that appears to have a clean signal such as shown in Figure 11.5, is meaningless unless the spatial extent of the relationship is known. Figure 11.6 shows the distribution of the correlation of the surface pressure at Darwin with surface pressures around the globe (8). The domain of significant correlations is extremely large and indicates a grand scale of coherent surface pressure variation connected to the TI shown in Figure 11.5.

Troup's correlations essentially matched those described earlier and rather succinctly by Walker as:

> We can best sum up the situation by saying that there is a swaying of pressure on a big scale backwards and forwards between the Pacific Ocean and the Indian Ocean . . . (10)

---

* The TI is a standardized surface pressure anomaly difference between Darwin and Tahiti which was calculated as follows. First the long-term average surface pressure for Darwin and Tahiti was calculated for each month (see Table 11.1). The anomaly for a given month was found by subtracting the long-term monthly mean surface pressure from the year-by-year monthly mean surface pressures. Then the monthly anomalies of Darwin and Tahiti were subtracted from each other. This difference was then standardized by dividing it by the standard deviation of pressure difference for the months listed in Table 11.1. The TI values in the time series show the difference in terms of the standard deviation which allows the significance of the index to be emphasized and its importance gauged relative to statistical noise in the data (9).

TABLE 11.1. The Monthly Mean Tahiti and Darwin Surface Pressure (mb) and the Tahiti–Darwin Surface Pressure Difference Over the 40-Year Period 1941–1980.[a]

| | Surface Pressure (mb) | | Surface Pressure Difference (mb) (Tahiti − Darwin) |
	Darwin (12.4°S, 130.9°E)	Tahiti (17.5°S, 149.6°W)	
January	1006.4 (1.2)	1010.8 (1.2)	4.4
February	1006.3 (1.3)	1011.1 (1.2)	4.8
March	1007.4 (1.0)	1011.7 (1.0)	4.3
April	1009.5 (0.9)	1011.8 (0.7)	2.3
May	1010.9 (1.0)	1012.5 (0.6)	1.6
June	1012.4 (0.8)	1013.7 (0.7)	1.3
July	1013.1 (1.0)	1013.9 (1.0)	0.8
August	1012.7 (0.8)	1014.6 (0.9)	1.9
September	1011.9 (0.9)	1014.4 (1.0)	2.5
October	1010.6 (1.0)	1013.7 (0.9)	3.1
November	1008.7 (0.9)	1011.8 (0.8)	3.1
December	1007.2 (1.2)	1011.0 (1.1)	3.8
DJF	1006.7 (1.0)	1010.9 (0.9)	4.2
MAM	1009.3 (0.7)	1012.0 (0.6)	2.7
JJA	1012.7 (0.8)	1014.1 (0.7)	1.4
SON	1010.4 (0.8)	1013.3 (0.7)	2.9

[a] Values in parentheses are the standard deviations of the surface pressure (9).

Walker noted that relationships also exist between other geographically remote regions:

> There are *swayings* on a much smaller scale between the Azores and Iceland and between the areas of high and low pressure in the North Pacific Ocean . . . (10)

and that all appeared to be related:

> There is a marked tendency for the highs of the last two *swayings* to be accentuated when the pressure in the Pacific is raised and that in the Indian Ocean is lowered . . . (10)

Thus in his search for statistical precursors of the state of the Indian summer monsoon, Walker had uncovered a global system of interaction involving the near-equatorial and extratropical circulations as well as the monsoons.

Despite his discoveries, given his objectives in compiling the SOI, Walker's exercise was only marginally successful. The reason for the weakness in the predictive value of the index for the Indian monsoon may be guessed from Figure 11.7 which

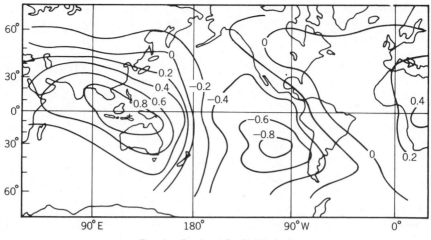

Figure 11.6. The correlation coefficient distribution of surface pressure variations relative to Darwin (from reference 8). Pattern indicates that when the pressure is anomalously high over Indonesia, it is low over the central and southeast Pacific Ocean, and vice versa. This seasaw of surface pressure defines the Southern Oscillation. It also shows why the Troup Index is a keen measure of the Southern Oscillation.

shows the departures from the normal surface pressure for extreme values of the TI, here taken as one standard deviation above the long-term average (8)*.

Figure 11.7 shows a considerably different pressure pattern for each season during the extreme phases of the TI. In JJA (Fig. 11.7a) pressure departures over Indonesia and India are relatively small and both regions lie very close to the zero pressure departure line. Maximum pressure variations occur in the winter hemisphere over Australia and the southeastern Pacific Ocean. Although small in magnitude, the 1.6-mb pressure difference between Indonesia and the central Pacific Ocean will drive an easterly surface wind anomaly of about 5 m/sec which is about the same magnitude as the climatological surface wind! In DJF, the maximum changes in pressure associated with the anomalous SOI occur in the precipitating region of the winter monsoon over North Australia and Indonesia.

Figure 11.7 shows that the rainfall anomalies associated with the extreme values of the TI (i.e., the hatched and stippled regions) are not located over the Indian subcontinent in summer. Thus, by implication, variations in the Indian rainfall which do occur may well be related to influences other than the variations of sea surface temperature in the eastern Pacific Ocean, the key correlate with the TI as we had noted earlier (Figs. 11.4 and 11.5). During the winter (Fig. 11.7b) regions of rainfall abundance and deficit appear over the winter monsoon region in Indo-

---

* Values of the surface pressure for extreme values of negative SOI or TI have the same distribution shown in Figure 11.7, but with reversed signs. Correspondingly, we would expect the regions of enhanced and deficient rainfall to be reversed as well.

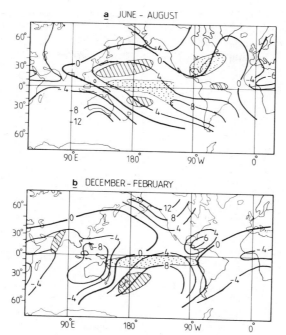

Figure 11.7. The seasonal departures of surface pressure for (*a*) JJA and (*b*) DJF (tenths of mb) from normal for extreme positive values of the Southern Oscillation Index. Pressure derivations for equally anomalous negative values of the SOI are identical except for a sign reversal (from reference 8). Areas of maximum (significant) correlation of the SOI with rainfall enhancement and deficiency are hatched and stippled, respectively (from reference 11).

nesia and Australia. Table 11.2 shows correlations between rainfall and the SOI (11). Generally, the correlations are weaker for the Indian monsoon during summer than for the winter monsoon, as noted previously.

The correlations between the index and the monsoon precipitation are at zero lag, that is, variations in the SOI or TI are correlated with *simultaneous* variations in monsoon precipitation. Of course, Walker was interested in forecasting monsoon precipitation for future seasons; that is, if the SOI is known for a previous month or season, what can be said about the monsoon precipitation in the following June, July, and August? Table 11.2 also lists these *lag correlations* between the TI of some previous season and the precipitation at a number of monsoon stations during either JJA or DJF (11). The predictive skill of the index increases somewhat and, although still small and only accounting for a relatively small part of the total variability of the rainfall, the trend is encouraging.

In summary, there is some evidence that a relationship exists on the very long time scale between the TI and monsoon precipitation. As we noted, the TI is intimately related to the Walker Circulation, and the precipitation in the monsoon regions to the monsoon circulation. Therefore one can infer that there is some inter-

TABLE 11.2. Relationship Between Seasonal Rainfall and the Preceding Seasonal Southern Oscillation Index (referred to as the Troup Index) and the Corresponding Seasonal Index for Selected Stations, Luanda (8°S, 13°E), Allahabad (26°N, 81°E), Madras (13°N, 80°E), and Darwin (13°S, 131°E).[a]

| | DJF Rainfall (mm) | Correlation of DJF Rainfall with SOI of: | | | |
| | | Previous | | | Current |
		MAM	JJA	SON	DJF
Allahabad	35	−.04	−.17	−.20	−.22
Madras	171	−.08	−.08	−.06	−.08
Darwin	927	+.31	+.39	+.32	+.23
Luanda	95	−.01	−.18	−.13	−.21

| | JJA Rainfall (mm) | Correlation of JJA Rainfall with SOI of: | | |
| | | Previous | | Current |
		DJF	MAM	JJA
Allahabad	696	+.34	+.39	+.41
Madras	263	+.19	+.33	+.35
Darwin	7	+.14	−.01	−.05
Luanda	1	+.19	+.22	+.19

[a] Underlined values significant at greater than 5% level (11).

action (albeit weak) between the near-equatorial flow and the monsoon circulation. Additionally, either other circulation features influence the monsoons or they possess internal, self-instigated modulations that account for the fairly large unexplained variance.

1.2.2 Influence of Monsoons on Extratropical Circulations.

A number of very specific influences appear to exist between the equatorial regions and the higher latitudes. Figure 11.7 indicates that the winter hemisphere is most affected during times of anomalous tropical conditions. A number of studies (12–16) support this contention from both observational and theoretical viewpoints.

But even when there are no anomalous conditions in the tropics (i.e., during normal, non-El Niño periods), there is considerable influence by low-latitude heating on the extratropics, again mainly in the winter hemisphere. Specifically, observations suggest that the monsoon heating affects the location and magnitude of the winter jet streams of the middle latitudes. This is an extremely important interaction because the extratropical jet streams are the producers of weather activity for the winter middle latitudes. Conceivably then, any variations in the monsoon may create changes in the source regions of weather in the extratropics.

Other influences also exist. In conjunction with the near-equatorial heating, monsoon heating produces distinct regions of westerlies and easterlies along the equator as seen in Figure 11.3. Regions of westerlies appear to allow extratropical waves to cross the equator (17–19) producing communication between the hemispheres. If the strength or location of the westerlies were to change, propagation may be blocked or modified.

In the following paragraphs we will discuss some of these interactions in more detail.

The Monsoon–Winter Jet Stream Connection. In Chapter 1 (see Section 4.3 and Figs. 1.9 and 1.10), we noted that the action of the earth's rotation causes strong westerlies in the winter hemisphere by forcing the upper-level air returning from the summer hemisphere to flow eastward. This may be seen in more detail in Figure 11.8 which shows the 200-mb meridional and zonal winds in the longitude–latitude plane for both JJA and DJF. The major convective zones (stippled) have been transcribed from Figure 11.1. The black areas show the outlines of the major orographic features: the Alps, the Rockies, the Himalayas, and the Andes.

From Figure 11.8 one's first impression is that the zonal wind maxima of the extratropics are all downstream of the orographic features. Specifically during DJF, middle latitude wind maxima exist downwind of the three Northern Hemisphere mountain ranges. This observation has led many to theorize that it is the perturbing effect of the mountains *alone* which produces the winter jet streams (20, 21). However, a further and equally important fact emerges from this figure: *All winter jet streams are poleward of either the near-equatorial or monsoonal heating centers* (22, 23). We also note that, in the JJA case, no corresponding orographic feature accounts for the Australian winter jet stream. On the contrary, this jet appears connected to a strong cross-equatorial flow which emanates from the highly convective monsoon region of South Asia as shown by the strong 200-mb mean meridional velocity component maximum. In fact, *all* the major winter jet streams in Figure 11.8 are on the poleward side of substantial meridional flows that originate in either the monsoon or the near-equatorial convection.

Figure 11.8 suggests that the large-scale, low-latitude heating, in conjunction with the influence of the orographic barriers, determines the location and the magnitude of winter hemisphere jet streams. Because both the areas of heating and the mountains are finite in lateral extent, so are the responses they produce in the extratropics; that is, the resultant jet streams have a finite longitudinal scale.

Figure 11.9 shows height and latitude cross sections of circulation through the most active convective regions of the summer and winter monsoon regions, through 90°E in JJA (Fig. 11.9a) and 130°E in DJF (Fig. 11.9b). In the bottom panel (Fig. 11.9c) are representations of the relative magnitudes of the various diabatic heating components as functions of latitude.

In DJF, the maximum precipitation (and so latent heating) occurs between the equator and about 15°S and coincides with the cloudy region to the south of the equator. Weak net radiational heating extends through the summer hemisphere while strong cooling, caused by a much diminished solar radiation dominates in the winter

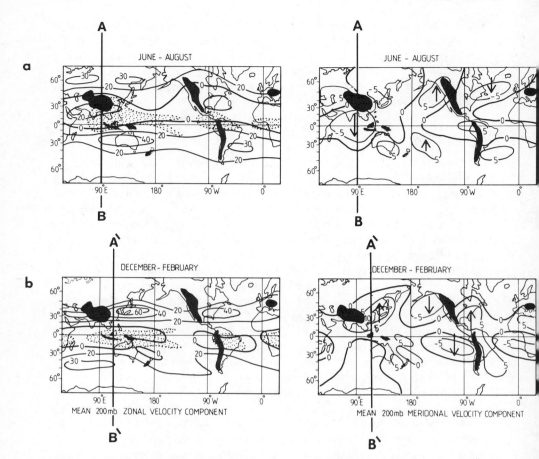

Figure 11.8. Latitude—longitude distribution of the mean seasonal zonal velocity component $\bar{u}$ and meridional velocity component $\bar{v}$ (in m/sec) for (*a*) JJA and (*b*) DJF. In the zonal wind plots the darkened areas denote the major mountain ranges and the stippled zones show the regions of maximum mean convection (from Fig. 11.1, the area within the 250°K IR isotherm). The arrows in the meridional velocity component diagram indicate the direction of the wind component.

hemisphere (see Chapter 1, Fig. 1.3). The sum of these two produces a strong net heating gradient between approximately 10°S and the winter hemisphere extratropics (Fig. 11.9*c*). Rising air associated with the monsoon convection is forced at high levels toward the equator by the resulting pressure gradient force. The flow is first turned westward by the Coriolis force to produce a weak easterly jet (the E in Fig. 11.9*b*) and then eastward after crossing the equator, producing the winter hemisphere westerly jet (J).

During JJA (Fig. 11.9*a*), the situation is almost reversed except that the major precipitating region is located further poleward than during the winter monsoon

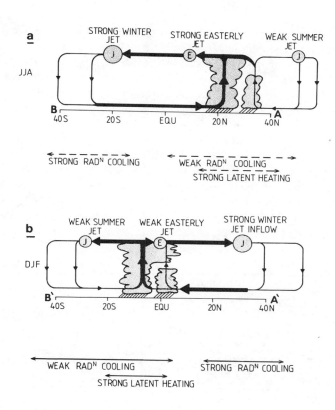

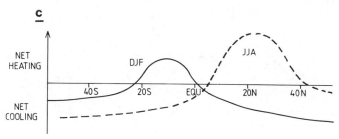

Figure 11.9. Schematic representation of the latitudinal structure of the local meridional circulations through (*a*) 90°E (for JJA) and (*b*) 130°E (for DJF). (See Fig. 11.8 for geographic reference as marked by arrows of A, B and A′, B′ cross sections.) Ⓙ represents the locations of the extratropical westerly jet streams. Ⓔ indicates the near-equatorial easterly jet streams. (*c*) shows the total heating averaged through the column (radiational and latent heating) for each season (solid line DJF, dashed line JJA). Maximum cooling occurs in the winter hemisphere and largest heating in the monsoon regions. Note that the strongest flow in the upper troposphere in (*a*) and (*b*) corresponds to the largest total heating gradient in (*c*).

because of the differences in the land–sea distribution between the two hemispheres. Thus the strongest heating in JJA exists further from the equator than in DJF. The result is a very strong easterly jet stream (E) located at about 5°N in contrast to the much weaker easterly jet stream in DJF. Again after the air flows across the equator, a westerly jet stream is produced. This is the jet stream that resides over Australia during the Southern Hemisphere winter (Fig. 11.8).

Easterly Barriers and Westerly Ducts. Figure 11.10 shows the upper tropospheric (200 mb) structure of the equatorial belt between 48°N and 48°S for JJA and DJF averaged over 11 years (19). The solid lines indicate the strength of the zonal wind component. Winds stronger than 30 m/sec (the extratropical jet streams) or less than 0 m/sec (easterlies) are shaded. The zonal wind structure is essentially the same as in Figure 11.8. Superimposed on the mean wind structure is a measure of the kinetic energy (dashed lines) of shorter time scale weather events and atmospheric

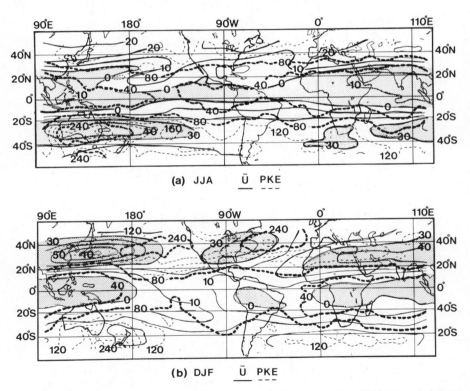

Figure 11.10. Longitude–latitude distribution at 200 mb of (*a*) JJA and (*b*) DJF mean zonal (east–west) wind component $\overline{U}$ (m/sec) plotted as solid lines (from reference 19). Regions with winds greater than 30 m/sec and less than 0 m/sec (easterlies) are shaded. Dashed lines show the perturbation kinetic energy, PKE. Large values of PKE are found in regions of high-frequency weather events and waves. Note the maxima associated with the winter hemisphere jet streams and the equatorial westerlies. Regions of easterlies, on the other hand, correspond to absolute minimum PKE.

waves. This is the perturbation kinetic energy or PKE (19). Regions of maximum PKE indicate regions of maximum synoptic or weather-scale activity.

Two major regimes of maximum PKE are apparent. In the extratropics, especially in winter, very large PKE values are over and downstream of the jet streams. These maxima are associated with the extratropical disturbances emanating from the jet streams (24, 25). Recalling the relationship between the location of the jet streams and the monsoon convection (Fig. 11.8), the colocation of the jet streams and the PKE maxima clearly suggests a connection between the monsoons and extratropical weather, at least in the winter hemisphere.

A second PKE maximum (at least relative to the rest of the tropics) exists in the region of the equatorial westerlies of the upper troposphere. Regions with easterly winds, on the other hand, have less PKE, in fact the lowest globally.

To test the visual correlation between PKE and the zonal wind in low latitudes suggested by Figure 11.10, the PKE was plotted with the corresponding time mean zonal velocity component, point by point, in the latitude–longitude plane between 5°N and 5°S. These relationships, shown in the scatter diagrams of Figure 11.11, suggest that the stronger the westerlies the greater the PKE. The same is not true for easterly winds where the values of PKE lie between 20 and 40 m^2/sec^2, irrespective of the magnitude of the easterlies.

For most weather systems in the tropics, the winds are very light. For example, easterly waves and weather disturbances have anomalous wind speeds of about 5 to 10 m/sec which corresponds to PKE values in the range noted in the tropical easterlies. Furthermore, a comparison of the location of the mean convection (Fig. 11.1) and the location of the equatorial maximum in PKE (Fig. 11.10) shows little relationship. Consequently, we can assume that the higher values of PKE in the equatorial upper tropospheric westerlies do not reflect convective disturbances and must be accounted for in other ways.

For the moment, assume that the high PKE values along the equator result from disturbances originating in the extratropics and either propagating into the deep tropics or through the tropics into the other hemisphere. This hypothesis suggests that the upper-level westerlies act as *ducts* or *channels*, through which wave propagation can occur, whereas the easterlies act as *barriers* to wave propagation. Furthermore, since the monsoonal and near-equatorial heating influence the strength and position of the upper-level winds, it follows that they also affect the propagation of weather systems from one hemisphere to the other.

One way to test the validity of the relationship between the PKE and the zonal velocity is to see if it holds when the large-scale wind field near the equator changes. Large changes occur during El Niño periods (see Fig. 11.4) when the region of equatorial westerlies over the eastern Pacific Ocean become weaker or even change sign and become easterlies. Figure 11.12 shows time sections of the 200-mb zonal wind field (Fig. 11.12*a*) and the corresponding section of PKE (Fig. 11.12*b*). The westerly winds and PKE values greater than 40 m^2/sec^2 are shaded. From Figure 11.5 we note that 1968–1969, 1972–1973, and 1977 were El Niño (marked by X in Fig. 11.12) or negative TI years and 1970–1971, 1973–1974, and 1975–1976 were non-El Niño (Y) or strongly positive TI years. By comparing the two panels

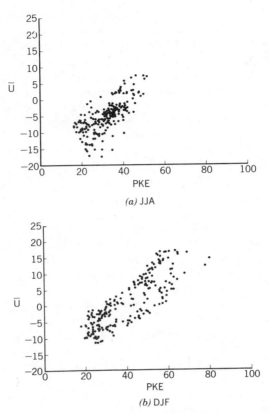

Figure 11.11. Scatter diagrams of perturbation kinetic energy (PKE) and the zonal wind component in the equatorial channel between 5°N and 5°S at 200 mb for (*a*) JJA and (*b*) DJF (from reference 19). Data were taken point-by-point from Figure 11.10. Note that the stronger the westerlies, the stronger and larger the PKE.

of Figure 11.12, we can see that the El Niño years are characterized by very weak westerlies (or even easterlies) with small values of PKE in the eastern Pacific Ocean, the non-El Niño years with strong westerlies and quite large values of PKE. Thus these data from years in which the TI was extreme tend to confirm the correlation noted in Figures 11.10 and 11.11.

In Figure 11.3 we saw a reversal of the zonal velocity with height, such that upper-level easterlies lie over lower tropospheric westerlies, and vice versa. Potentially, then, extratropical waves could penetrate the tropics and even cross the equator in certain regions of the tropics in the lower troposphere as well. Some evidence for such a transmission from the middle to low latitudes has been documented for the summer monsoon (26). In Section 1.2.4 we will show how this may occur over the South China Sea during the winter monsoon.

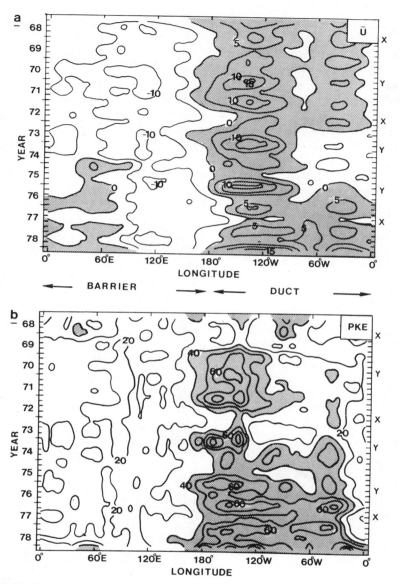

Figure 11.12. Time–longitude sections of (a) the zonal wind component in m/sec and (b) the perturbation kinetic energy in m²/sec² at 200 mb along the equator for the years 1968 through 1978 (from reference 19). Westerly winds and PKE which is greater than 40 m²/sec² are shaded. Note that the zonal wind and the PKE wax and wane together. The domains of the easterly barrier and the westerly duct are indicated. Years corresponding to extremes in the TI of Figure 11.5 are indicated as X (extreme positive TI) and Y (extreme negative TI).

1.2.3 Intraseasonal Monsoon Variability. Rather than a continual deluge, the monsoon is made up of a series of discrete events, both pluvial and dry. The main agent of precipitation is the propagating monsoon disturbance that usually lasts for a few days and produces torrential rainfall along its path. An example of the rainfall associated with a monsoon depression is shown in Figure 11.13. Forming in the Bay of Bengal, this depression progressed westward along the Ganges valley to the Arabian Sea. The path, spatial scale, and longevity of the depression are typical of both the summer and winter monsoon disturbances. Periods during which there are a succession of disturbances and precipitation are referred to as *active* periods of the monsoon.

There are also periods during the monsoon summer when no disturbances occur at all in a particular region. Such periods, in which there is little or no monsoon precipitation, are called monsoon *breaks* and sometimes, depending upon their timing and duration, are the harbingers of regional drought.

Breaks appear on two time scales, either 10–20 days or 40–50 days. In Figure 11.14, the average daily rainfall rate (cm/day) for the west coast region of Peninsular India is plotted for the summers of 1963 and 1971. A 15–20 day variation in precipitation is clearly evident and there is a suggestion of an even longer variation as well (27).

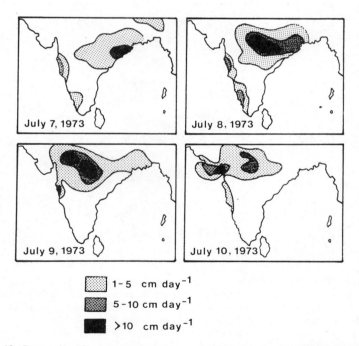

Figure 11.13. Progression of a monsoon depression up the Ganges valley from July 7–10, 1973 during an active phase of the summer monsoon (from reference 1). The isopleths (smoothed) indicate the rainfall rates of 1, 5, and 10 cm/day. An active phase of the monsoon is usually made up of a number of such disturbances.

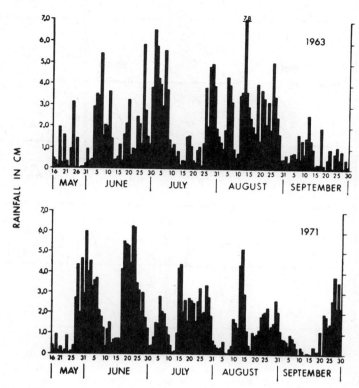

Figure 11.14. The daily rainfall (cm/day) along the western coast of India incorporating the districts of Kunkan, Coastal Mysore, and Kerala for the summers of 1963 and 1971 (from reference 27). The peaks show the active periods of the monsoon that may be made up of a succession of weather disturbances such as those shown in Figure 11.13. The active periods are separated by lulls known as the monsoon breaks.

10- to 20-Day Variations of the Monsoon. Figure 11.14 depicts monsoon rainfall over a particular region. As such, it does not relate the variation in conditions occurring simultaneously in other regions. Sikka and Gadgil (28) tried to find a broader scale relationship. Figure 11.15 shows the latitude of maximum cloudiness obtained from satellite data and averaged across the Indian Ocean as a function of time. From Section 1.1 we know that the maximum cloudiness is related to maximum precipitation. Over the five-year period, we see a succession of orderly and northerly propagating cloudiness (precipitation) zones starting near the equator and ending in the foothills of the Himalayas. In all cases the progression is from south to north. The numbers mark the longevity of the events in days; the average lifetime is about 15 days. Once or twice during the summer, in addition to their fairly regular northward propagation, the maximum cloudiness zones appear to become locked in position near the Himalayas; these are the long or *extended* break periods. Notice that when the maximum cloudiness is in the foothills of the Himalayas, a second

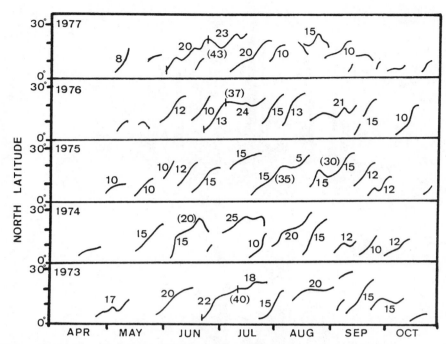

Figure 11.15. The mean latitudinal position of the maximum cloudiness zone and the 700-mb trough (which correspond in location) in the Indian Ocean during the Northern Hemisphere summer for the five years, 1973–1977. The numbers refer to the longevity of a particular zone. Usually the zones propagate from north of the equator to just south of the Himalayas in about 10–20 days. When the zone reaches its northern limit, it is replaced by a new zone in the south. Occasionally, the zone remains stationary in the far north for an extended period; this is the monsoon break period. (From references 28 and 29).

cloud maximum starts to form just north of the equator. Eventually, this zone also propagates northward over the land mass.

Figure 11.15 suggests that the monsoon is made up of a series of very orderly northerly progressions of precipitation events with an occasional extended break. We will show later that the orderliness of the progression provides some clues regarding the physical basis of the monsoon transitions.

To provide a visual perspective of the monsoon over South Asia during the extremes of the active and break monsoon sequence, latitude–height cross sections (Fig. 11.16) along 90°E are shown for conditions in late June and early July of 1976 in Figure 11.15. Figure 11.16*a* illustrates the active monsoon where the maximum precipitation zone is located over central India. Because the latent heating occurs some distance from the equator, the easterly jet stream (Ⓔ) is very strong. The summer extratropical jet is north of the Himalayas (see Fig. 11.9). Figure 11.16*b* shows the circulation when the precipitation zone has moved to the foothills of the Himalayas and subsidence is occurring over the entire Peninsular India.

Because a secondary precipitation zone has developed very close to the equator, the easterly jet is weaker than it was during the active monsoon period in late June.

So far, we have discussed variations relating to the summer monsoon, but we should not ignore the possibility of similar events occurring during the winter monsoon. Although less observational work has been done on the convective region of the winter monsoon, it appears that similar distinct variations in the monsoon circulation do occur (30). However, it seems that the latitudinal variability noted in the Northern Hemisphere is either masked by the complexity of the orography of the Indonesian region or exists on a longer time scale. Furthermore, the observed variability takes place in the winter monsoon regions in the *longitudinal* direction (30).

40- to 50-Day Variations of the Monsoon. As suggested in Figure 11.15, sometimes the monsoon appears to remain locked into a break configuration for an extended period of time. Mounting evidence shows that these extended break periods are related to a pronounced 40- to 50-day periodicity first discovered in the early 1970s by Madden and Julian (31). Their pioneering investigation resulted in the schematic representation shown in Figure 11.17. A series of longitude-height sections of the tropospheric circulation are shown along the equator (Fig. 11.17a). Regions of maximum convection appear as clouds. Longitudinal variations of the sea level pressure are shown along the abscissa of each section. For reference, the average sea surface temperature along the equator is shown in Figure 11.17b. The numbering

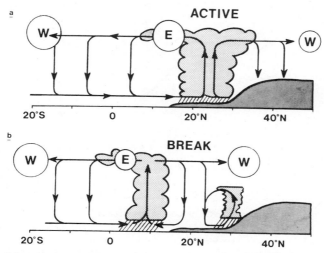

Figure 11.16. Schematic representation of (*a*) an active period and (*b*) a break period during the Northern Hemisphere summer as might be indicative of late June to late July 1976 (see Fig. 11.15). In the active period, the convection is in mid-India and a strong easterly jet exists equatorward. During the break period with convection near the equator and in the foothills of the Himalayas, subsidence exists over most of Peninsular India and the easterly jet is weaker.

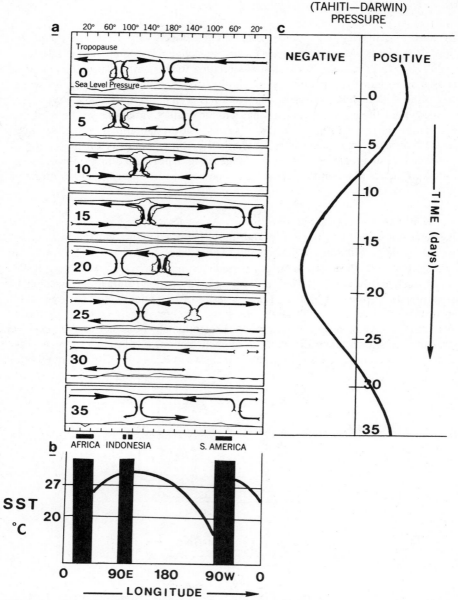

Figure 11.17. (*a*) Schematic representation of the time and space variations of the disturbance associated with the 40–50 day variation along the equator (from reference 31). The times of the cycles (days) are shown to the left of the panels. Clouds depict convection. The curve along the base of each panel shows the departure from the average surface pressure. The low-pressure anomaly accompanies the convection. The circulation on days 10–15 is quite similar to the Walker Circulation shown in Figure 11.4; (*b*) shows the mean annual sea-surface temperature distribution along the equator. The 40–50 day wave appears strongly convective when the sea-surface temperature is greater than 27°C as in panels 2–5 in (*a*); (*c*) shows the variations of the pressure difference between Darwin and Tahiti. The swing is reminiscent of the Southern Oscillation but with a time scale of tens of days rather than years.

along the axis of Figure 11.17*a* roughly refers to the phase of the cycle in days. Starting at an arbitrary day 0, the maximum cloudiness zone lies over the warm water of the eastern Indian Ocean. After 10 to 20 days the zone and the low-pressure area have moved eastward to the Indonesian archipelago and the western Pacific Ocean. At this time in the cycle, the circulation is very similar to the mean Walker Circulation of figure 11.4*a*. With further progression to the east over the cold water of the eastern Pacific Ocean, the convection weakens and eventually disappears. With time, convection is re-established in the eastern Indian Ocean and the cycle continues with an eastward propagation like that shown in the top section of Figure 11.17*a*. Figure 11.17*c* shows the pressure difference between Tahiti and Darwin during the 40–50 day cycle. The variation is indicative of a very large-scale pressure progression with a 40–50 day period.

A number of subsequent investigations (32–38) have corroborated the Madden–Julian observations and have found that the wave completes a full circuit of the globe after the convection diminishes in the eastern Pacific Ocean in about 40 to 50 days, although in something of a weaker state, until it reaches the warmer water of the Indian Ocean where once again the convection becomes active.

The 40–50 day wave appears to have a distinct influence on the character of the monsoon (36). The influence may affect both the timing of the extended-period monsoon breaks and also the timing of the monsoon onset.*

Figure 11.18 shows the eastward propagation of the 40–50 day wave in terms of its velocity potential† (a measure of the divergent part of the wind) in the Eastern Hemisphere (37). The four panels, each separated by five days, show the distributions of the velocity potential at 850 mb. The center of the ascending (descending) region of the wave is denoted by A (B). Because mass is conserved, the divergent part of the wind must flow between the rising and sinking parts of the wave, in the same way as shown for the Walker Circulation in Figure 11.4 and the monsoon circulation in Figure 11.9. Such flow occurs in a direction given by the gradient of velocity potential (the arrows in Fig. 11.18) with the strongest flow corresponding to the maximum gradient (stippled regions). Figure 11.18 shows the lower-level divergent motion field of the schematic Figure 11.17 from 0° to 180°E and 40°N to 40°S.

As the wave moves eastward, it intensifies as shown by the increased gradient. Furthermore, and very important, as center A moves from southwest of India (which lies between 70°E and 90°E) to the east of India, the direction of the divergent wind over India changes from easterly to westerly. Notice also that as center A

---

* From Figure 11.14 we see that the monsoon precipitation in the coastal region of Peninsular India commences each year in the time window between late May and late June. This is the *onset* of the monsoon and its timing is critical because of its social and economic impacts. Figure 11.15 shows that in the north the monsoon onset is later in the summer than in the south but also at different dates each year, that is, interannual variability is endemic to the monsoon onset in all parts of the monsoon lands.
† As in all fluid flow, the atmospheric circulation may be broken down into two components: a rotational (or nondivergent) and an irrotational (or divergent) part. These components are represented by the stream function and the velocity potential, respectively. Thermally driven circulations such as the Walker Circulation and the monsoons are very divergent and the divergent part of the wind field, represented by the velocity potential, is a major component of the flow.

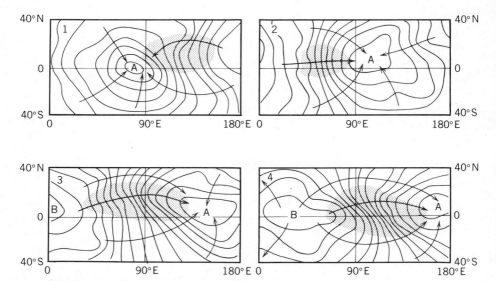

Figure 11.18. The latitude–longitude structure of the 40–50 day wave in the Eastern Hemisphere in terms of the velocity potential at 850 mb (from reference 37, units: per sec $\times$ 10^{-6}). The arrows denote the direction of the divergent part of the wind and the stippled region the locations of maximum speed associated with the wave. Depicted by A and B, the centers of velocity potential of the wave propagate east. Center A may be thought of as a region of rising air and B a region of subsidence.

moves across the Indian region that the velocity potential gradient intensifies to the north as indicated by the movement of the stippled regions in panels 2 and 3. Thus, depending on where the centers A and B are located relative to the monsoon flow, *the strength of the monsoon southwesterlies flowing toward the heated Asian continent will be strengthened or weakened.* Thus we can see the importance of the phase of the Madden–Julian 40–50 day wave on the mean monsoon flow.

Based on the observational studies reported previously, we can now pose a hypothesis that may explain how the 40–50 day wave influences the onset and active–break periods of the summer monsoon.

1. *Effect on the Monsoon Onset.*   It is hypothesized that the onset of the summer and winter monsoons is determined by two factors:

a. *The annual march of solar insolation.*   As a result of the geographic distribution of land and sea, there is an annual variation in radiational heating and a corresponding variation of the pressure gradient force (see Chapter 1) that constitute the basic driving force of the monsoon. Model studies (e.g., 39–41) have shown that the pressure gradient force will produce low-level monsoon winds of about 10 m/sec in the direction indicated in Figure 11.1. If there were no other influences, the date of the onset of the monsoon would probably be within a much narrower time window than observed (27, 40, 41).

b. *The phase of the Madden–Julian 40–50 day wave.* Calculations using the velocity potential fields of Figure 11.18 show that the divergent wind field associated with the Madden–Julian wave varies from 5 m/sec to -5 m/sec depending on the phase of the wave. The divergent wind speed is about half of the background mean flow resulting from the differential heating of land and sea (37). Figure 11.19 shows the magnitude of the divergent part of the wind field for Bombay during the period from April 30 to August 18, 1979 (36, 37). As the 40–50 day wave progressed eastward, the lower tropospheric divergent wind (850 mb) at Bombay changed sign. The onset of the monsoon accompanied the change in sign. The effect of the wave is to either enhance or to reduce the flow of the moist air to the heated continent which is necessary for the release of latent heat (see Chapter 1). Depending on the wave phase, the onset will either be early or late. The phase of the wave at the beginning of summer in 1979 may have been responsible for the late onset of the monsoon in the south of India. Prior to June 20, the wave contributed easterlies to the low-level monsoon flow, reducing the strength of the monsoon southwesterlies; this may have delayed the onset well past its climatological norm. Only when the monsoon westerlies were enhanced by the wave did the onset occur.

We have postulated that the interaction of two phenomena determine the date of the onset of the monsoon. First, the insolation over the continental regions has to be sufficiently intense for the onset of the monsoon to occur. Second, the 40–50 day wave has to have the correct phase so that the southwesterlies are enhanced. Together these phenomena produce a window within which the monsoon may commence. The monsoon rains cannot occur before mid-May, regardless of the phase of the Madden–Julian wave, since the basic heating gradient has not been established at that time. On the other hand, even with the worst possible phase of

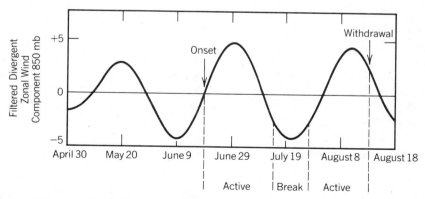

Figure 11.19. The 850 mb divergent, zonal component of the wind associated with the 40–50 day wave for Bombay during the summer of 1979. The labels indicate the major active and break periods and the monsoon onset and withdrawal at Bombay. The 40–50 day wave can be seen to be associated with westerly (positive) and easterly (negative) divergent winds of about 5 m/sec in magnitude or about 50% of the winds of the background monsoon flow in the lower troposphere (from references 36 and 37).

the wave, it is likely that the monsoon would commence during the second half of June based simply on the strength of the seasonal heating gradient.

2. *Effect on the Extended Active and Break Periods of the Monsoon.* It would appear that the same arguments used to relate the phase of the Madden–Julian wave and the monsoon onset are quite relevant to the timing of the monsoon active and break periods indicated in Figure 11.15. With continuous progression around the equator, the 40–50 day wave returns to the Indian Ocean periodically strengthening and weakening the monsoon southwesterlies and modulating the flow of moist air toward the continent. Figure 11.19 supports this hypothesis, as we can see that the major extended break in the monsoons (mid-July) occurs when the 40–50 day wave enhances the monsoon flow. Similarly, the withdrawal of the monsoon (the term used to signify the cessation of organized monsoon precipitation) occurs as the 40–50 day westerly contribution turns to easterly, and of course, the solar insolation diminishes.

What is appealing about the hypothesis is that it explains why there may be such a large interannual variation in the time of the monsoon onset and why there are extended breaks in the monsoon flow. It is also valuable as a forecasting tool for these important monsoon phenomena, especially since the Madden–Julian 40–50 day wave appears to exist throughout the year, with its maximum amplitude occurring during the Northern Hemisphere summer. The phase of the wave is easily determined at any time of the year and its movement is predictable.

1.2.4 Relationships and Interactions on the Synoptic Time Scale. In Section 1.2.2 we showed that the monsoonal and near-equatorial heating have a distinct effect on the location and strength of the extratropical upper-tropospheric jet streams. The winter jet streams, of course, are sources of extratropical weather disturbances. We would now like to go one step further and see if these weather disturbances complete the circuit and, in turn, influence the monsoon flow.

Figure 11.20 (42, 43) provides three examples of the low-level monsoon flow measured by three USSR research ships in the South China Sea during December 1978.* The panels show isopleths of wind speed (m/sec) in the lower half of the troposphere. The arrows below each panel show the direction of the average wind between 1000 and 900 mb. The geographic coordinates of each ship are given at the upper right of each panel. Five distinct wind maxima or surges can be seen in the wind field of the lower troposphere between December 6 and 31, 1978†. In each panel these are referred to as S_2, S_3, . . . S_6.

---

* During December 1978 and January 1979, the Winter Monsoon Experiment (part of the Global Weather Experiment) took place near the South China Sea. Full details of the experiment are given in reference 44.
† We refer to the low-level wind maxima, generically, as *surges* or *cold surges* even though by the time the surges reach the low latitudes there is very little temperature drop associated with them. The term *cold surge* is more relevant in the maritime areas of South China (e.g., Hong Kong) where the relatively warm winter temperatures drop rapidly with the passage of a surge.

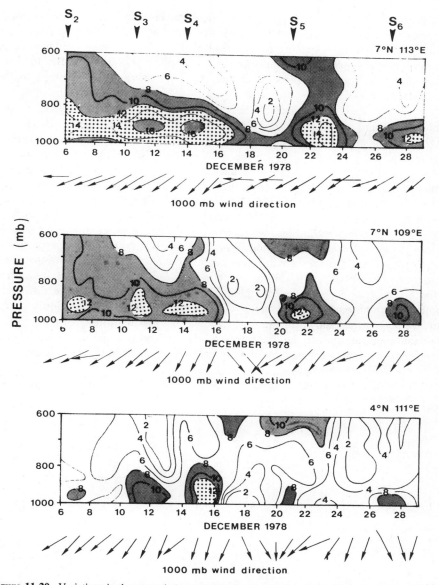

Figure 11.20. Variations in the strength (m/sec) and direction of the wind in the South China Sea in the lower troposphere during December 1978 (from reference 42). Data are from three research ships located as indicated at the upper right of each panel. Maxima in wind strength occurred around December 7, 12, 14, 21, and 28 with speeds of up to 16 m/sec at 7°N and 12 m/sec at 4°N. These are the surges of the northeast monsoon. The direction of the average flow between 1000 and 900 mb is shown by arrows below each section. Note that with the surge the winds become more northerly (backing winds) and with the lulls more easterly (veering winds).

With each surge the wind speed nearly doubles. Furthermore, the wind changes direction becoming more northerly with the surge and more easterly during the lulls. This successive *backing* and *veering* of the low-level wind is indicative of some important physical mechanisms that we will discuss later. Finally, for later reference, we note in Figure 11.20 that the wind maxima are confined to the boundary layer of the monsoon flow.

The most intriguing aspect of the surges, and the one most germane to our interests, is that they can be traced back to other variations at higher latitudes, in particular, the Siberian high-pressure region and the East Asian upper-tropospheric jet stream. Respectively, they are the most intense high pressure system and the strongest jet stream that exist in the atmosphere at this time of the year. We advance the hypothesis that the interaction of these two strong features results in surges of the low-level wind field of the monsoon. At the same time, we recall that the East Asia jet stream is intimately connected with monsoonal and near-equatorial heating.

To establish this relationship, we start by studying the pressure difference between Hong Kong and a point over China at 30°N, 111°E (marked by HK and X, respectively, on Fig. 11.21*a*). The pressure differences are shown in Figure 11.21*b* for the entire month of December, 1978, during which they varied by as much as 20 mb in five discrete events. For each event, there is a rapid buildup of pressure difference followed by an equally rapid relaxation. By comparing the time section with the low-latitude winds shown in Figure 11.20, we see that the timing of the pressure relaxation corresponds to the five surges S_2–S_6, indicating a one-to-one correlation between surges near the equator and the pressure gradient at higher latitudes.

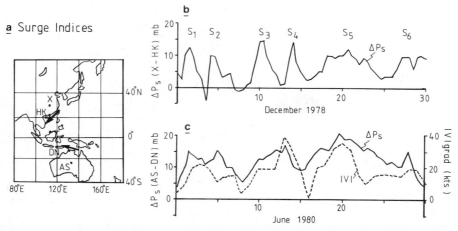

Figure 11.21. An index used to forecast and measure the intensity of the cold surge is the surface pressure difference, ΔP_s, between Hong Kong (HK) and 30°N, 100°E (X). Note the buildup of the pressure difference, or gradient, prior to the surges noted in Figure 11.20. For the Southern Hemisphere winter, the Darwin–Alice Springs (DN–AS) pressure difference is used. These two indices are plotted for December 1978 and June 1980. The near-surface wind speed at Darwin (marked V and in knots) is shown by the dashed curve (from reference 43).

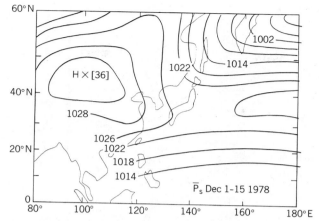

Figure 11.22. The average surface pressure distribution $\overline{P}_s$(mb) of the first 15 days of December 1978, featuring the Siberian High which is the strongest and most persistent high-pressure system in the atmosphere. The average central pressure for this period is 1036 mb (marked by X) which is somewhat lower than is typical during winter. The high pressure lies beneath the climatological position of the East Asian jet stream and in the descending (subsiding) branch of the winter hemisphere convection cell shown in Figure 11.9b.

To examine further the interrelationships between the East Asia jet and the Siberian high-pressure system, we look at the mean surface pressure distribution for the first two weeks of December 1978 (Fig. 11.22). The variations of the daily surface pressure from the average are plotted in Figure 11.23 as latitude-time sections along Figure 11.23a, 40°N and Figure 11.23b, 30°N. In each section, strong pressure variations are seen to propagate in an orderly manner from west to east. Figures 11.20 and 11.21, show that the surges S_1, S_2, S_3 and S_4 are all part of coherent variations associated with the intense Siberian High*.

Figure 11.24 completes the picture by connecting the propagating surface variations with events occurring in the upper troposphere. For brevity, we will limit our discussion to surge S_2, although all surges seem to have similar relationships. We note that prior to the surge on December 5, the upper tropospheric jet was relatively weak with its core, the region of most intense winds (stippled in the diagram), extending well west of the coast of China. By December 6, the same time as the relaxation of the pressure gradient in Figure 11.21, the jet core had intensified and moved eastward. The intensification and migration was accompanied by the eastward propagation of an upper tropospheric trough (dashed line) that was located to the west of the jet on December 5. The interaction of the trough and jet resulted in an increase in the confluence at the entrance to the jet stream. The jet intensification and the eastward propagation of the jet core and trough continued on December 7 when the surge reached its peak as seen at the left panels of Figure 11.20. This

---

* The ships commenced collecting data on December 6. Therefore we have no special low-latitude measurements of the first surge S_1. It is included here for completeness.

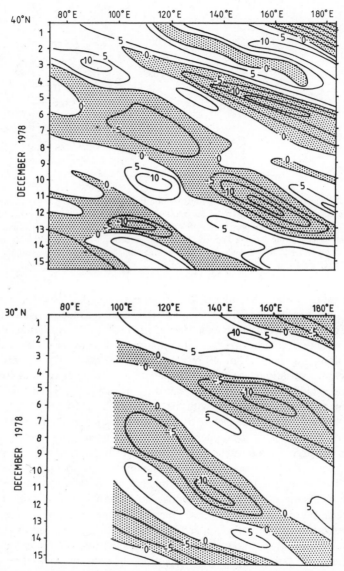

Figure 11.23. Time–longitude sections from December 1–15, 1978, along 30°N and 40°N, of the difference between the daily pressure and the average pressure (i.e., pressure deviations) shown in Figure 11.22. The positive and negative (shaded) regions sloping down from left to right indicate a systematic propagation of weather events from west to east (from reference 43).

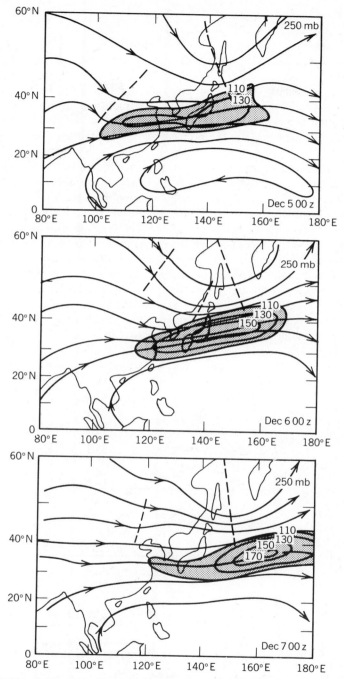

Figure 11.24. The upper tropospheric flow (250 mb) during the surge S$_2$ (December 5–7, 1978). Solid lines with arrows indicate the direction of the upper-level winds. Stippled regions denote jet stream core with speeds greater than 110 knots (roughly 55 m/sec). Dashed lines show position of the trough (from reference 45).

sequence of events is repeated in all other cases studied. Later we will discuss the physical processes that link the upper and lower troposphere during the generation of the surges.

A number of studies (46, 47) show very direct relationships between the surges and the major precipitation regions of the winter monsoon in particular. It appears that quite often a surge will trigger the formation of a disturbance in the South China Sea (see Chapter 18) or the propagation of a disturbance along the equator (45).

An example of the latter phenomenon is shown in Figure 11.25 where the surface pressure variation is plotted as a function of time at a number of stations lying roughly along the 105°E meridian. The plot shows a pressure pulse starting near Nanning (23°N, 107°E) on January 14, 1979 and arriving near the equator at Singapore (1°N, 104°E) just a day or so later. The pressure rise, associated with a strong surge out of the Siberian High, traversed over 2000 km in 24 hours or at a speed of about 30 m/sec or 100 km/hr! Such a speed is nearly a factor of five greater than the average wind speeds in the region or a factor of three greater than the winds associated with the surges themselves. This observation eliminates the possibility of the effects of the surge being advected by the wind field itself. Rather it is better to think of the cold surges spreading their influence by the propagation of a pressure pulse in the form of a gravity wave (46).

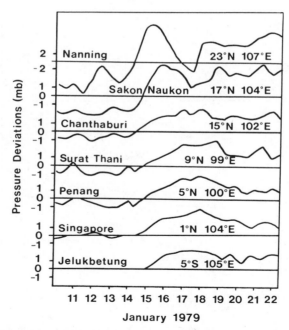

Figure 11.25. The surface pressure deviation (mb) at a number of stations arranged in order of decreasing latitude as a function of time. Following the surge of January 14 at Nanning, the pressure increased rapidly and successively toward the equator but at speeds a factor of four greater than the prevailing wind (from reference 45).

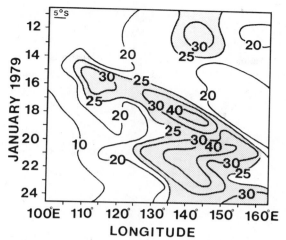

Figure 11.26. A longitude–time section of the wind speed (m/sec) along 5°S for the period January 11–25, 1979. Note the rapid wind acceleration and propagation toward the east after the pressure pulse (see Fig. 11.25) reaches the equator (from reference 45).

The low-latitude response to the cold surge is seen in Figure 11.26 which shows a longitude–time section of wind speed along 5°S for the period January 11 to January 25, 1979. This is roughly the same period used in Figure 11.25. Starting at Telukbetung (5°S, 105°E), we see that from the time of arrival of the pressure pulse at the equator (i.e., January 15 from Fig. 11.25), a considerable wind acceleration occurs which propagates eastward to about 140°E in six days and then to 150°E in another four days; or at propagation speeds of 10 m/sec and 5 m/sec, respectively. Figures 11.25 and 11.26 represent the first observational evidence of a wave propagating along the equator that can be related directly to extratropical forcing (45).

While we have chosen as our example the Northern Hemisphere winter monsoon, it should be pointed out that similar events occur elsewhere. For example Figure 11.21c shows the surface pressure difference between Darwin, Australia, (DN on the map) and Alice Springs (AS) together with the strength of the southeast trades at 900 mb at Darwin during June 1980. The variations correspond to surges from the Southern Hemisphere into the summer monsoon of southeast Asia. Other studies show that further to the west there appear to be similar relationships between the Mascarene High (the major subtropical high-pressure zone in the South Indian Ocean) and the cross-equatorial flow into the southwest monsoon of the Indian region.

2 A PHYSICAL BASIS FOR THE INTERACTIVE MONSOON

Perhaps the most interesting aspect of the observations discussed so far is the multiplicity of the interaction modes between the monsoons and other weather and climate systems. We have seen that the monsoon circulations, restricted basically to the Eastern Hemisphere, are related to low-latitude motions that encircle the equator and appear to exert a significant influence on the general circulation of the adjacent winter hemisphere. The significance of this influence is that it determines, to a large degree, the location of the cyclogenesis region of the winter hemisphere. To complete the cycle of influence, we have seen that the winter hemisphere jet streams feed back to the monsoon system by altering the return low-level flow and possibly produce propagating equatorial waves. The study of the forced equatorial waves by Williams (45), discussed previously, shows the degree of interplay between the monsoons and other atmospheric systems, illustrating the final act within a sequence of events which links near-equatorial and monsoon circulations to those of higher latitudes.

In the following sections, we will examine the physical nature of these interactions from a fundamental viewpoint. We will find that many of the observations described in the previous section can be explained by:

1. the manner in which the atmosphere responds to the annual variation of insolation;
2. the latitudinal dependence of the atmospheric response to forcing imposed at different latitudes;
3. the type of forcing, such as orographic barriers, low-level heating, or deep convective heating; and
4. the manner in which the atmosphere feeds back upon itself to alter the initial forcing.

2.1 The Fundamental Modes of Tropical and Subtropical Circulations

If the earth were not rotating, energy would be transferred from one location to another almost solely by gravity waves and density currents; exchanges of heat between regions of deficit and abundance would be accomplished by a series of direct circulations that would rapidly reduce temperature differences. Except for having a different heat balance, the equatorial regions would have a circulation no different from the polar regions in its basic dynamic response to forcing. Energy would radiate away from a point in the same fashion irrespective of the location of the source.

But the atmosphere does rotate, and because the earth is spherical in shape, air parcels at the equator and at the middle latitudes are located at different distances from the axis of rotation. Thus there is a distinct latitudinal structure in the manner in which the energy is transmitted away from the energy source.

The reasons for the latitudinal dependence of the atmospheric response to forcing are quite straightforward. All motions are subject to conservation laws which govern how energy is created or dissipated and how the mass of the atmosphere behaves subject to various forcing mechanisms. The total energy and total mass of the atmosphere must each remain constant. However, in terms of motion on a rotating sphere, the most important conservation law is that of absolute (or total) angular momentum. The law states that a fluid parcel moving from one latitude to another is subject to a restoring force such that its absolute angular momentum is conserved. The magnitude of the angular momentum at any point depends on the distance of a parcel from the earth's axis of rotation. Thus the magnitude of the restoring force on a moving parcel of air is a function of its starting and ending points. The restoring force imposed by the rotation of the earth on a parcel moving from the equator to 10°N, which brings the parcel only slightly closer to the axis of rotation, is much less than on a parcel moving from 60°N to 70°N, where it has moved significantly closer to the axis because of the spherical shape of the earth.*

In addition to the latitude of the forcing, the direction of the flow within which the forcing is imbedded is also of great importance. That is, the response of an atmosphere with basically easterly winds will be quite different from that of an atmosphere dominated by westerly winds, because relative to the forcing, the atmosphere is rotating either faster or slower than the earth. With these factors in mind, we can make the following generalizations:

1. At low latitudes in the easterly wind regime.
   a. Low-frequency waves (periods of a few days to a season) tend to be confined close to the equator and are restricted to an easterly propagating equatorial wave (the so-called Kelvin wave, see Chapter 9) and westward propagating waves (the Rossby wave, see Chapter 9). The waves decay rapidly away from the equator because they are equatorially "trapped" by the rotation of the earth; that is, the Coriolis force deflects motion to the right on the northern side of the equator, but to the left of the southern side, thus confining the wave motion to near the equator.
   b. Waves tend to be divergent and occupy the entire troposphere, with inflow in the lower troposphere and outflow in the upper troposphere. The depth of the waves is affected by moist processes (see Chapter 1). Latent heat release in rising moist air allows a parcel to maintain its buoyancy and it may continue to rise until its moisture is expended. If the air were dry and the atmosphere only heated at the surface, then atmospheric motions would only be 2 to 3 km deep rather than the observed 12 to 15 km. The near-equatorial and monsoon flows have this deep convective structure.
   c. Waves tend to be of an extremely large horizontal scale, for example, the scale of the Eastern Hemisphere.
   d. Asymmetric motions with respect to the equator can be produced by asymmetric forcing distributions such as strong heating to the north of the equator during the summer (e.g., in the monsoon regions) and cooling in the winter hemisphere.

* A *restoring force* is intrinsic to the system and differs in character from an *external force* applied to the system. Using the analogy of a spring, the external force is used to initially extend the spring. The tension in the spring is the restoring force. Thus, in the atmosphere, heating may produce a body force (the external force) that causes a latitudinal movement. The change in the distance from the axis of rotation is proportional to the restoring force, for example.

2. At higher latitudes, in the westerly wind regime.

    a. Low-frequency, Rossby waves usually radiate equatorward and poleward from the point of excitation depending, to some extent, on the scale of the forcing and the strength of the basic westerlies. The Rossby waves owe their existence to oscillations of fluid parcels under a restoring force produced by the change in angular momentum as an air parcel moves latitudinally and, thus, closer to or further away from the earth's axis of rotation.

    b. The middle-latitude disturbances propagate only as far as their phase speed remains greater than the magnitude of the basic wind. Thus, waves propagating equatorward may encounter a "critical latitude" at which their energy may be either reflected or absorbed (see Chapter 9) because of a change in the background wind field. If the "critical latitude" were to stretch completely around the equator the low latitudes would be insulated from the extratropics, at least from the relatively low-frequency Rossby waves. High-frequency gravity waves, however, may pass through these latitudes as we have seen with those associated with cold surges (see Section 1.2.4). However, because the basic flow varies considerably in longitude as well as latitude (see Fig. 11.3), we will see that energy can infiltrate the equatorial regions in specific areas and even propagate to the other hemisphere. We noted examples of this behavior in Section 1.2.2.

2.2 The Effect of the Latitude of the Forcing: The Production of Near-Equatorial and Monsoon Circulations

The model response of the upper- and lower-tropospheric JJA mean circulation to large-scale heating is different when the same initial heating field is at the equator (Fig. 11.27) or away from it (Fig. 11.28) at 24°N. The model used is a two-layer system (14, 15) in which the initial heating field, a region of anomalously warm sea-surface, is applied to the mean JJA winds shown in Figure 11.2. Although the heating represents the effects of condensation and takes into account the modification of the heating by the motion, the model does not contain an explicit hydrologic cycle. The model is iterated until an equilibrium or steady-state perturbation circulation is reached. The final heating field and the steady-state circulation are shown in Figures 11.27a–d and 11.28a–d, respectively.

With the initial heating at the equator, a symmetric response is produced with the largest response confined to the plane along the equator (Fig. 11.27). The final heating field shows the modifications made to the initial heating fields by the circulation-induced release of latent heat. Convergence occurs in the vicinity of the heating in the lower level with divergence aloft produced by the strong vertical motion. Note the similarity in form to the Walker Circulation shown in Figure 11.4. Technically the flow is almost completely determined by the Kelvin wave (4, 6).

When the initial heating field is moved to 24°N (Fig. 11.28), there is strong low-level flow into the heating region (Fig. 11.28c). This is connected to strong

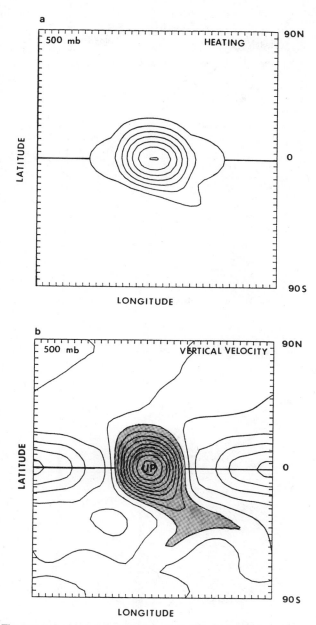

Figure 11.27. The response of a model atmosphere for the Northern Hemisphere winter (JJA winds of Fig. 11.2b) to heating at the equator. (a) The final heating distribution; (b) the vertical velocity distribution (upward motion shaded). (*Figure 11.27 continues on page 310.*)

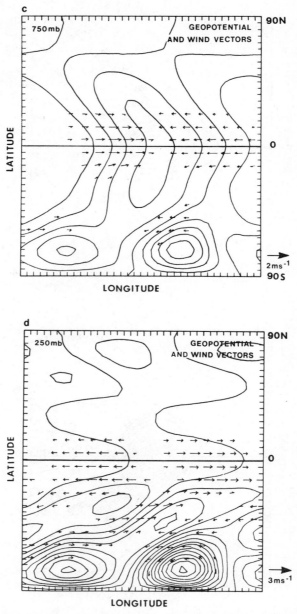

Figure 11.27. (*continued*) (*c*) the lower-level (750 mb) tropospheric flow; and (*d*) the upper-level (250 mb) tropospheric flow. Solid contours on the 250- and 750-mb charts indicate the height of the pressure levels and the arrows indicate the flow field. Magnitudes are relative (from references 14 and 15).

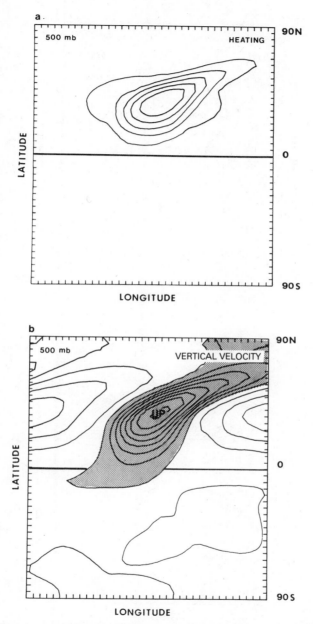

Figure 11.28. The same as a Figure 11.27 but with heating at 24°N. Even though the magnitude and shape of the initial heating is the same, there is a vast difference in the response of the atmosphere, compared to Figure 11.27. (*Figure 11.28 continues on page 312.*)

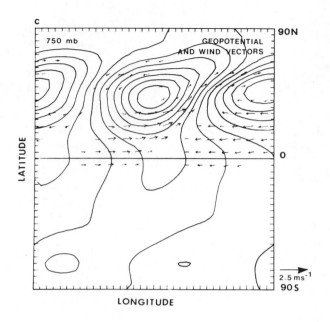

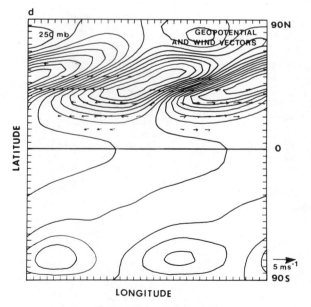

Figure 11.28. (*continued*)

outflow aloft by the intense vertical motion. However, the response that was previously confined to the zonal-height plane along the equator has been replaced by a swirling circulation, which spans many latitudes, induced by the rotation of the earth (see Chapter 1). The result is the production of lower-level southwesterlies moving toward the heat source and strong easterlies aloft on the equatorial side of the heating. This is the easterly jet stream (the E of Figs. 11.9 and 11.10) of the summer monsoon. Thus the effect of the earth's rotation is to trap equatorial motion within the boundaries of equatorial latitudes, on one hand, and to produce a large swirling circulation which crosses latitude circles, on the other. The difference of course, is the location of the forcing.*

An interesting aspect of Figure 11.27 is the rather large response at the high latitudes of the *winter* hemisphere (the Southern Hemisphere, in this case) to forcing at the equator. The heating at the equator excites Rossby waves in the winter hemisphere westerlies which extend into low latitudes. Such a wave propagation is probably a major vehicle of influence during such climate events as El Niño (48, 14, 15, 16) and explains why we can expect global anomalous climatic responses to accompany changes in geographically remote regions of heating.

2.3 The Importance of the Type of Surface and of Moist Processes

The three principal factors that determine the monsoon structure are, as described in Chapter 1: the annual march of the sun, the land and sea distribution and the closeness of atmospheric temperatures to the triple point of water. It would seem, then, that a meaningful monsoon model should include:

1. an *interactive ocean*, the temperature of which can change by solar heating and mixing by wind stress;
2. *moist processes* which include the complete hydrologic cycle from the evaporation of sea water to the release of latent heat in clouds during precipitation processes;
3. an *interactive land surface* of sufficient detail to provide both an accurate representation of the heating of the atmosphere at the lower boundary and of the modification of that heating by the evaporation of ground moisture; and
4. *sufficient internal complexity* to allow the atmosphere to respond to the forcing and to have feedback processes 1–3 above.

Few such models have been developed and those that have usually include some simplifying assumption [e.g., no variation in the longitudinal direction so that the model is only two-dimensional (40, 41)]. In one attempt to model the Indian summer monsoon, the land and sea boundary was set at 18°N with the continental cap to the north and ocean to the south (40, 41). The model was time-dependent with the

---

* Note that the flow in the upper troposphere is confined mainly to the Northern Hemisphere. This is because the strong radiational cooling in the winter hemisphere was not included in the experiment so that the heating gradient across the equator was absent.

solar radiation input following the annual cycle. Like all models, parameters could be changed at will. For example, the land mass could be removed to produce an oceanic globe and hydrologic processes either introduced or omitted.

The objective of the model simulations was to understand the importance of the type of lower boundary and of moisture on the monsoon circulation. Toward this end, the following experiments were performed:

1. the earth covered by ocean with a full hydrologic cycle (the vertical velocity and temperature response are shown as the solid lines in the upper panels of Fig. 11.29);
2. an ocean–land complex (as described above) but with *no* hydrology cycle, that is no release of latent heat or precipitation permitted (the dashed lines in the lower panels); and
3. an ocean–land complex with a full hydrology cycle (solid lines in the lower panels).

The distribution of land and sea are shown on the bottom axes of the lower panels of Figure 11.29. The resulting seasonal averages of sea-surface temperature and vertical velocity are shown in Figure 11.29. The effects of land and sea, and of moisture can be seen clearly. With the earth covered by ocean, the intense vertical velocity (in response to the model's annual solar heating cycle) is confined quite close to the equator and corresponds to the region of maximum sea surface temperature. Here the air parcels in the boundary layer are moistest and the air is the most unstable,* making this the region of most efficient and largest release of latent heat. The effect of the continent in the "dry" ocean–land experiment (Fig. 11.29c and d, dashed curves) is to add a seasonal distortion to the vertical velocity and the surface temperature fields. The maximum upward motion occurs over the Northern Hemisphere continent during summer and corresponds to the broad and very wide surface temperature maximum. However, when moist processes are added in the ocean–land system (Fig. 11.29c and d, solid curves), a mixture of the two previous experimental results is obtained: strong upward motion to the south of the equator during DJF and over the continent during JJA. The maximum vertical velocity over the land does not correspond to the maximum land temperature but to the slightly lower temperature region to the north of the coast (c.f. Fig. 1.1). The lower temperature (cf. with the dry case) is a result of precipitation moistening the soil and thus increasing the evaporative cooling. We shall see later that this process causes substantial subseasonal variation of the monsoon.

From the results of the simulations shown in Figure 11.29, we can see a greater complexity in the physics of the monsoon compared to that of the near-equatorial circulations. The latter circulations may be explained in terms of the sea-surface

---

* 10 mb/sec ($\times$ 10^{-4}) is equivalent to about 1 cm/sec. At this vertical speed, a parcel would rise from the surface to the upper troposphere in about 15 days. This seemingly slow vertical drift represents the large-scale, or spatial-average motion. In a convective area, the upward motion takes place at a rapid rate in the clouds, but is offset by downdrafts inside and outside the clouds that nearly compensate for the upward motion. Thus the area average vertical velocity is quite small although the upward velocity within the convective clouds may be two to three orders of magnitude larger.

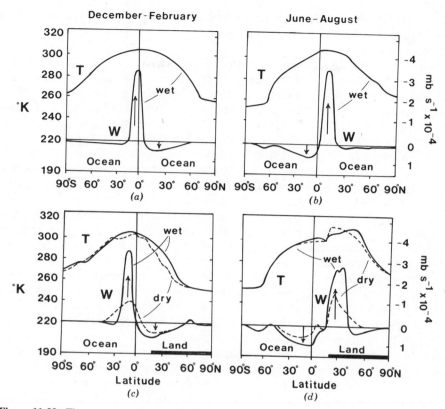

Figure 11.29. The seasonal average circulation of a monsoon model (from references 40 and 41). The DJF surface temperature (*T*, degrees K) and the vertical velocity (*w*, mb/s $\times$ 10^{-4}) are shown in (*a*) and (*c*) and the JJA fields in (*b*) and (*d*). Three cases are shown for which the globe is assumed to be purely oceanic (solid lines, *a*, *b*), to have a continent north of 18°N but with no moist processes (dashed lines, *c*, *d*, labeled dry) and to have moist processes over the continent (solid curves, *c*, *d*, labeled wet). The extent of the model continent is shown along the abscissa in *c* and *d*.

temperature distribution, and, in particular, the dominance of the warm sea surface temperatures in the eastern Indian Ocean and the western Pacific Ocean. The resultant Walker Circulation (or Kelvin wave) appears to be forced by the very efficient release of latent heat associated with the warm sea surface temperature pools. When the SST distribution changes, as in the case of El Niño, the main ascending branch of the Walker Circulation follows the warm water eastward as seen in Figure 11.4*b*. In the case of the monsoon circulations (simulated by Fig. 11.29), the effect of ground hydrologic processes is to produce complex circulation patterns which add rich variation to the monsoon structure.

2.4 Interactions

Armed with some knowledge of the important physical processes involved, we can now approach an understanding of certain observed interactions between the monsoon and other circulation features. Many of the relationships are still not understood and therefore, some of the explanations are speculative.

2.4.1 Monsoons and Walker Circulations.

Viewed as independent entities, we can go a long way in understanding the physical mechanisms that produce the monsoons and the Walker Circulations. But they are not independent; interactions have been observed, or rather, sensed through statistics which relate the monsoon and near-equatorial (Walker) circulations. The precise manner in which these two major circulation systems interact is, unfortunately, not well understood.

If we view the schematic circulation features of Figures 11.27 and 11.28, it is not difficult to envisage some form of interaction between these two large-scale atmospheric circulation patterns. The Walker Circulation cells impose strong convergence and divergence patterns in the same locations where the monsoonal cross-equatorial flow occurs. Thus a variation in the location of the surface heating (say, during El Niño) will change the location of the convergence and divergence patterns and, in turn, precipitation patterns, of the Walker Circulation. How might this be related to the monsoon circulation?

Discussion relative to Figure 11.9 indicated that the monsoon systems are driven by the *gradient* in heating (i.e., the spatial difference in heating and cooling; see also Chapter 1). With this in mind, we can provide a tentative explanation of why the winter monsoon precipitation over Indonesia and northern Australia may be affected by changes in the Walker Circulation during El Niño to a much greater extent than the summer monsoon precipitation over southeast Asia. During DJF, over Indonesia and Australia, the regions of ascending motion of the winter monsoon circulation and the Walker Circulation are both fairly close to the equator and almost coincident (see Figs. 11.1 and 11.9). Hence, a variation in location of the ascending branch of the Walker Circulation will cause a large change in the longitudinal heating gradient of the winter monsoon, even if the latitudinal heating gradient were to remain the same. On the other hand, during JJA the monsoon and Walker Circulations ascending regions are fairly well separated geographically. Summer monsoonal heating is centered well north and to the west of the ascending region of the Walker Circulation in both "normal" and El Niño periods.

2.4.2 Interhemispheric Interactions; Barriers and Ducts.

In our discussion on the fundamental nature of tropical and subtropical circulations it was stated that wave-like motions, or modes, forced at high latitudes, generally would not affect the tropics because they encounter a critical latitude which acts as a barrier to waves. This means that if the mean state of the atmosphere were perfectly zonally symmetric, having no variability in the longitudinal direction and possessing only bands of easterlies and westerlies, the tropical regions would be well insulated from energy sources outside the tropics (5, 49, 50, among others). But Figures 11.10 and 11.11 suggest that the mean zonal winds are not zonally symmetric (they contain easterlies and westerlies along the equator) and that energy does propagate from the extratropics to very low latitudes and possibly to the other hemisphere. The physical mechanisms responsible for this propagation can be studied by considering the propagation of extratropical wave energy in a realistic mean flow that allows for time-mean easterlies *and* westerlies along the equator (18).

Figures 11.30 and 11.31 show a model simulation of waves in the upper troposphere propagating through two different basic flows (upper panels) from identical energy sources (hatched) that are located in the extratropics. The two basic fields represent, respectively, periods of weak and strong upper-level equatorial westerlies which are, in fact, observed along the equator. The model response to the forcing is shown in the lower panels of Figures 11.30 and 11.31. Energy radiates out of the middle latitude source region both to higher latitudes and toward the equator in the form of waves. These are indicated by the alternating negative and positive velocity (u') values of the response.

It should be noted that the wave path is only through the region of equatorial westerlies. These are the so-called westerly ducts of the tropical upper troposphere. In the model simulation, it is only during periods of strong, mean equatorial westerly

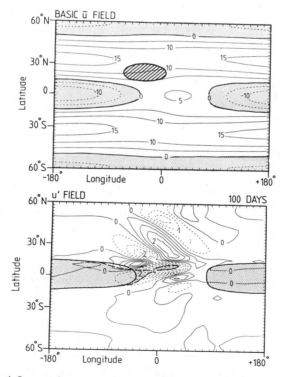

Figure 11.30. The influence of upper tropospheric, extratropical forcing (hatched region) on the low latitudes represented by a basic state $\overline{U}$ with weak equatorial westerlies and easterlies (shaded). Lower panel shows the response u' (zonal velocity component) after 100 days of model integration with solid lines representing westerly speeds and dashed lines representing easterly speeds (units m/sec). Shaded region in the lower panel shows the region of basic easterlies (from reference 18). Note that the propagation is into the region of westerlies.

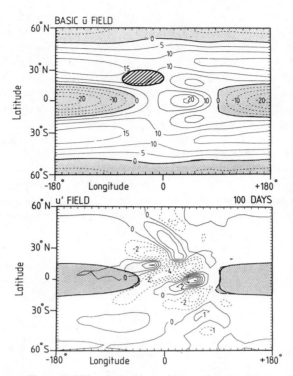

Figure 11.31. Same as Figure 11.30 but the basic state has strong equatorial westerlies and easterlies. Note that the propagation in this case is *through* the equatorial westerlies across the equator and into the Southern Hemisphere (from reference 18).

flow that energy propagates freely from one hemisphere to another. However, at all other times, whenever equatorial westerlies exist, wave energy from middle latitudes propagates much further equatorward than it would if easterlies existed everywhere in the tropics (18, 19).

2.4.3 Subseasonal Variations and Interactions: Surface and Moisture Effects.

We noted earlier that there is a predominance of variability in the monsoon at periods of 10–20 days and 40–50 days (see Section 1.2.3). Using a model, we can simulate the behavior of the monsoon on these time scales in order to learn more about the physical mechanisms that produce this variability.

Figure 11.32 shows the mean monthly vertical velocity distributions in latitudinal sections between 45°S and 45°N calculated using a model of the monsoon that contains both an interactive ocean and a hydrologic cycle (40, 41). The panels show the monthly progression of the field (six months on each panel) for Figure 11.32*a*, the purely oceanic and Figure 11.32*b*, the ocean–land cases (see Fig. 11.29).

Considering first the pure ocean case, we find a relatively orderly march of the vertical velocity maximum (corresponding to the Intertropical Convergence Zone), with the most northerly extent occurring in September and the most southerly in March. The movement of the field is out of phase with the position of the sun by more than two months, a manifestation of the lag between isolation and sea surface temperature caused by the manner in which heat is stored in the ocean (see Chapter 1). The ocean–land case, especially for the six months encompassing summer, is much more complicated. Here the advance of the maximum vertical velocity to the land area north of 18°N occurs very rapidly between April (A) and May (M). However, as summer progresses, there is no systematic colocation of the solar heating maxima and the rising motion as we found in the ocean case, even with the two-month phase lag.

Figure 11.33 shows a time–latitude section of the vertical velocity calculated with the model over one annual cycle. The insets labelled A, B, and C show the detailed vertical velocity and surface temperature structure. This detail explains why the ocean–land case is characterized by a chaotic fluctuation of the position of the mean monthly vertical velocity fields. Obviously a monthly mean is an inappropriate sampling interval given this strong submonthly variability. Indeed, depending upon the start of the averaging period, the mean monthly vertical velocity maximum during summer could occur anywhere within a broad latitudinal belt over

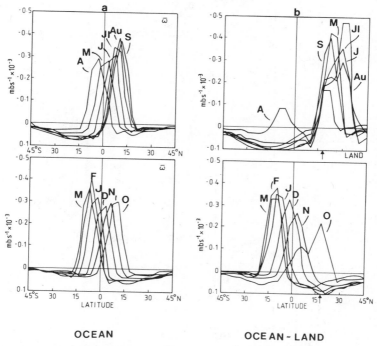

OCEAN   **OCEAN - LAND**

Figure 11.32. The mean monthly vertical velocity distributions between 45°S and 45°N for (*a*) ocean simulation of the Webster–Chou model and (*b*) the ocean–land simulation. Months are indicated by letter (e.g., A, April; M, May; etc.). Units are 10^{-3} mb/sec (from references 40 and 41).

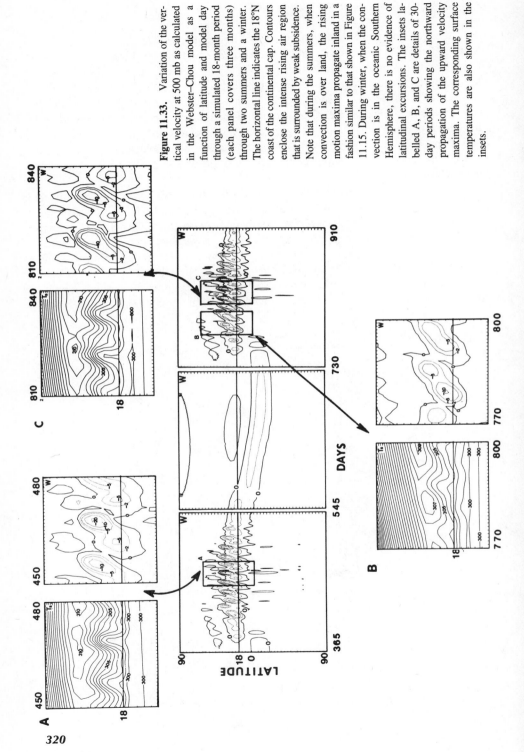

Figure 11.33. Variation of the vertical velocity at 500 mb as calculated in the Webster–Chou model as a function of latitude and model day through a simulated 18-month period (each panel covers three months) through two summers and a winter. The horizontal line indicates the 18°N coast of the continental cap. Contours enclose the intense rising air region that is surrounded by weak subsidence. Note that during the summers, when convection is over land, the rising motion maxima propagate inland in a fashion similar to that shown in Figure 11.15. During winter, when the convection is in the oceanic Southern Hemisphere, there is no evidence of latitudinal excursions. The insets labelled A, B, and C are details of 30-day periods showing the northward propagation of the upward velocity maxima. The corresponding surface temperatures are also shown in the insets.

320

the land area. Also one should note that the propagation of the vertical velocity maxima from south of the "coast line" northward over the land mass in successive waves is quite similar to the observed movements of the maximum cloudiness zone shown in Figure 11.15 with much the same period (41, 46). Note also that during the winter, the rising motion maximum moves to the south of the equator but does not show any migratory behavior. Obviously the reason for the migrations rests with the lower boundary over land and its interaction with the heating.

It has been argued (28, 29) that the propagation of the cloud zone results from a radiational heating gradient on either side of the cloud band caused by the variation of ground and cloud albedo. This radiation differential causes surface heating variations which, in some manner, cause migrations of the cloud band. However, these migrations are explainable in terms of a complementary argument (29) which proposes that the surface heating variations are caused instead by the precipitation that alters the ground moisture over the land area. This, in turn, causes the cloud zone to propagate. A schematic of the mechanism is shown in Figure 11.34. It is argued (29) that the sensible heating ($\dot{Q}_{sen}$ in Fig. 11.34) is approximately proportional to the surface temperature (see Fig. 11.29). As the land heats with the coming of summer, rising motion (w) is induced by the sensible heat input into the atmosphere. As the moist ocean air is drawn over the heated land, latent heating results from condensation ($\dot{Q}_{cv}$). The resulting precipitation (PPT) causes the ground to cool because much of the solar heat goes into evaporation. Correspondingly, the sensible and the convective instability decreases, so that locally the total heating ($\dot{Q}_{tot}$) decreases. This causes the distribution of total heating of the atmosphere (i.e., latent plus sensible and radiational heating) to move slightly poleward. In the monsoon latitudes, the vertical velocity is almost exactly proportional to the total heating (27). Thus the vertical velocity follows the total heating distribution inland. The progression inland continues as the precipitation modifies the heating field, but as the near-coast land (moistened by the last convective zone) dries out, the sensible heat and the convective instability are re-established near the coast and a new zone of rising motion occurs. This quickly overwhelms the inland convective zone by starving it of moist air. As the inland zone diminishes in intensity, the process of propagation inland continues as the new convective zone moves northward. In Figure 11.34, the progression of the zone inland is seen in the sequence t_1, t_2, t_3.

The relatively simple interaction of the components of the total heating may account for some aspects of the observed 10–20 day variation of the monsoon. However, it probably does not account for the longer period 40–50 day oscillation which we have seen to be part of a longitudinal propagation around the entire tropics. Whether or not this longer period variability is driven by a feedback of heating components affecting the sea-surface temperature remains to be seen.

2.4.4 Synoptic-Scale Influences from the Extratropics.

We described earlier how the positions of the equatorial and monsoonal heating affect the establishment of the finite, strong, winter jet stream. We recall that the position and magnitude of the jet is variable and that this variability is associated with orderly surface pressure migrations across East Asia and with surges in the near equatorial wind

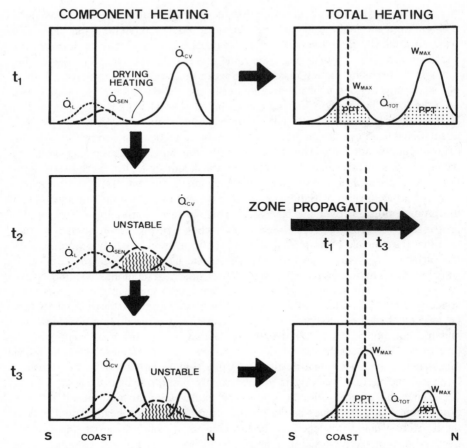

Figure 11.34. Schematic diagram of the interaction of the sensible heating ($\dot{Q}_{sen}$) and the convective latent heating ($\dot{Q}_{cv}$) of the atmosphere via the ground hydrology. $\dot{Q}_L$ is latent heating due to nonconvective processes. Right panels show total heating $\dot{Q}_{tot}$; maximum vertical velocity is associated with $\dot{Q}_{tot}$ maximum. Sensible heating is approximately proportional to the ground temperature. Rising motion caused by the sensible heating causes moist air to condense and precipitation to fall. The moistening of the ground lowers the temperature (evaporative cooling) and reduces the sensible heating so that the total heating maximum ($\dot{Q}_{tot}$) moves slightly poleward. Thus the propagation of the zone of ascent continues until the coastal land dries and warms causing another ascending zone which quickly dominates the old zone lying inland by starving it of moist air. The summer continues with a succession of poleward propagations until the insolation decreases sufficiently so that the influence of the lower boundary on the convection becomes less important. The zones of upward vertical motion then move over the ocean where, without the complexity of the ground hydrology, they are very steady (from reference 29).

fields. The variability is enhanced by periodic interference of eastward propagating upper tropospheric troughs across Asia (see Fig. 11.23). We also recall that the finite jet stream is the most active breeding ground for winter storms and in this sense is the most unstable region of the winter hemisphere. The disturbances that develop in the core of the jet stream produce wave trains which may look something like the wave disturbances shown in Figures 11.30 and 11.31.

A finite jet stream has a region of mass confluence in the entrance of the jet and difluence in the exit. Such a configuration in a rotating atmosphere cannot exist as a steady state in usual circumstances. Figure 11.35 presents, schematically, the finite winter jet stream. An air parcel which is in geostrophic balance, with velocity

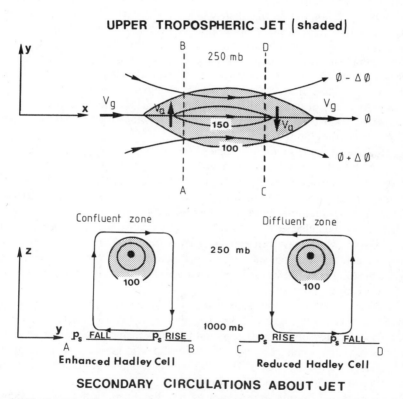

Figure 11.35. Upper panel shows schematic representation of an upper-tropospheric extratropical finite westerly jet steam (stippled region). Solid lines represent the pressure surface heights (labeled φ). Solid, closed contours indicate the wind speed (m/sec). V_g denotes that part of the wind which is in balance with the pressure gradient and the Coriolis force (the geostrophic wind) and V_a the wind deviation caused by the acceleration and deceleration at the entrance and exit regions of the jet, or the ageostrophic component. Lower panel shows two cross-sections in the vertical plane traversing the jet (AB and CD) with the circulation induced by the ageostrophic motions in the jet (stippled). These ageostrophic secondary circulations either enhance the Hadley Circulation (at the entrance of the jet) or reduce it (at the exit of the jet). Regions of rising and falling surface pressure, P_s, relative to the secondary circulations, is indicated (from reference 43).

V_g, upstream of the jet stream accelerates on entering the jet stream. In order to maintain a balanced state, it is forced to move across the isobaric surfaces. The cross-isobaric or ageostrophic flow (V_a) enhances the circulation seen in the vertical section AB behind the jet. The enhanced divergence on the south side of the jet and enhanced convergence to the north produces a change in the mass of the column which results in changes in surface pressure; rising pressures to the north (B) and falling pressures to the south (A). The meridional circulation is direct (i.e., descending cold air and rising warm air) and can be thought of as an enhancement of the local Hadley (or monsoon) Circulation shown in the schematic sections of Figure 11.9.

Downstream of the jet, we find the reverse. An air parcel decelerates as it leaves the jet. The ageostrophic flow is equatorward in the upper troposphere causing divergence to the north and convergence to the south. The resulting surface pressure changes (increasing at C and falling at D of Fig. 11.35) are such that the local Hadley cell is opposed by the indirect, meridional circulation along CD.

The changes in surface pressure produced by the accelerations in the jet stream described above, produce an *isallobaric* wind. This wind is proportional to the pressure change and flows from regions of rising pressure to regions of falling pressure. As a rule of thumb, a 10-mb change in surface pressure in 24 hours over a distance of 1000 km will produce an isallobaric wind of about 10 m/sec toward the region of falling pressure.

The regions of rising pressure (B and C) and falling pressure (A and D) are fixed relative to the jet stream and the magnitude of the pressure changes is directly related to the acceleration and deceleration in the jet's confluent and difluent regions. With no outside influences, a small isallobaric flow is produced in the sense shown in Figure 11.35 with a magnitude of a few meters per second. However, the effect is magnified by the influence of eastward propagating troughs in the upper troposphere. These troughs which pass through the jet (as seen in Fig. 11.24), enhance the confluence and, in so doing, increase the isallobaric wind. With the trough causing the jet to accelerate and move eastward, a stronger isallobaric wind and a strengthened Hadley Cell move eastward.

The fields of isallobars (pressure change per 24 hr) for a three-day period about surge S_2 are shown in Figure 11.36. Together with Figure 11.24, it can be seen that the acceleration of the jet stream and the movement eastward correspond to a strengthening (and a change in sign) of the isallobaric gradient. Typical gradients of 10 mb/day over a distance of 1000 kilometers produce isallobaric winds of the same order as the background geostrophic flow.

A schematic of the presurge isallobars and the surge isallobars are shown in Figures 11.37 and 11.38, respectively. The direction of the background or mean geostrophic wind field (V_g) may be inferred from the mean isobars plotted on Figure 11.22. Prior to the surge, the isallobaric wind V_a is weak and onshore (westward) causing the surface wind V to turn onshore. However, with the arrival of the trough, and with the subsequent increase in magnitude and eastward propagation of the jet, the surface winds increase in strength and turn more directly offshore with the addition of the strong offshore (eastward) isallobaric wind. We note that such changes in the surface winds were evident in the time sections of near-equatorial

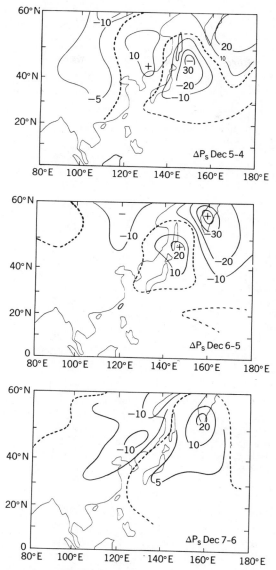

Figure 11.36. Patterns of the 24-hr pressure change (mb) (i.e., 24-hr isallobars) between the 5th and 4th, the 6th and 5th, and the 7th and 6th of December, 1978. Dashed line represents the zero pressure change.

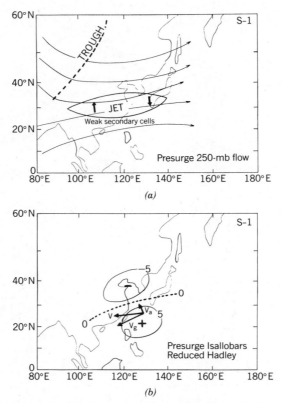

Figure 11.37. Schematic representation of the isallobaric distribution (*a*) at 250 mb and (*b*) at the surface when the jet stream is weak and well inland. The secondary circulation reduces the monsoon or local Hadley circulation. The resultant surface wind V is towards the coast. V_g is the geostrophic wind and V_a the isallobaric wind. The contours at 250 mb are streamlines of flow and at the surface are pressure changes in mb/24 hr.

wind shown in Figure 11.20 during periods of surge and nonsurge conditions. From this evidence we infer a connection between extratropical high-frequency variability and a response in the near-equatorial and monsoon flow.

One may question why the surges always occur in roughly the same location. Obviously they must occur near the location of the East Asia jet stream since the variation of the jet is of primary importance in our arguments. However, there may be another reason. If the interaction of the jet and the troughs, and the resultant isallobaric wind, are thought of as providing an impulsive force to the atmosphere, the surge may be thought of as a gravity wave (see Fig. 11.26). In fact, the recognition of the surge as having gravity wave characteristics has been made (46). But there is no preferential direction for gravity wave dispersion of energy; however, there is ample evidence (51) that suggests that the eastern Himalayas may duct the gravity waves to the southwest. In that sense, the surges may be thought of as

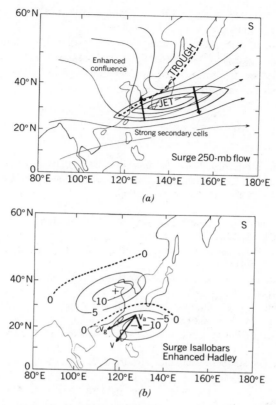

Figure 11.38. Same as Figure 11.37 but when the jet stream is strong and has moved eastward. The secondary circulation associated with the exit region of the jet enhances the monsoon circulation and the surface wind is strong and directed toward the equator (from reference 43).

"edge-waves" or gravity waves trapped by rotation against the sides of the mountains. Because of this, the surges may be smaller-scale cousins of the Walker Circulation which, we recall from Section 2.1, are trapped about the equator by rotation. Of course, the "edge" for the Walker Circulation is the equator where the sign of the Coriolis force changes.

3 CONCLUSIONS

We have presented considerable evidence of the interactive and variable nature of the monsoon circulation. Emanating from vast and concentrated heating in the summer hemisphere tropics and subtropics, the monsoon system is robust and vigorous and few regions of the globe are unaffected by its variations. But perhaps the most intriguing feature of all is that the effects exist on all scales of motion, large and small alike.

Important lessons may be learned from the examination of the multiscale monsoon. To forecast either monsoon weather or weather events in the extratropics for more than a few days in advance, it is necessary to use a prediction model that encompasses the whole globe. Likewise, to gauge the influences of major climate anomalies (such as El Niño) on the monsoons, or even the reverse interaction, we need to consider data sets that are also global in nature.

REFERENCES

1. P. J. Webster, Monsoons, *Scientific American*, **245**(2), 108–118 (1981).
2. B. Liebman and D. L. Hartmann, Interannual variation of outgoing longwave radiation: Associations with tropical circulation during 1974–78, *J. Atmos. Sci.*, **39**, 1152–1163 (1981).
3. T. N. Krishnamurti, Tropical east–west circulations during the northern summer, *J. Atmos. Sci.*, **28**, 1342–1347 (1971).
4. P. J. Webster, Response of the tropical atmosphere to local steady forcing, *Mon. Wea. Rev.*, **100**, 518–540 (1972).
5. P. J. Webster, Temporal variations of low-latitude zonal circulations, *Mon. Wea. Rev.*, **101**, 803–816 (1973).
6. A. Gill. Some simple solutions for heat-induced tropical circulation, *Quart. J. Roy. Meteor. Soc.*, **106**, 447–462 (1980).
7. J. Berlage, The southern oscillation and world weather, *Med. en Verhanddelingen*, n. **88** (1965).
8. A. J. Troup, The southern oscillation, *Quart. J. Roy. Meteor. Soc.*, **91**(390), 490–506 (1965).
9. K. Trenberth, Signal versus noise in the southern oscillation, *Mon. Wea. Rev.*, **112**, 326–332 (1984).
10. G. T. Walker, Correlation in seasonal variations of weather: IX, A further study of world weather (World Weather II), *Memoirs of India Meteor. Dept.*, **24**, 275–332 (1924).
11. P. B. Wright, "The Southern Oscillation—Patterns and Mechanisms of the Teleconnection and the Persistence," Technical Report HIG-77-13, Hawaii Institute of Geophysics, University of Hawaii, Honolulu, 1977, 107 pp.
12. J. Horel and J. M. Wallace, Planetary-scale atmospheric phenomena associated with the interannual variability of sea surface temperature in the equatorial Pacific, *Mon. Wea. Rev.*, **109**, 813–829 (1981).
13. J. D. Opsteegh and H. M. van den Dool, Seasonal differences in the stationary response of a linearized primitive equation model: Prospects for long-range weather forecasting?, *J. Atmos. Sci.*, **37**, 2169–2185 (1980).
14. P. J. Webster, Mechanisms determining the atmospheric response to large-scale sea surface temperature anomalies, *J. Atmos. Sci.*, **38**, 554–571 (1981).
15. P. J. Webster, Seasonality in the local and remote atmospheric response to sea surface temperature anomalies, *J. Atmos. Sci.*, **39**, 41–52 (1982).
16. B. J. Hoskins and D. Karoly, The steady linear response of a spherical atmosphere to thermal and orographic forcing, *J. Atmos. Sci.*, **38**, 1179–1196 (1981).
17. T. Murakami and M. S. Unninayer, Atmospheric circulation during December 1970 through February 1971, *Mon. Wea. Rev.*, **105**, 1024–1038 (1977).

18. P. J. Webster and J. R. Holton, Cross-equatorial response to middle-latitude forcing in a zonally varying basic state, *J. Atmos. Sci.*, **39**, 722–733 (1982).

19. P. A. Arkin and P. J. Webster, Annual and interannual variability of tropical-extratropical interaction: An empirical study, *Mon. Wea. Rev.* **113**, 1510–1523 (1985).

20. J. M. Wallace, "The Climatological Mean Stationary Waves: Observational Evidence," in B. J. Hoskins and R. P. Pearce, Eds., *Large-Scale Dynamical Processes in the Atmosphere*, Academic, New York, 1983, pp. 27–52.

21. I. M. Held, "Stationary and Quasi-Stationary Eddies in the Extratropical Troposphere: Theory," in B. J. Hoskins and R. P. Pearce, Eds., *Large-Scale Dynamical Processes in the Atmosphere*, Academic, New York, 1983, pp. 127–167.

22. T. N. Krishnamurti, "The Subtropical Jet Stream of Winter," Research Report Contract No. N6, NR 082-120, University of Chicago, 1959.

23. P. J. Webster, "Interhemispheric Interaction" Proceedings of the Conference on Southern Hemisphere Meteorology, Sao Paulo, Brazil, American Meteorological Society, Boston, 1983.

24. J. Frederiksen, The effect of long planetary waves on the regions of cyclogenesis: Linear theory, *J. Atmos. Sci.*, **36**, 2320–2335 (1979).

25. M. J. Blackmon, J. M. Wallace, N. G. Lau, and S. Mullen, An observational study of the Northern Hemisphere wintertime circulation, *J. Atmos. Sci.*, **34**, 1040–1053 (1977).

26. T. N. Krishnamurti and H. H. Bhalme, Oscillation of a monsoon system. Part I: Observational aspects, *J. Atmos. Sci.*, **33**, 1937–1954 (1976).

27. P. J. Webster, "Large-Scale Structure of the Tropical Atmosphere," in B. J. Hoskins and R. P. Pearce, Eds., *Large-Scale Dynamical Processes in the Atmosphere*, Academic, New York, 1983, pp. 235–275.

28. D. R. Sikka and S. Gadgil, On the maximum cloud zone and the ITCZ over Indian longitudes during the south-west monsoon, *Mon. Wea. Rev.*, **108**, 1840–1853 (1980).

29. P. J. Webster, Mechanisms of monsoon low-frequency variability: Surface hydrological effects, *J. Atmos. Sci.*, **40**, 2110–2124 (1983).

30. J. L. McBride, Satellite observations of the Southern Hemisphere monsoon during winter MONEX, *Tellus*, **35**, 68–76 (1983).

31. R. A. Madden and P. R. Julian, Description of global scale circulation cells in the tropics with a 40–50 day period, *J. Atmos. Sci.*, **29**, 1109–1123 (1972).

32. T. Murakami, Analysis of the deep convective activity over the western Pacific and southeast-Asia, Part II, Seasonal and intraseasonal variations during northern summer, *J. Meteor. Soc. Japan*, **61**, 60–76 (1983).

33. T. Murakami, and T. Nakazawa, Tropical 45-day oscillations during the 1979 Northern Hemisphere summer, *J. Atmos. Sci.*, **42**, 1107–1122 (1985).

34. T. Murakami, T. Nakazawa, and J. He, "40–50 Day Oscillations During the 1979 Northern Hemisphere Summer," UHMET 83-02, Department of Meteorology, University of Hawaii, Honolulu, 1983.

35. T. Murakami, T. Nakazawa, and J. He, On the 40–50 day oscillations during the 1979 Northern Hemisphere summer, Part I: Phase propagation, *J. Meteor. Soc. Japan*, **63**, 250–271 (1984).

36. A. C. Lorenc, "The Evolution of Planetary Scale 200 mb Divergences During the FGGE Year," Meteorological Office 20, Technical Note II/210, Meteorological Office, Bracknell, Berkshire, England, 1984.

37. T. N. Krishnamurti and D. Subrahmanyan, The 40–50 day mode at 850 mb during MONEX, *J. Atmos. Sci.*, **39**, 2088–2095 (1982).

38. T. Yansunari, A quasi-stationary appearance of 30–40 day period in the cloudiness fluctuations during the summer monsoon over India, *J. Meteor. Soc. Japan*, **58**, 225–229 (1980).

39. T. Murakami, R. V. Godbole, and R. R. Kelkar, "Numerical Experiment of the Monsoon Along 80 E Longitude," Scientific Report No. 62. India Meteorological Department, Poona, 1968.

40. P. J. Webster and L. Chou, Seasonal structure of a simple monsoon system, *J. Atmos Sci.*, **37**, 354–367 (1980).

41. P. J. Webster and L. Chou, Low-frequency transitions of a simple monsoon system, *J. Atmos. Sci.*, **37**, 368–382 (1980).

42. P. J. Webster and G. L. Stephens, Tropical upper-tropospheric extended clouds: Inferences from winter MONEX, *J. Atmos. Sci.*, **37**, 1521–1541 (1980).

43. P. J. Webster, "Mechanisms Determining the Mean and Transient Structure of the Large Scale Winter Monsoon: Cold Surges," Proceedings International Conference on Early Results of FGGE and Large Scale Aspects of Its Monsoon Experiment, Tallahassee, Florida, World Meteorological Organization, Geneva, 1981.

44. R. S. Greenfield and T. N. Krishnamurti, The winter monsoon experiment—Report of the December 1978 field phase, *Bull. Amer. Meteor. Soc.*, **60**, 439–444 (1979).

45. M. Williams, "Interhemispheric Interaction During Winter MONEX," Proceedings International Conference on Early Results of FGGE and Large Scale Aspects of Its Monsoon Experiments, Tallahassee, Florida, World Meteorological Organization, Geneva, 1981.

46. C. P. Chang and K. M. Lau, Northeasterly cold surges and near-equatorial disturbances over the winter MONEX area during 1974, Part II: Planetary scale aspects, *Mon. Wea. Rev.*, **108**, 293–312 (1980).

47. K. M. Lau and M. T. Li, The monsoon of east Asia and its global associations—A survey, *Bull. Amer. Meteor. Soc.*, **65**, 114–123 (1984).

48. H. Lim and C. P. Chang, A theory of mid-latitude forcing of tropical motions during winter monsoons, *J. Atmos. Sci.*, **38**, 2378–2392 (1981).

49. M.-K. Mak, Laterally driven stochastic motions in the tropics, *J. Atmos. Sci.*, **26**, 41–64 (1969).

50. J. G. Charney, A further note on large scale motions in the tropics, *J. Atmos. Sci.*, **26**, 182–185 (1969).

51. P. J. Webster and J. M. Fritsch, Edge waves: Ubiquitous, multiscale atmospheric phenomena, *Quart. J. Roy. Meteor. Soc.* (in press).

12

Orography and Monsoons

Takio Murakami
Department of Meteorology
University of Hawaii
Honolulu, Hawaii

INTRODUCTION

The Asiatic continent is dominated by the Tibetan Plateau, one of the most complex geographical features in the world. At an average elevation of about 4 kilometers with individual peaks climbing to more than twice that altitude, these high mountains not only act as a mechanical barrier to the air currents, but also serve as an effective recipient of heat for atmospheric motions. These are two of the most important effects of the Tibetan Plateau upon the regional weather and monsoonal climate over South and East Asia. The mechanical barrier effect appears to be important for the Asiatic winter monsoon circulation, while the thermal effect is primarily responsible for the establishment of the summer monsoon over Asia.

Edmund Halley (1) in the seventeenth century described the Asiatic monsoon circulation as resembling a giant sea breeze caused by the Eurasian continent–Indian Ocean heat contrast at the earth's surface. Since the Tibetan Plateau is a huge massif of extremely high elevation, it receives a large amount of solar radiation which effectively heats the mountain surface and the air above it, causing a strong heat contrast with the surrounding free atmosphere. This heat contrast at the mid-tropospheric level (4 to 5 km altitude) may be even more pronounced than the continent–ocean contrast at the earth's surface. Thus it is only natural for recent observational investigators to find that Halley's classical theory, which considers only the continent–ocean heat contrast, is far from complete. Throughout the summer monsoon season, substantial cloud development and rainfall occurs over the southeastern Tibetan Plateau, causing condensational heating. This latent heating can be as important as the sensible heat source at the mountain surface. Section 1 discusses the importance of the thermal effect of the Tibetan Plateau in enhancing the Asiatic summer monsoon circulation.

During winter, the Tibetan Plateau, because of its height and extent, exerts strong mechanical barrier effects on the prevailing upper tropospheric westerly winds. These winds tend to flow up as they approach the western periphery of the Tibetan Plateau, and down its eastern slopes. These downward motions enhance the downdraft leg of the north–south Hadley circulation to the east of the Tibetan Plateau during winter. A phenomenon unique to this area is the frequent occurrence of cold surges, that is, strong surges of cold northerly air which burst out of Siberia, through central China, to as far south as the equatorial South China Sea and the Indonesian Seas. The effect of the Tibetan Plateau is to enhance the northerly cold surges during winter and contribute to the equatorial rainfall over the South China Sea–Indonesian Sea region. These points will be elaborated on in Section 2.

The meteorology of the Tibetan Highlands was unknown until the beginning of the twentieth century principally because of a lack of data. During the late 1950s, a network of surface and upper-air stations was established by the military of the People's Republic of China. Most of these stations were situated above an altitude of 3000 m. During the next 10 years (1961–1970) data coverage and quality continued to improve under the Chinese National Project on the meteorology of the Tibetan Plateau region. This project was reported on by Yeh and Gao (2), and the main results of the project were summarized in excellent review papers by Gao et al. (3) and Yeh (4). The observational network was further upgraded with several new and some temporary stations during a special 1978–1979 Global Weather Experiment, making the data distribution over the Eastern Tibetan Plateau, as shown in Figure 12.1, comparable to the coverage over the Rocky Mountain region of the United States. Utilizing this additional data, researchers have already improved our understanding of the orographic influence of the Tibetan Plateau upon monsoon circulations.

There are mountains, such as the Abyssinian Highlands of East Africa, the western Ghats of India and the Arakan Yona Mountains of western Burma, which

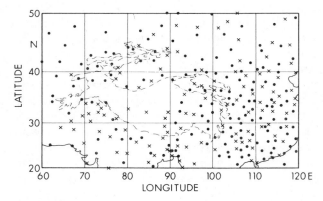

Figure 12.1. Distribution of Tibetan region upper-air stations: rawinsonde (●) and pilot balloon (×) for June 1979.

influence monsoon circulations. These regional features have been covered by others in this volume. This chapter concentrates on the influence of the Tibetan Plateau. While the Tibetan Highlands produce a variety of atmospheric circulation systems with different time and space scales, discussion will be limited to the planetary scale aspects of the monsoon (i.e., spatial scales of 10,000 km and temporal scales of one month and longer).

1 INFLUENCE OF THE TIBETAN PLATEAU UPON THE SUMMER MONSOON CIRCULATION

Flohn (5) postulated that the Tibetan Plateau acts as a sensible heat source at the mid-tropospheric level, and is instrumental in generating a heat low near the mountain surface and a warm core anticyclone (high) in the upper troposphere. Using a diagnostic model with the diabatic heat source prescribed as an external parameter, Murakami (6) demonstrated the importance of sensible heat supply in exciting the upper-tropospheric Tibetan high. Later Flohn (7,8) and Riehl (9) speculated that the latent heating, rather than the sensible heating, is the primary factor for the maintenance of the Tibetan high. However, more recent measurements by Yeh and Gao (2) indicate that sensible heating is as important as, or more important than, latent heating.

1.1 Observed and Analyzed Fields

Figure 12.2 (top) depicts longitude–height sections of seasonal mean temperature departures (anomalies) at 33.75°N during the summer. These anomalies are obtained by subtracting the zonal mean (average temperature around a latitude circle) values from the three-month seasonal average temperatures at different longitudes and heights. During summer, surface temperatures above the Tibetan Plateau are higher than the zonal mean temperatures for the free atmosphere at the same height, that is, the anomalous temperature is positive and exceeds +4.0°C at the top of the Plateau. A maximum temperature anomaly of +6.5°C is found near 230 mb at 70°E. These anomalous temperature fields, which are the largest found anywhere in the world at these altitudes, are reflected in the pressure distribution: prominent upper-level anticyclonic cells are found near 150 mb and low-level cyclonic cells between 700 and 850 mb to the east and west of the Tibetan Plateau. The upper anticyclonic system is generally referred to as the Tibetan high, while the lower cyclonic system is called the monsoon trough. Compared to the Tibetan Plateau region, the temperature fields over the low-elevation African continent are of a different character. Here the temperature anomalies are positive but substantial only in the lower troposphere below about 700 mb. Similarly, over the North American continent, temperature anomalies exceed +4°C only in the lower troposphere below 600 mb. In contrast, over the Atlantic and Pacific Oceans, temperature anomalies are negative throughout the troposphere.

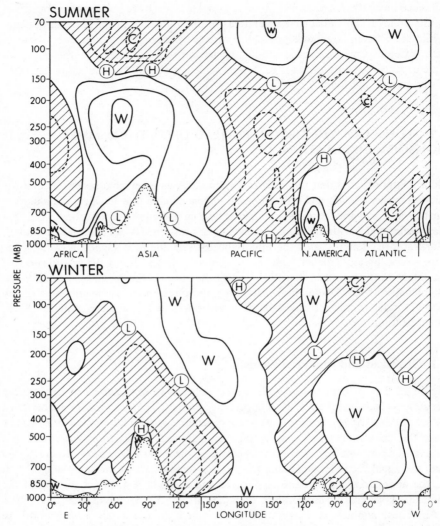

Figure 12.2. Mean temperature anomalies; longitude–height sections of summer (June–August 1979, top) and winter (December 1978–February 1979, bottom) at 33.75°N (2°C interval). Negative anomalies are shaded. Major warm (cold) centers denoted by W (C). Major high (H) and low (L) pressure centers also shown.

The apparent thermal effect of the Tibetan Plateau upon the summer monsoon circulation was questioned by Rangarajan (10), who analyzed the 500-mb temperature field over Eurasia and found an east–west elongated band of high temperatures extending along latitudes 25°–30°N from Egypt, through northern India, into southern China. He concluded that the Tibetan Plateau is not directly responsible for high temperatures over Eurasia. Rather he espoused a more classical theory, similar to

Halley's (1), for explaining the maintenance of the upper-tropospheric monsoon anticyclonic system, namely, that the Tibetan high is a part of the upper-tropospheric monsoon anticyclonic system which extends along 25°N–30°N over the entire Eurasian continent. While the exact role of the Tibetan Plateau upon the planetary-scale monsoon circulation is still unknown, these high mountains undoubtedly contribute to the enhancement of local circulation around the entire highland complex. Let us examine this point.

Figure 12.3 shows the two components of the mean wind for July 1979 as a function of height and latitude along 90°E. The zonal (east–west) component is shown at the top; the meridional (north–south) component at the bottom. The low-level southerlies are seen to be strongest near the southern margin of the Himalayas and extend upward along the south slope. Flow against a barrier causes forced ascending motions. Similarly, there are forced ascending motions to the north of the Tibetan Plateau, as the northerlies flow up the north slope. Where these two meridional flows converge corresponds to the location of the upper-tropospheric Tibetan high. To the south of the Himalayas, above 200 mb, northerly return flow prevails. Its maximum is 4 m/sec near 150 mb at 17°N. This northerly flow represents the upper-level branch of the monsoon meridional circulation along 90°E. Note that it is associated with strong zonal easterlies with a maximum of about 25 m/sec near 100 mb at 17°N (Fig. 12.3, top). Koteswaram (11) defined these strong easterlies as the tropical easterly jet. Below this easterly jet are the low-level westerlies, which exceed 10 m/sec at 850 mb near 13°N. These opposite currents indicate that the tropospheric temperature decreases southward from near the Tibetan Plateau to the equatorial region [see the thermal wind Eq. (10), Section 2.4 in Chapter 9].

The latitude–height cross sections for July shown in Figure 12.3 are representative of the mean winds when the summer monsoon is at its peak. Let us next investigate how these winds are established during the onset of the summer monsoon. In Assam and the southeastern corner of Tibet, summer rains start much earlier than in central India. The frequency and amount of rainfall during April and May in Assam is remarkable, generally exceeding 200–500 mm of rain for these two months alone. The combined effect of the direct heating of the elevated surface and the release of latent heat in the ascending air, causes upper-tropospheric warming. This upper-tropospheric warming and its associated thermal wind balance result in a weakening of the upper westerlies which circulate around the southern periphery of the Tibetan Plateau during winter and spring. By early June, the continual upper-tropospheric warming causes the westerly jet to disappear and to be replaced by the easterly jet. In general, the establishment of this upper-level easterly jet occurs nearly simultaneously with the onset of monsoon rains over central India.

The shift from the westerly to easterly jet takes place in an abrupt manner. Figure 12.4 shows the rapid northward jump of the westerly jet across the Himalayas from mid-May to early June, and the establishment of the easterly jet and the low-level westerlies to the south of the Plateau in early June. Yin (12) claims that the thermal effects of the Tibetan Plateau are responsible for these sudden changes in the zonal wind fields. Following this study, other authors have postulated different views.

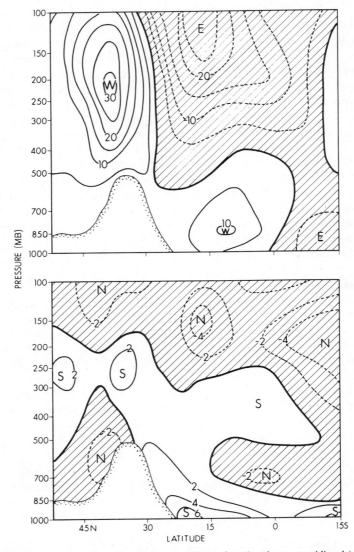

Figure 12.3. Mean zonal (east–west) wind (10 m/sec interval, top) and mean meridional (north–south) wind (2 m/sec interval; bottom) along 90°E for July 1979.

Sutcliff and Bannon (13) confirmed that the northward retreat of the westerly jet over the Middle East occurs nearly simultaneously with the northward jump of the westerly jet around the Tibetan Plateau. Staff members of the Academia Sinica (14) and Yeh et al. (15) also found similar northward jumps of the westerly jet at locations some distance away from the Tibetan Plateau region. Yeh et al. postulated that as the elevation of the sun increases northward from winter to summer in the Northern

Hemisphere, the temperature contrast from pole to equator decreases until it reaches a threshold value, at which time a type of instability in the atmosphere appears, causing an abrupt change in the upper-air circulation which is not directly related to mountain effects.

The 1979 summer monsoon over central India commenced on June 19. Murakami and Ding (16) computed differences in the mean temperature fields (ΔT) between the pre- and post-onset phases. (Here, the term "onset" refers to the monsoon onset over central India. As mentioned earlier, there is a marked regional difference in the timing of the monsoon onset.) From Figure 12.5, it is immediately evident that at both 700 and 300 mb there are well-organized east–west oriented bands of positive temperature differences ΔT at approximately 30°N-40°N across Eurasia with three distinct centers: near Japan, the western end of Tibet, and the Saudi Arabian desert. Hence, these temperature changes between the pre- and post-monsoon onset periods represent a very large-scale phenomenon. The largest temperature increase occurs over the western Tibetan Plateau where the northward retreat of the

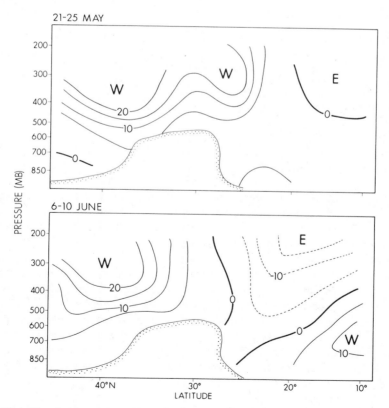

Figure 12.4. Five-day mean zonal (east–west) wind (5 m/sec interval) along 90°E for 1956, May 21–25 (top) and June 6–10 (bottom) (14).

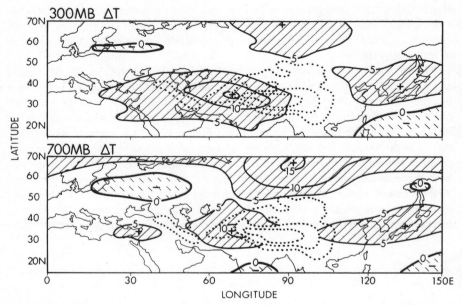

Figure 12.5. Mean temperature difference between the pre-onset (May 15–30) and post-onset (June 20–30) phases of the 1979 summer monsoon as a function of latitude and longitude at 300 mb (top) and 700 mb (bottom) with 5°C intervals. Topographical contours (1.5 km interval) are shown by the dotted lines. Areas of ΔT greater than 5°C are hatched and ΔT less than 0°C are indicated by dashed-hatching.

westerly jet is most prominent (Fig. 12.4). In contrast, over the eastern Plateau where atmospheric warming occurs much earlier than the monsoon onset over central India, the temperature increase is small and the associated northward jump of the westerly jet is not clearly defined. Murakami and Ding's study indicates the importance of the thermal effect of the Eurasian continent as a whole. However, the western Tibetan Plateau is instrumental in causing a local enhancement of the temperature increase during the monsoon onset phase.

Insight as to how this local enhancement of temperature comes about may be gained by investigating the changes in the heat balance over the Tibetan Plateau before, during and after the monsoon onset over central India. In an effort to measure the net heat source from the earth's surface to the 100-mb level, Luo and Yanai (17) applied objective analysis techniques to Global Weather Experiment data for the 40-day period from May 26 to July 4, 1979. They evaluated an apparent heat source Q_1 as a residual in the thermodynamic equation. The computations summarized in Table 12.1 reveal significant changes in Q_1 over the western Tibetan Plateau from 208 units in Period I (pre-onset) to -2 units (or cooling) in Period IV (post-onset). Over the eastern Tibetan Plateau, Q_1 is positive and substantial throughout the four periods, with the largest value of 348 units during Period IV. Luo and Yanai also found large diurnal changes in the surface air temperature on the Plateau.

TABLE 12.1. Large-Scale Heat Sources (Q_1) in cal/(cm²day), Vertically Integrated from the Earth's Surface to 100 mb

	Period	Western Tibet	Eastern Tibet
I	(5/26–6/4)	208	236
II	(6/5–6/14)	152	205
III	(6/15–6/24)	104	168
IV	(6/25–7/4)	−2	349

[a] Data are from over the western and eastern Tibetan Plateau (17).

A deep, mixed boundary layer with nearly constant potential temperature was observed in the evening. They postulated that dry thermal convection originating near the heated surface in the afternoon hours may reach the upper troposphere, depositing sensible heat there. The heating above the Plateau is pronounced in the upper-tropospheric layer between 200 and 500 mb with a mean heating rate of 3°C per day.

Based on ten years (1961–1970) of station data, Yeh and Gao (2) computed various components of the long-term mean tropospheric heat balance over the Tibetan Plateau as follows:

$$Q = Q_{rad} + Q_{sens} + Q_{lat} \tag{1}$$

where Q represents the net heat source, Q_{rad} is the net radiational heating or cooling, Q_{sens} is the sensible heat transfer from the earth's surface and Q_{lat} is the latent heat released by condensation (refer to Eq. (6), Section 2.3 in Chapter 9). The net radiational effect Q_{rad} can be evaluated from the sum of the effective long-wave radiation from the earth's surface, the absorption of solar radiation, and the outgoing long-wave radiation from the tropopause. Note that the vertically integrated Q_1, as measured by Luo and Yanai (17), corresponds to the difference $(Q - Q_{rad})$ in the above equation.

In Figure 12.6, only the monthly mean values of Q_{sens}, Q_{lat}, and Q are shown. (Q_{rad} is nearly constant, about −200 units, throughout the year and thus not reproduced here.) One immediately notes an extremely large sensible heat flux over the arid western Tibetan Plateau with a maximum (450 units) in June. This is about twice the magnitude of Flohn's (8) estimate of 250 units. Because of the height of the Plateau, this sensible heat supply is used directly to warm up the middle and higher troposphere. Compared to western Tibet, sensible heating, Q_{sens}, over eastern Tibet is much less pronounced; yet it exceeds latent heating, Q_{lat}, until June. This contradicts Riehl's (9) speculation that latent heating, rather than sensible heating, is the primary factor in atmospheric heating sources over the eastern Tibetan Plateau. (The monthly mean rainfall over the eastern Plateau is slightly more than 100 mm in July. This amount is approximately 10 times as large as the July mean rainfall over the western Tibetan Plateau. However, it should also be noted that it is less

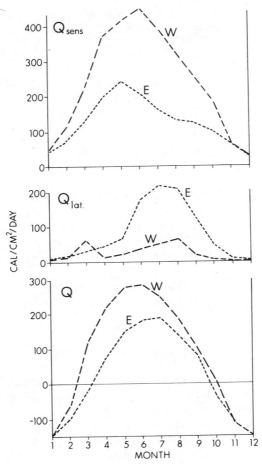

Figure 12.6. Ten-year (1961–1970) means for sensible heat flux, in cal/cm²/day, at the surface (Q_{sens}), latent heat of precipitation (Q_{lat}), and net atmospheric heat source (Q), over the western (W) and eastern (E) Tibetan Plateau (2).

than one-fourth the July mean rainfall over central India). It is only during the height of summer (July and August), that Q_{lat} becomes slightly larger than Q_{sens} over eastern Tibet. Because of the predominance of Q_{sens}, the net heating Q, is quite large over the western Tibetan Plateau, resulting in this region contributing most to the net heat balance over the entire Tibetan Plateau. The net contribution from the eastern Tibetan Plateau is less, particularly from March through June.

Although computed heating components obtained by Yeh and Gao (2) provide useful information on the heat budget over the Tibetan Plateau, some of the values used to calculate heating are only estimated and may be of questionable accuracy due to observational errors and/or assumptions made to do the calculations.

An important feature of the Tibetan Plateau is the strong diurnal variations. Nitta (18) showed the large diurnal variations in the mean tropospheric heating rate Q over the eastern Tibetan Plateau. Using Japanese satellite infrared observations taken eight times a day during the 1979 summer, M. Murakami (19) investigated the diurnal variation of convective activity over and around eastern Tibet. He found intense diurnal variations over the southern part of Tibet, with the minimum activity at 0900 LST and the maximum intensity at 1800 LST. Significant diurnal variations are also observed in the wind, temperature and pressure fields [e.g., (8) and (2)]. Figure 12.7 represents the 10-year (1961-70) mean 600-mb geopotential height

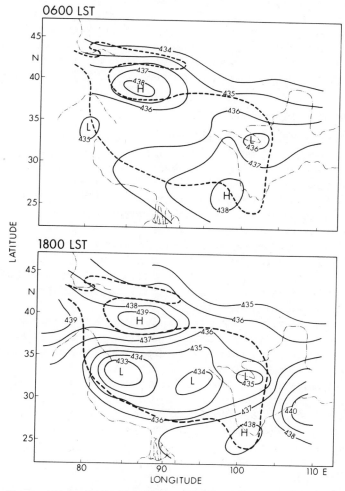

Figure 12.7. Ten-year (1961–1970) mean 600-mb geopotential height, in decameters, at 0600 LST (top) and 1800 LST (bottom) for July. Major rivers are indicated by thin chain lines and the Tibetan Plateau is outlined by a dashed curve (2).

fields at 0600 and 1800 LST in July. At 1800 LST, a low-pressure system occupies almost the entire Plateau, while in the morning, a weak high-pressure system prevails. Consequently, the daily mean pressure at the 600-mb level is lower over the Plateau than it is over the surrounding area. This means that the effect of daytime heating is not offset by nighttime cooling. Continental-scale heating could also play a substantial role in the continued maintenance of the daytime circulation through the night. Further study is needed to investigate a possible interaction between diurnal variations around Tibet and the large-scale summer monsoon circulation.

So far we have shown that our knowledge gained by observation is insufficient to adequately describe the exact role of the Tibetan Plateau in maintaining the summer monsoon circulation. Despite the efforts of many observational meteorologists, there continues to be no unified and detailed description. This may be due partly to the fact that the observed data reflect the effects of processes on many different time and space scales making it difficult to isolate the orographic influence. To overcome this difficulty, several scientists employ other approaches, such as numerical model experiments and laboratory simulation or so-called "dish pan" experiments.

1.2 Numerical Model Experiments

An effective method of identifying the role of mountains in the monsoon circulation is to perform numerical simulation experiments. The experiments are performed both with and without the mountains and the results compared. Several such experiments have been conducted. Murakami et al. (20) used an eight-layer, two-dimensional model to study the summer monsoon. The primary purpose of the model experiment was to study the influence of the Himalayas on the zonal wind components—the low-level westerlies and the upper-level easterly jet—of the monsoon circulation. The model extends from the equator to the North Pole, at 80°E, with the Asiatic continent lying poleward of 15°N and the Indian Ocean lying equatorward of 15°N. The model includes the effects of radiation and condensational (latent) heating. The land-surface temperature is determined from the calculated heat balance, while the sea surface temperature is kept constant at 27°C.

The model was started with a completely calm and dry atmosphere using the observed July mean temperature distribution. The experiment was performed for a period of 80 days with 10-minute time steps. Without mountains, the low-level westerly wind maximum was less than 5 m/sec and the upper-level easterly wind maximum was less than 10 m/sec. When the mountains were included, the low-level westerly jet exceeded 10 m/sec and the upper easterly jet became stronger than 35 m/sec (Fig. 12.8). These model computed westerly and easterly jets are slightly stronger than the corresponding observed speeds in July along 90°E (Fig. 12.3), but generally the results confirm the strong impact of Tibet on the monsoon circulation.

In the Murakami et al. experiment, the meridional motions first responded to the heat contrast between the continent (including mountains) to the north and the ocean to the south. The zonal motions were then induced via the Coriolis force (the earth's rotation). In the real atmosphere, disturbances (i.e., traveling cyclones

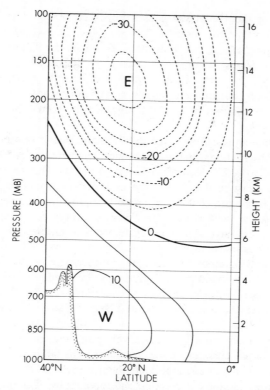

Figure 12.8. Height–latitude section of simulated zonal wind (5 m/sec interval) at 80°E for the summer season (20).

which cannot be accounted for explicitly in a two-dimensional model) also play a role, carrying energy away from the source region. Thus in this model, the role of the meridional motions is overemphasized. To overcome this difficulty, a three-dimensional numerical model is needed.

Hahn and Manabe (21) performed numerical experiments using a three-dimensional, global general circulation model to study the effect of the Himalayas on the establishment of the monsoon circulation. Experiments were conducted with no mountains (hereafter called the NM model) and with mountains (M model). In the M model, surface southeasterly winds diverged out of the subtropical anticyclones of the Southern Hemisphere, becoming southerly winds as they crossed the equator. They then converged into the vicinity of the south Asian low-pressure belt around 25°N. Other important low-level characteristics of the south Asian monsoon were also simulated. For example, along the Abyssinian Highlands off the east coast of Africa, a strong concentrated southerly flow, commonly called the Somali jet, was simulated by the M model. In comparison, the NM model produced only an ill-defined wind field in that vicinity.

The presence of the Himalayas causes important changes in flow patterns at levels much higher than the mountains themselves. The M model simulates the rapid northward progress of the subtropical jet across the Tibetan Plateau during the monsoon onset. This is consistent with Yin's (12) finding discussed earlier. Without mountains, the subtropical jet does not abruptly jump northward to its summertime position, but rather it slowly moves northward throughout May and June, stabilizing its position in July at a latitude approximately 10° farther south than in the M model.

Maximum temperatures found over the Tibetan Plateau in the M model agree with Flohn's (8) observations. In the NM model, temperatures at 500 mb over Tibet are approximately 10°–12°C lower than in the M model. This clearly indicates that the mountains, at least in the model, act as a mid-tropospheric heat source and contribute to the formation of the warm core low pressure area near the surface of the Tibetan Plateau as well as the warm core high pressure system aloft. The M model and NM model have significant differences in the heating fields. In the M model, the latent heating tends to be most important over Tibet while in the NM model, sensible heating tends to dominate. Furthermore, in the M model, larger amounts of rainfall are found particularly along the southern slope of the Tibetan Plateau. In short, the M model appears to overemphasize the effect of latent heat of condensation over Tibet as compared to Yeh and Gao's (2) observations shown in Figure 12.6.

More recently, Kuo and Qian (22) investigated the influence of the Tibetan Plateau on the diurnal changes of the meteorological fields in July. They used a five-layer, global, primitive equation model which included the effects of solar and longwave radiation, cumulus convection and topography. This model created pronounced diurnal variations. Over the central portion of the Tibetan Plateau, it produced a maximum diurnal temperature change of 12°C, while in the surrounding area, diurnal changes remained less than 2°C. Analysis of the various heating processes in the model revealed that the large net heating over the Tibetan Plateau is mainly due to absorption of solar radiation by the plateau surface and sensible heat transfer to the atmosphere, in agreement with Yeh and Gao's (2) observations. Kuo and Qian also showed that latent heat released by cumulus convection and by large-scale flow is concentrated along the southern edge of the plateau and in the region to the south.

Figure 12.9 shows the distributions of the total nighttime and daytime net heating rates at 90°E. Over the plateau, all levels of the atmosphere lose heat during the night, while to the south over the plain, the atmosphere gains heat from condensation. During the daytime, the atmosphere gains heat almost everywhere below 100 mb. Here the strong heating on the south side is largely due to convective precipitation, while heating on the north slope is mainly from solar radiation.

The nighttime and daytime meridional circulations along 90°E are illustrated in Figure 12.10. During the night, descending motion prevails below 300 mb over the plateau as well as along the northern and southern slopes. During the day, these descending motions are replaced by strong updrafts. Further south of the mountains, strong low-level southerlies are capped by prominent upper-level northerlies during

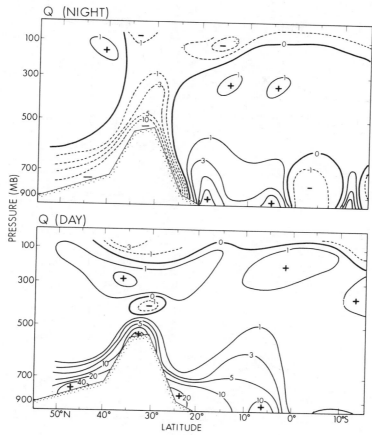

Figure 12.9. Computed heating rate, Q (°C/day) along 90°E at nighttime (top) and daytime (bottom) (22).

both day and night. The daily mean fields (not shown) compare well with the observed July mean meridional circulation of Figure 12.3 (bottom).

Using a different approach, Webster and Chou (23) performed an experiment which included the Asiatic continent and adjacent oceans but not the Tibetan Plateau, thereby focusing on the heat contrast between the Asiatic continent and adjacent cooler oceans rather than on the topography. This experiment used a two-level primitive equation model in which a continental cap north of 18°N is surrounded by interactive and mobile oceans. Experiments were conducted with three variations of the model; a dry ocean-continent model (DOC), a moist ocean–continent model (MOC) and a moist ocean model (MO) where the entire earth is covered by ocean. The MOC includes hydrologic processes but is otherwise identical to the DOC. Figure 12.11 depicts the changes in the zonal mean winds at 250 mb during the one-year period from day 1095 (January 1) to 1460 (December 31). In the MOC

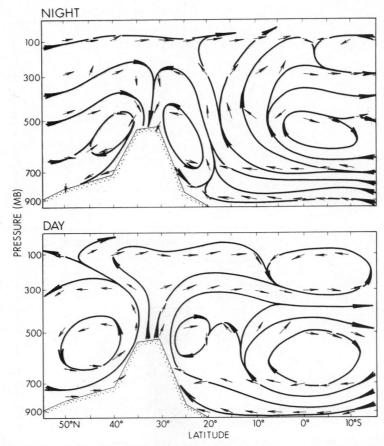

Figure 12.10. Schematic nighttime (top) and daytime (bottom) mean meridional circulations at 90°E with vertical velocity magnified 100 times (22).

model, the upper-level zonal winds underwent substantial interseasonal changes, from 37 m/sec (westerly) at 30°N during winter to -25 m/sec (easterly) at 10°N in summer even without the effects of the Tibetan Plateau. The annual changes are slightly reduced in the DOC model and much reduced in the MO model. Thus the interactive nature of the oceans adjacent to the Asiatic continent appears to be a crucial element in determining the magnitude and spatial variation of the upper-level circulation. (See Webster's discussion of these and related experiments in Chapter 11).

Although the theoretical information obtained by Webster and Chou is instructive, their experiments would have to be extended to determine the relative importance of the orographic influence. In the MOC model, the annual changes over the Southern Hemisphere are substantially smaller than the corresponding changes in the Northern Hemisphere. In fact, the MOC model, which has no land mass south of the equator, fails to produce the normally observed pattern of easterly winds over the Southern Hemisphere tropics during the Southern Hemisphere summer (Northern Hemisphere winter). This suggests that topography does play a role in producing the annual variations of the upper-level winds over the Southern Hemisphere summer monsoon region. Another interesting feature of the MOC model is that the westerly jet near 30°N during the Northern Hemisphere winter is comparable to the corresponding westerly jet near 30°S during the Southern Hemisphere winter. In the real atmosphere, the Northern Hemisphere westerly jet is much stronger than its Southern Hemisphere counterpart over the Indian Ocean. As will be shown in

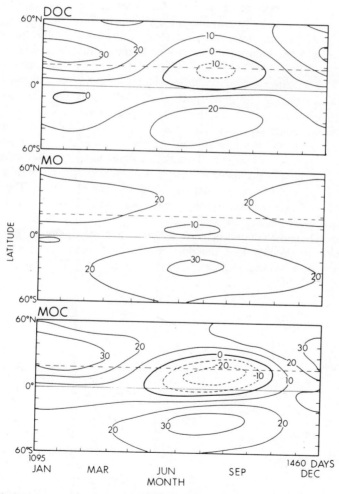

Figure 12.11. Latitude–time sections of 250-mb zonal wind component (10 m/sec interval, negative values denote easterlies). The dashed line at 18°N separates ocean from continent (see text) (23).

Section 2, the Tibetan Plateau plays an important role in the intensification of the world's strongest westerly jet over East Asia during the Northern Hemisphere winter.

1.3 Rotating Annulus Experiments

Experimental physical simulation of the atmospheric circulation using laboratory apparatus began in the mid-nineteenth century. It was recognized at that time that the large, planetary-scale features of the atmospheric circulation are controlled principally by differential heating between the equator and the poles and by the rotation of the earth. Other features, such as land–ocean contrasts and mountains were viewed as secondary. The first significant work was carried out by Fultz et al. (24) and Hide (25), both of whom performed a series of atmospheric simulation experiments in differentially heated rotating cylinders of fluid. Recently a group of Chinese scientists (among them Yeh and Chang, 26) successfully simulated the summer monoon circulation over the Eastern Hemisphere. They used a rotating multichamber annulus, 1 m in diameter (Fig. 12.12). This annulus (A) was filled to 10 cm with a working medium which consisted of a mixture of glycerine and water with small white, neutrally buoyant, plastic balls as tracers. The tracers were illuminated by a horizontal beam of light which could be raised or lowered to show the fluid motion at various levels. The results were recorded by a motion picture camera. The cold source was a mixture of ice and water in the inner chamber (C) and the heat source was a hot-water bath with automatic temperature control (B). The Tibetan Plateau was simulated by a half ellipsoidal block (D) with major and minor axes of 7.0 and 4.8 cm, respectively, and a height of 3.0 cm. The southern half of the block had a resistance wire which could be electrified (and heated) to simulate the summer heating of the Plateau.

The basic state in the annulus is the flow field before the "plateau" is immersed. The basic state can be altered by changing the rate of rotation and/or the temperature difference between the outer cylinder and the inner cylinder. Two basic states were

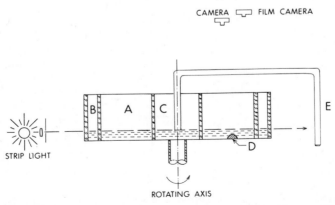

Figure 12.12. Schematic diagram of the Yeh and Chang (26) annulus experiment.

investigated. In the first, A, there were many irregular small vortices and the average zonal speed of the fluid was nearly zero. The second basic state, B, was a purely zonal flow which was produced by rotating the annulus at about the same rate as for A but increasing the temperature difference by nearly a factor of 10.

Into each basic state, the "plateau" is immersed at the appropriate geographical position. Experiments were conducted with an unheated plateau and then the plateau was heated at one of three different rates, namely, 2, 5, and 8 watts. Without heating, the effect of the plateau is purely mechanical with temperatures in the annulus gradually decreasing from the outer cylinder wall to the inner cylinder wall. When heat is applied at 8 W, the temperature above the plateau (T_g) is slightly higher than the temperature at the wall of the (hot) outer cyclinder (T_o). At the low heating rate (2 W), T_g is slightly lower than T_o. In the real atmosphere, the temperature above the plateau is higher than that at the same height over India. Thus, this pattern is duplicated in the annulus when T_g is higher than T_o. The effect of the plateau differs considerably for the different basic states.

The near surface flow in state A consists of many irregular small vortices and the mean flow is almost zero. When the nonheated plateau is added to the annulus, the motion is still irregular; however, the vortices become larger. When heat is added to the plateau, there is a conspicuous change in the circulation. First, the irregular small vortices in the upper layer in the vicinity of the plateau gradually organize into a large anticyclone. At a heating rate of 2 W, a closed anticyclone forms in the upper layer centered over the southeast part of the plateau and comparable in size to the plateau (Fig. 12.13, top left). With higher rates of heating, the anticyclone becomes larger, and extends northward (Fig. 12.13, bottom left). In the lower layer, the flow circulates cyclonically around the plateau and covers a much larger area when the rate of heating is increased.

The dynamics of this circulation are quite straightforward. When it is heated, convergent motion is induced toward the plateau in the middle and lower layers. In the rotating annulus, the tracers were observed to climb up the plateau in almost all quadrants. Therefore, in the upper layer, there is compensating divergent flow outward from the plateau. The convergent flow into the plateau generates the lower and middle layer cyclonic circulation, and the divergence above creates upper-layer anticyclonic circulation.

The circulation in the annulus appears similar to the observed meridional circulation shown in Figure 12.3. In the annulus, the vertical motions associated with the meridional circulation cannot be observed; however, the tracer balls do confirm that in the lower layers, the flow converges into the plateau as it circulates cyclonically around it. This lower-level flow is southerly, climbing up the plateau. To compensate, there is a return northerly flow in the upper layers of the annulus which descends further south. In basic state A, the medium- and small-scale convective systems are effective elements transporting heat from the plateau to the high "atmosphere."

The basic state B is a purely zonal flow and has a large horizontal temperature gradient. According to Hide (25), the rotation of the annulus produces a vertical temperature gradient that is directly proportional to the imposed horizontal temperature gradient. Consequently in basic state B, the fluid is very stable vertically (e.g.,

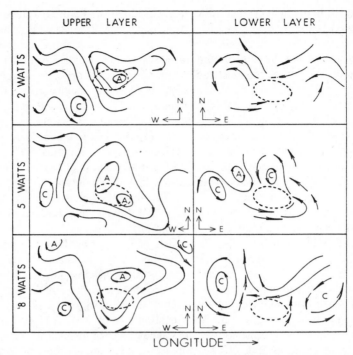

Figure 12.13. Streamline charts in the upper (left) and lower (right) layer, obtained in Yeh and Chang's (26) annulus experiments with the heating rate of the Plateau at 2, 5, and 8 W, respectively, and basic state A. The dashed curve represents the simulated Tibetan Plateau. North is toward the top of each panel. A's and C's mark anticyclonic and cyclonic circulations, respectively.

warm fluid overlying cooler fluid). Also, as a result of thermal wind balance, the fluid has large vertical wind shear resulting in strong upper-level westerlies. In this basic state, there is relatively little effect from increasing the north–south differential heating because the vertical stability prevents the development of vertical motion and associated heat transport.

When heat is added to the plateau, a major change occurs in the circulation. As expected from thermal wind considerations, the strong upper-layer westerlies over the plateau weaken. Also, a relatively stable trough forms at about 30° longitude upstream from the plateau with an upper-level ridge to the southeast. As the rate of heating is increased, the ridge becomes stronger. But even with a heating rate of 14 W, the upper-level anticyclonic circulation does not become closed off. In basic state A, a closed high appeared over the southeastern part of the plateau with only 2 W of heating. The strong temperature gradient and high velocities in basic state B probably prevents the formation of closed circulations even at 14 W. A steep vertical temperature gradient (high vertical stability) hinders the development of vertical velocity and upper-level divergence, which is the mechanism for the formation of highs. Another conspicuous change caused by the heated plateau is a clear northward displacement of the belt of strong westerlies in the plateau region. This agrees with the observation that a relatively abrupt northward shift of the westerly jet over Tibet occurs with the onset of the summer monsoon (Fig. 12.4).

The annulus experiments by the Chinese scientists clearly indicate the importance of heating of the Tibetan Plateau during summer. However, the thermal effect of the Tibetan Plateau is probably overemphasized in their annulus experiments, which do not include heating over the vast Asiatic continent. Furthermore the heating applied to the model plateau is constant in time, and thus corresponds to perpetual summer. The influence of annual cycles of heating–cooling should be investigated. During the winter, the Plateau may act as a cold source (except the southeast slopes which remain as a warm source). Furthermore, the basic zonal flows for the annulus experiments are ideal patterns. In the future, another basic state which more closely resembles the summer (winter) situation should be used to investigate the effect of a heated (cooled) plateau. The influence of diurnal cycles of heating and cooling should also be investigated.

2 INFLUENCE OF THE TIBETAN PLATEAU UPON THE WINTER MONSOON CIRCULATION

We know that with or without mountains there would be strong temperature contrasts between continents and adjacent oceans that reverse sign from summer to winter. This is, in fact, the primary reason we have monsoons (see Chapters 1 and 9). To what extent mountain complexes enhance these contrasts and affect the monsoon circulation as regional heat sinks in winter is not well understood. We do know that in winter the Tibetan Plateau acts as a heat sink for the monsoon circulation (Fig. 12.6) but, although it is not certain, it appears that the thermal effect is not significant during winter. On the other hand, a great deal of research has been done on the mechanical influence of the Plateau.

2.1 Observed and Analyzed Fields

The Himalayan complex is a huge massif of extremely high elevation, therefore, it probably represents an important mechanical influence upon winter westerly flows. Chinese meteorologists (14, 27, 28) have shown that air currents tend to flow around, rather than up and over, the massif. They also have shown that the low-level westerlies split at the western end of the Tibetan Plateau into northern and southern branches and that the upper-level westerly jet stream is strongest, not above the Plateau, but rather along its southern periphery. Although these findings are based on relatively sparse data for the period of 1955 to 1957, they agree with a more recent analysis by Yeh and Gao (2), in which 10 years (1961–1970) of station data were used.

Except for the Chinese studies, there has been little observational information on the orographic influence of the Tibetan Plateau upon either the winter mean or daily circulation patterns. There remain many unanswered questions. Recently Murakami (29–32) examined several problems related to the Tibetan Highlands using temperature, wind and geopotential height data for the 1978–1979 (Global Weather Experiment) winter. He investigated the extent to which the Tibetan Plateau controls the winter mean circulation, and its role in altering the structure (horizontal and vertical tilt, phase speed and intensity) of wave perturbations as they pass over the high mountains. He also investigated the mechanisms responsible for initiating strong low-level cold surges in the lee of the elevated highlands.

The 1978–1979 winter mean wind and vorticity fields at 700 mb are shown in Figure 12.14. The 700-mb mean wind splits into two major streams as it encounters the western end of the Tibetan Plateau. The northern branch is directed northeastward approximately along the 2 km height contours, then turns eastward into the saddle area between the Tibetan Plateau and the Altai Mountains. The southern branch, which is slightly weaker than its northern counterpart, flows eastward nearly parallel to the southern periphery of the Tibetan Plateau. To the east of the Plateau, these branches tend to converge into a single major stream near 120°E. Of particular

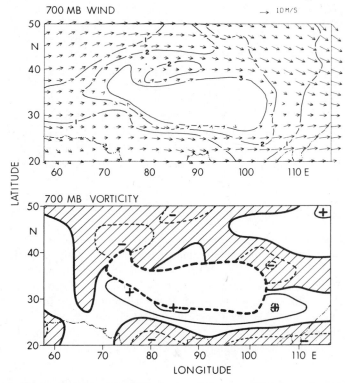

Figure 12.14. Wind and vorticity. Top: Winter mean wind vectors (10 m/sec unit vector). Full lines are smoothed topographic contours with 1 km interval. Bottom: Winter mean vorticity (1×10^{-5} sec^{-1} interval). Negative (anticyclonic) vorticity shaded. Heavy dashed line is 3 km contour of the plateau. Fields are at 700 mb, averaged from December 1978–February 1979 (27).

interest is the zone of extremely weak winds which extends approximately 1000 km downstream from the eastern slope. Associated with this wind minimum is a vorticity pair, with anticyclonic vorticity to the north and cyclonic vorticity to the south (Fig. 12.14, bottom). A similar feature, although not as well defined, is observed near the western end of the Tibetan Plateau. Interestingly, these features, called Karman vortices, resemble the flow fields found in laboratory experiments in which a fluid moves around a sharp-edged obstacle. Hirota and Miyakoda (33) were able to simulate such a Karman vortex street behind a circular cylinder using a mathematical model for the vorticity of a two-dimensional, incompressible fluid.

At 500 mb (Fig. 12.15, top), the splitting of the main westerly flow is still seen at both 25°N and 45°N between about 70°E and 120°E. An interesting feature is that the winter mean winds at 500 mb are weakest (less than 10 m/sec) directly over the Plateau. These extremely weak winds may be an indication of the frictional effects of the Tibetan massif. By comparing observed against calculated geostrophic wind velocities, Murakami (29) estimated the average height of the planetary boundary layer to be about 1.5 km above the Plateau. Thus, over the Plateau, the top of the planetary boundary layer may reach up to about 6 km above sea level (500 mb).

Winter mean winds at 300 mb (not shown) are also weak over the Plateau, indicating that the mechanical effect of the Tibetan Plateau may extend that high. Taylor (34) demonstrated in the laboratory that a rotating homogeneous fluid not only flows around isolated obstacles, but also flows around the imaginary fluid columns which extend upward from the obstacles. These imaginary fluid columns are now referred to as "Taylor columns" after Hide (35). The 300-mb wind field suggests that the Taylor column phenomenon may occur above the Tibetan Plateau. Nakamura (36, 37) simulated flow fields over and around a prescribed obstacle using a six-level primitive equation model. His simulated flows exhibited features that were similar to the Taylor column phenomenon; namely, (a) lower-tropospheric winds tended to flow around the obstacle, and (b) upper-tropospheric flows were characterized by weak winds above the obstacle. In an annulus experiment with a nonheated plateau, Yeh and Chang (26) found splitting of the westerlies with a point of diffluence upstream and a point of confluence downstream. This splitting phenomenon diminishes sharply upward, which is consistent with the observational results of Murakami (Figs. 12.14 and 12.15).

We have so far shown that low-level winds generally flow along the borders of the east–west oriented Tibetan Plateau, with relatively weak winds over the high mountains. In fact, even at 200 mb (Fig. 12.15, bottom), winds are strongest not above, but along the southern periphery of the Himalayas. This characteristic feature of the winter mean flow cannot be explained by the conservation theorem of absolute potential vorticity (see Chapter 9, Section 2.5). According to the theory, if air encountering an obstacle rises, acquiring anticyclonic curvature as the depth of the air column decreases on its windward side, the wind speed should increase markedly above and to the north of the barrier (38). This is not the case in the observed winter mean upper-tropospheric flows for 200 mb.

One also should note the distinct eastward acceleration of the jet stream south of the Himalayas at the 200-mb level. East of 70°E, this eastward acceleration is

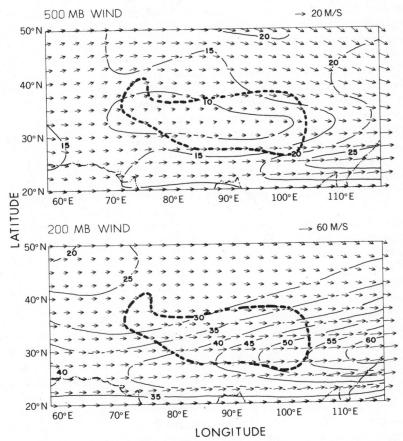

Figure 12.15. Winter (December 1978–February 1979) mean wind vectors with 5 m/sec isotachs for 500 (top) and 200 (bottom) mb. Heavy dashed line denotes the smoothed 3 km topographic contour (27).

associated with prominent southerly flows from the tropics. By estimating the value of each term in the time-averaged, zonal momentum equation, Murakami (29) confirmed that the jet stream and its substantial eastward acceleration are maintained primarily by southerly nongeostrophic winds. This southerly flow may correspond to an upper branch of the thermally direct, local Hadley circulation which is associated with the heat contrast between the cooler continent and high mountains to the north and the warmer ocean and equatorial rainfall to the south. This reasoning was corroborated by Chinese scientists (28) using a quasi-geostrophic model with prescribed diabatic heat sources. They demonstrated the importance of a heat sink over the Tibetan Plateau in inducing a strong jet stream to its south.

To further investigate the characteristic features of local Hadley circulations over the Asiatic continent, the winter mean (December 1978–February 1979) velocity potential patterns were calculated from observed divergence fields. Figure 12.16 gives an indication of strong local orographic effects acting on the winter mean divergent circulation. Near the western end of the Tibetan Plateau, a marked divergent center with strong 200-mb outflow in all directions is located at 43°N, 68°E, while at 700 mb there is a region of dominant inflow. Thus upward motion is implied over the southern Soviet Union. (Vertical wind observations from the good data coverage shown in Fig. 12.1 support this idea.) The substantial snowfall that occurs over this region during winter is further evidence of upward motion. This upward motion is associated with significant eastward deceleration of the 500-mb westerlies (Fig. 12.15, top) as they approach the western end of the Tibetan Plateau. Presumably this easterly deceleration contributes to the enhancement of upward motion in the mid-troposphere, with upper-level divergence above and low-level convergence

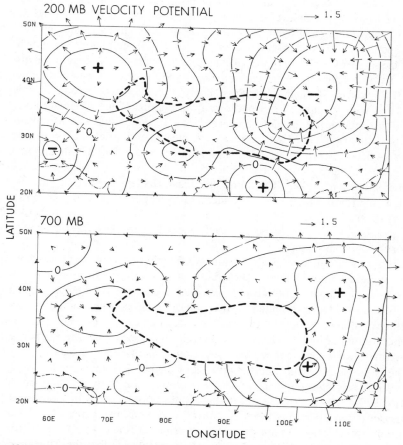

Figure 12.16. Winter (December 1978–February 1979) mean velocity potential (full lines, 2.5 × $10^5 m^2$/sec interval) and divergent wind (arrows, 1.5 m/sec unit) fields for 200 (top) and 700 (bottom) mb (27).

below 500 mb. As noted earlier, a large portion of 500-mb winds tend to flow around the western end of the Tibetan Plateau. However, a substantial part of the 500-mb winds does flow upward, above and parallel to the steep western slope of the Plateau.

South of the updraft region over the southern Soviet Union, the 200-mb divergent flows are northerly, while there are southerly 700-mb divergent flows, over Afghanistan, Pakistan, northwest India and the northern Arabian Sea, implying downdrafts and persistent dry weather during winter. Therefore, the north–south vertical overturning in this region west of Tibet is thermally indirect.

To the east of the Tibetan Plateau, there is a thermally direct, north–south oriented vertical circulation with a dominant downdraft center at around 40°N, 105°E. Here, a dominant inflow center at 200 mb nearly overlies a prominent outflow center at 700 mb, resulting in mid-tropospheric descending motions. Near the eastern end of the Tibetan Plateau, 200-mb divergent flows are predominantly southerly and emanate from an upper-level divergent center near 22.5°N, 95°E in the northern Bay of Bengal. As mentioned earlier, these southerly, upper-level divergent winds are responsible for the eastward acceleration of the 200-mb jet stream which flows along the southern periphery of the Tibetan Plateau. Interestingly, upper-level divergent flows are also southerly over southern China and the South China Sea region east of 100°E between 20° and 40°N. Presumably, these southerly divergent winds are associated with the upper-level outflows from regions of heavy equatorial rainfall (Malaysia, Indonesia, and northern Australia). In fact, Sumi and Murakami (39) computed the 1978–1979 winter mean velocity potential fields over an extensive region (40°E–110°W, 30°S–45°N) and found that the 200-mb divergent center during the Northern Hemisphere winter, is located over the Southern Hemisphere summer monsoon region. Therefore, the local Hadley circulation east of about 100°E is thermally direct, and characterized by upper-level divergent flows from regions of heaviest equatorial rainfall, and a vigorous, lower-tropospheric return flow (monsoonal surges) with orographically enhanced subsidence near the northeastern end of the Tibetan Plateau.

A phenomenon unique to this area is the frequent occurrence of low-level cold surges bursting out of Siberia. These cold surges correspond to an intensification of the low-level branch of the Hadley circulation described in the paragraph above. The intensified Hadley circulation then enhances the convective activity over the maritime continent (Malaysia, Borneo, Indonesia, and Australia). Our main concern in what follows is how the cold surges develop in association with the passages of mid-latitude wave disturbances over and around the Tibetan Plateau.

The winter mean wind fields (Figs. 12.14 and 12.15) undoubtedly exert some control in changing the intensity and phase propagation of wave disturbances over East Asia. Figure 12.17 shows the standard deviation patterns of meridional wind fluctuations with a period of 3.5 to 5.5 days. At 200 mb, two bands of high standard deviations (greater than 3 units) occur along about 30°N and 50°N, with a minimum zone over the Tibetan Plateau. The distribution suggests that the Tibetan Plateau influences wave perturbations even at the 200-mb level. Members of the Academia Sinica (28) found that extended north–south troughs split into two parts when they

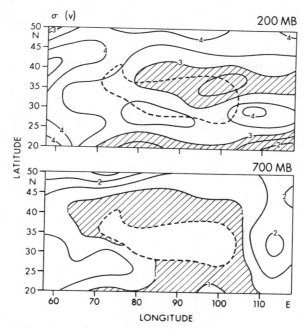

Figure 12.17. Standard deviation (0.5 m/sec interval) of 3.5–5.5 day filtered, meridional wind data for the 1978–1979 winter, at 200 (top) and 700 (bottom) mb. Shading indicates regions of standard deviations less than 3 (1) units at 200 (700) mb (31).

approach the western end of the Tibetan Plateau. The northern part continues propagating eastward along about 50°N. The southern part moves very slowly eastward along the southern periphery of the Tibetan Plateau. In conducting laboratory annulus experiments with an unheated and a heated simulated plateau, Yeh and Chang (26) also found that eastward moving troughs and ridges split near the western end of the Tibetan Plateau (Fig. 12.18, left). They reported that the splitting is primarily due to mechanical effects of the plateau, with the thermal effects contributing little.

At 700 mb (Fig. 12.17, bottom), standard deviations are smallest (less than 1 unit) to the immediate north and south of the Tibetan Plateau. East of the Tibetan Plateau, a region of high standard deviations (greater than 2 units) is found over China. At 500 mb, the standard deviation pattern does not exhibit this characteristic, but more reflects the conditions at 200 mb. The large variation in the winds at 700 mb is probably related to monsoon cold surge activity below 500 mb. Frequently, these cold northerly surges are associated with lee cyclogenesis. Chung et al. (40) provided detailed descriptions of the frequency, paths and structures of these lee cyclones and recently, Yeh and Gao (2) made an extensive study of 700-mb cyclogenesis in the lee of the Tibetan Plateau.

The association of northerly winds with low temperatures in monsoon surges produces a northward transport of sensible heat. This is shown in Figure 12.19

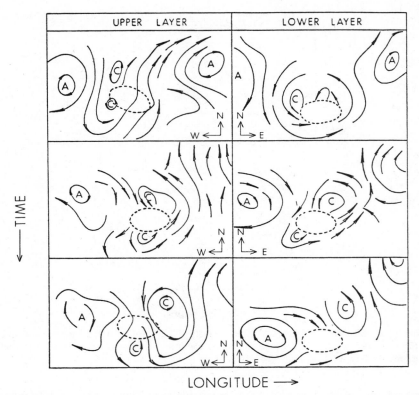

Figure 12.18. A deep trough passing through the plateau region in Yeh and Chang's (26) annulus experiment with a 5 W heating rate. Left (right): Streamline charts for the upper (lower) layer. The sequence of events is from top to bottom.

where a large northward sensible heat flux is found in the lee of the Tibetan Plateau. Figure 12.19 also indicates the strong orographic influence (mechanical blocking) of the Tibetan Plateau upon sensible heat transports. There is no indication of a direct northward heat flux from the tropics (India and the Bay of Bengal), across the Tibetan Plateau, to the Mongolia–Siberia region. This situation is favorable for the development of a pronounced surface high-pressure cell near Siberia. However, substantial heat fluxes originating in the Bay of Bengal–South China Sea region cross over the southeastern Tibetan Plateau and turn northward over central China. Part of this northward heat flux is then shifted northwestward along the northeast corner of the Tibetan Plateau and reaches as far west as the Takla Makan Desert (40°N–45°N, 80°E).

Further study is needed to determine the coupling mechanism between eastward propagating waves that move north and south of the Tibetan Plateau and, in particular, to determine what causes them to move with different phase speeds. Other problems which must be investigated include: the relationship, if any, between upper-level wave disturbances and low-level lee cyclones over central China; the relation of these systems to low-level monsoonal surges; and the determination of whether or not the orographic effects are essential to the development of cold surges in the lee of the Tibetan Plateau. In the next section, a numerical model is introduced to investigate those problems which are unique to the winter monsoon over East Asia.

2.2 Numerical Model Experiments

During the winter, the circulation over East Asia is governed by a dominant high-pressure system over Siberia, which is maintained primarily by radiative cooling. There are frequent monsoonal cold surges bursting out of Siberia which spread over the China continent, and propagate southward as far as the South China Sea and the Indonesian Seas. As mentioned earlier, these cold surges represent the intensification of the low-level branch of the Hadley circulation which, in turn, contributes to the increase of equatorial rainfall over the maritime continent. The question arises as to what mechanisms are responsible for cold surges over East Asia.

Lim and Chang (41) as well as Lau and Lim (42) carried out numerical experiments on cold surges in order to examine the response of the tropical wind field to prescribed mass sources in the mid-latitudes, which they regarded as cold air pools separated from the Siberian high pressure system. Their results showed that subgeostrophic northerly wind surges occur south of the prescribed mass sources and propagate equatorward reaching as far as the Southern Hemisphere summer monsoon region (5°S–10°S). However, a fundamental problem still remains unsolved. What mechanisms are responsible for the initial mass sources and the imbalance between pressure and wind fields?

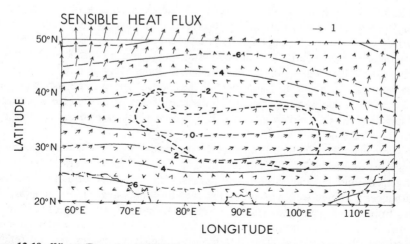

Figure 12.19. Winter (December 1978–February 1979) mean, vertically integrated (50 mb–surface), sensible heat flux (arrows, 10⁴ °C mb m/sec per unit as shown at upper right) with temperature field shown by full lines (2°C interval) (27).

The frequent occurrence of cold surges and lee cyclogenesis, unique to East Asia, is probably related to the orographic influence of the Tibetan Plateau. Nakamura and Murakami (43), and Murakami and Nakamura (44) examined this possibility with a numerical experiment using a 10-layer dry model. They did not assume an initial imbalance between the pressure and wind fields, that is, initially there are no disturbances and the zonal mean state is purely in geostrophic balance. The numerical integration is performed as a marching problem with respect to time and a train of wave perturbations is excited by the prescribed mountain. Eventually, the model produced prominent low-level cold surges and subsequent lee cyclogenesis on the east and southeast sides of a facsimile of the Tibetan Plateau. These phenomena were preceded by a sequence of events that occurred along the western, northern, and eastern periphery of the prescribed mountain. In the model simulation, when a major ridge system approaches the western end of the mountain, cold air advected from the north accumulates in the lower troposphere due to the barrier effect of the mountain's east-west orientation. The cold air pool is then advected rapidly eastward along the northern periphery of the mountain by the prevailing low-level westerly jet. Associated with this cold air pool is the development of a small-scale (mountain) edge anticyclone trapped below 700 mb near the northeastern side of the mountain (Fig. 12.20, top left).

This orographically induced low-level anticyclone then moves southward (clockwise) along the eastern periphery of the mountain. Its southward movement is faster than the movement of the edge cyclone which has already passed the eastern periphery, and has become nearly stagnant at the southeast corner of the mountain (Fig. 12.20, top left). The phase speed difference between the edge anticyclone to the north and the edge cyclone to the south, results in a sudden increase in the pressure gradient with prominent subgeostrophic northerlies east of the mountain. These flow down the pressure gradient and are responsible for the strong advection of cold air from the north. The advected cold air pool takes a form similar to a gravity-type Kelvin wave (Fig. 12.20, top right). The edge cyclone begins to shift eastward, advected by the prevailing southwesterlies near the southeast corner of the mountain. Subsequently, the edge cyclone is amalgamated with the major trough which has traveled from north of the mountain at about 50°N on day 6 to northeast of the mountain at about 45°N on day 8. The merged lee cyclone intensifies into a major mid-latitude weather system. To the rear of this intensified lee cyclone are widespread northerly cold surges, which penetrate southward to the South China Sea (Fig. 12.20, bottom left).

In summary, Murakami and Nakamura's numerical experiment confirms that the topographic influence of the Tibetan Plateau has a dominant effect on the lee cyclogenesis and associated northerly surges. The simulated structures of these disturbances compare well with those observed by Murakami (31). In general, cold surges from Siberia propagate southward along the eastern periphery of the Tibetan Plateau, reaching as far south as Malaysia and Indonesia, where subsequent cumulus activity becomes enhanced. This is a region of heavy equatorial rainfall (energy source) during the Northern Hemisphere winter. The relationship between cold monsoon surges from the Tibetan Plateau region, and the development of equatorial disturbances and their associated convection is not yet completely understood.

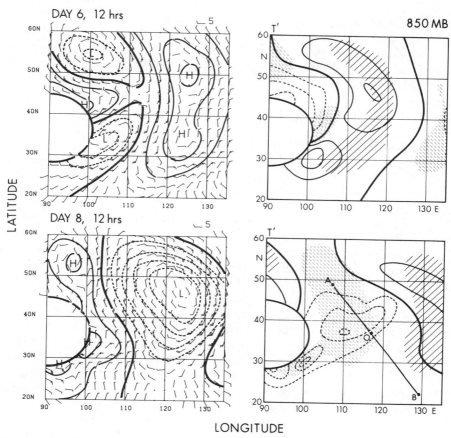

Figure 12.20. Top left: 850-mb wind (5 m/sec unit) and geopotential height (20 m intervals) at 1200 LST on day 6 of Murakami and Nakamura's model experiment (44). Bottom left: As in top left, for 1200 LST on day 8. Top right: 850-mb temperature (4°C interval; zero isotherm is thick contour, and negative values are dashed contours) for 1200 LST on day 6. Southerlies (northerlies) exceeding 5 m/sec indicated by low-density hatching (high-density hatching). Bottom right: As in top right, for 1200 LST on day 8. H and L denote, respectively, high (anticyclonic) and low (cyclonic) circulations. The outline of the model plateau is shown at left center. For our purpose, disregard the cross section denoted by A-B.

In other numerical simulations using a global general circulation model, Manabe and Terpstra (45) investigated the role of mountains in maintaining the wintertime general circulation. Their experiments were performed with no mountains (NM model) and with mountains (M model), and orography was found to play a significant role in the maintenance of the Siberian surface high pressure cell in the model atmosphere. In the NM model, an anticyclone is centered at 30°N, while in the M model, it is located near 45°N which corresponds well with the observed location of the Siberian high. Furthermore, the high is more intense in the M model than in the NM model. Manabe and Terpstra also found that cyclogenesis in the model atmosphere increases significantly in the presence of high mountains. Unfortunately, however, they did not investigate the role of the Tibetan Plateau in the development of low-level cold surges nor the relationship of cold surges to the changes in equatorial rainfall over the maritime continent. It would be quite interesting to examine these problems using data obtained from global general circulation model experiments.

3 SUMMARY AND CONCLUSIONS

This chapter has discussed recent research activities concerned with the orographic influence of the Tibetan Plateau upon the planetary-scale summer and winter monsoon circulations.

During the summer monsoon, it appears that the thermal effect of the Tibetan Plateau at the mid-tropospheric level (about 5 km) is as important as the Eurasia continent–Indian Ocean heat contrast at the earth's surface in influencing the planetary-scale summer monsoon circulation. Over the western and central Tibetan Plateau where the July mean rainfall amounts only to about 10 mm, the sensible heat supply at the mountain surface contributes the most to the net heat balance. In these regions there is a deep, mixed boundary layer with nearly constant potential temperature. This indicates the possible occurrence of dry thermal convection which originates near the heated surface and transports sensible heat into the upper troposphere. Over the southeastern Tibetan Plateau, the July mean rainfall exceeds 100 mm, and latent heating is as important as, or even more important than sensible heating. When considering the entire Tibetan Plateau, however, it is very difficult to determine the relative importance of sensible and latent heating on the basis of observations alone. Numerical experiments with dry and moist atmospheres may help to solve this problem.

During winter, the thermal effect becomes less important than the mechanical (blocking) effect of the Tibetan Plateau. In the lower troposphere, the winter mean winds tend to flow around the Tibetan Plateau with two major streams, one to the north and one to south of these high mountains. At 500 mb, winds are light over the Plateau due to planetary boundary layer friction. A pronounced eastward acceleration of 500 mb westerlies near the eastern end of the Plateau results in mid-tropospheric descending motions which enhance the downdraft leg of the thermally direct north–south Hadley circulation along 100°E-110°E. The low-level divergent winds of this orographically enhanced Hadley circulation are directed southward as northerlies. This favors the occurrence of low-level northerly cold surges bursting out of Siberia and penetrating equatorward, eventually enhancing convective rainfall over the maritime continent. The upper-level southerly return flows contribute to the eastward acceleration of the world's strongest jet stream near Japan.

As outlined previously, much progress has been made on the monsoon meteorology over and around the Tibetan Plateau: observationally, based on the 1978–1979 Global Weather Experiment data and numerically, by regional and/or global simulation models. Since the Global Weather Experiment observations represent the best data coverage and quality ever obtained over the Tibetan Plateau, an effort should be made to maximize their use. For example, the observations may be extremely useful for further investigating the structural features and the mechanisms through which edge disturbances and Tibetan vortices develop and propagate across these mountainous regions. Since it is unlikely that we in the foreseeable future will acquire observational data with better coverage and quality than the Global Weather Experiment data, a numerical experimentation program can be a useful alternative to observational studies. These numerical experiments can provide model data of extremely fine time and space resolution, providing a very effective means for identifying the exact role of mountains in the formation and propagation of synoptic-scale, model generated disturbances over and around the Himalayan massif.

REFERENCES

1. E. Halley, *Phil. Trans., Roy. Soc. London*, **16**, 153 (1686).
2. T. C. Yeh and Y. X. Gao, *Meteorology of Tibetan Plateau*, Scientific Publication Agency, Beijing, 1979, p. 278 (in Chinese).
3. Y. X. Gao, M. C. Tang, S. W. Luo, and Z. B. Shen, *Bull. Amer. Meteor. Soc.*, **62**, 31 (1981).
4. T. C. Yeh, *Bull. Amer. Meteor. Soc.*, **62**, 14 (1981).
5. H. Flohn, *Erdkunde*, **12**, 294 (1958).
6. T. Murakami, *J. Meteor. Soc. Japan*, **36**, 239 (1958).
7. H. Flohn, *Austral. Meteor. Mag.*, **49**, 55 (1965).
8. H. Flohn, "Contributions to a Meteorology of the Tibetan Highlands," Department of Atmospheric Science, Colorado State University, Fort Collins, 1968, 120 pp.
9. H. Riehl, "Southeast Asia Monsoon Study," Department of Atmospheric Science, Colorado State University, Fort Collins, 1967, 33 pp.
10. S. Rangarajan, *Austral. Meteor. Mag.*, **42**, 24 (1963).
11. P. Koteswaram, *Tellus*, **10**, 43 (1958).
12. M. T. Yin, *J. Meteor.*, **6**, 393 (1949).
13. R. C. Sutcliffe and J. K. Bannon, *Sci. Proc. International*, Association of Meteorology, **317** (1954).
14. Staff Members, Academia Sinica, *Tellus*, **9**, 432 (1957).
15. T. C. Yeh, S. H. Dao, and M. T. Li, *The Atmosphere and the Sea in Motion, Rossby Memorial*, The Rockefeller Institute Press, New York, 1959, pp. 249–267.
16. T. Murakami and Y. H. Ding, *J. Meteor. Soc. Japan*, **60**, 183 (1982).
17. H. Luo and M. Yanai, *Mon. Wea. Rev.*, **112** (1984).
18. T. Nitta, *J. Meteor. Soc. Japan*, **61**, 590 (1983).
19. M. Murakami, *J. Meteor. Soc. Japan*, **61**, 60 (1983).
20. T. Murakami, R. V. Godbole, and R. R. Kelkar, Proceedings of the Conference on the Summer Monsoon of Southeast Asia, Navy Weather Research Facility, Norfolk, 1970, pp. 39–51.
21. D. G. Hahn and S. Manabe, *J. Atmos. Sci.*, **32**, 1515 (1975).
22. H. L. Kuo and Y. F. Qian, *Mon. Wea. Rev.*, **109**, 2337 (1981).
23. P. J. Webster and L. C. Chou, *J. Atmos. Sci.*, **37**, 354 (1980).
24. D. Fultz, R. Long, G. Owens, W. Bohan, R. Taylor, and J. Weil, *Meteor. Monograph*, **21**(4), American Meteorological Society, Boston (1959).

25. R. Hide, *J. Atmos. Sci.*, **24**, 6 (1967).
26. T. C. Yeh and C. C. Chang, *Scientia Sinica*, **20**, 631 (1977).
27. Staff Members, Academia Sinica, *Tellus*, **10**, 58 (1958).
28. Staff Members, Academia Sinica, *Tellus*, **10**, 299 (1958).
29. T. Murakami, *J. Meteor. Soc. Japan*, **59**, 40 (1981).
30. T. Murakami, *J. Meteor. Soc. Japan*, **59**, 66 (1981).
31. T. Murakami, *J. Meteor. Soc. Japan*, **59**, 173 (1981).
32. T. Murakami, *J. Meteor. Soc. Japan*, **59**, 201 (1981).
33. I. Hirota and K. Miyakoda, *J. Meteor. Soc. Japan*, **43**, 30 (1965).
34. G. I. Taylor, *Proc. Roy. Soc. London*, **104**, 213 (1923).
35. R. Hide, *Nature*, **190**, 895 (1961).
36. H. Nakamura, *J. Meteor. Soc. Japan*, **56**, 317 (1978).
37. H. Nakamura, *J. Meteor. Soc. Japan*, **56**, 341 (1978).
38. R. B. Smith, *J. Atmos. Sci.*, **36**, 2385 (1979).
39. A. Sumi and T. Murakami, *J. Meteor. Soc. Japan*, **59**, 625 (1981).
40. Y. S. Chung, K. D. Hage, and E. R. Reinelt, *Mon. Wea. Rev.*, **104**, 879 (1976).
41. H. Lim, and C.-P. Chang, *J. Atmos. Sci.*, **38**, 2377 (1981).
42. K. M. Lau and H. Lim, *Mon. Wea. Rev.*, **110**, 336 (1982).
43. H. Nakamura and T. Murakami, *J. Meteor. Soc. Japan*, **61**, 524 (1983).
44. T. Murakami and H. Nakamura, *J. Meteor. Soc. Japan*, **61**, 547 (1983).
45. S. Manabe and T. B. Terpstra, *J. Atmos. Sci.*, **31**, 3 (1974).

13

The Indian Ocean: Interaction with the Monsoon

Robert A. Knox
Scripps Institution of Oceanography
University of California, San Diego
La Jolla, California

INTRODUCTION

The broad subject of the chapters in Part V is how monsoons interact with other natural features and geophysical systems, both within and outside the atmosphere. This chapter will treat some of what is known, and some of what is not yet known, about one of the most fascinating and complex of these interactions, that with the circulation of the underlying ocean. Two boundaries on the discussion are needed both to limit the text and to sharpen the focus. The first is a geographical limitation to the Indian Ocean. While there are monsoon winds over both of the other oceans [see Chapter 1 of this volume and Chapter 2 of (1) for a good summary of the geographical reach of monsoon winds], the areas involved are not large and the effects on those oceans are generally secondary and local. But over the Indian Ocean north of about 10–20°S the large scale wind field is dominated by the Asian monsoon, and the oceanic response is unique and clearly monsoon-driven.

The second, looser boundary will be to concentrate heavily but not exclusively on the influences of the atmosphere on the ocean, rather than the reverse. This is perhaps the simpler branch of the interaction to pursue, it is the one which distinguishes this chapter from others in the volume, it fits my scientific competence, and the chief ingredient is clear: the seasonal variation of the monsoon winds. In some ways, however, the reverse interaction, driven primarily by the effects of sea surface temperature as a lower boundary condition for the atmosphere, is the more important one for questions affecting the lives of people, for example, improvement of rainfall prediction for monsoon regions. Other chapters will deal with this subject more fully and more expertly. It is hoped that in the foreseeable future much of what has been written here will be supplanted by better observations and deeper understanding of interactions in both directions within the fully coupled atmosphere–ocean system.

1 THE WIND FIELD

Without repeating more detailed meteorological treatment of the monsoon wind field structure, it is useful to review some of the elementary features. It is important to bear in mind that the *surface* wind is what matters as a forcing function for the ocean; winds aloft, even at low elevations, can be different. For large-scale views of the surface wind patterns, various climatic atlases are helpful. Adaptations from a particularly useful one (2) are given in Figure 13.1. The January chart shows a wintertime high pressure zone over southern Asia. A generally northerly or north-

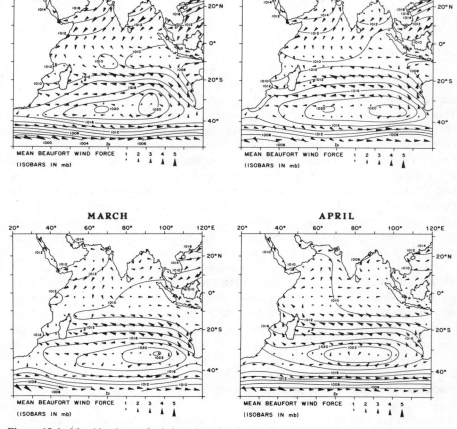

Figure 13.1. Monthly charts of winds and sea-level pressure over the Indian Ocean. From reference (2). Pressure is in mb. Beaufort numbers 1–7 correspond to wind speed ranges (kts) 1–3, 4–6, 7–10, 11–16, 17–21, 22–27, 28–33.

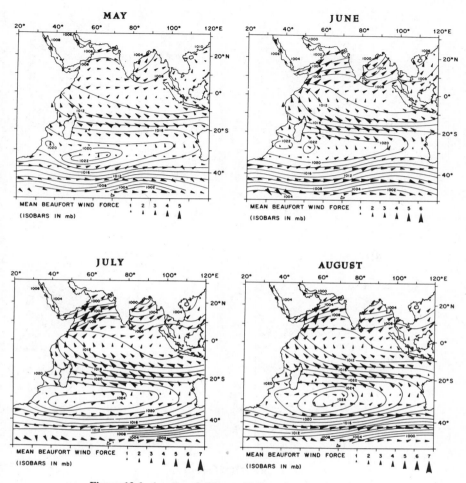

Figure 13.1. (*continued*) (*Figure 13.1 continues on page 368.*)

easterly* flow reaches across the northern Indian Ocean to a convergence with the southeast trade winds at 10–20°S. This is the northeast monsoon, and it is at this season that the tropical Indian Ocean looks most like the other two oceans in terms of surface wind patterns: two trade wind systems, each with a significant easterly component in much of the tropical zone.

As the year progresses the high pressure center over Asia weakens, the northeast monsoon decays, the Southern Hemisphere high pressure center intensifies and moves northward, and the southeast trades penetrate farther northward. By July the

---

* Both meteorological and oceanographic directional conventions are observed. An airflow from the north is a northerly; an ocean current in the same direction is a southward current.

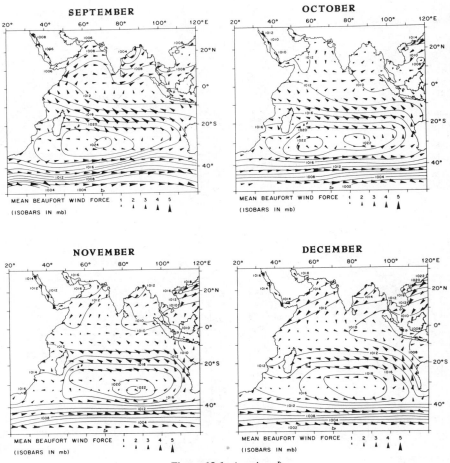

Figure 13.1. (*continued*)

southwest monsoon is fully developed. Southeasterlies from the Southern Hemisphere reach across the equator as southerlies or southwesterlies. Along this path they acquire the moisture which brings the monsoon rains to the Indian subcontinent. The southwest monsoon involves stronger winds than the northeast. In January and February there are only restricted regions of winds in excess of Beaufort force 3 (3.4–5.4 m/sec) and virtually none in excess of force 4 (5.5–7.9 m/sec) north of the equator, whereas in June, July, and August there are very few areas with mean winds less than force 3.

The annual cycle is completed in the fall of the year by the decay of the southwest monsoon. Asian high pressure rebuilds, the southwesterly winds retreat, and northerly flow is re-established in the Northern Hemisphere.

Based as they are upon heavy averaging of scattered observations from many years, atlases of mean monthly fields tend to obscure features of scale smaller than basinwide and, of course, say nothing about differences of patterns between particular years. One important such structure in the wind field is the low-level jet studied by Findlater (3). This is a pronounced intensification of the southwest monsoon airstream near the African coast, and results in part from the barrier of high ground in Kenya and Ethiopia. Figure 13.2a shows the location of the jet maximum at the 1-km level month by month in the first half of the year. We see it move toward the African coast and shift northward, following the general advance of the southeast trades noted above. By May the jet flows across the coast line, is blocked by the highlands, and runs parallel to them, recrossing the coast near Ras Asir (Cape Guardafui). The fully developed July pattern at 1 km is shown in Figure 13.2b. The intensification is best seen in cross sections of the jet. Figure 13.3b shows the southwest monsoon situation; the asymmetry of the flow, bunched against the highlands, is obvious. A similar but weaker concentration takes place in the northeast monsoon (Fig. 13.3a). Recent observations from research aircraft (5) confirm this

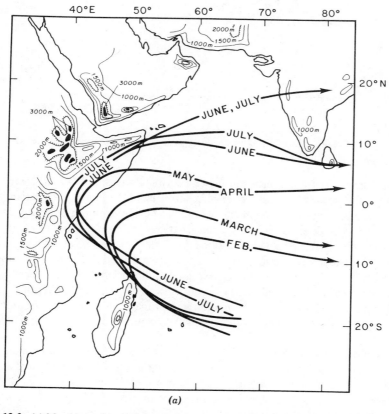

(a)

Figure 13.2. (a) Monthly positions of the maximum velocity at 1 km in the low-level jet. From reference (3). (*Figure 13.2 continues on page 370.*)

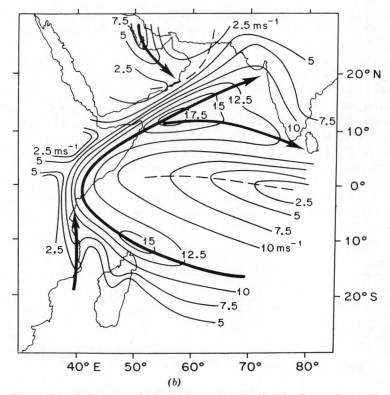

Figure 13.2. (*b*) Mean flow in the low-level jet at 1 km in July. From reference (4).

general pattern. As the jet turns out to sea it increases in speed (4), with the result that the surface winds beneath the jet core northeast of the Horn of Africa are some of the strongest and steadiest in the world during the northern summer. In July (Fig. 13.1*b*) this zone has a *mean* wind of Beaufort force 6 (10.8–13.8 m/sec), and 40% of the observed winds are force 7 (13.9–17.1 m/sec) or higher.

This jet-related concentration of the surface wind field into intense flows near the western boundary of the Indian Ocean has important consequences for the oceanic response, as we shall see. Certain aspects of the Somali Current system, notably its initial onset and the formation and propagation of eddies within it, are thought to be closely related to details of the local winds near the East African coast.

A second feature of the wind field, smaller than basinwide, which emerges on close inspection of atlas data, is that near the equator, during the spring and autumn transitions between monsoons, the wind becomes westerly and increases in strength. An alternative view of this phenomenon is given in Figure 13.4, which shows time series of wind *stress* at Gan, Maldives (0° 42′S, 73° 10′E), from Knox (6). (In this figure an eastward stress results from a wind toward the east—a westerly.) Note

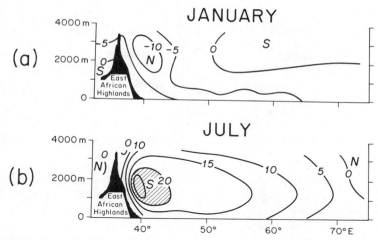

Figure 13.3. Cross sections through the low-level jet along the equator in (*a*) January, and (*b*) July. From reference (3). Speed contours in m/sec.

that in almost all years there is a pronounced pulse of eastward wind twice, once in spring, and once in fall. The associated wind speeds are not extraordinary, but the oceanic response [the so-called equatorial jet (7)] is large, due to the efficiency with which zonal winds can accelerate zonal currents in the equatorial zone.

The concept of the onset, or sharply defined beginning, of the southwest monsoon recurs in the scientific literature,* but remains somewhat elusive. For oceanographers the important onset is that of the winds over the sea, while meteorologists and residents of monsoon regions tend to apply the same term to the start of heavy rains over the land. We deal here with the oceanographic meaning, which by itself leaves room for discussion. Fieux and Stommel (9), by examining surface wind

---

* It occurs in other writing, too. Nehru, for example, writes:

I have been to Bombay so many times, but I had never seen the coming of the monsoon there. I had been told and I had read that this coming of the first rains was an event in Bombay; they came with pomp and circumstance and overwhelmed the city with their lavish gift. It rains hard in most parts of India during the monsoon and we all know this. But it was different in Bombay, they said; there was a ferocity in this sudden first meeting of the rain-laden clouds with land. The dry land was lashed by the pouring torrents and converted into a temporary sea. Bombay was not static then; it became elemental, dynamic, changing.

So I looked forward to the coming of the monsoon and I became a watcher of the skies, waiting to spot the heralds that preceded the attack. A few showers came. Oh, that was nothing, I was told; the monsoon has yet to come. Heavier rains followed, but I ignored them and waited for some extraordinary happening. While I waited I learned from various people that the monsoon had definitely come and established itself. Where was the pomp and circumstance and the glory of the attack, and the combat between cloud and land, and the surging and lashing sea? Like a thief in the night the monsoon had come to Bombay, as well it might have done in Allahabad or elsewhere. Another illusion gone. (8)

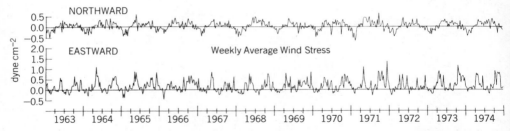

Figure 13.4. Time series of wind stress (positive for eastward, northward winds) at Gan, Maldives.

reports from merchant vessels along two well-traveled shipping lanes (Red Sea–Ceylon, and Persian Gulf–Capetown), concluded that in most years an abrupt onset does occur, and that its timing is remarkably constant near the intersection of these lanes (12°N, 54°E): May 24, give or take a few days. In 16% of the years studied, the onset was "gradual," that is, building to full strength in about one month. The average midpoint of the buildup was May 20. In 24% of the cases there were multiple onsets, that is, one or more intermediate slackenings of the wind before reaching its full strength. The midpoint in these cases was May 24. The regularity of the abrupt onsets, comprising 60% of the years studied, is the most remarkable result of this analysis.

The shipping lane intersection is almost exactly under the offshore extension of the low-level atmospheric jet axis (Fig. 13.2*b*), and it seems likely that the onset studied there reflects the timing and intensity of the northward penetration of the jet over the Arabian Sea.* Elsewhere in the data studied by Fieux and Stommel the picture is somewhat more complicated. Farther south, near the equator on the Persian Gulf–Capetown lane, onset seems to occur earlier though the data are sparse. We shall discuss this point further in Section 2.2 in connection with the spinup of the Somali Current. Winds at the eastern end of the Red Sea–Ceylon lane during their example sudden-onset year of 1954 are already westerly in early May, and of a strength not very different from that in late June. The observations at Gan (Fig. 13.4), south of the eastern end of this lane, also show the appearance of strong westerlies well before the end of May in most years. The strengthening of these westerlies seems less sudden, too. The variability in Figure 13.4 appears to the eye to be dominated by a few low frequency components, and indeed a spectrum of these data (10) shows a strong semiannual (annual) peak in zonal (meridional) wind.

We are left with the picture that in the high wind region northeast of Ras Asir, in a majority of years, the onset is sudden, but that elsewhere over the monsoonal Indian Ocean the change from winter to summer winds takes place less abruptly and/or at different times. When interannual variations are considered, specifying the detailed space–time variation of the winds becomes complex indeed, and the

---

* It would be most interesting to examine data on the strength and position of the jet on a year-by-year basis rather than as a sequence of monthly mean values, to see the connection to the different patterns of onset identified in the marine data by Fieux and Stommel (9). I am unaware of any such published studies.

simple analytical forms used in many models of ocean response must be viewed with some reservation. Certainly the advent of satellite-based routine wind observations over fine time and space increments will revolutionize such model building in the years to come.

2 THE OCEAN RESPONSE

We now turn to a discussion of the oceanic response to monsoon winds. Following a short overview of the general surface current pattern, the discussion is organized by regions. This reflects the limited scope of observational programs, which have of necessity studied one selected area or another, not the entire ocean.

2.1 The Climatological Surface Circulation

As in the case of the surface wind field, atlases of mean monthly conditions are useful for presenting the gross patterns of surface currents in the Indian Ocean. Adaptations from such an atlas (11) are given in Figure 13.5. The January and February charts show the northeast monsoon situation. Since at this time of year the wind field most resembles the trade wind systems of the other two oceans, it is perhaps not surprising to find a rough correspondence in the currents as well. There is a generally westward flow, the Northeast Monsoon Current, in the northern ocean, extending to about 2°S. It deflects into a moderate, generally southwestward, current along the Somali coast. An eastward equatorial countercurrent exists in the Southern Hemisphere, between 2°S and 8°S. South of 8°S we find the westward South Equatorial Current. Apart from the fact that the countercurrent lies in the Southern Hemisphere (because in the Indian Ocean it is the northeast, not southeast trades which reach across the equator at this season), this alternate westward–eastward pattern is rather like that of the tropical Atlantic or Pacific.

The chart for April, the month of transition between monsoons, shows the most dramatic event in the equatorial surface circulation, the establishment of a strong eastward jet within a few degrees of the equator in the central and eastern portions of the ocean. This arises in direct response to the moderate equatorial westerlies of the transition period, as noted by Wyrtki (7). The strong ocean response to these moderate winds is due to the efficiency with which zonal winds drive zonal currents near the equator, where the Coriolis force is weak. Further discussion of this portion of the circulation is given in Section 2.3.

In summer, as exemplified by the July and August charts, the fully developed southwest monsoon drives a generally eastward flow in the northern ocean with appropriate detours for land barriers. This is the Southwest Monsoon Current. Along the east African coast, the extremely swift Somali Current flows northeastward (see Section 2.2). Immediately at the equator, winds are lighter than in April, and are more southerly than westerly. Currents there are less coherent than in the April jet, and may even revert to westward.

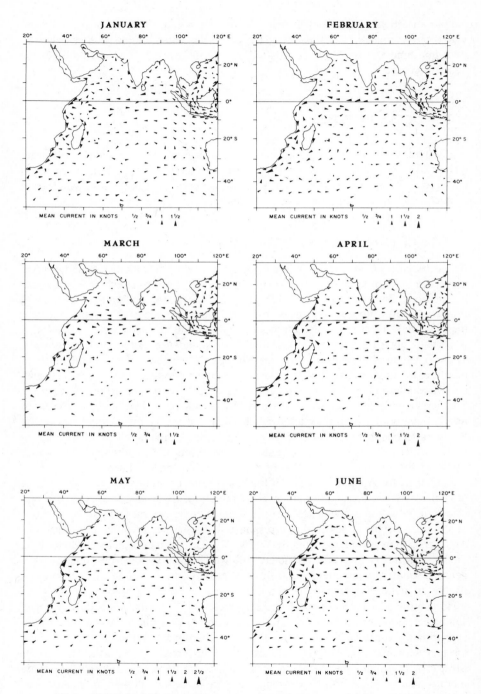

Figure 13.5. Monthly charts of Indian Ocean surface currents. From reference (11).

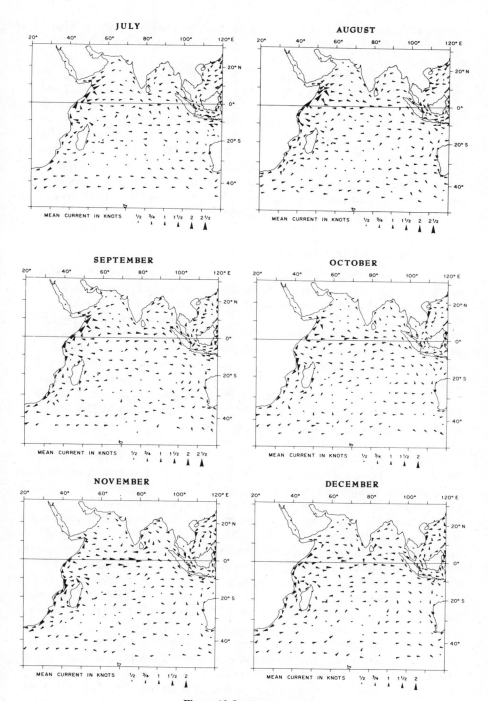

Figure 13.5. *Continued.*

Autumn (October, November) brings a repetition of the westerly winds and eastward equatorial jet, similar to April. In the northern ocean interior, currents are highly variable.

This rather broad-brush description is a fair representation of the state of this aspect of oceanography until very recent times. Chapter 6 of Schott (12) and the associated charts exemplify this lack of fine detail for want of observations. In the 1960s investigations of the International Indian Ocean Expedition (IIOE) added materially to our understanding of the circulation, especially in the Somali Current system and near the equator. Other research since then, notably INDEX (Indian Ocean Experiment), the oceanographic component of MONEX (Monsoon Experiment), carried out in 1979, has continued to address these problems and to discover the patterns and dynamics of the subsurface flows. The following subsections attempt summaries of where our knowledge stands today.

2.2 The Somali Current System

By far the most spectacular event in the annual march of oceanic response to monsoon winds is the "reversal" (quotation marks to be clarified below) of the Somali Current along the coasts of Kenya and Somalia. It is the most strongly time-varying major current in any of the oceans, and it achieves some of the highest surface velocities (3.7 m/sec) (13). Two modern reviews, one by Düing (14) on observations and one by Anderson (15) on theoretical aspects, are very useful, and much of what follows is drawn from them. Subsequent important work on numerical models has been carried out by Cox (16), by Anderson (17), and by Philander and Delecluse (18). Extensive observations from the southwest monsoon of 1979 have been presented in groups of papers by INDEX scientists in the August 1980 issue of *Science* and the December 1982 issue of *Journal of Physical Oceanography*. Individual references to these and other papers are made below.

Prior to the IIOE, the general picture of summertime (southwest monsoon) circulation was of a strong boundary current flowing more or less uniformly along the coast from south of the equator to a separation from the coast at 9–10°N. Atlas presentations such as Figure 13.5 tend to support this view, though the spatial resolution is poor. The existence of very cold upwelled water inshore and north of this separation, in the vicinity of Ras Hafun, had been known for years, and was correctly thought to be due to the baroclinic structure of the offshore-turning current and/or to wind-driven upwelling, in some undetermined combination. In 1979 for example, sea surface temperatures in July just south of Ras Hafun were 18°C or less, while at the same time offshore values were roughly 28°C (19). The downstream recirculation of the current in a "great whirl" (20) or "prime eddy" was also well known. Figure 13.6, from Bruce (21) shows this offshore turning and circulation; something resembling it has always been found during the southwest monsoon.

IIOE investigations tended by and large to confirm this picture. Emphasis was given to exploring the fully developed current of August–September 1964. The two key interpretations of the circulation at this time are one based on water mass properties (22) and one based on direct current measurements coupled with geostrophic

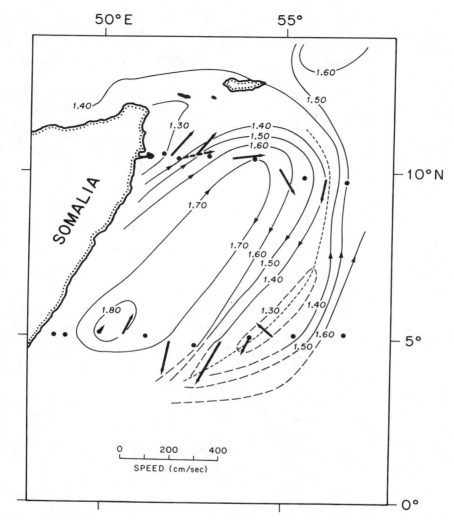

Figure 13.6. Surface currents (bold arrows, speed scale at bottom) and dynamic topography contours (solid lines, units of dynamic meters) composited by Bruce (21) from observations in three consecutive southwest monsoons. Dynamic topography is analogous to atmospheric pressure in the sense that large-scale flow tends to be parallel to contours with high (low) values of topography on the right in the Northern (Southern) Hemisphere. This expected sense of flow is indicated by arrowheads on contours. Solid circles denote hydrographic stations used to make the map.

calculations (23). Both papers indicate a continuous flow along the coast to the separation point, although both mention the evidence, inconclusive in the sparse data set, for a meander near 4°N. This seems to be the first hint of a feature which has prompted much subsequent research, the eddies or meanders that can occur within the northward-flowing Somali Current.

An aspect of the flow brought to light for the first time by the IIOE work is the clockwise eddy to the north of the great whirl (Fig. 13.7). In the IIOE data the northern or Socotra eddy is clearly distinguishable from the Somali Current by its properties; it contains warm, salty Arabian Sea water rather than the cooler, fresher water transported north by the main current. Work since the IIOE has confirmed

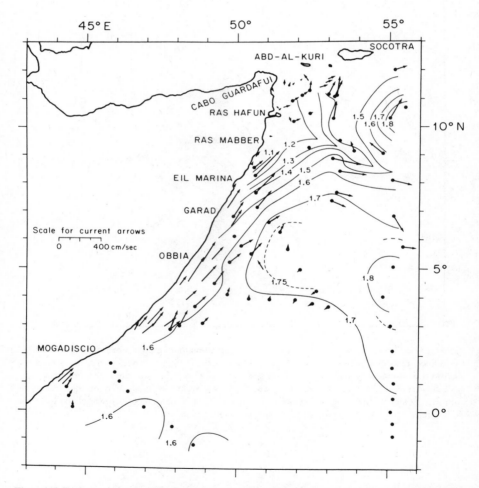

Figure 13.7. Currents at 10 m depth and dynamic topography contours during August and September 1964, using the same graphic conventions as in Figure 13.6. From reference (23). Note offshore turning of flow between Ras Mabber and Ras Hafun, forming boundary between great whirl and Socotra eddy.

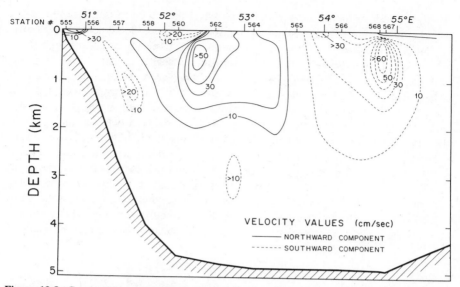

Figure 13.8. Geostrophic currents for the winter of 1964–1965 on a section at 9°N, with reference velocity determined by neutrally buoyant floats. From reference (26).

that the great whirl and the northern eddy are repeatable elements of the circulation, recurring each summer. Bruce (24), in an important paper describing four years of repeated temperature sections through this region using expendable bathythermographs (XBTs), documents formation of these two eddies in each summer from 1975 to 1978, and notes that in all available earlier southwest monsoon observations a similar picture emerges. A valuable companion data report by Bruce (25) gives fuller details. He also demonstrates that the great whirl persists for some months after the summer monsoon winds begin to slacken. In some years it may persist, albeit beneath the surface and with decreased intensity, throughout the winter monsoon and reintensify during the subsequent summer. In these cases the winter circulation along the coast consists of the southward near surface current atop the northward-flowing remnant of the previous summer flow. 1975–1976 was probably such a winter; the XBT sections of Bruce (25) are marginally deep enough to show this. There was a clearer case (26) in 1964–1965: available then were both deep hydrographic observations for calculating geostrophic shear and neutrally buoyant floats for establishing a reference velocity. Figure 13.8 shows the 1964–1965 measurements, and it is obvious that the northward and southward cores at about 500-m depth correspond well to the position of the great whirl of summer at this latitude (Fig. 13.6). Figure 13.8 also shows the weak nature of the near-surface southward flow near 52°E, that is, the surface circulation response to the northeast monsoon. Current atlases (Fig. 13.5) also show that the winter surface flow is far weaker and less coherent than that of summer. In all, to speak of an annual reversal of the Somali Current, in the sense of equal amplitudes in opposite directions, is a bit misleading.

To the south of the prime eddy or great whirl the structure of the flow is even more complex. Late in the summer monsoon (August, September) the Somali Current usually appears to flow fairly continuously along the coast to the offshore turning at 9–10°N. 1964 (see Fig. 13.7) was such a year. But earlier in the season, for example in 1970 (27), 1976 (24), and 1979, the year of the INDEX observations (13), a southern eddy often develops in which the Somali Current turns offshore at a lower latitude (4–6°N), and a separate clockwise gyre north of this point constitutes the great whirl. In 1979 (Fig. 13.9) the contrast between the southern eddy and the great whirl was particularly well resolved, both by direct current measurements and by the salinity difference; the low salinity water transported north from equatorial regions turned offshore at 4°N, and the coastal limb of the great whirl from 4°N to 10°N was composed of distinctly saltier Arabian Sea water. The frequent but not universal (28) occurrence of the southern eddy during the early months of the southwest monsoon is generally ascribed to the detailed structure of the wind field (13). The particular aspects of importance have not been isolated

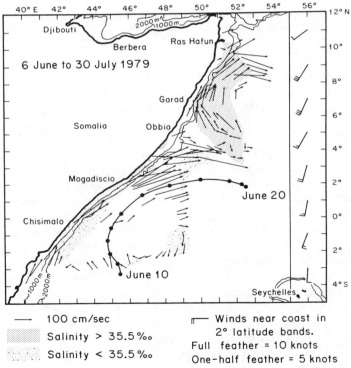

Figure 13.9. Map of surface currents (arrows) and surface salinity patterns for June–July 1979, showing separate southern eddy and great whirl. Winds in the near-coastal region averaged over 2 degree latitude bands are indicated at the right margin. The track of a drifting buoy (solid circles at one day intervals) also indicates the southern eddy.

completely, but some useful hints are contained in the model results discussed below.

To this point we have treated the several eddies of the Somali Current system as if they were locked into fixed geographical positions. This is only partially true; variations in positions and sizes, particularly of the southern eddy, have been observed. Attempts to understand the time dependence and possible longshore translation of eddies have been bound up in attempts to understand the dynamics of the spinup of the Somali Current as a whole, to which we now turn.

A key paper, the goad and the inspiration for much subsequent theoretical, numerical, and observational work in the area, was Lighthill's (29) attempt to explain the onset of the Somali Current using the simplest possible linear ocean model. There were two crucial ingredients. First was proper inclusion of the equatorially trapped wave modes,* which allow much more rapid low-latitude baroclinic current response to time-dependent, large-scale forcing than is possible at mid-latitudes (31). Second was to model the forcing as due to winds over the ocean interior, (i.e., the open ocean seaward of the boundary current), rather than to those near the coast. These were taken to have a simple step-function time dependence. The current resulted from the impingement upon the coast of signals propagated at the high equatorial wave speeds from the interior forcing region. From this scheme Lighthill deduced a time scale for Somali Current spinup of about one month following the monsoon onset over the interior.

Leetmaa's (32, 33) observations off the Kenya coast soon showed that at least the southern portion of the Somali Current began to flow northward well before the change of winds over the ocean interior and that the detailed time dependence of the flow in this region was closely coupled to the *local* winds. Lighthill's mechanism was still potentially important in the subsequent intensification of the current but was not responsible for the initial onset. Anderson and Rowlands (34) further clarified the distinction between response to local and remote forcing, again in the context of a linear analysis. The locally driven response is proportional to time, the remotely driven response to (time)^2, and so local effects can be expected to predominate at early stages. All of this would be easier to unravel, of course, if the coastal and interior winds had the same simple time of (abrupt) onset, but in fact the wind field is structured in both space and time, as shown previously. In particular, the low-level jet (Fig. 13.2) moves toward the Kenya coast a month or so before it penetrates to the strong wind region northeast of the Horn of Africa, and so it is perhaps not surprising to find the current onset off Kenya in advance of the stronger monsoon onset over the Arabian Sea as defined by Fieux and Stommel (9).

How is the current system, including the various eddies, established along the rest of the coast; what are the details of the time evolution? These questions have prompted much of the recent research, including the 1979 INDEX investigations. To study them adequately requires nearly synoptic observations along the entire coast, some 10 degrees of latitude, and the difficulty of doing this under the extreme

---

* Although he missed the equatorial Kelvin wave, an oversight remedied by Moore (30).

wind and sea conditions which occur during the southwest monsoon cannot be overstated. Success has been partial. Nevertheless, some useful generalizations can be made. The first is that in any given year the circulation seems to develop either with or without a southern eddy, and this structure persists, though perhaps with modest shifts of the eddy boundaries, until late in the southwest monsoon. The idea of eddies moving smoothly northward along the coast throughout the development of the current, found in some of the literature, is based on overextrapolation of the first evidence for northward motion of eddy structure. These observations (but not the extrapolations) were made by Bruce (27), using data from the summer of 1970. In that year the main current turned offshore at a low latitude (6°N), that is, both a southern eddy and the great whirl occurred. In late August and early September, however, this offshore flow at the boundary between the southern eddy and great whirl moved northward at 9–27 cm/sec, gradually coalescing with the great whirl to form a single clockwise circulation. Subsequent to this work, numerical models appeared [e.g., Hurlburt and Thompson (35)] in which northward-moving eddies arose soon after the onset of the winds and continued to move north. But in Bruce's 1970 observations, the northward motion was of the zone of separation, not an entire eddy, and it occurred only late in the summer, after the winds had begun to decrease. Similar behavior, judged from satellite-derived sea surface temperature observations of the cold-water wedges inshore of points where eddy flows turn eastward, occurred in 1979 (19).

Cox (16) has published numerical model results which seem to include and explain a number of the observed properties of the Somali Current and its eddies. It is a three dimensional, fully nonlinear, stratified ocean model driven by an applied wind stress. The stress is imposed abruptly (zero to maximum in six days), within about 1000 km of the coast; remote interior wind forcing is excluded. The model current responds rapidly to these local winds, as is observed in nature. Other important points are:

1. The geometry of the western boundary is crucial. Whereas most earlier models, for reasons of analytical simplicity, used a western boundary oriented north–south, Cox has run a comparison case using a boundary oriented NE–SW, roughly like the Somali Coast. The result is to stall the northward motion of eddies. A further refinement toward geographical realism—inclusion of a slight bend in the coast at 6°N and of an embayment to model the Gulf of Aden—yields an even more realistic eddy system. There seems little doubt in the context of this model that the northward motion of eddies, or lack of it, is very dependent upon the coastal geometry.
2. The strength and distribution of the wind forcing strongly influences the motion and location of eddies. An overall increase in the specified wind maximum value leads to a southward displacement of the offshore turning of the main current. Actual wind stresses in the region can vary by a factor of 2 from year to year (36), so this is a potentially important mechanism for explaining interannual differences in the eddy field structure. Similarly, applying a more realistic stress pattern which has maximum value increasing from 1

to 3 dyne/cm^2 northward along the coast instead of a constant everywhere moves the separation point northward; the current penetrates farther north under the weaker low-latitude winds before being forced offshore. These model responses may explain why the southern separation point, in years when it occurs, seems to stay approximately stalled during the peak of the monsoon but then can migrate north and coalesce with the great whirl under the lighter winds of August–September.

The vorticity balance in this flow is complex, as Cox (16) has pointed out, and all terms—planetary advection, vortex stretching, nonlinear advection, and vertical friction—contribute importantly to the evolution of the field. Horizontal friction appears to be less critical. Wind stress curl, a source of vorticity, is excluded from the region of the model current by the form of the specified wind stress (no variations normal to the coast until 1000 km offshore), but Cox notes that it too is potentially important. Certainly the observed values of wind stress curl in this region [up to 1.2×10^{-7} dyne/cm^3 in July northeast of Ras Asir under the low-level jet (22)] are comparable to the model planetary advection term (about 2.3×10^{-7} dyne/cm^3 in Cox's Figure 3, not shown) and to oceanic observations of the same term. For example, a typical northeastward transport of 50×10^6 m^3/sec between the surface and 200 m depth across a 500 km section (23, 37) gives a value for the advection term of 1.6×10^{-7} dyne/cm^3.

Cox's conclusion that realistic basin geometry is important has since been criticized by Philander and Delecluse (18) as unwarranted. They point out that changing the western boundary from north–south to northeast–southwest, while retaining the model wind forcing in a strip parallel to the coast, has the effect of changing the wind stress curl distribution (there is a line of nonzero curl along the seaward edge of the wind strip) over the basin, and hence of changing the structure of the basinwide steady flow toward which the model must tend during its spinup. One therefore cannot ascribe differences in the boundary current development between the two cases solely to the change in geometry. Their own model, however, lacks the wind stress curl vorticity source and thus exhibits some other unrealistic features. In particular, a two-gyre system does not form in response to their forcing.

Anderson (17) has used a model with a north–south coast but more realistically structured wind forcing, with elements intended to represent (a) the northeast monsoon, (b) the southwest monsoon, including intensification near the coast, and (c) Southern Hemisphere easterly trade winds. The model is driven initially by (a) and (c) until a steady northeast monsoon ocean flow is set up. Then (a) is switched to (b), while (c) remains. The results are pleasingly realistic in many ways. A two-gyre Somali Current develops, with separation points at the proper latitudes. The model has not been extended farther in time to simulate later developments such as the coalescence of the gyres.

This short tour of the Somali Current may be summarized as follows: under the influence of the southwest monsoon winds, and especially the near-coastal winds at first, a system of clockwise eddies with swift northward flow near the coast is established. In all years the great whirl and the northern or Socota eddy appear. In most years there is a third or southern eddy. Eddy boundaries, of which the most obvious are the nearshore wedges of cold upwelled water associated with offshore turnings of the coastal current, do change and move during the monsoon but substantial northward displacement of the southern separation point, when it happens, seems to take place after the winds begin to decrease. None of the models cited does a uniformly good job of simulating all of these ocean responses and of exposing the aspects of the forcing and boundary conditions which are responsible. All the models proceed by spinning the ocean up to equilibrium with one [or two, in the case of Anderson (17)] specified steady pattern of wind forcing, so that no account is taken of the continuous evolution of the wind field over the full annual cycle. Work in progress (Luther and O'Brien, personal communication) on a model that drives the Indian Ocean with a full annual wind cycle does seem to capture most of the important responses. This same work also indicates that the basin boundary geometry is a critical element for a realistic simulation, a point that was left somewhat uncertain by the earlier results.

Because of the sensitivity of the Somali Current system to the winds, a fruitful avenue for future research will be to explain interannual differences in that system in terms of interannual wind field differences. The *sine qua non*, a long-time series of routine detailed surface wind observations over the ocean, is not yet in hand, but advances in remote sensing will one day change this situation. Coupled with modest methods for monitoring the current systems, some of which exist [satellite sea surface temperature data show the cold wedges; Bruce (24) has already demonstrated the usefulness of repeated XBT sections along the main tanker route and these could be revived], we can at least hope for real progress in this area in the imaginable, if not the near, future.

We should also consider the effect of the ocean on the atmosphere, principally the heat exchange at the surface. Through a combination of upwelling and offshore advection of cold water, the currents play a substantial role in the heat budget of the Arabian Sea, a point explored more fully in Section 2.4. Clearly it is of more than purely oceanographic interest to improve our understanding of the interannual wind-driven variations in the current system given the associated changes in the amplitudes of these two heat transport processes.

So far we have concentrated primarily on the relation between the monsoon and the surface or near-surface flows in the Somali Current system. Apart from the observations referred to above on the longevity of the subsurface portion of the great whirl, much less is known about the monsoon–subsurface circulation connections. This is because observations are so sparse; there is no subsurface data base at all comparable to the accumulated surface current data represented by Figure 13.5, for example. Instead there is tantalizing fragmentary evidence. Leetmaa, Rossby, Saunders, and Wilson (38) made subsurface current measurements along the African coast as part of INDEX in 1979. The flow was strongly time-dependent. During May and early June, as the surface Somali Current was developing, a pre-existing southward flow at 700-m depth was found to weaken and, north of the equator, to reverse. At the same time, a westward jet at the same depth was found along the equator. The connection between the westward jet and the flow along the coast is a complex one, both in the direct current measurements and in terms of water mass properties (39) and it is not possible yet to discern the "typical" annual cycle of this system.

2.3 Equatorial Circulation

Over the tropical Atlantic and Pacific Oceans, easterly trade winds prevail. As a result the ocean surface flow has a mean westward component. Water accumulates near the western boundaries, and sea level stands higher there; consequently a pressure gradient force toward the east exists in the shallow layers from the thermocline (layer of rapid decrease of temperature with depth) to the surface (see Fig. 13.10a). The easterly wind also drives a poleward Ekman flow in the surface layers of both

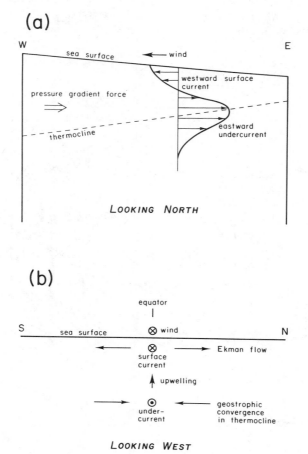

Figure 13.10. Schematic vertical sections, (a) zonal and (b) meridional, of the mean circulation in the Atlantic and Pacific equatorial undercurrents. Symbols indicate eastward ⊙ or westward ⊗ flowing undercurrent, surface current, and wind.

hemispheres, creating a divergence near the equator which is compensated by upwelling. The upwelled water is in turn replaced by a meridionally convergent flow in the thermocline, in geostrophic balance with the eastward pressure gradient force. Near the equator this geostrophic balance breaks down, and a jetlike eastward current in the thermocline arises; the pressure gradient is balanced not by the Coriolis force due to meridional velocity but by vertical diffusion or vertical advection of momentum involving the zonal flow in the jet. This jet is the equatorial undercurrent, a major permanent feature of the Atlantic and Pacific circulations. A schematic picture is given in Figure 13.10b. The undercurrent was first noted in the Atlantic by Buchanan (40, 41), forgotten, and rediscovered in the Pacific in 1952 (42). Figure 13.11 shows a typical section of the zonal velocity in the central Pacific; it is in accord with the qualitative description of Figure 13.10b.

The key ingredients of the Atlantic and Pacific undercurrent circulations are the ocean boundaries which support the zonal pressure gradient and the easterly trades which drive the surface flow to sustain this gradient and force the meridional and

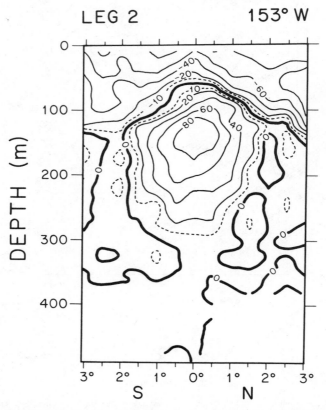

Figure 13.11. Typical vertical section of zonal velocity in cm/sec (positive eastward) in the Pacific at 153°W. From reference (43).

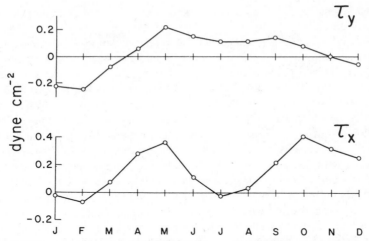

Figure 13.12. Monthly mean wind stress components (τ_y positive northward, τ_x positive eastward) from Gan, Maldives, based on same data as Figure 13.4.

vertical flow. The tropical Indian Ocean has the requisite coasts, but the winds, and thus the ocean circulation, are profoundly different. Figure 13.4 shows the strong time variation and direction reversals of the zonal wind at Gan, due to the monsoons. Direction reversals in the central Pacific are very rare and the variance is much less. Figure 13.12 shows the sequence of monthly rather than weekly means for the same data as Figure 13.4. The strong semi-annual cycle in the zonal wind stress is obvious, as is the fact that the *mean* zonal wind is *westerly*, the opposite of the other oceans. The annual march of the zonal wind in this record can best be described as spring and fall periods of strong westerlies, with interludes of weak or easterly winds between. The oceanic response to such forcing is very different from that to the nearly steady easterlies of the Atlantic or Pacific.

The IIOE coincided with a surge of interest in the undercurrent. By this time the initial surveys had been completed in the Pacific (44) and Atlantic (45), and the 1960 issue of *Deep Sea Research* with its collection of theoretical models had appeared. An obviously important next step, incorporated into IIOE physical ocean-ography plans, was to find out whether, and when, a similar current existed in the Indian Ocean. Answers appeared in the classic monograph of Taft and Knauss (46); important additional observations from a different year were reported by Swallow (47). The basic result was that an eastward undercurrent in the thermocline with weak or westward near-surface flow above, similar to the Atlantic/Pacific pattern, existed all along the Indian Ocean equator only briefly during the late stages of the northeast monsoon, but not during the southwest. Swallow's observations showed a stronger undercurrent lasting later in the year. The oceanwide, single-ship surveys of Taft and Knauss took three months to complete, thereby mixing space and time variations of the flow field significantly. Thus it is unwise to interpret their results too closely for synoptic detail. Both sets of observations, however, established the fact that a strong seasonal, monsoon-related variation of the undercurrent exists in the Indian Ocean.

The second major component of the equatorial circulation is the eastward surface jet which develops in response to the spring and fall westerlies, as mentioned in Section 2.1. Wyrtki (7) was the first to notice this in available climatic data, to relate it clearly to the winds, and to point out the associated cycle in sea level measurements, which indicate that the near-surface layer of warm water is drained near Africa and filled near Sumatra as the jet flows east. The change of surface elevation is about 20 cm and the compensating variation in the thermocline depth is about 40 m; these changes are roughly consonant with a simple two-layer model of the flow, with the thick lower layer (below the thermocline) at rest.

With assistance from many people, I was able to obtain a long-time series* of quasi-weekly current observations near Gan (48) which is complementary, in terms of sampling strengths and weaknesses, to the measurements mentioned. The time evolution of the surface jet and of the undercurrent appears clearly, albeit at a single location. Figure 13.13 shows wind stresses at Gan and currents averaged over a surface layer (u_s, 0–20 m), and over a layer at the thermocline/undercurrent core (u_{th}, 60–80 m) from this work. The spring and fall westerlies are obvious, as is the rapid response of the currents in both layers. Cross-correlations (not shown) demonstrate that the zonal wind and zonal current in both layers are highly correlated (0.7–0.8) with no detectable lag; the semiannual period is also obvious in this analysis. Only in the spring of 1973 was there a "classical" Atlantic/Pacific undercurrent structure, with eastward flow in the thermocline beneath weak or westward surface flow. This situation is shown in more detail by the sequence of current profiles in Figure 13.14. The March 29 profile shows the development of the eastward surface jet, which by April 5 has strengthened and deepened so that eastward velocity increases upward throughout the profile.

Cane (49) has constructed a model of equatorial ocean response to changing winds which accounts for most of the important features of the Gan data. A three-layer ocean in a bounded basin centered on the equator is assumed. Wind stress driving is modeled as a body force in the surface layer. An intermediate layer of the same density as the surface layer, analogous to but thicker than the thermocline layer of Figure 13.13, is also in motion, connected to the surface layer by friction and by vertical velocity. The denser bottom layer is at rest. The calculation is fully nonlinear, and this is the crucial property in considering the differing responses to (a) switch-on of easterly winds, (b) switch-on of westerlies, (c) switch-off of easterlies, and (d) switch-off of westerlies. For a linear system, responses to (a) and (b) would

---

* The full record, of which the majority is shown in Figure 13.13, probably still stands as the longest time series of Indian Ocean equatorial currents: 2½ years. That so paltry a record length can sustain such a claim must strike my meteorological colleagues, accustomed as they are to daily observations all over the globe, year in and year out, as proof of the primitive nature of oceanographic data sets. They are right. We lack the oceanic analogues of aviation requirements, forecasting needs, and so on, to propel us into the major leagues in this respect. We try to compensate with sharpened skill—some might view it as poetic license—in extracting general patterns from limited measurements.

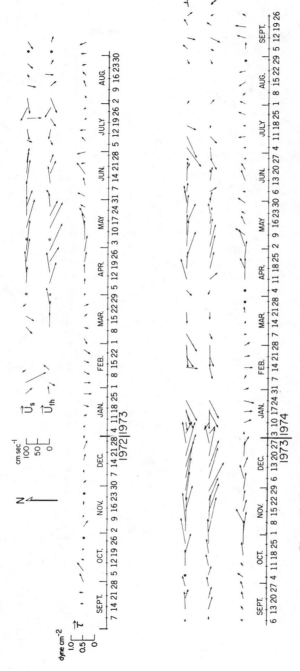

Figure 13.13. Time series of wind stress, τ (weekly averages), and of currents near the surface ($\mathbf{u}_s$; 0–20 m average) and the thermocline ($\mathbf{u}_{th}$; 60–80 m average) at Gan. Open circles denote unuseable data, solid circles are zero values. From reference (48).

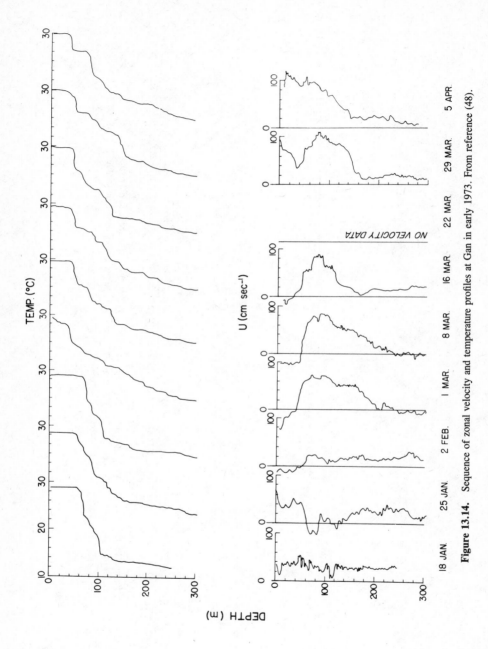

Figure 13.14. Sequence of zonal velocity and temperature profiles at Gan in early 1973. From reference (48).

be related by a simple change of sign, as would responses to (c) and (d), but the nonlinear results are very different.

Consider first the switch-on of winds over a resting ocean, cases (a) and (b). Early in the response, before coastal effects have had time to propagate to the basin interior, the zonal pressure gradient is not yet established and can be ignored. Easterlies put westward momentum into the surface layer, and it is diffused downward, but they also drive a poleward meridional Ekman flow which advects zonal momentum away from the equator. In response to the meridional divergence in the surface layer caused by the poleward Ekman flow, the equatorial vertical velocity is upward. This vertical motion brings water with less zonal momentum to the surface, that is, vertical advection acts to weaken the surface layer flow. With westerlies, the surface layer flow is meridionally convergent near the equator, and the vertical velocity is downward. Thus the early development of the flow in response to easterlies is slower since it represents a competition between the strengthening effect of vertical momentum diffusion and the weakening effects of the advections. With westerlies the process is much faster, since both diffusion and advection work to accelerate the flow in the surface and intermediate layers. The fast response to westerlies, in both active layers of the model, is consistent with the observations in Figure 13.13.

Starting at the onset of the winds, changes in the model zonal pressure gradient or lower layer interface slope, initially zero, begin to propagate eastward (westward) from the western (eastern) boundaries at very nearly the speed of the equatorial Kelvin (fundamental Rossby) wave of the system.* How these effects modify the evolution of the flow at any given point on the equator depends on those wave speeds, on the forms of both wave signals, and on the location of the point. But eventually, a steady zonal pressure gradient is set up. For easterlies, it overcomes the intermediate-layer westward flow due to the wind forcing and drives the classic eastward undercurrent. For westerlies the flow in both layers continues to be eastward, though the westward pressure gradient force decreases the intermediate-layer current somewhat.

Similarly, when winds are switched off allowing pre-existing steady circulations to relax, we find that responses to (c) and (d) above differ by more than a simple sign change. Consider again the early stages, before the arrivals of coastal effects. For switch-off of easterlies, the surface current is reversed by the action of the pressure gradient, but the intermediate current, already nearly in balance with this gradient, hardly changes. For westerlies, however, switch-off breaks the downward advection of zonal momentum which had been maintaining the eastward intermediate flow against the adverse pressure gradient. Consequently, deceleration of both layers is immediate and substantial. Note in Figure 13.13 that the Gan currents relax and even become westward almost simultaneously with the decay of the strong westerlies.

---

* Cane's model has only one baroclinic mode, since his surface and intermediate layers have equal densities. But readers of his paper should note that in scaling his model results to the real ocean he argues for using the phase speed of the oceanic 2nd baroclinic mode. In fact it is a subtle question, pursuit of which would take us too far afield, as to why the linear baroclinic wave speeds should appear so prominently in results from a highly nonlinear model.

The actual reversal (as opposed to a simple decrease) of the eastward currents in the model does not result from switching off the westerlies, but does result if the wind is changed from westerly to northerly or southerly, as happens at Gan (Fig. 13.13). The meridional wind shifts the whole current system upwind, bringing a band of westward flow to the equator. Details are given by Cane (49).

After coastal effects propagate to the longitude of interest, the relaxation sequence is altered. For easterlies, arrival of the Kelvin wave signal from the west begins to decrease the zonal pressure gradient, leading to a fairly smooth decay of the model intermediate current. For westerlies, the superposition of the Kelvin and Rossby signals leads to a more complex behavior, and at a central longitude (e.g., Gan), the initially westward pressure gradient force is reversed after a time by the interplay of these waves. The timing is about right for the appearance of an eastward pressure gradient force at Gan some four months after the decay of the westerlies. Thus the model yields a transient undercurrent in the central basin which is dynamically similar (response to eastward pressure gradient force) to the Atlantic and Pacific ones, but which is really a temporary aspect of response to the decay of the westerlies. In this mechanism Cane has found an explanation for the presence of a classic undercurrent in 1973 but not in 1974. In 1972 the autumn westerlies began to relax in October, and winds were weak until the following spring. This allowed time enough for the late stages of relaxation to evolve as described above; the temporary undercurrent which resulted was seen at Gan in March, 1973. But in autumn 1973 the westerlies lasted into December; there was not enough time for the late-stage relaxation to set up before arrival of the next round of westerlies in spring 1974, and so no undercurrent was seen in early 1974. Cane (49) points out that the autumn preceding Swallow's (47) observations in 1964 was one in which the westerlies died early, as in 1972. This may explain why Swallow found a stronger, longer-lasting undercurrent than did Taft and Knauss (46).

In principle the same sort of relaxation process should occur in summer, following the spring westerlies. Certainly the immediate decrease and reversal of current in both layers (the early-stage response) is seen in Figure 13.13. The record is less conclusive regarding a late-stage temporary undercurrent. Data return was spotty in summer 1974. In August 1973 there are two profiles that indicate westward surface flow with eastward current in the thermocline. In all, the available data are quite consistent with Cane's (49) analysis.

The importance of Cane's (49) work is to clarify the fundamentally different equatorial ocean responses to easterlies and westerlies, due to nonlinear effects, and to drive his model ocean with an equatorial wind that incorporates the basic features of the observed field over the central Indian Ocean: autumn and spring westerlies, with weak, largely meridional winds in between. Obvious elaborations for the future will include taking account of more detail in the wind field, such as the tendency for westerlies to begin sooner in the east, and the differences in strength at various locations. This will change details of the calculated response, both the fast local effects and the slower propagating changes. Interannual differences, such as that between 1973 and 1974, need fuller exploration. In all of this work better long-term wind data will be essential to improved understanding of the real ocean response.

What of equatorial currents beneath the thermocline; is there a monsoon signal at depth? Until very recently, data with which to address such questions have been lacking. Available theoretical considerations, mainly derived from the approach of Lighthill (29), predicted a simple structure, composed of a few low order baroclinic modes. This is the result if the wind forcing is treated as a body force distributed uniformly over an upper mixed layer and this forcing is then projected onto the (standing) baroclinic modes of a flat-bottomed ocean. But in 1976, in one of the most significant and reassessment-inspiring observational contributions to equatorial oceanography of recent years, Luyten and Swallow (50) published continuous full-depth profiles of currents in the Indian Ocean showing that the vertical structure was far more complex than previously suspected. Figure 13.15 shows some of their results. The profiles of zonal velocity have many zero crossings in the vertical and thus must involve many baroclinic modes, if such a representation is used. From the dates of the profiles it is clear that the field changes slowly over the span of these observations; individual cores of high velocity can be traced from profile to profile. The most energetic ones are westward flows at 200 and 700 m. In the wake of this discovery, similar measurements have been made in the Pacific by Eriksen (51), and similarly rich vertical structure has been found. It is clearly an equatorial phenomenon; poleward of 1−2 degrees of latitude the structure reverts to a simpler low-mode profile typical of mid-latitudes.

Wunsch (52) has offered an instructive calculation which shows how complex vertical structure might be forced by the long-period variations of the wind. The basic linear dynamical equations for an equatorial beta plane ocean can be solved in two ways. The first is Lighthill's (29) approach. The vertical problem is separated and solved subject to boundary conditions at the ocean surface and bottom; this

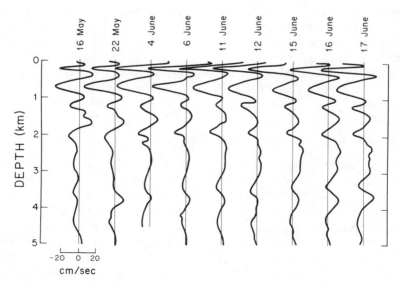

Figure 13.15. Profiles of zonal velocity on the Indian Ocean equator at 53°E. Dates in May and June of 1976 are indicated. From reference (50).

leads to the vertically standing barotropic and baroclinic modes. Wind forcing is projected onto these modes, and as already noted only a few low-order ones are importantly excited. The remaining inhomogeneous (forced) problem for the horizontal and time dependence of each vertical mode is then solved. The second approach, which is more common in meteorology, admits vertically *propagating* solutions. Forcing and the dynamical variables are expanded in terms of the meridionally trapped equatorial wave modes, and the problem is separated, with the inhomogeneity now occurring in the vertical problem. Upward- and downward-traveling forced waves result. Wunsch (52) argues that it is reasonable to accept the solutions with group velocity (energy propagation) away from the surface, since the vertical propagation for forcing periods of interest (annual, semiannual) is so slow that they would never be reflected from the bottom unattenuated. In effect, this approach treats the ocean as infinitely deep for purposes of this class of waves.

Wunsch (52) presents solutions obtained by this technique, having made particular choices of several parameters of the problem, in particular a single period and wavenumber of the forcing. In a further simplification the forcing is taken to consist of an *ad hoc* simple field of vertical velocity at the top of the deep ocean rather than the complicated combination of wind stress components and gradients thereof which arise if the problem is attacked directly. All of this means that it is not possible to take Wunsch's model literally as a representation of ocean response to actual winds. It is rather a simple example that shows how forced, vertically propagating motions at the equator can give rise to a complicated vertical structure in the deep ocean; it gives a valuable context for the interpretation of results from more complicated models and calculations.

Now good observational evidence exists (53) that much of the variability in the deep equatorial Indian Ocean takes the form of vertically propagating signals with semiannual period, and so it is, presumably, monsoon-driven. In particular, the westward jet at 750 m, which was mentioned in Section 2.2 as possibly connecting to longshore flows in the Somali Current system, has a clear semiannual variation. In principle the (linear) wave which Luyten and Roemmich (53) have detected extends to the surface, and so also accounts for some of the variability in the shallow layers, such as the Wyrtki jet. Cane's (49) work indicates, however, that a linear scheme cannot be a complete one for the fast shallow currents, as discussed above.

2.4 Arabian Sea

The most remarkable feature of the Arabian Sea interior is that the surface layer cools in summer, by 1–4°C in the region west of 75°E, and warms in the winter. One might suppose that this is due to direct evaporative heat exchange driven by the strong winds of the southwest monsoon, but Düing and Leetmaa (54) have shown that this is not the case, except perhaps locally. Over the Arabian Sea as a whole, which they take to be bounded by 75°E and the equator, they compute an upper layer heat budget composed of three terms: total heat exchange with the atmosphere through the surface, heat flux across the equator due to the northward flow of cooler Somali Current water, and heat flux due to wind-driven upwelling along the African and Arabian coasts. Calculation of the terms is necessarily subject to large errors, given the available data. Fluxes in the surface layer across 75°E and across the equator east of the Somali Current region are ignored, but are probably small. Nevertheless, the clear result is that the surface exchange transfers heat from the atmosphere into the ocean during the summer, despite the strong winds, but this is more than offset by the other two processes, resulting in a net summer cooling. Ocean circulations are therefore of the essence in determining the temperature and heat content of the Arabian Sea surface layer, at least in summer. Less is known about the relative importance of ocean heat transports and direct surface heat exchange in winter.

The detailed processes which bring cold water to the surface are, of course, not sorted out in such an integrated budget, but remain a fascinating subject for further study. Presumably upwelling is an important element. We have noted in Section 2.2 the cold wedges that appear inshore of places where the Somali Current leaves the coast. Both upwelling, in the sense of wind-driven Ekman transport to the right of the wind, and uplifting of isotherms in geostrophic balance with the strong current are driven by the same longshore wind, which is primarily responsible for the development of the current (16). Distributions of properties show the importance of the upwelling component. Warren, Stommel, and Swallow (22) discuss observations of surface water south of Ras Hafun with dissolved oxygen values at only 60% of saturation. This water must have been upwelled to the surface quite recently, since prolonged exposure to surface mixing by the strong southwest monsoon would have increased the oxygen content to 100% of saturation. Along the Arabian coast the current is less pronounced, but upwelling is still very strong, as distributions of temperature and of nutrients in the surface layer show (54). Recent satellite views of sea surface temperature (55) have disclosed a fabulously complex collection of sea-surface temperature structures in this region, especially long cold streamers and eddies emanating from the several capes of this coast. There is evidence of seasonal variation in some of these features. They must be important pieces of the machinery which moves cold upwelled water into the Arabian Sea, but clarifying their dynamics and quantifying their role in the overall cooling process remain unsolved problems.

3 CONCLUDING REMARKS

If this chapter has given readers some sense of the richness and uniqueness of the wind-driven ocean circulations in the monsoonal Indian Ocean, and has pointed to some of the principal papers needed to begin serious study of the literature, it has served its purpose. That the monsoon winds are the forcing which drives these circulations and makes them very different from patterns in other oceans should be clear. It follows inescapably that among the prerequisites for new advances in understanding and modeling these circulations, better knowledge of the actual surface wind field stands high. This is true for attempts to understand the typical annual cycle, and even more so for attacks on the harder but more important question of understanding interannual variability. It is interannual changes in the typical monsoon cycle that disturb the lives of people in this region. It is unfortunate that the scientific problem is difficult and will take a long time to solve, but it would be even more unfortunate not to make the attempt.

ACKNOWLEDGMENTS

I am indebted to Henry Stommel for suggesting that I write this chapter, for introducing me to the fascinations of Indian Ocean problems, and for making possible my first research efforts there. It is a pleasure to recall the kind cooperation I received from men of the British Meteorological Office and of the Royal Air Force at the former RAF base on Gan. My work in the Indian Ocean was supported for many years by the Office of Naval Research. Preparation of this chapter was supported in part by the National Science Foundation under Grant OCE8024369.

REFERENCES

1. C. Ramage, *Monsoon Meteorology*, Academic, New York, 1971.
2. Meteorological Office, *Monthly Meteorological Charts of the Indian Ocean*, HMSO, London, 1949.
3. J. Findlater, *Geophysical Memoirs*, **16**, 115 (1971).
4. J. Findlater, "An Experiment in Monitoring Cross-Equatorial Airflow at Low Level over Kenya and Rainfall of Western India During the Northern Summers," in J. Lighthill and R. P. Pearce, Eds., *Monsoon Dynamics*, Cambridge University Press, Cambridge, 1981, Ch. 20.
5. G. V. Rao and H. M. E. van de Boogaard, "Structure of the Somali Jet Deduced from Aerial Observations Taken During June-July, 1977," in J. Lighthill and R. P. Pearce, Eds., *Monsoon Dynamics*, Cambridge University Press, Cambridge, 1981, Ch. 21.
6. R. A. Knox, "Wind Stress Data from RAF Gan," unpublished manuscript, Scripps Institution of Oceanography, La Jolla, California, 1976.
7. K. Wyrtki, *Science*, **181**, 262 (1973).
8. J. Nehru, *The Unity of India; Collected Writings 1937–1940*, Lindsay Drummond, London, 1942; quoted in (1).
9. M. Fieux and H. Stommel, *Mon. Wea. Rev.*, **105**, 231 (1977).
10. D. S. Luther, "Observations of Long Period Waves in the Tropical Oceans and Atmosphere," unpublished doctoral dissertation, Woods Hole Oceanographic Institution–Massachusetts Institute of Technology Joint Program in Oceanography and Ocean Engineering, Massachusetts Institute of Technology, Cambridge, 1980, Ch. 2.
11. Deutsches Hydrographisches Institut, *Monatskarten für den Indischen Ozean*, Report #2422, Hamburg, Federal Republic of Germany, 1960.
12. G. Schott, *Geographie des Indischen und Stillen Ozeans*, C. Boysen, Hamburg, 1935, Ch. VI.
13. W. Düing, R. L. Molinari, and J. C. Swallow, *Science*, **209**, 588 (1980).
14. W. Düing, "The Somali Current: Past and Recent Observations," in FINE Workshop Proceedings, Nova/N.Y.I.T. University Press, Dania, Florida, 1978, Ch. 5.

15. D. L. T. Anderson, "Low Latitude Western Boundary Currents," in FINE Workshop Proceedings, Nova/N.Y.I.T. University Press, Dania, Florida, 1978, Ch. 6.
16. M. D. Cox, *J. Phys. Oceanogr.*, **9**, 311 (1979).
17. D. L. T. Anderson, *Ocean Modelling*, **34**(6), Department of Applied Mathematics and Theoretical Physics, Cambridge University, Cambridge, England, 1981.
18. S. G. H. Philander and P. Delecluse, *Deep-Sea Res.*, **30**, 887 (1983).
19. O. B. Brown, J. G. Bruce, and R. H. Evans, *Science*, **209**, 595 (1980).
20. A. G. Findlay, *A Directory for the Navigation of the Indian Ocean*, Richard Holmes Laurie, London, 1866.
21. J. G. Bruce, *Deep-Sea Res.*, **15**, 665 (1968).
22. B. Warren, H. Stommel, and J. C. Swallow, *Deep-Sea Res.*, **13**, 825 (1966).
23. J. C. Swallow and J. G. Bruce, *Deep-Sea Res.*, **13**, 861 (1966).
24. J. G. Bruce, *J. Geophys. Res.*, **84**, 7742 (1979).
25. J. G. Bruce, Variations in the Thermal Structure and Wind Field Occurring in the Western Indian Ocean During the Monsoons, U.S. Naval Oceanographic Office Report TR-272, 1981.
26. J. G. Bruce and G. H. Volkmann, *J. Geophys. Res.*, **74**, 1958 (1969).
27. J. G. Bruce, *Deep-Sea Res.*, **20**, 837 (1973).
28. J. C. Swallow and M. Fieux, *J. Mar. Res.*, **40** (supplement), 747 (1982).
29. M. J. Lighthill, *Phil. Trans., Roy. Soc.*, **265**, 45 (1969).
30. D. Moore, "Planetary-Gravity Waves in an Equatorial Ocean," unpublished doctoral dissertation, Harvard University, Cambridge, Massachusetts, 1968.
31. G. Veronis and H. Stommel, *J. Mar. Res.*, **15**, 43 (1956).
32. A. Leetmaa, *Deep-Sea Res.*, **19**, 319 (1972).
33. A. Leetmaa, *Deep-Sea Res.*, **20**, 397 (1973).
34. D. L. T. Anderson and P. B. Rowlands, *J. Mar. Res.*, **34**, 295 (1976).
35. H. E. Hurlburt and J. D. Thompson, *J. Phys. Oceanogr.*, **6**, 646 (1976).
36. J. G. Bruce, *J. Geophys. Res.*, **83**, 963 (1978).
37. J. G. Bruce, *Deep-Sea Res.*, **16**, 227 (1969).
38. A. Leetmaa, H. T. Rossby, P. M. Saunders, and P. Wilson, *Science*, **209**, 590 (1980).
39. D. R. Quadfasel and F. Schott, *J. Phys. Oceanogr.*, **12**, 1358 (1982).
40. J. Y. Buchanan, *Proc. Roy. Geogr. Soc.*, **8**, 753 (1886).
41. J. Y. Buchanan, *Scottish Geogr. Mag.*, **4**, 177 (1888).
42. T. Cromwell, R. B. Montgomery, and E. D. Stroup, *Science*, **119**, 648 (1954).
43. E. Firing, C. Fenander, and J. Miller, Profiling Current Meter Measurements from the NORPAX Hawaii to Tahiti Shuttle Experiment, Hawaii Institute of Geophysics report HIG-81-2, Honolulu, Hawaii, 1981.
44. J. A. Knauss, *Deep-Sea Res.*, **6**, 265 (1960).
45. W. G. Metcalf, A. D. Voorhis, and M. C. Stalcup, *J. Geophys. Res.*, **67**, 2499 (1962).
46. B. A. Taft and J. A. Knauss, *Bull. Scripps Inst. Oceanogr.*, **9**, 163 pp. (1967).
47. J. C. Swallow, *Nature*, **204**, 436 (1964).
48. R. A. Knox, *Deep-Sea Res.*, **23**, 211 (1976).
49. M. A. Cane, *Deep-Sea Res.*, **27**, 525 (1980).
50. J. R. Luyten and J. C. Swallow, *Deep-Sea Res.*, **23**, 999 (1976).
51. C. C. Eriksen, *J. Phys. Oceanogr.*, **11**, 48 (1981).
52. C. Wunsch, *J. Phys. Oceanogr.*, **7**, 497 (1977).
53. J. R. Luyten and D. H. Roemmich, *J. Phys. Oceanogr.*, **12**, 406 (1982).
54. W. Düing and A. Leetmaa, *J. Phys. Oceanogr.*, **10**, 307 (1980).
55. B. J. Cagle and R. Whritner, Arabian Sea Project of 1980—The Development of Infrared Imagery, Office of Naval Research Report ONRWEST 81-3, 1981.

14

Interannual Variability of Monsoons

J. Shukla
Center for Ocean–Land–Atmosphere Interactions
Department of Meteorology
University of Maryland
College Park, Maryland

INTRODUCTION

The basic question addressed in this chapter is why one monsoon season is different from another. This question is not any different than asking why some summers are warmer than others, and some winters are colder than others. This is fundamentally a more difficult question than why summers are warmer than winters and vice versa. For brevity, I first describe the processes that determine the mean seasonal monsoon climate itself and the mechanisms that are responsible for day-to-day weather changes. I then attempt to address the question of the interannual variability of the seasonal averages. This chapter will focus almost exclusively on the Indian summer monsoon for which we have an extensive historical data set.

Let us begin by asking what appears to be a simple question: Why weather on a given date (say, January 1 of one year) is so different from the weather on the same date in another year? We know that the amount and distribution of solar energy, the most important driving force for the atmosphere, remains the same on both dates, as does the rotation of the earth, the composition of the atmosphere, and the distribution of oceans and continents. This would suggest that the causes for different weather on the same date for two different years do not lie outside the atmosphere, but most probably inside the atmosphere. To provide a reasonable answer to the above question, we have to address some additional fundamental questions. For example, what causes weather, and what causes weather to change from one day to another? An understanding of the processes that are responsible for changes in weather from one day to another is a prerequisite to understanding the mechanisms for changes in seasonal mean weather from one year to another. Since seasonal mean weather, which is sometimes referred to as the short term climate, is a consequence of the average of daily weather over the length of a season, it is natural to think that day-to-day changes in weather in a given season might be responsible, at least in part, for producing changes in the seasonal mean from one year to another.

The basic driving force for atmospheric motions is the uneven solar heating of the earth–atmosphere system due to the near-spherical geometry of the earth's surface and the revolution of the earth around the sun. The actual rates of heating vary with height, latitude and longitude, and the magnitudes are determined primarily by the composition of the earth's atmosphere, and the time of day and day of the year. The earth's equatorial regions receive more energy from the sun than they lose to space, but the reverse is the case for the polar regions. The net heating of the warmer equatorial regions and the net cooling of the colder polar regions is a source of energy for the motion of air particles. Other important quasi-stationary forces are provided by the asymmetric heat sources of land and ocean and by mountains which act as mechanical barriers and produce quasi-stationary circulation features. Solar energy heats the oceans and evaporates the water that later condenses into the atmosphere, often at some distance from where it evaporated, providing another very important energy source, the latent heat of condensation. The three-dimensional structure of this heat source is determined in part by the motion field itself, producing one of the most complex feedback loops (i.e., nonlinearities) of atmospheric dynamics. The mean circulation produced by the forcing functions described above can be considered, hypothetically, as the seasonal mean climate of the earth–atmosphere system. This seasonal mean climate, however, is characterized by horizontal and vertical gradients of wind, temperature, and moisture, that are favorable for the growth of thermodynamic and hydrodynamic instabilities. The day-to-day weather fluctuations are produced by transient weather disturbances which are the manifestations of the growth, decay and propagation of these instabilities. The disturbances can derive their enery from the mean circulation and thereby change the mean circulation itself. The observed, quasi-equilibrium mean circulation, to be referred to as the mean climate, is produced by interactions among the stationary and quasi-stationary forcing functions, and transient disturbances.

Since changes in the weather occur due to instabilities of the atmospheric state, and since the internal atmospheric dynamics are intrinsically nonlinear, atmospheric behavior is aperiodic and, therefore, at long ranges, unpredictable. This inherent aperiodicity of the weather at all time scales also produces aperiodicity of monthly and seasonal averages. Therefore, even if there were no changes in the external forces, it would be reasonable to expect seasonal averages to be different from one year to another. Actually, small changes do occur; for example, although the solar energy input for a given season remains nearly the same from one year to the next, there is some variation due to the interannual variability of seasonal mean cloudiness. The height of the mountains remains constant, but there may be large changes in the circulation because the wind impinging on the mountains may be quite different. Another plausible reason for the interannual variability of seasonal mean climate appears to be the interannual variability of boundary parameters such as seasonal mean sea surface temperature (SST), soil moisture, and sea ice and snow. Based on these considerations, the mechanisms responsible for interannual variability of seasonal mean climate fall into two categories: "internal dynamics" and "boundary forcing" (1, 2). These will be discussed further in Section 3.

It is reasonable to expect that the nature of the interannual variability of seasonal averages will depend upon the spatial and temporal domains for which the averages are calculated. For example, if we average rainfall over the whole Afro–Asian monsoon region, it may not show large interannual variability, but the average over the Indian subcontinent, or part of it, might show large interannual variability. Similarly, monthly means might display large interannual variability, whereas seasonal means might not. This suggests that in order to get a meaningful description of the interannual variability of any atmospheric phenomenon, we need a good understanding of the dominant space and time scales of atmospheric anomalies. Part of the interannual variability may be due to changes in the intensity of the mean atmospheric circulation systems, and part of it may be simply due to shifts in location and timing of those circulation systems. It would appear, therefore, more appropriate to study the question of interannual variability on a global scale. However, due to insufficient long-term, global data records, we are constrained to study the interannual variability of regional phenomena.

In Section 1, 81 years of monsoon rainfall data is used over different subdivisions of India to describe the observed structure and interannual variability of monsoon rainfall. The interannual variability of rainfall is described here because of its social–economic importance and also because reliable records are available for a long period. Discussion of observed variability will be limited to the summer monsoon season. In Sections 2 and 3, the basic mechanisms for intraseasonal and interannual variability of the monsoon circulation and rainfall are described and the results of some of the key studies which document possible relationships between monsoon and other global circulation features are summarized. It will be shown that the year-to-year changes in the Indian summer monsoon are related to a global scale atmospheric circulation feature called the Southern Oscillation, and ocean temperature anomalies in the tropical Pacific called El Niño. A summary of several numerical experiments with global climate models, and results of observational studies is presented to show that the slowly varying boundary conditions have significant influence on the fluctuations of seasonal averaged monsoons.

1 INTERANNUAL VARIABILITY OF SUMMER MONSOON RAINFALL OVER INDIA

After the great famines of 1877 and 1899, the India Meteorological Department established an extensive and efficient network of rain gauge stations over India. This network has provided a great wealth of monsoon rainfall data for systematic studies of interannual variability which have been carried out by a large number of investigators. Some of the earlier work will be summarized and then an analysis will be presented of the structure and space–time variability of monsoon rainfall over India for the period 1901–1981.

1.1 Previous Studies

There are a large number of scientific publications, especially from Indian scientists, describing various aspects of the interannual variability of Indian monsoon rainfall and its relationship with regional and planetary-scale circulations. It is not possible to summarize here the results of all such papers; I therefore refer only to those works that examine sufficiently long time series. This allows some definitive statements to be made about the nature of variability and its possible causes. Case studies using limited data samples are indeed illuminating but insufficient to distinguish between a genuine signal and random noise because of sampling problems.

Parthasarathy and Dhar (3) studied 60 years (1901–1960) of annual rainfall for 31 subdivisions over India (Fig. 14.1). Their data were derived from approximately 300 rain gauge stations. The analysis showed that rainfall in most of the subdivisions fits a normal distribution. Positive trends (increasing rainfall) were found in many

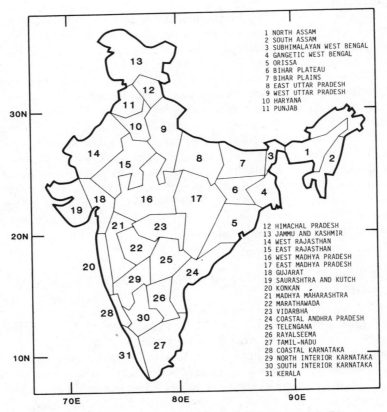

Figure 14.1. Locations and names of 31 subdivisions over India.

subdivisions, but these have been reversed by the large number of years with deficient rainfall in the 1960s and 1970s. Power spectrum analyses did reveal some evidence of a 2- to 3.5-year cycle in some subdivisions. Dhar et al. (4) computed monthly mean rainfall for the whole of India from corresponding subdivisional data for 1901–1960. Annual and seasonal mean rainfall statistics were computed from these monthly values. The authors showed that a direct relationship usually exists between the rainfall amount and the number of cyclonic storms crossing India. Pareek and Ramaswamy (5) analyzed the summer monsoon rainfall over Burma for the period 1907–1938 and concluded that droughts are not a serious problem in Burma. Parthasarathy and Mooley (6) studied seasonal rainfall for the whole of India. The data, covering the period 1841–1977, were derived from rain gauge stations. Two series were constructed; one for prepartition India and Burma covering the period 1841–1935; another for Indian rain only, 1901–1977. From these two series a homogeneous series was constructed for 1866–1970. Some data were discarded because of unreliability and sparse coverage (1971–1977). They found no trend in the rainfall data, but pointed out that the period 1931–1960 had higher than normal rainfall. A power spectrum analysis showed evidence of a possible cycle with a period of between 2 and 3 years.

Bhalme and Mooley (7) used monthly mean percentage departure rainfall data for the period 1891–1975 for 31 subdivisions to construct indices of flood and drought intensity. They found that large-scale floods (heavy rainfalls) and droughts each occur about 15 times per century. They also found a periodicity of about 20 years in the general flood index. Frequent large-scale droughts were found to appear during the periods 1891–1920 and 1961–1975, while there were few droughts between 1921–1960. They also studied composites of circulation patterns for flood and drought years and indicated that departures from normal of the monsoon rain may be foreshadowed by circulation anomalies in the upper troposphere during May. Mooley et al. (8) considered a time series from 1871–1978 of annual rainwater volume falling on the whole of India, derived from a network of 306 rain gauge stations. Time series of the area covered by excess and deficit rain were prepared to show the extreme years. The authors point out that in the period 1921–1950 there was only one excess and one deficit (as per their definition) year of rainfall. They also commented on the economic impact of extreme rainfall. Mooley and Pant (9) studied the historical data for 1771–1977 and classified 32 major droughts. These years did not necessarily correspond to the largest negative departures in the time series for seasonal rain because they considered the annual rainfall.

1.2 Variability of Monsoon Rainfall over India, 1901–1981

The observed interannual variability of monsoon rainfall over India based on 81 years of data is presented here. The percentage departure from normal rainfall for the months of June, July, August, September, and the seasonal mean, for 81 years (1901–1981), for 31 subdivisions of India (Fig. 14.1) was obtained from the India Meteorological Department (the seasonal mean is given in Table 14.1). Various

Table 14.1. Seasonal Rainfall Anomalies (percentage departure from normal) for 31 Subdivisions

Year	Subdivision Number															
	1	2	3	4	5	6	7	8	9	10	11	12	13	14	15	16
1901	−4	−3	−18	−5	−27	−15	−26	−16	−5	−25	−20	−11	−14	−53	−36	−10
1902	15	2	24	−9	−3	−4	−2	−5	−4	−4	−25	−35	−26	−31	−9	−17
1903	5	4	−3	−14	0	−28	−17	−5	−16	−11	11	−6	45	12	4	−3
1904	−6	−2	−33	0	6	9	−2	6	8	−1	−40	−17	−16	−51	16	−18
1905	2	10	14	5	−15	0	23	−13	−44	−49	−32	−30	3	−59	−62	−23
1906	−4	4	3	−14	−11	−15	1	4	31	21	11	49	24	−16	−3	14
1907	0	9	2	−9	1	8	−9	−40	−39	−32	−22	−53	−20	19	−27	−30
1908	−9	1	−31	12	10	−10	−41	−22	4	45	48	1	34	112	43	1
1909	−7	−7	−1	23	−1	13	20	19	13	36	57	25	52	40	6	−14
1910	1	9	22	−6	4	−1	16	2	8	15	23	12	23	11	3	0
1911	12	−1	9	−8	−2	12	15	1	−20	−15	−51	−34	−31	−63	−35	−23
1912	1	−14	−6	−14	−5	−20	−16	−10	−2	11	−6	−2	−21	−2	5	−12
1913	−16	−3	1	27	5	11	22	−26	−44	−28	−18	−24	−10	−27	−41	−19
1914	−7	−19	−14	−8	12	−15	−11	1	12	31	36	36	35	2	8	−8
1915	3	14	−18	−17	−17	−23	−1	32	−5	−37	−45	−6	−22	−72	−57	−22
1916	−9	−8	27	5	−5	−4	21	27	34	31	18	2	32	45	34	21
1917	8	−6	−5	5	5	11	1	18	30	82	90	48	67	139	82	26
1918	28	26	13	11	−13	−5	21	−35	−43	−57	−49	−53	−40	−76	−58	−40
1919	−4	−13	−6	9	11	17	4	5	5	4	−10	−7	14	3	23	27
1920	−5	−11	17	−8	1	10	−2	−13	−14	−26	−31	−22	−47	−28	−16	−30
1921	5	5	16	−2	−5	1	13	15	48	−5	−33	−7	−2	−20	−13	−9
1922	−7	−3	20	41	10	22	32	42	33	14	0	30	−4	−11	5	−5
1923	−6	1	−11	−5	−23	7	−22	6	−1	−3	22	−15	−9	−4	12	16
1924	1	−3	13	−6	−27	4	25	18	27	27	0	8	−5	4	41	4

Year																
1925	-8	-18	5	-25	27	-1	1	19	21	12	26	8	12	-32	-27	-20
1926	-3	-5	3	23	3	7	-6	0	-1	18	8	12	22	56	22	4
1927	10	12	4	-17	4	-14	-17	-11	2	-9	-17	24	-4	16	-1	-18
1928	-1	-3	10	19	4	-6	-12	-37	-35	-43	-29	-26	-9	-4	-30	-14
1929	-2	4	-11	-1	6	0	-7	-4	-25	-43	-27	-11	9	10	-18	-12
1930	1	4	-25	0	-2	5	-6	28	1	10	5	-21	-18	-3	-8	-5
1931	10	-12	12	-11	-10	-8	-7	-1	3	-1	10	-7	-11	40	9	9
1932	10	0	-2	-18	-8	-13	-28	-26	-2	-8	-9	18	-4	-3	-16	3
1933	-8	1	-6	26	27	3	9	-21	20	82	65	24	21	27	43	18
1934	8	4	1	-20	11	-6	2	10	9	0	-22	2	-15	19	20	28
1935	15	13	16	-21	-6	-7	9	-7	-14	-6	-12	-10	-15	-12	0	-4
1936	-5	-4	-2	7	25	17	27	53	49	16	11	14	-1	-10	-15	-6
1937	-9	-2	-17	6	5	0	-9	-1	-17	-16	-15	7	-38	-6	-1	9
1938	7	7	22	-12	-13	-9	26	40	-4	-46	-27	-16	-13	-29	-20	3
1939	-2	-7	-2	23	2	12	5	2	-12	-36	-41	-12	-23	-59	-36	-4
1940	-11	-1	-11	-5	17	-17	-14	-16	-12	-22	-2	-8	-10	-2	-14	8
1941	-7	3	-1	23	8	4	6	-18	-36	-35	-11	-20	-2	-19	-26	-26
1942	-2	4	-23	19	-1	25	7	0	28	63	53	47	28	16	48	24
1943	-1	-6	12	-1	32	13	-6	12	16	-11	-25	47	0	10	7	-3
1944	-6	-1	-2	4	-3	-4	-5	-4	-23	-15	-5	-2	14	87	25	32
1945	-3	9	-6	-24	-3	-11	-16	-13	18	30	49	0	-18	28	30	15
1946	-3	-13	-13	11	19	4	3	-6	3	-4	-20	11	-6	-18	27	15
1947	3	19	-14	-13	-14	-14	-10	3	2	-9	19	9	8	-5	4	15
1948	4	5	1	-8	-13	-2	4	32	17	15	3	-4	15	-23	1	19
1949	7	4	-2	-3	-20	2	10	7	10	5	-4	-22	-31	-15	-17	-2
1950	1	-8	1	13	-7	13	-6	-8	15	11	91	28	46	9	13	-7
1951	-12	-18	-3	-18	-17	-21	-29	-22	-25	-45	-26	-17	-44	-43	-47	-27
1952	-2	-5	3	-8	2	5	2	7	4	2	0	-9	-12	4	-7	-7

(Table continues on p. 406.)

Table 14.1. (*Continued*)

Year	1	2	3	4	5	6	7	8	9	10	11	12	13	14	15	16
									Subdivision Number							
1953	−6	−3	−6	16	−5	18	21	35	7	13	10	10	71	10	−17	−16
1954	−4	−18	24	−12	−18	−14	−16	−6	−12	−23	19	9	−14	−13	−11	15
1955	5	−12	22	−18	2	−21	15	40	13	16	15	5	4	31	20	35
1956	−13	−14	−1	14	22	10	12	22	6	0	27	−2	44	29	13	3
1957	−4	−23	−18	−13	−19	−11	−17	−4	40	10	12	0	95	−21	3	−17
1958	−10	−17	18	−20	10	−14	0	−2	40	50	67	9	41	−5	23	8
1959	−10	−5	−14	17	−15	−4	−28	−28	−14	−10	27	24	45	22	0	11
1960	8	18	−4	−6	19	−1	4	9	21	40	10	−4	−12	−16	−7	−1
1961	−11	−4	−18	−11	36	11	−16	3	26	46	33	19	53	40	38	45
1962	52	−5	−15	−19	−13	−20	−5	3	−3	−2	47	9	8	−7	1	−4
1963	0	1	−12	−11	−1	−15	0	−4	18	17	−6	19	−9	−35	−11	−4
1964	1	5	23	−5	13	−2	−14	2	21	63	−34	6	8	22	−6	2
1965	−3	15	33	0	−15	−21	−6	−34	−24	−27	−38	−41	−39	−23	−36	−37

Year																
1966	6	36	-6	-27	-15	-32	-46	-29	-3	8	18	11	21	-14	-34	-29
1967	-13	-18	-16	2	-1	-7	-15	5	29	27	-15	24	1	2	6	-4
1968	-5	1	-1	17	0	1	-10	-20	-13	-41	-19	-30	7	-45	-20	-9
1969	-7	-6	-7	-3	1	-13	12	2	6	-6	-38	-15	-25	-54	0	19
1970	7	-7	3	20	-6	-4	-12	14	-7	-6	13	13	15	33	0	11
1971	-9	-12	-7	39	-2	39	15	32	12	23	17	60	-26	-30	23	16
1972	-3	-21	-26	-3	-5	-16	-35	-25	-16	-20	-28	-28	-30	-23	-32	-22
1973	1	-9	-10	17	3	8	-1	-8	14	16	21	3	30	51	49	49
1974	61	14	16	4	-34	-7	10	-9	-18	-22	-36	-24	-6	-48	-21	-11
1975	-9	-21	-8	0	-2	4	-5	24	21	65	44	7	95	115	33	9
1976	15	8	-10	-15	-14	-1	8	1	-1	39	62	-12	46	64	15	12
1977	15	-14	-8	19	0	28	-4	-16	11	28	26	11	15	41	38	7
1978	-8	-11	-24	38	-4	21	-1	7	28	45	24	23	15	40	27	10
1979	-24	10	-12	-11	-18	-26	-20	-47	-50	-31	-34	-46	-23	-23	-23	-50
1980	5	-26	-7	5	18	10	6	64	2	8	32	-8	10	-22	10	11
1981	-7	-28	-8	26	-7	-5	-3	14	-18	9	-19	-28	-3	-32	-8	-14
Mean	0	-2	-1	1	-1	-2	-2	1	2	3	3	0	6	0	0	-1
S.D.	12	12	14	16	14	14	16	21	22	30	32	23	35	40	27	19

(*Table continues on p. 408.*)

TABLE 14.1 (*Continued*)

Year	Subdivision Number														
	17	18	19	20	21	22	23	24	25	26	27	28	29	30	31
1901	0	−36	−55	1	−10	8	2	−24	−20	−36	10	−5	−9	−11	−3
1902	−22	4	0	−7	−15	−20	−27	4	−15	−10	−5	13	−11	−16	9
1903	−10	0	17	5	−6	27	12	39	37	20	33	−1	21	24	6
1904	−12	−51	−58	−19	−26	−9	−27	−33	−21	−51	−26	3	−11	−15	6
1905	−17	−16	−26	−31	−33	−28	−10	−16	−14	−18	−24	−19	−33	−31	−15
1906	0	0	12	−11	−3	12	16	22	10	11	2	−11	11	10	−13
1907	−11	9	7	5	−11	−26	−7	−14	−5	−31	−19	14	33	23	24
1908	14	10	15	2	8	2	21	7	49	−15	−20	22	7	−5	−6
1909	−16	7	27	10	3	27	−7	9	−7	26	3	−3	13	13	−8
1910	0	4	34	−9	18	56	22	30	1	31	8	−5	31	19	−10
1911	4	−56	−72	−23	−29	−11	−15	−15	−22	−34	−21	−21	−21	−5	−8
1912	−7	23	56	0	−10	−35	−7	11	−15	−16	−24	6	23	29	15
1913	−21	29	46	−2	−4	−14	2	−17	−19	−33	−23	−16	−17	1	−18
1914	4	24	37	33	41	39	15	25	50	−6	−15	26	50	6	−1
1915	−4	−52	−55	−11	19	14	−5	16	18	3	10	−21	16	9	7
1916	−3	0	6	15	12	47	25	27	22	37	20	1	17	19	5
1917	16	47	41	16	0	40	13	34	35	16	21	2	9	13	−9
1918	−2	−59	−72	−45	−55	−50	−33	−21	−30	−40	−50	−45	−41	−44	−47
1919	20	6	−5	4	11	−7	0	−7	−10	12	4	−14	7	12	−1
1920	−22	−32	4	−29	−24	−51	−47	−38	−39	−36	0	−13	−13	3	26
1921	2	33	34	2	−7	−23	5	14	12	65	4	−4	−27	5	−8
1922	−4	−3	−23	1	−17	−12	−1	−22	−7	−55	−11	4	−37	−19	9
1923	6	−37	−51	4	−11	3	−7	−17	−5	−33	−21	24	7	39	40
1924	−14	0	−22	−10	−6	−13	−19	14	−9	−6	35	15	13	47	64

Year															
1925	19	-32	-26	-23	-20	-30	-20	6	-2	-23	-25	9	-19	4	-5
1926	12	37	83	20	7	-6	-9	-5	-23	-13	-9	-3	7	17	11
1927	-1	52	39	-3	12	-15	2	16	0	0	-2	-2	12	13	6
1928	-19	-6	-1	13	5	10	16	-1	5	-18	-25	0	57	-19	-22
1929	7	-21	-8	-16	-16	-26	-20	-10	-24	-25	-6	9	-13	1	13
1930	-8	5	6	7	10	-18	-12	-11	-22	-25	-25	-14	-14	-19	-17
1931	-8	14	-28	26	20	22	14	3	11	-18	-11	19	15	15	23
1932	-3	2	-1	-3	10	0	6	-11	-7	-18	-9	-7	10	14	-14
1933	6	25	43	11	28	42	29	-1	22	-15	11	17	19	23	25
1934	16	8	-6	13	3	15	17	-2	2	-32	-37	-3	-11	-22	-21
1935	-1	-15	20	-9	-8	16	14	-8	-2	16	-10	-9	-7	-3	-20
1936	19	-33	-21	-6	-21	-23	14	6	3	-19	-5	-1	-22	1	-8
1937	8	28	26	10	4	-20	7	-21	-15	-24	-15	1	-30	-13	-9
1938	6	-12	-25	10	20	38	31	29	31	44	12	0	28	5	-13
1939	5	-30	-52	-4	-2	-26	-18	-20	-31	-18	-17	-17	9	-10	-7
1940	-3	-14	-20	11	3	12	21	-5	0	-16	-4	11	-5	5	4
1941	-36	10	-9	-35	-10	-34	-20	-11	-29	-28	7	-22	-9	-2	-2
1942	16	40	-13	17	16	28	24	-11	10	-22	-4	10	-9	2	-2
1943	10	-4	14	5	-2	-5	-15	-7	3	-25	-11	-4	-3	-9	1
1944	3	43	60	0	20	-9	21	0	-15	3	12	-13	-7	-13	-29
1945	5	48	56	16	2	-13	13	1	10	-10	-18	0	-8	-14	-16
1946	12	27	16	12	21	0	1	-10	-7	-17	-5	22	18	24	18
1947	12	-11	12	4	5	12	5	18	19	17	17	9	24	8	15
1948	1	-54	-58	0	4	4	-3	-6	-3	-26	-10	11	-2	5	6
1949	-6	-7	17	14	5	18	25	32	10	38	19	0	4	-11	0
1950	-9	28	6	2	6	-14	-29	3	-4	-21	-6	20	17	9	12
1951	-15	-49	-36	-7	-19	-13	-6	39	-4	-22	-2	-4	5	-11	-35
1952	-2	-14	-12	-9	-15	-17	-26	-30	-32	-43	-38	-16	-25	-25	-32

(Table continues on p. 410.)

TABLE 14.1 (*Continued*)

Year	Subdivision Number														
	17	18	19	20	21	22	23	24	25	26	27	28	29	30	31
1953	-7	23	64	20	4	16	4	7	23	14	23	9	28	26	-15
1954	-11	40	40	42	16	26	10	28	14	-2	-4	21	10	5	-10
1955	1	3	-10	33	4	51	14	6	38	18	-7	-5	20	-18	-13
1956	9	31	61	22	17	16	-3	34	30	23	19	2	31	4	-20
1957	-20	-28	-2	9	-7	6	-6	15	15	-3	-17	-8	17	-12	-2
1958	-3	25	20	49	23	33	4	32	30	3	-17	14	1	15	-7
1959	16	53	102	27	30	43	49	26	40	8	-13	38	26	52	30
1960	-8	-28	-18	11	4	-10	-11	15	-2	2	2	-2	15	-6	-7
1961	38	2	91	31	5	-1	33	11	17	-14	26	62	5	42	50
1962	-26	-13	-19	15	-4	18	-3	21	23	-16	6	13	8	9	2
1963	-10	7	-25	31	-3	47	2	-11	18	-19	3	-6	-11	-16	-15
1964	8	11	35	12	12	15	2	39	10	46	6	-2	54	22	-3
1965	-40	-31	11	2	-2	15	-28	-4	8	2	-8	-9	11	-18	-30

Year															
1966	−27	−16	−21	−21	−11	−13	−14	7	−3	17	29	−21	−3	−15	−26
1967	7	−3	40	−1	11	−1	−11	1	7	−3	−7	5	0	−14	−9
1968	−20	−22	−31	−26	−7	−7	−13	−30	−26	−21	−3	5	−9	−4	31
1969	−9	−6	−37	13	31	15	−6	−17	2	−15	−26	4	6	−5	−13
1970	11	53	79	25	9	45	21	15	27	20	−4	26	11	−8	−14
1971	4	−10	13	−4	−21	−22	−29	−18	−39	−29	10	2	−17	19	5
1972	−12	−40	−55	−31	−47	−45	−38	−26	−34	−31	9	−26	−14	−7	−14
1973	−4	38	−18	5	13	20	0	−26	−4	−5	17	0	2	15	−15
1974	−27	−67	−72	10	−5	−34	−34	−13	−25	4	16	14	−5	2	8
1975	15	26	−4	13	24	50	13	27	14	9	43	59	41	24	19
1976	−15	67	−1	9	31	−10	−9	−6	17	−3	15	−7	6	−38	−49
1977	5	43	10	9	−12	−20	−4	−17	−29	3	14	5	−6	0	−16
1978	−2	−4	−5	1	−12	−4	12	47	40	38	2	18	19	−1	−4
1979	−52	−25	60	−38	1	12	−3	−28	−18	−15	13	−19	9	32	1
1980	33	22	61	0	15	24	5	17	−9	−15	−32	−1	2	12	2
1981	−13	4	1	15	24	6	14	10	18	22	41	12	46	16	24
Mean	−3	1	4	3	1	3	0	2	1	−6	−2	2	5	3	−2
S.D.	15	30	39	18	17	25	18	20	21	24	18	17	20	18	19

statistics have been calculated from these data.* The percentage departure from normal is the difference (expressed as a percentage) between the average monthly mean rainfall for all the stations in a subdivision for which data is available for that particular month and year and the normal rainfall (the average, over all years, for that month) for the same stations. The percentage departure of seasonal rainfall is calculated in the same way. The seasonal mean percentage departure is not identical but close to the average of the percentage departures for the four months.

It is important to note that the number of stations varies with month and year, that is, the same stations are not available for each year. Thus a straightforward averaging of rainfall for all the available stations would show some interannual variation solely due to the sampling error that arises from using different stations each year. If the total number of stations is not large, and/or the rainfall in a given subdivision has large geographical variability, the spurious interannual variability arising from the sampling error could be quite large and might mask the real interannual variability. We consider the percentage departure from normal as a more appropriate parameter than the actual rainfall to study interannual variability.

We have also examined the interannual variability of the percentage departure of rainfall for several large regions of India. A regional average is defined as the area weighted mean of all the subdivisions in a given region. The "rainfall anomaly" is defined as the deviation of the percentage departure of rainfall from its long-term mean and the "normalized rainfall anomaly" is defined as the ratio of rainfall anomaly and its standard deviation. The coefficient of variability is defined as the ratio of the standard deviation and the mean rainfall.

Walker (10) suggested that rainfall anomalies over several subdivisions of India should be grouped together to define areal averages for large homogeneous regions. The basis for Walker's groupings was the uniformity of correlation coefficients between the anomalies and geographically distant atmospheric parameters. The two regions (Northwest and Peninsula) which were first defined by Walker and which are still being used by the India Meteorological Department for seasonal forecasts are shown in Figure 14.2 and defined in Table 14.2. The interannual variability of the seasonal rainfall for the subdivisions in the Northeast region was found to be small, and the correlation with the distant parameters was not significant, so Walker did not use these subdivisions for seasonal prediction.

It can be argued that the criteria chosen by Walker to define homogeneous regions of India were not quite appropriate because they were based on correlations with distant parameters. A more appropriate procedure might have been to examine the structure of the dominant mode of variability of the Indian rainfall data itself.

We have attempted to determine spatially homogeneous regions of India by the following three methods: (a) correlation among all the subdivisions, (b) correlation

---

* Percentage departure from normal, $P = 100(\widetilde{R} - \widetilde{\widetilde{R}})/\widetilde{\widetilde{R}}$, where $\widetilde{R}$ is the monthly or seasonal mean rainfall, and $\widetilde{\widetilde{R}}$ is the "normal" monthly or seasonal mean rainfall for 50 years (1901–1950).

Rainfall anomaly, $P' = (P - \widetilde{\widetilde{P}})$, where $\widetilde{\widetilde{P}}$ is time mean of P for 81 years.

Normalized rainfall anomaly, $P'' = P'/\sigma$ where σ is the standard deviation of P' for 81 years.

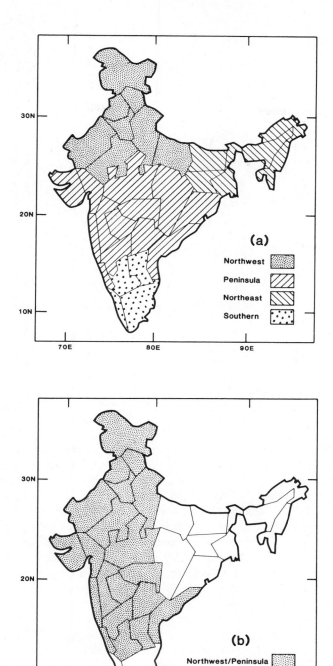

Figure 14.2. Locations and names of regions used for spatial averaging: (*a*) Northwest, Peninsula, Northeast, Southern, (*b*) Northwest–Peninsula. (*Figure 14.2 continues on page 414.*)

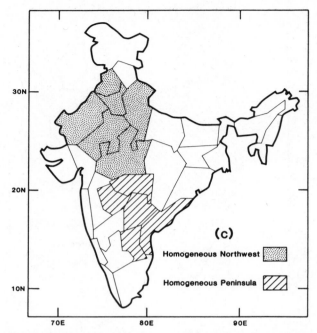

Figure 14.2. (*continued*) (*c*) Homogeneous Northwest, Homogeneous Peninsula.

between the subdivisional rainfall anomaly and the Southern Oscillation, and (c) empirical orthogonal functions (EOFs). Table 14.3 gives the correlation coefficients among rainfall anomalies for all the subdivisions. Correlations of 0.7 or higher are found between adjacent subdivisions as shown in Figure 14.3a. Six subdivisions in Northwest India (Western Uttar Pradesh, Haryana, Punjab, Western Rajasthan, Eastern Rajasthan, and Western Madhya Pradesh) and five subdivisions in Peninsula India (Marathawada, Vidarbha, Coastal Andhra Pradesh, Telengana, and Rayalseema) are the only two contiguous groups of highly homogeneous rainfall anomalies. We will refer to them as the Homogeneous Northwest and Homogeneous Peninsula regions. A similar map for the threshold correlation coefficient of 0.5 is shown in Figure 14.3b. This defines a much larger number of nearly homogeneous subdivisions which can be used for spatial averaging of Indian summer monsoon rainfall. Figure 14.3c shows the correlation coefficients between the normalized winter to spring pressure tendency for Darwin (which can be considered to be an index of the Southern Oscillation phenomenon), and seasonal rainfall anomalies for each subdivision using 81 years of data.

There is a marked similarity between the homogeneous subdivisions of Figure 14.3b and subdivisions with large correlation coefficients (significant at the 95% confidence level) in Figure 14.3c. This suggests that the Southern Oscillation influences a large number of subdivisions of India and, therefore, the separate groupings of

TABLE 14.2 Names of Subdivisions Constituting the Eight Regions

Region

1. *Northwest India*: East Uttar Pradesh (8), West Uttar Pradesh (9), Haryana (10), Punjab (11), Himachal Pradesh (12), Jammu and Kashmir (13), West Rajasthan (14), East Rajasthan (15)
2. *Peninsula India*: Orissa (5), West Madhya Pradesh (16), East Madhya Pradesh (17), Gujarat (18), Saurashtra and Kutch (19), Konkan (20), Madhya Maharashtra (21), Marathawada (22), Vidarbha (23), Coastal Andhra Pradesh (24), Telengana (25), Coastal Karnataka (28), North Interior Karnataka (29)
3. *Southern India*: Rayalseema (26), Tamil-Nadu (27), South Interior Karnataka (30), Kerala (31)
4. *Northwest India*: North Assam (1), South Assam (2), Subhimalayan West Bengal (3), Gangetic West Bengal (4), Bihar Plateau (6), Bihar Plains (7)
5. *Homogeneous Northwest*: West Uttar Pradesh (9), Haryana (10), Punjab (11), West Rajasthan (14), East Rajasthan (15), West Madhya Pradesh (16)
6. *Homogeneous Peninsula*: Marathawada (22), Vidarbha (23), Coastal Andhra Pradesh (24), Telengana (25), Rayalseema (26)
7. *Northwest–Peninsula*: West Uttar Pradesh (9), Haryana (10), Punjab (11), Himachal Pradesh (12), Jammu and Kashmir (13), West Rajasthan (14), East Rajasthan (15), West Madhya Pradesh (16), Gujarat (18), Saurashtra and Kutch (19), Konkan (20), Madhya Maharashtra (21), Marathawada (22), Vidarbha (23), Coastal Andhra Pradesh (24), Telengana (25), Rayalseema (26), Coastal Karnataka (28), North Interior Karnataka (29), South Interior Karnataka (30)
8. *Whole India*: All 31 Subdivisions

Northwest India and Peninsula India, as suggested by Walker, are not appropriate to study the variability and predictability of Indian summer monsoon rainfall in relation to the Southern Oscillation. This is mainly because large-scale rainfall anomalies over Northwest India and Peninsula India are not independent. This point is further supported by Figure 14.4 which shows the composite rainfall anomaly maps for years with heavy and deficient monsoon rainfall for both Northwest and Peninsula regions. Figures 14.4a and 14.4b show the composite anomaly maps for the Northwest region for the years with normalized rainfall anomalies equal to or greater than 1.0, and less than or equal to −1.0, respectively. Figures 14.4c and 14.4d show similar composites for the Peninsula region.* It can be seen that the years with heavy or deficient rainfall over one of the regions is accompanied by anomalies of the same sign (although not of the same magnitude) for a large number of subdivisions over India.

* Composite anomaly for the years 1908, 1909, 1916, 1917, 1933, 1942, 1953, 1956, 1957, 1958, 1961, 1973, 1975, 1976, and 1978 is shown in Figure 14.4a; for 1901, 1905, 1911, 1913, 1915, 1918, 1920, 1939, 1951, 1965, 1972, and 1979 in Figure 14.4b; for 1914, 1917, 1933, 1955, 1956, 1958, 1959, 1961, 1964, 1970, 1975, and 1980 in Figure 14.4c; and for 1901, 1904, 1905, 1911, 1918, 1920, 1941, 1951, 1965, 1966, 1968, 1972, 1974, and 1979 in Figure 14.4d.

TABLE 14.3 Correlation Coefficients Among All Subdivisions for Seasonal Mean Rainfall Anomalies

							Subdivision Number								
	1	2	3	4	5	6	7	8	9	10	11	12	13	14	15
1	1.00	.34	.22	−.23	−.33	−.13	.13	−.00	−.10	−.12	−.07	−.14	−.11	−.12	−.10
2		1.00	.15	−.23	−.24	−.24	.03	−.29	−.26	−.22	−.23	−.23	−.26	−.17	−.31
3			1.00	−.12	−.08	−.06	.43	.07	−.05	−.18	−.21	−.16	−.20	−.12	−.20
4				1.00	.18	.67	.32	.09	−.03	.10	.07	.16	.07	.12	.17
5					1.00	.35	.12	.21	.28	.32	.17	.29	.09	.24	.30
6						1.00	.49	.35	.25	.28	.18	.26	.10	.17	.34
7							1.00	.53	.19	.06	−.05	.01	−.05	−.15	.03
8								1.00	.57	.33	.21	.36	.18	.09	.27
9									1.00	.75	.50	.66	.44	.31	.60
10										1.00	.72	.62	.49	.62	.77
11											1.00	.59	.59	.64	.67
12												1.00	.40	.42	.58
13													1.00	.50	.44

14	1.00	
15		.75
16		1.00
17		
18		
19		
20		
21		
22		
23		
24		
25		
26		
27		
28		
29		
30		
31		

(Table continues on p. 418.)

TABLE 14.3 (*Continued*)

	Subdivision Number															
	16	17	18	19	20	21	22	23	24	25	26	27	28	29	30	31
1	-.05	-.14	-.17	-.33	-.01	-.16	-.13	-.20	.01	-.07	.08	.00	-.11	-.11	-.21	-.13
2	-.24	-.26	-.22	-.17	-.32	-.16	-.14	-.06	-.13	-.10	.06	-.00	-.34	-.20	-.22	-.18
3	-.11	-.19	-.03	.00	.02	-.06	.06	-.08	.14	-.05	.25	-.04	-.16	.01	-.06	-.10
4	.09	.08	.19	.13	-.00	-.02	-.06	.01	-.19	-.14	-.06	.05	.08	-.03	.05	.10
5	.25	.50	.23	.29	.14	.13	.02	.15	.00	.08	-.14	-.16	.16	-.01	.06	.05
6	.23	.37	.17	.07	-.01	-.12	-.21	-.11	-.23	-.28	-.16	-.03	.14	-.21	.10	.17
7	.09	.05	-.03	-.14	-.03	-.06	.04	-.04	-.05	-.09	.10	-.01	-.18	-.20	-.10	-.07
8	.37	.47	.02	-.01	.25	.14	.24	.14	.29	.21	.22	.13	.15	-.00	.07	.09
9	.49	.45	.21	.10	.41	.21	.29	.24	.36	.39	.30	.18	.33	.06	.17	.16
10	.51	.39	.43	.31	.45	.36	.39	.32	.36	.43	.27	.29	.42	.25	.28	.14
11	.48	.30	.54	.37	.47	.37	.42	.35	.39	.48	.18	.22	.46	.22	.27	.08
12	.51	.48	.43	.35	.41	.28	.35	.33	.33	.36	.16	.20	.35	.13	.28	.12
13	.27	.19	.33	.30	.33	.31	.39	.30	.39	.46	.31	.28	.35	.35	.25	.09

	16	17	18	19	20	21	22	23	24	25	26	27	28	29	30	31
14	.52	.35	.65	.50	.42	.41	.37	.44	.34	.43	.30	.31	.43	.32	.25	.04
15	.74	.46	.56	.37	.47	.40	.38	.36	.20	.34	.17	.24	.49	.20	.34	.26
16	1.00	.62	.47	.24	.59	.47	.44	.54	.20	.34	.21	.17	.44	.14	.24	.14
17		1.00	.28	.26	.33	.22	.21	.44	.23	.25	.08	−.09	.38	.02	.21	.14
18			1.00	.72	.54	.58	.33	.47	.27	.36	.30	.16	.33	.25	.22	−.03
19				1.00	.41	.45	.30	.46	.37	.34	.36	.14	.32	.32	.39	.08
20					1.00	.67	.62	.57	.45	.63	.34	.14	.58	.45	.24	.09
21						1.00	.64	.59	.39	.56	.39	.24	.45	.63	.35	.13
22							1.00	.70	.51	.71	.46	.27	.37	.54	.29	.03
23								1.00	.51	.68	.45	.19	.42	.39	.31	.07
24									1.00	.73	.71	.35	.34	.57	.31	.03
25										1.00	.56	.25	.46	.56	.21	.02
26											1.00	.54	.12	.49	.24	−.08
27												1.00	.24	.37	.41	.26
28													1.00	.41	.53	.59
29														1.00	.49	.23
30															1.00	.71
31																1.00

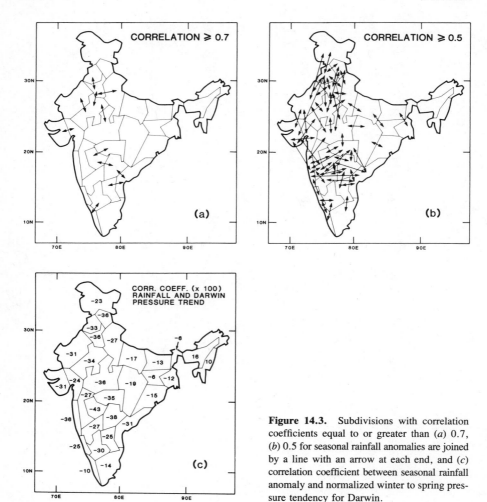

Figure 14.3. Subdivisions with correlation coefficients equal to or greater than (*a*) 0.7, (*b*) 0.5 for seasonal rainfall anomalies are joined by a line with an arrow at each end, and (*c*) correlation coefficient between seasonal rainfall anomaly and normalized winter to spring pressure tendency for Darwin.

Figure 14.5 shows the structure of the empirical orthogonal functions (EOFs) for the normalized seasonal rainfall anomaly. EOFs describe large-scale patterns which represent maximum variability. The percentage departure from normal rainfall is normalized by dividing it by its standard deviation. The first EOF explains 31.5% of the total variance while the second, third, and fourth functions explain only 11.7%, 7.6%, and 6.3% of the variance, respectively. It is again seen that the large values for the first function in Figure 14.5 are confined to nearly the same subdivisions which show large correlation coefficients with the Southern Oscillation in Figure 14.3*c*. Results shown in Figures 14.3, 14.4, and 14.5 suggest that the most dominant pattern of the spatial homogeneity of the Indian summer monsoon rainfall is determined

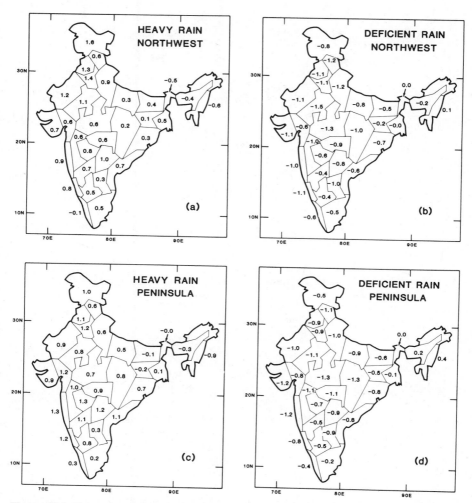

Figure 14.4. Composite maps of normalized seasonal rainfall anomaly for: (*a*) heavy rainfall over Northwest India, (*b*) deficient rainfall over Northwest India, (*c*) heavy rainfall over Peninsular India, (*d*) deficient rainfall over Peninsula India.

by its relationship with the Southern Oscillation, which simultaneously affects a large number of subdivisions over India. We, therefore, propose to examine regional averages over a larger area which would be referred to as the combined Northwest–Peninsula region (see Fig. 14.2*b* and Table 14.2).

As mentioned earlier, the Northwest and Peninsula regions were defined by Walker and are currently used by the India Meteorological Department for long-range forecasting of monsoon rainfall. The Southern and the Northeast regions are simply the remaining subdivisions. Homogeneous Northwest and Homogeneous

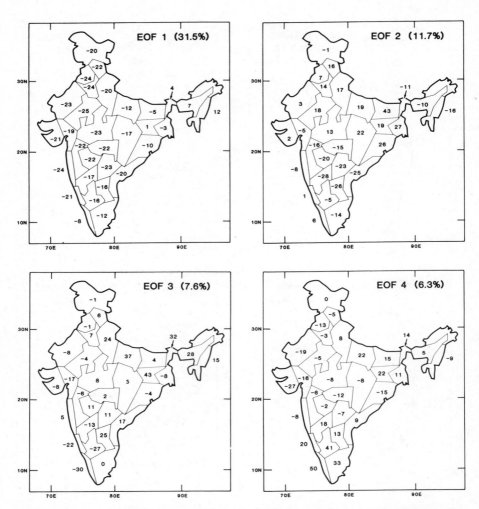

Figure 14.5. First four empirical orthogonal functions (EOFs) for normalized seasonal rainfall anomalies. Percentage of variances explained is indicated at the top.

Peninsula are defined on the basis of high correlation coefficients for seasonal rainfall anomalies (see Fig. 14.3a). The combined region of Northwest–Peninsula is defined on the basis of significant correlation with the Southern Oscillation. Table 14.4 gives the rainfall anomalies for the eight regions (listed in Table 14.2) for the period 1901–1981.

Table 14.5 shows the correlation coefficients of the seasonal rainfall anomalies among the eight regions. It is seen that the correlation coefficient between the Northwest and Peninsula regions is 0.71, which is perhaps too high to consider these regions as independent. It should also be noted that the correlation coefficient

TABLE 14.4 Seasonal Rainfall Anomalies (percentage departure from normal) for the Eight Regions in Table 14.2

Year	Region Number							
	1	2	3	4	5	6	7	8
1901	−24	−15	−7	−10	−26	−14	−21	−16
1902	−18	−12	−7	3	−16	−14	−15	−11
1903	10	8	25	−7	0	28	13	8
1904	−12	−21	−24	−2	−16	−28	−22	−15
1905	−33	−21	−24	8	−44	−16	−29	−20
1906	11	6	4	−4	7	14	11	6
1907	−22	−5	−5	1	−19	−14	−10	−9
1908	37	13	−13	−10	41	17	26	14
1909	31	0	9	5	15	6	16	11
1910	12	12	14	5	7	24	15	11
1911	−32	−21	−18	7	−36	−19	−32	−20
1912	−6	0	−3	−11	−3	−11	0	−4
1913	−27	−5	−18	5	−30	−15	−14	−12
1914	16	21	−6	−12	6	27	21	12
1915	−27	−8	8	−3	−40	10	−19	−12
1916	32	13	21	1	32	30	25	18
1917	71	22	14	3	71	28	48	33
1918	−51	−34	−46	18	−54	−33	−48	−32
1919	8	8	7	0	13	−3	7	7
1920	−26	−25	−4	−3	−24	−42	−27	−20
1921	1	2	15	5	−3	14	3	3
1922	11	−8	−20	13	4	−17	−6	0
1923	0	−7	1	−5	7	−11	−3	−4
1924	14	−8	33	4	16	−7	6	5
1925	1	−7	−14	−10	−12	−12	−10	−6
1926	21	9	0	1	20	−12	15	11
1927	1	5	4	−2	−3	1	5	3
1928	−22	1	−22	−1	−21	3	−8	−9
1929	−7	−10	−6	−1	−13	−21	−12	−7
1930	−2	−6	−22	0	−3	−17	−8	−5
1931	7	4	−1	−4	15	7	8	3
1932	−8	0	−5	−7	−4	−6	−1	−4
1933	24	21	11	4	31	16	26	18
1934	5	10	−30	0	16	1	4	3
1935	−10	−1	−4	5	−8	6	−4	−3
1936	13	−1	−7	6	4	−1	−3	4
1937	−13	3	−16	−4	−4	−14	−7	−5
1938	−10	9	14	6	−15	34	0	3
1939	−27	−12	−14	4	−28	−23	−23	−14
1940	−10	2	−3	−9	−4	3	−3	−4
1941	−19	−18	−4	4	−26	−24	−18	−13

(*Table continues on page 424.*)

TABLE 14.4. (*Continued*)

Year	Region Number							
	1	2	3	4	5	6	7	8
1942	28	12	− 6	8	31	7	20	15
1943	8	4	− 12	− 1	4	− 8	1	3
1944	19	13	− 2	− 3	29	0	18	11
1945	10	11	− 15	− 7	24	2	12	5
1946	− 2	11	3	− 2	4	− 6	5	4
1947	3	9	14	− 1	5	14	8	6
1948	7	− 6	− 7	2	4	− 6	− 3	− 1
1949	− 11	4	12	4	− 5	23	0	0
1950	21	− 2	− 3	1	11	− 12	11	6
1951	− 36	− 15	− 13	− 18	− 35	0	− 24	− 22
1952	0	− 13	− 35	− 2	1	− 30	− 12	− 9
1953	23	8	18	6	− 3	13	17	13
1954	− 9	12	− 2	− 11	− 3	15	7	0
1955	19	15	− 6	− 4	25	25	16	11
1956	23	20	11	− 1	13	20	21	16
1957	46	− 7	− 11	− 14	− 1	6	21	8
1958	23	15	− 3	− 11	19	21	22	11
1959	10	29	15	− 8	7	35	29	16
1960	0	− 1	− 1	6	1	− 1	− 3	0
1961	34	31	25	− 7	38	12	32	25
1962	3	− 4	2	5	0	10	2	0
1963	− 7	− 1	− 9	− 4	− 8	7	− 4	− 4
1964	10	15	18	− 1	10	20	15	11
1965	− 32	− 15	− 11	− 1	− 31	− 2	− 19	− 18
1966	− 6	− 16	8	− 7	− 17	− 2	− 8	− 9
1967	8	4	− 8	− 12	6	− 1	5	2
1968	− 19	− 15	− 3	0	− 23	− 20	− 17	− 13
1969	− 18	1	− 16	− 4	− 9	− 4	− 8	− 7
1970	12	21	− 1	0	10	25	19	13
1971	3	− 7	4	9	6	− 29	− 6	0
1972	− 26	− 27	− 6	− 16	− 23	− 34	− 29	− 23
1973	26	9	8	1	40	− 4	20	13
1974	− 22	− 26	9	21	− 25	− 21	− 22	− 14
1975	60	16	28	− 8	46	21	39	27
1976	29	5	− 11	5	27	0	18	11
1977	19	− 1	4	6	24	− 15	10	7
1978	24	8	8	3	26	28	18	12
1979	− 33	− 19	11	− 13	− 37	− 12	− 18	− 19
1980	11	18	− 12	− 2	2	3	8	9
1981	− 11	4	28	− 6	− 17	14	0	0
Mean	2	1	− 2	− 1	0	0	2	1
S.D.	22	14	15	7	23	18	18	12

TABLE 14.5 Correlation Coefficients for Seasonal Mean Rainfall Anomalies Among Eight Regions

Region	1	2	3	4	5	6	7	8
1 Northwest India	1.00	.71	.42	−.06	.93	.55	.93	.93
2 Peninsula India		1.00	.44	−.19	.76	.79	.90	.90
3 Southern India			1.00	−.06	.38	.55	.54	.55
4 Northeast India				1.00	−.05	−.18	−.16	−.03
5 Homogeneous Northwest					1.00	.51	.91	.91
6 Homogeneous Peninsula						1.00	.75	.72
7 Northwest–Peninsula							1.00	.98
8 Whole India								1.00

between the combined Northwest–Peninsula region and whole India is 0.98, which suggests that any statement about the combined Northwest–Peninsula region would generally be valid for the whole of India.

1.3 The Observed Variability

Table 14.6 gives the seasonal mean rainfall ($\bar{R}$) in mm (column 1), standard deviation of seasonal mean rainfall (σ_R) in mm (column 2), and coefficient of variation (CV) calculated from 70 years (1901–1970) of rainfall data for 31 subdivisions and eight regions (column 3). As pointed out earlier, the number and location of stations used for calculating subdivisional mean rainfall is not same for each year and, therefore, part of the variability shown could be due to sampling of different stations. Only 70 years of data were used for these calculations because rainfall reports from numerous stations had not yet been received for the period after 1970. The seasonal normal rainfall given in column 1 is very similar to the rainfall based on 50 years (1901–1950) of data and used as normal by the India Meteorological Department. The differences are less than 5% for most of the subdivisions and less than 10% for a few subdivisions near high mountains. Column 4 gives the standard deviation of seasonal percentage departure from normal rainfall (σ_p) for 81 years (1901–1981). Column 5 gives the standard deviation of monthly percentage departure from normal (σ_m) for all the four months (June, July, August, September) combined, and column 6 also gives the standard deviation of monthly percentage departure from normal (σ_{ms}) for all the four months combined, except that for each year the seasonal percentage departure from normal is subtracted from the monthly percentage departure before calculating the standard deviation of monthly means. The standard deviations σ_m and σ_{ms} are not the same because seasonal mean anomalies contribute towards the interannual variability of monthly means. The area weighted average of seasonal normal monsoon rainfall for 31 subdivisions (whole India) is 890 mm and the coefficient of variation of area weighted seasonal rainfall is 9.5%.

TABLE 14.6. Seasonal Normal Rainfall ($\widetilde{R}$ in mm), Standard Deviation of Seasonal Mean Rainfall (σ_R) in mm and Coefficient of Variation (CV) Based on 70 Years (1901–1970) of Data.

	MEAN($\widetilde{R}$)	σ_R(mm)	CV(X100)	σ_p(%)	σ_m(%)	σ_{ms}(%)
1 North Assam	1531.6	139.8	9.1	12.3	27.4	24.7
2 South Assam	1819.5	209.3	11.5	11.7	22.4	19.4
3 Subhimalayan West Bengal	2222.4	358.3	16.1	14.3	30.2	26.8
4 Gangetic West Bengal	1069.8	159.6	14.9	15.9	35.8	31.7
5 Orissa	1123.1	153.4	13.7	13.7	30.3	26.8
6 Bihar Plateau	1101.3	142.2	12.9	13.7	31.7	28.1
7 Bihar Plains	1005.4	166.5	16.6	15.9	35.3	31.1
8 East Uttar Pradesh	886.3	175.1	19.8	21.4	42.9	36.3
9 West Uttar Pradesh	869.9	186.1	21.4	21.9	49.0	43.4
10 Haryana	460.2	138.0	30.0	30.4	66.2	59.3
11 Punjab	451.2	143.5	31.8	32.3	72.6	64.1
12 Himachal Pradesh	1347.5	356.8	26.5	23.2	61.2	54.9
13 Jammu and Kashmir	528.5	152.3	28.8	35.2	62.5	53.1
14 West Rajasthan	270.0	106.2	39.3	40.3	78.7	68.4
15 East Rajasthan	619.0	169.3	27.4	27.0	58.8	52.1
16 West Madhya Pradesh	932.8	173.6	18.6	19.0	43.0	38.4
17 East Madhya Pradesh	1197.3	175.3	14.6	15.3	32.5	28.3
18 Gujarat	929.9	266.0	28.6	30.2	63.9	55.9
19 Saurashtra and Kutch	488.3	179.3	36.7	39.1	86.4	76.8
20 Konkan	2756.2	481.8	17.5	17.9	37.7	32.7
21 Madhya Maharashtra	793.1	127.9	16.1	17.3	35.7	31.2

TABLE 14.6. (Continued) Standard Deviation of Seasonal Percentage Departure from Normal (σ_p), Standard Deviation of Mean Monthly Percentage Departure from Normal (σ_m) for June, July, August, September, and Standard Deviation of Monthly Percentage Departure (σ_{ms}) after Removing Seasonal Means Calculated from 81 Years (1901–1981) of Data

	MEAN($\widetilde{R}$)	σ_R(mm)	CV(X100)	σ_p(%)	σ_m(%)	σ_{ms}(%)
22 Marathawada	692.9	167.2	24.1	25.2	48.8	41.7
23 Vidarbha	936.4	174.6	18.6	18.3	37.6	32.3
24 Coastal Andhra Pradesh	583.1	109.8	18.8	20.0	34.9	28.7
25 Telengana	755.0	158.1	20.9	21.3	39.4	33.3
26 Rayalseema	376.2	99.6	26.5	23.7	48.8	42.8
27 Tamil-Nadu	344.7	62.8	18.2	18.5	36.0	31.1
28 Coastal Karnataka	2907.0	455.7	15.7	16.8	36.5	31.6
29 North Interior Karnataka	456.7	84.4	18.5	20.1	39.1	34.8
30 South Interior Karnataka	786.4	163.9	20.8	18.4	34.8	30.1
31 Kerala	1976.0	393.2	19.9	19.4	38.1	32.7
Region						
1 Northwest India	631.9	118.0	18.7	22.4	45.3	39.1
2 Peninsula India	944.3	117.4	12.4	13.5	28.9	24.9
3 Southern India	666.9	98.2	14.7	14.7	29.0	24.8
4 Northeast India	1394.1	90.9	6.5	7.2	18.3	16.9
5 Homogeneous Northwest	649.8	123.5	19.0	22.6	47.8	41.7
6 Homogeneous Peninsula	692.2	120.2	17.4	17.8	30.8	25.0
7 Northwest–Peninsula	729.4	112.7	15.4	17.8	34.3	29.0
8 Whole India	890.2	84.4	9.5	12.5	25.2	21.4

SEASON (JJAS)

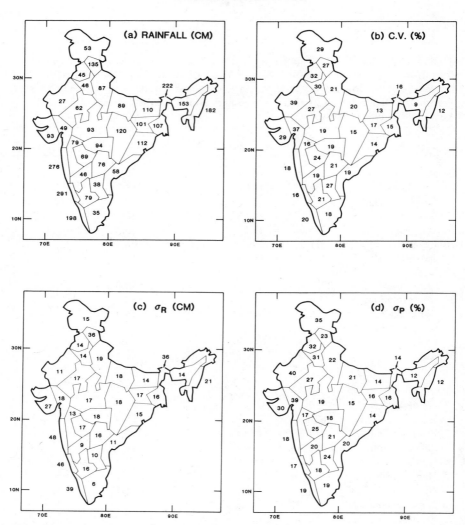

Figure 14.6. (*a*) Seasonal normal rainfall ($\bar{R}$) in cm, (*b*) coefficient of variation (CV) calculated from 70 years (1901–1971) of data, (*c*) standard deviation of seasonal mean rainfall (σ_R) in cm, and (*d*) standard deviation of seasonal percentage departure from normal (σ_p) calculated from 81 years (1901–1981) of data (see Table 14.6).

Figure 14.6 shows the maps of the first four quantities listed in Table 14.6. The large values of subdivisional mean rainfall occur along the west coast and over Northeast India. For both of these regions ocal orography plays an important role in determining the seasonal mean rainfall. The subdivisions in Northwest and Southeast

India get the least amount of rainfall. The coefficient of variation is largest for the subdivisions in Northwest India, primarily because normally the amount of rainfall is small but it can undergo large changes from year to year. The normal rainfall and coefficient of variation are both small for Southeast India. This can be seen more clearly from values of standard deviation of seasonal mean rainfall which are the smallest for Southeast India. Although the seasonal normal for subdivisions in Northeast India is quite large, the standard deviation of seasonal mean rainfall is relatively small giving rise to rather low values of coefficient of variation. The values of standard deviation of percentage departure of seasonal rainfall (shown in Fig. 14.6d) range from 12 to 40%, with the smallest values occurring over the astern part of Northeast India and the largest values over the western part of Northwest India. For most of the subdivisions over Central and Peninsula India, the values range from 15 to 25%. It can also be seen from Table 14.6 that the standard deviation of the percentage departure of seasonal rainfall ranges only from 13.5 to 22.6% for most of the large homogeneous regions of India, and the coefficient of variation is very similar to the standard deviation of percentage departure (Fig. 14.6).

Figures 14.7–14.10 show the mean rainfall, the coefficient of variation, the standard deviation of monthly mean rainfall, and the standard deviation of percentage departure for June, July, August, and September, respectively. In general, normal rainfall for July and August are comparable, but larger than that for June or September. The least rainfall occurs in June. The coefficient of variation is large for June (22 to 86%) and September (26 to 108%) and somewhat less for July and August (19 to 70%). As would be expected, the coefficient of variation for monthly mean rainfall is significantly larger than that for the seasonal mean rainfall. If the coefficients of variation were calculated for the monthly mean rainfall of individual stations, they would be even larger than those for subdivisions. The standard deviation of the percentage departure is comparable for July and August, and relatively less for June or September. The coefficient of variation is very similar to the standard deviation of percentage departure for each month. A more detailed discussion of intraseasonal variability as compared to the interannual variability will be presented later in this chapter. The standard deviation of percentage departure of combined July and August mean rainfall is comparable to that of seasonal mean rainfall for most of the subdivisions of India (Fig. 14.11).

The percentage departure of seasonal rainfall for 81 years for Northwest, Peninsula, Southern, and Northeast regions is shown in Figure 14.12, and for Homogeneous Northwest, Homogeneous Peninsula, combined Northwest–Peninsula, and whole India is shown in Figure 14.13. The years for which the normalized seasonal rainfall anomaly was equal to or greater than 1.0, or less than or equal to −1.0 are shown in Tables 14.7 and 14.8, respectively. It can be seen that during the 81 year period considered here, for the whole of India, there were 12 years of heavy rainfall (normalized anomaly equal to or greater than 1.0) and 14 years of deficient rainfall (normalized anomaly less than or equal to −1.0). A more detailed discussion of possible relationships between heavy and deficient rain and the Southern Oscillation is presented in Chapter 16.

JUNE

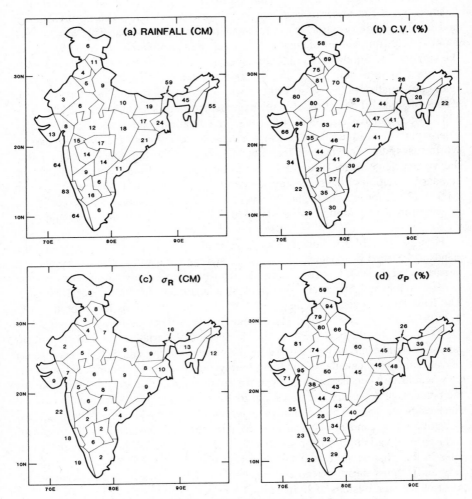

Figure 14.7. Same as Figure 14.6, except for June only.

JULY

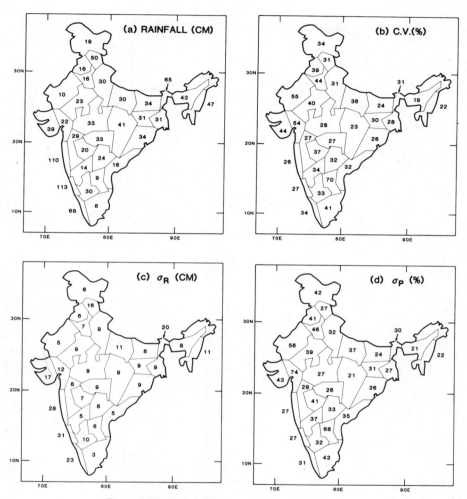

Figure 14.8. Same as Figure 14.6, except for July only.

AUGUST

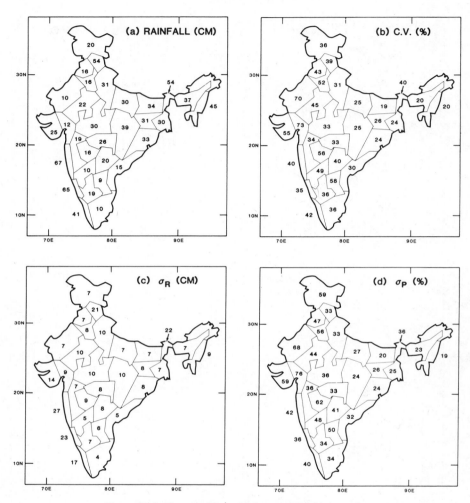

Figure 14.9. Same as Figure 14.6, except for August only.

SEPTEMBER

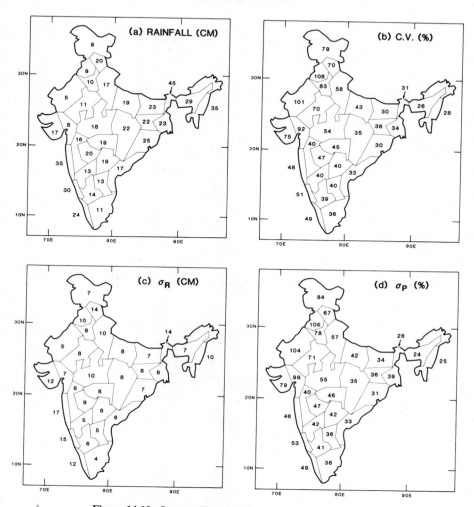

Figure 14.10. Same as Figure 14.6, except for September only.

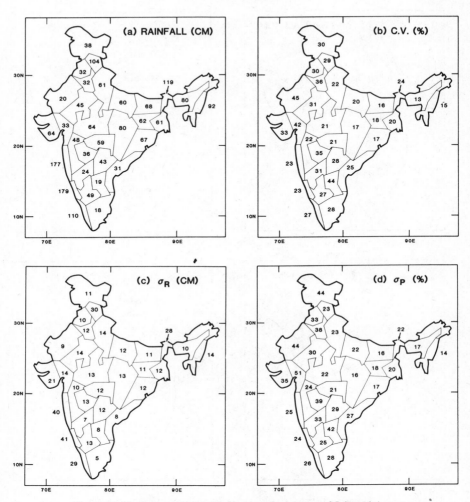

Figure 14.11. Same as Figure 14.6, except for average of July and August.

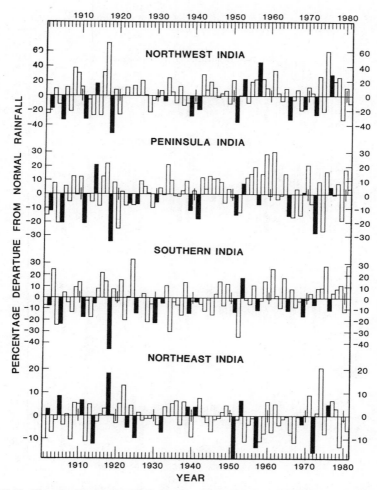

Figure 14.12. Normalized seasonal rainfall anomalies for Northwest, Peninsula, Southern, and Northeast regions of India (in cm). Solid bars denote the El Niño years (11).

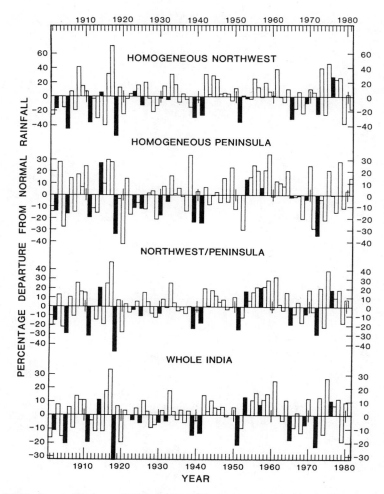

Figure 14.13. Same as Figure 14.12 for Homogeneous Northwest, Homogeneous Peninsula, Northwest–Peninsula, and Whole India (11).

TABLE 14.7 Years with Normalized Rainfall Anomalies Greater Than or Equal To 1.0 for the Eight Regions

Region															
1 Northwest India	08[a]	09	16	17	33	42	53	56	57	58	61	73	75	76	78
2 Peninsula India	14	17	33	55	56	58	59	61	64	70	75	80			
3 Southern India	03	16	21	24	53	61	64	75	81						
4 Northeast India	05	18	22	42	71	74									
5 Homogeneous Northwest	08	16	17	33	42	44	45	55	61	73	75	76	77	78	
6 Homogeneous Peninsula	03	10	14	16	17	38	49	55	56	58	59	64	70	75	78
7 Northwest–Peninsula	08	14	15	17	33	42	44	56	57	58	59	61	70	73	75
8 Whole India	08	16	17	33	42	53	56	59	61	70	73	75	76	78	

[a]Year 1908 is written as 08, etc.

TABLE 14.8 Years with Normalized Rainfall Anomalies Less Than or Equal To −1.0 for the Eight Regions

Region																	
1 Northwest India	01[a]	05	11	13	15	18	20	39	51	65	72	79					
2 Peninsula India	01	04	05	11	18	20	41	51	65	66	68	72	74	79			
3 Southern India	04	05	11	13	18	22	28	30	34	37	45	52	69	74			
4 Northeast India	01	03	08	12	14	25	40	51	54	57	58	59	66	67	72	75	79
5 Homogeneous Northwest	01	05	11	13	15	18	20	39	41	51	65	68	72	74			
6 Homogeneous Peninsula	04	11	18	20	29	39	41	52	68	71	72	74	79				
7 Northwest–Peninsula	01	04	05	11	15	18	20	39	41	51	65	72	74				
8 Whole India	01	04	05	11	18	20	39	41	51	65	68	72	74	79			

[a]Year 1901 is written as 01, etc.

There is considerable intraseasonal variation in the data. The standard deviation (σ_m) of monthly mean anomalies is twice that of the standard deviation of the seasonal mean anomalies (σ_p); see columns 4 and 5 of Table 14.6. It is of interest to note also that during the heavy or deficient rain seasons, the individual months also show heavy or deficient rain. In the next section we describe the mechanisms of intraseasonal variability.

Table 14.9 gives the normalized rainfall anomalies for the summer season and for the individual months of June, July, August, and September for the years of heavy rain and deficient rain for four regions. For most of the years, at least three of the four months have anomalies of the same sign. This indicates that, in spite of the large intraseasonal variability in general, the particular seasons of heavy and deficient rain have significant temporal and spatial coherence (Fig. 14.4). The temporal coherence is higher for deficient years compared to heavy rainfall years.

2 INTRASEASONAL VARIABILITY OF THE INDIAN SUMMER MONSOON

In spite of the highly periodic nature of the planetary-scale monsoon circulation, there are large variations in the circulation and rainfall within the monsoon season. The intensity of the seasonal mean monsoon is influenced by the nature of variability within a monsoon season. Dhar et al. (12) have examined the possible association between the monsoon rainfall over three west coast subdivisions of India (Kerala, Coastal Karnataka, and Konkan) and dates of onset of the monsoon over the respective regions. Despite the fact that the dates of onset can fluctuate by more than 30 days (see Table 14.10), they found the interesting but counterintuitive result that the rainfall anomaly for the month of June as well as for the whole monsoon season is not related to the date of onset. This also indicates that the intraseasonal variability of monsoon rainfall is quite large. This result may, at least in part, be due to the arbitrariness in defining the dates of onset. A season of highly deficient monsoon rainfall and severe drought does not imply an absence of rainfall for the whole season; rather, it can occur due to prolonged periods of reduced rainfall. These periods are usually referred to as the break monsoon conditions by the India Meteorological Department.

2.1 Break Monsoons

Ramamurthy (13) has studied the climatology of break monsoon conditions over India. The phrase "break in the rains" can be traced back as far as the Indian Daily Weather Reports of 1888, and even now it refers only to the situations with reduced or no rainfall. The systematic study of break conditions has shown that although a reduction in rainfall occurs over most of the Indian subcontinent, it is accompanied by increased rainfall over extreme northern India (near the Himalayan foothills) and extreme southern India. This suggests that the break monsoon is actually a spatial redistribution of the monsoon rainfall. During break monsoon conditions, the monsoon trough shifts to its extreme northern position, the surface pressure

TABLE 14.9 Normalized Monthly and Seasonal Anomalies for Heavy and Deficient Monsoon Years

Northwest (Heavy Rain)

Year	Season	June	July	Aug	Sept
1908	1.7	−0.7	1.6	2.4	−0.4
1909	1.4	1.1	1.0	−0.7	1.1
1916	1.4	1.2	0.2	1.6	0.5
1917	3.2	2.2	0.6	1.4	3.5
1933	1.1	1.7	−0.5	1.1	0.6
1942	1.3	−0.2	1.2	0.9	0.7
1953	1.0	0.6	1.1	0.8	−0.2
1956	1.0	0.2	2.2	0.3	−0.6
1957	2.0	−0.1	1.6	2.0	0.6
1958	1.0	−0.9	0.6	0.1	2.2
1961	1.5	0.7	0.4	1.0	1.8
1973	1.2	0.1	−0.1	3.0	−0.2
1975	2.7	1.2	2.4	1.8	1.1
1976	1.3	0.4	−0.1	2.2	0.4
1978	1.1	1.3	1.7	0.3	−0.3

Northwest (Deficient Rain)

Year	Season	June	July	Aug	Sept
1901	−1.1	−1.2	−0.7	−0.1	−0.9
1905	−1.5	−1.2	−1.2	−1.7	0.7
1911	−1.4	0.3	−2.6	−1.3	0.9
1913	−1.2	1.5	−1.0	−1.2	−1.1
1915	−1.2	−0.4	−1.5	−0.6	−0.2
1918	−2.3	−0.3	−2.4	−1.0	−1.3
1920	−1.2	0.6	0.4	−1.7	−1.2
1939	−1.2	0.2	−1.0	−1.4	−0.2
1951	−1.6	−0.5	−1.7	−0.6	−0.9
1965	−1.4	−1.3	−0.1	−1.3	−0.9
1972	−1.1	−0.5	−1.2	−0.3	−0.6
1979	−1.5	−0.2	−0.7	−1.1	−1.1

Peninsula (Heavy Rain)

Year	Season	June	July	Aug	Sept
1914	1.5	1.0	1.0	0.1	1.4
1917	1.6	1.2	−0.6	1.2	1.7
1933	1.6	0.9	−0.2	1.3	1.5
1955	1.1	0.6	−1.5	1.8	1.8
1956	1.5	0.5	1.8	0.5	0.2

Peninsula (Deficient Rain)

Year	Season	June	July	Aug	Sept
1901	−1.1	−1.1	−0.8	0.6	−1.3
1904	−1.5	−0.0	−1.3	−1.5	−0.4
1905	−1.5	−1.5	−0.7	−0.9	−0.1
1911	−1.5	0.5	−2.2	−0.9	−0.6
1918	−2.5	0.1	−2.7	−0.9	−1.6

(*Table continues on p. 440.*)

TABLE 14.9. (Continued)

	Peninsula (Heavy Rain)				
Year	Season	June	July	Aug	Sept
1958	1.1	−1.0	1.3	1.1	1.3
1959	2.1	0.0	2.0	0.6	1.6
1961	2.3	0.2	1.8	0.4	2.7
1964	1.1	0.1	−0.0	1.4	0.7
1970	1.5	1.4	−1.1	2.5	0.9
1975	1.2	0.8	−0.4	1.0	1.2
1980	1.3	2.9	−0.8	0.7	−1.1

	Peninsula (Deficient Rain)				
Year	Season	June	July	Aug	Sept
1920	−1.8	−0.1	−0.6	−1.5	−1.4
1941	−1.3	−0.7	−0.5	−0.8	−1.0
1951	−1.1	−0.5	0.0	−0.4	−1.3
1965	−1.1	−1.3	0.4	−0.7	−0.7
1966	−1.2	−0.4	−0.1	−1.5	−0.4
1968	−1.1	−1.0	−0.3	−0.7	−0.4
1972	−2.0	−0.6	−1.9	−0.6	−1.2
1974	−1.9	−1.0	−1.7	−0.3	−1.2
1979	−1.4	−0.7	0.9	−1.3	−1.1

	Whole India (Heavy Rain)				
Year	Season	June	July	Aug	Sept
1908	1.1	−0.8	1.1	1.6	−0.2
1916	1.4	1.2	0.3	1.4	0.7
1917	2.7	1.9	−0.2	1.4	3.1
1933	1.5	1.3	−0.5	1.7	1.0
1942	1.2	−0.1	1.2	1.0	0.5
1953	1.1	0.3	1.0	1.3	−0.1
1956	1.3	0.7	2.1	0.3	−0.2
1959	1.3	−0.4	1.5	0.1	1.5
1961	2.0	0.6	1.4	0.8	2.3

	Whole India (Deficient Rain)				
Year	Season	June	July	Aug	Sept
1901	−1.3	−1.3	−1.1	0.2	−1.1
1904	−1.2	−0.3	−0.5	−1.2	−0.9
1905	−1.6	−1.6	−1.2	−1.1	0.3
1911	−1.6	0.6	−2.9	−1.2	0.3
1918	−2.6	0.0	−3.3	−0.9	−1.7
1920	−1.6	0.2	−0.2	−1.9	−1.3
1939	−1.1	−0.3	−1.0	−0.5	−0.6
1941	−1.1	−0.1	−1.3	−0.7	−0.5
1951	−1.7	−0.6	−1.1	−0.8	−1.3

Year	Season	June	July	Aug	Sept
1970	1.0	1.3	-1.4	1.7	0.9
1973	1.0	-0.2	-0.1	2.4	0.2
1975	2.2	0.8	1.8	1.4	1.4

Year	Season	June	July	Aug	Sept
1965	-1.4	-1.4	-0.0	-1.0	-0.9
1968	-1.0	-0.8	0.3	-1.0	-1.1
1972	-1.8	-0.6	-2.1	-0.9	-0.8
1974	-1.1	-0.4	-0.3	-0.8	-1.0
1979	-1.5	-0.5	0.1	-1.6	-1.0

Northwest–Peninsula (Heavy Rain)

Year	Season	June	July	Aug	Sept
1908	1.4	-0.8	1.4	1.7	-0.1
1914	1.2	1.0	1.6	-0.3	1.0
1916	1.4	0.9	0.4	1.5	0.7
1917	2.7	2.0	-0.1	1.6	3.1
1933	1.4	1.5	-0.4	1.5	1.1
1942	1.1	0.1	1.3	1.0	0.1
1944	1.0	-0.4	1.6	1.2	-0.8
1956	1.2	0.3	2.6	0.4	-0.4
1957	1.2	0.2	0.7	1.7	-0.2
1958	1.2	-0.9	1.1	0.9	1.9
1959	1.6	-0.3	1.9	0.3	1.5
1961	1.8	0.6	1.5	0.6	2.0
1970	1.1	1.3	-1.3	2.0	0.8
1973	1.1	-0.1	0.2	2.5	-0.1
1975	2.2	1.0	1.4	1.4	1.4
1976	1.0	0.6	0.4	1.9	-0.3
1978	1.0	1.5	1.1	0.9	-0.6

Northwest–Peninsula (Deficient Rain)

Year	Season	June	July	Aug	Sept
1901	-1.2	-1.1	-0.8	0.1	-1.2
1904	-1.2	-0.6	-0.7	-1.2	-0.6
1905	-1.6	-1.3	-1.3	-1.2	0.1
1911	-1.8	0.3	-2.4	-1.3	-0.0
1915	-1.1	0.1	-1.5	-0.9	-0.3
1918	-2.7	-0.6	-2.9	-1.1	-1.5
1920	-1.5	0.6	-0.6	-1.7	-1.3
1939	-1.3	-0.5	-1.2	-0.4	-0.6
1941	-1.0	-0.3	-1.1	-0.6	-0.4
1951	-1.4	-0.5	-0.7	-0.7	-1.1
1965	-1.1	-1.3	0.4	-0.8	-1.0
1972	-1.6	-0.3	-1.8	-0.7	-1.0
1974	-1.3	-0.1	-0.8	-0.9	-1.0

TABLE 14.10 Dates of Onset of the Monsoon over Kerala, the Southern Tip of India (Range: May 11–June 18)

Year	Date	Year	Date	Year	Date	Year	Date
1901	June 7	1921	June 2	1941	May 23	1961	May 18
1902	June 6	1922	May 31	1942	June 10	1962	May 17
1903	June 12	1923	June 11	1943	May 29	1963	May 31
1904	June 7	1924	June 2	1944	June 3	1964	June 6
1905	June 10	1925	May 27	1945	June 5	1965	May 26
1906	June 13	1926	June 6	1946	May 29	1966	June 1
1907	June 8	1927	May 27	1947	June 3	1967	June 9
1908	June 11	1928	June 3	1948	June 11	1968	June 8
1909	June 2	1929	May 29	1949	May 23	1969	May 17
1910	June 2	1930	June 8	1950	May 27	1970	May 26
1911	June 6	1931	June 4	1951	May 31	1971	May 27
1912	June 8	1932	June 2	1952	May 20	1972	June 18
1913	June 2	1933	May 22	1953	June 7	1973	June 4
1914	June 4	1934	June 8	1954	May 31	1974	May 26
1915	June 15	1935	June 12	1955	May 29	1975	May 30
1916	June 2	1936	May 19	1956	May 21	1976	May 31
1917	May 31	1937	June 4	1957	June 1	1977	May 30
1918	May 11	1938	May 26	1958	June 14	1978	May 29
1919	June 3	1939	June 5	1959	May 31		
1920	June 3	1940	June 14	1960	May 14		

departure from normal is positive over most of the country and maximum over central India, and negative over extreme northern and southern parts of the country. The easterly jet at the upper levels is stronger and shifts northward. The meridional component of the wind over the northern parts of India is from the north in the middle troposphere and from the south in the upper troposphere, just the opposite of the meridional wind directions during active monsoon conditions.

The duration of break conditions ranges from 3 to 21 days. During the 80 year (1888–1969) period examined by Ramamurthy, there were 56 cases of break monsoon conditions in July of which 38 lasted for 3–5 days, 14 for 6–10 days and 4 for 11–20 days; of 57 cases in August, 31 lasted for 3–5 days, 19 for 6–10 days and 7 for 11–20 days. The total number of the break days were 306 for July and 380 for August. Although the number of break periods are about the same in July and August, they tend to last longer in August.

The observed rainfall distribution and the circulation patterns support the idea that break monsoon conditions are associated with a northward shift of the normal meridional circulation and the occurrence of enhanced convective activity near the southern tip of India. Most of India is therefore under the descending branches of the two thermally forced meridional circulations with ascending branches near the foothills of the Himalayas and the extreme southern tip of India. The northward shift and the increase in the speed of the easterly jet stream during the monsoon breaks are consistent with the possibility that an enhanced meridional circulation accelerates the zonal flow by deflection of the winds by the Coriolis force.

2.2 Factors Causing Intraseasonal Variability

It is possible to identify certain phenomenological factors that produce variability within a monsoon season on time scales of a few days to a few weeks. They fall into four broad categories: synoptic-scale disturbances, monsoon troughs, quasi-periodic oscillations, and mid-latitude effects.

2.2.1 Synoptic-Scale Disturbances (Lows, Depressions, Storms).

An examination of daily weather charts and daily rainfall amounts over India suggests that the rainfall distribution on any given day is generally related to the presence of synoptic-scale disturbances, quasi-symmetric zones of convergence, and the interaction of strong monsoon flow with orographic barriers. In particular, rainfall depends upon the frequency, intensity, life cycle, and propagation characteristics of the synoptic disturbances that influence a particular region. Dhar and Rakhecha (14) found that in the absence of tropical disturbances affecting the Indian subcontinent, the July and August rainfall over the northern Indian plains was reduced by 19 and 14%, respectively. This study did not include the effects of low-pressure areas which can also produce large amounts of rainfall. Sikka (15) examined changes in the monsoon rainfall for five years of high rainfall and deficient rainfall. He also found that the number of monsoon depressions as well as the number of depression days were quite similar for both heavy and deficient rain years. The most striking difference was found in the number of low-pressure areas and the number of days with low-pressure areas. The ratio of the number of lows for heavy and deficient rain years was 1.6. Based on these results, Sikka concluded that the higher number of monsoon lows is a manifestation of greater instability of the monsoon trough.

2.2.2 The Monsoon Trough.

As mentioned earlier, the northward shift of the monsoon trough is accompanied by break monsoon conditions over most of the central Indian regions, and enhanced rainfall near the southern tip of India. A weakening of the monsoon trough over North India is associated with the strengthening of the convergence zone near the southern tip of India. We are not aware of any physical explanation for this behavior of the monsoon trough. The northward shift of the monsoon trough could be related to changes in the large-scale circulation in middle latitudes, and to changes in the intensity of the near-equatorial trough. The latter could be due either to air–sea and/or interhemispheric interaction over the Indian Ocean or the formation of tropical disturbances. A partial explanation for the observed fluctuations of the monsoon trough and the equatorial convergence zone is provided by the recently documented northward propagation of cloudiness described in the following text.

2.2.3 Quasi-Periodic Oscillations.

Yasunari (16), Sikka and Gadgil (17), and Krishnamurti and Subrahmanyam (18) have presented observational evidence for a possible northward propagation of the convergence zone and cloudiness. Fluctuations in the intensity of the monsoon trough could be interpreted in terms of the phase of this northward propagating convergence zone. There is a large body of observational evidence for 15-day oscillations during the monsoon season [for references see the paper by Krishnamurti and Ardanuy (19)]. From spectral analysis of digital cloud data, Yasunari (16) showed dominant periodicities near 15- and 40-day periods. Both oscillations showed a tendency for northward phase propagation. From the analyses by Yasunari (20, 21) and Krishnamurti and Bhalme (22), there appears to be a significant relationship between the phase of 15-day oscillations and the formation of monsoon depressions. However, this relationship by itself does not clarify whether these oscillations are responsible for the formation of monsoon depressions by creating a favorable large-scale environment, or whether the oscillations are merely a consequence of a regular formation of depressions at 15-day intervals. The 15-day oscillations may also be due to instabilities produced by interaction of the zonal flow with mountains and diabatic heat sources (23). The 40-day oscillation is of much larger scale and not necessarily unique to the monsoon flow (24). Physical mechanisms responsible for these fluctuations are not yet understood, and more observational studies are needed to describe their structure and origin. Webster and Chou (25) conducted numerical experiments with a simple model to study low frequency transitions of a monsoon system and found that the model fluctuations are very sensitive to the parameterizations of land surface processes and treatment of soil moisture. Goswami and Shukla (26) showed that a zonally symmetric general circulation model of the atmosphere exhibited quasi-periodic oscillations with periods of 15–40 days due to the interaction of the motion field and moist convection. If this interaction was eliminated by prescribing the diabatic heating, the oscillations disappeared. Webster (see Chapter 11) has shown that the feedback mechanisms between soil moisture changes and atmospheric circulation play an important role in northward propagation of the monsoon trough.

2.2.4 Mid-Latitude Effects. A large number of observational studies indicate a possible relationship between the mid-latitude circulations of both hemispheres and the summer monsoon circulation and rainfall (see Chapter 11 for an extended discussion). Ramaswamy (27) has suggested that intrusions of large-amplitude troughs from the mid-latitudes of the Northern Hemisphere are associated with break monsoon conditions over India. The preponderance of westerly winds over northern India is favorable for the propagation of mid-latitude influences to the monsoon region. Due to lack of upper air data over the southern Indian Ocean, it has not been possible to examine the upper air circulation over the Southern Hemisphere. Future studies with the special 1979 Global Weather Experiment data and the global analyses produced from these data will be required to gain a better understanding of such relationships.

3 MECHANISMS OF INTERANNUAL VARIABILITY OF MONSOONS

As mentioned earlier, a convenient framework for understanding and describing the mechanisms of interannual variability is to isolate the factors associated with the atmosphere's internal dynamics and with its lower boundary conditions. It is important to understand the relative contributions of these two factors to the observed interannual variability of the Indian monsoon. This separation is only an idealization of the real atmosphere where the internal dynamics and boundary conditions continuously interact. The recent research has been summarized under these two categories.

3.1 Internal Dynamics

Even if the external forcing by solar radiation and boundary conditions at the earth's surface were constant in time, the atmospheric circulation would exhibit interannual variability due to the inherent aperiodic nature of the system. The combined effects of dynamical instabilities (manifested as synoptic-scale disturbances—see Section 2.2.1), nonlinear interactions among various scales of motion, thermal and orographic forcing, tropical–extratropical interactions, and so on, can be considered as examples of internal dynamical processes that can produce interannual variability. The orography and the land–sea distribution at the earth's surface are fixed with time; however, their interactions with fluctuating winds can produce large changes and, therefore, they are to be considered as part of the variability associated with internal dynamics. It should be noted that if all the observed interannual variability were due to internal dynamical processes alone, the prospects for long-range forecasting would be rather limited because of the inherent limits of predictability of internal dynamics (1). A number of observational studies suggest possible relationships between the monsoon circulation and other features of the global circulation.

Bannerjee et al. (28) found a significant correlation between the average position of the ridge at 500 mb over India during April and summer monsoon rainfall. If the ridge position is to the south (north) of its climatological position, the summer monsoon rainfall is deficient (excessive). This result is substantiated by Rao (29). Parthasarathy and Mooley (6) concluded from a time series of Indian summer monsoon rainfall that the rainfall is random and normally distributed. This does not necessarily imply that it is unpredictable; realizations of a nonlinear deterministic system can also have certain statistical properties which are similar to that of a random process. Verma (30) examined the monthly mean anomalies of 300–100 mb thickness, which is a measure of the upper-tropospheric temperature anomaly, for 10 years (1968–1977) for selected stations over India. He found that anomalies in April and May tend to persist for the whole monsoon season, and that negative (positive) thickness anomalies in the pre-monsoon months are associated with negative (positive) anomalies of Indian summer monsoon rainfall. It is difficult to explain this long persistence as an internal dynamical process of the atmosphere; it seems likely that it is related to some slowly varying boundary forcing at the earth's surface (for example, snow over Eurasia or SST anomalies over the tropical oceans). Joseph et al. (31) found highly significant correlation between monthly mean meridional wind during May at selected Indian stations and summer monsoon rainfall over India for 15 years (1964–1978) of data. However, the correlation coefficients drop abruptly for the meridional winds in both April and June.

Tanaka (32) examined the monthly mean rainfall and height fields for June, July, August and September over the Asian monsoon region, and the wind speed of the tropical easterly jet above 10°N at 150 mb for 17 years (1964–1980). He found that a strong tropical easterly jet at 10°N is associated with a strong low-level monsoon circulation and heavy monsoon rain over India and predominantly zonal circulation near 50°N. In contrast, a weaker tropical easterly jet at 10°N is associated with deficient monsoon rainfall over India and a blocking high to the north of the Caspian Sea. Based on these observations, Tanaka concluded that the interannual fluctuations of the summer monsoon are strongly influenced by the middle latitude circulation of the Northern Hemisphere. This conclusion seems tenuous; the causal mechanism is not clear. He has not considered the possibility that the changes in diabatic forcing due to changes in the rainfall could have produced anomalous tropical and middle latitude circulations. Moreover, in examining the wind speed only at 10°N, Tanaka may have missed the jet core since the latitudinal position of the jet maximum varies from one year to the next. In fact, Ramamurthy (13) has shown that during the break-monsoon situations, when most of India experiences deficient rainfall, the tropical easterly jet shifts northward and is stronger than normal. Raman and Rao (33) suggested that blocking ridges over East Asia are also associated with prolonged breaks in monsoon rainfall.

3.2 Influence of Global Surface Boundary Conditions

There is a growing body of modeling and observational evidence that suggests that the slowly varying boundary conditions of sea surface temperature (SST), soil moisture, and sea ice and snow at the earth's surface can influence the interannual variability of the atmospheric circulation.

The variations in these surface boundary conditions can influence the location and intensity of diabatic heat sources that drive the atmospheric circulation. Anomalous boundary conditions can be more effective in producing circulation anomalies in the tropics than in the mid-latitudes because the tropical circulation is dominated by the planetary-scale Hadley, Walker, and monsoon circulations, and changes in the boundary conditions can alter the locations and intensity of these systems. They can also influence the amplitude and phase of planetary waves in mid-latitudes which, in turn, can influence the tracks and intensity of cyclone-scale disturbances. The physical mechanisms responsible for such influence are rather complex (see Chapter 16) and depend upon the nature of the boundary forcing. Charney and Shukla (2) suggested that the Asiatic monsoon is a dynamically stable circulation system, and, its interannual variability is largely determined by the slowly varying boundary conditions; therefore, monsoons are potentially more predictable than the mid-latitude circulations.

The following is a summary of selected observational and general circulation model studies of the relationships between these boundary conditions and monsoon circulation and rainfall.

3.2.1 Snow Cover.
As shown by Wiesnet and Matson (34), December snow cover for the Northern Hemisphere is a very good predictor of the snow cover for the

following January through March. Snow cover, therefore, is considered to be a slowly varying boundary condition useful for prediction.

Blanford (35) found that excessive winter and spring snowfall in the Himalayas was an indicator of the subsequent monsoon rainfall in India. The amount and time of occurrence of the cold weather (October–May) snowfall in the mountain districts adjacent to northern India was one of the important factors used by Blanford for monsoon rain forecasts which he started issuing in 1882. For the period 1880–1920, greater winter snowfall was found to be related to deficient monsoon rainfall, but for the subsequent 30-year period, reported snow accumulation showed very large variability and the relationship with the monsoon rainfall was opposite to what it was in the earlier four decades. After 1950 the India Meteorological Department dropped this factor as one of the predictors of monsoon rain.

Hahn and Shukla (36) examined 11 years of satellite-derived snow cover over Eurasia and summer monsoon rainfall over India, and found (Fig. 14.14) an apparent inverse relationship between the area extent of winter snow cover and Indian monsoon rainfall. This result, although based on a rather limited sample, supported the earlier

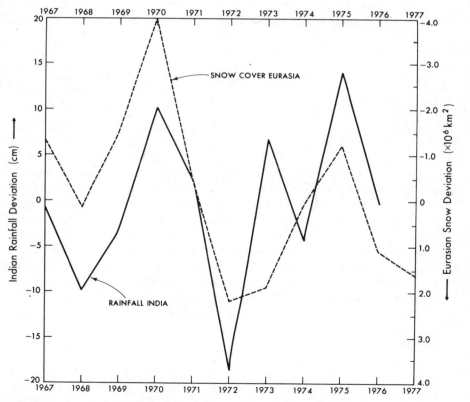

Figure 14.14. Area weighted summer monsoon rainfall anomaly over India (solid line) and winter snow cover departure over Eurasia (dashed line). Note the inverted scale for the latter.

findings of Blanford. Dickson (37) has extended this study to include data up to 1980 and found that the relationship still holds for the extended data period; however, the magnitude of the correlation between the summer monsoon rainfall and winter snow cover is less for the period 1967–1980, compared to 1967–1975. Dickson also found that the snow cover data for the period 1966–1974 did not include information on Himalayan snow cover and, in particular, the data for 1969 were considered to have a large error. He showed that if the data for 1969 are not included and the data for 1967–1974 are adjusted for bias, the correlation coefficient between snow cover and rainfall for the period 1967–1980 changes from -0.44 to -0.59, and for the period 1967–1975, it changes from -0.62 to -0.74. The statistical significance and practical utility of this relationship for predicting summer monsoon rainfall can be examined more systematically when a large sample size is available. In light of these recent results, we speculate that a lack of systematic relationship between Himalayan snowfall and monsoon rainfall during the period 1920–1950 (35) could have been, at least in part, due to the changes in the quality of reporting of the snow cover and snow depth over the Himalayas.

Dey and Bhanukumar (38) have examined the relationship between spring snow cover over Eurasia and the time taken by the Indian summer monsoon rainfall to advance from the southern tip of India to the northern border of India. They found that, for the period 1967–1978, when snow cover during the spring was greater than normal, the time taken by the monsoon to advance from south to north was also greater than normal, and vice versa. They also found a negative correlation between the amount of snow melt during spring and the advance period for the summer monsoon. When the difference of snow cover from March to May (snow melt) was above normal, the speed of the advance of the monsoon was slower than normal and vice versa. These results are not inconsistent with the earlier results of Hahn and Shukla because high snow melt is indicative of enhanced solar heating or shallow snow depth.

An inverse relationship between Eurasian snow cover and the summer monsoon is not implausible because large and persistent winter snow cover over Eurasia can delay and weaken the spring and summer heating of the land masses that is necessary for the establishment of the large-scale monsoon flow. During the spring and summer seasons following winters with excessive snow, most of the solar energy is used for melting the snow or evaporating from the wet soil. Systematic numerical experiments using global general circulation models (GCMs) with adequate treatment of the effects of albedo and ground hydrology can be used to understand the physical mechanisms that influence the atmospheric circulation due to excessive snowfall (39). Analysis of data has also shown (40) that for large (small) snow cover over the Tibetan Plateau, the arrival of the summer monsoon circulation over the plateau is late (early).

3.2.2 Sea Surface Temperature. Since 70% of the earth's surface is covered with water, and since changes in the sea surface temperature (SST) are much slower compared to the atmospheric fluctuations, it is natural to think that interannual variability of SST might contribute to the interannual variability of the atmospheric circulation and rainfall. During the last 30 years there have been numerous studies suggesting possible relationships between SST anomalies and atmospheric anomalies (see e.g., 41 and 42).

Arabian Sea Surface Temperature Anomalies. During the last 20 years, there have been several observational and GCM sensitivity studies that have suggested a possible relationship between anomalies of SST over the Arabian Sea and summer monsoon rainfall over India. These studies have produced, at times, conflicting results, and they have used different data sets and different analysis schemes. The first observational study on this topic was by Ellis (43) who showed that SST over the Arabian Sea was warm during 1920 when several parts of India experienced floods, and was cold during 1923 when several parts of India experienced droughts. In this study, data were examined for only these two years and therefore the results are of limited statistical significance. The characterization of 1920 as a flood year and 1923 as a drought year is also questionable. The concept, however, was physically appealing: warm SST can produce high evaporation and possibly larger rainfall.

In another observational study, Pisharoty (44) calculated the flux of water vapor across the equator into the Arabian Sea, and across the west coast of India from the Arabian Sea, and found that the latter was more than twice the former, and concluded that evaporation over the Arabian Sea is an important source of moisture for precipitation over India. Saha (45) and Saha and Bavadekar (46) repeated the calculations of Pisharoty but used additional stations in the western Arabian Sea where the northward flow is the strongest. They concluded that, contrary to the results of Pisharoty, the flux of water vapor across the equator is about 30% greater than evaporation over the Arabian Sea. They pointed out that owing to scarcity of data over the Arabian Sea, Pisharoty may have underestimated the cross-equatorial moisture flux and, consequently, overemphasized the role of evaporation over the Arabian Sea. Ghosh et al. (47) used observations taken during a monsoon field program called Monsoon-77 to calculate the water vapor budget over the Arabian Sea. They, like Pisharoty, found that the moisture flux across the west coast of India was more than twice the moisture flux across the equator from the Southern Hemisphere, thus supporting the idea that evaporation over the Arabian Sea is a significant moisture source for monsoon rainfall over India. A recent calculation by Cadet and Reverdin (48) has shown that most of the water vapor (about 70%) crossing the west coast of India comes from the Southern Hemisphere. This is in agreement with Saha. In summary, there is a large discrepancy among the calculations of different investigators. Part of this discrepancy can be attributed to the interannual variability of SST, evaporation, and wind speed over the Arabian Sea, but perhaps a much larger part is due to differences in the quality and density of data and the techniques used for analyzing the data.

Motivated by the earlier observational studies, Shukla (49) used a GCM to investigate the sensitivity of the model simulated monsoon rainfall over India to SST anomalies over the Arabian Sea. He ran the model for a period representing 60 days with and without prescribed negative (cold) SST anomalies (which were $-3°C$ near the African coast and $-1°C$ in the central Arabian Sea). He found that the monsoon rainfall over India was significantly reduced. Sikka and Raghavan (50) pointed out that the drastic reduction of rainfall over India could be due to limitations of the model parameterizations or to the choice of a large verification area which included the oceanic regions of the negative SST anomaly. Shukla (51) recomputed the rainfall over the Indian subcontinent for the two model runs and found that the earlier results did not change when the verification domain was changed. A similar GCM sensitivity study was carried out by Washington et al. (52) who also found a significant reduction of rainfall but only over the region of colder SST. Rainfall over India and adjacent regions also decreased but the decrease was not statistically significant. Washington et al. also studied the sensitivity of a warm SST anomaly over the eastern Arabian Sea and the central (equatorial) Indian Ocean. Only in the latter did they find any evidence of remote reponse. In order to resolve the conflicting results of the two GCMs, Shukla (41) repeated the same numerical experiment with a third climate model and found that the results were very similar to his earlier results. He pointed out that the primary reason for the differences in the results was the ability of each model to simulate the mean monsoon circulation in the control experiments. In the July mean simulation of the low-level monsoon flow by the model used by Washington et al., the southwesterly monsoon current did not reach the western Ghats of India.

Considering the large differences in the basic physical parameterizations of different GCMs, it is rather remarkable that the model simulated response of monsoon rainfall to SST anomalies over the Arabian Sea is so similar. It appears reasonable to conclude that any model with a good parameterization of moist convection and the boundary layer would produce a somewhat similar response. However, recent analyses of the observed SST anomalies over the Arabian Sea have suggested that the magnitudes of the SST anomalies used in these GCM experiments were too high to be realistic. Analyses of data along the ship tracks suggest that rarely does such a large anomaly occupy so broad an area in the Arabian Sea. In addition, SST anomalies change sign from the early part of the summer monsoon season to the later part. Thus it is not reasonable to assume that the anomalies persist for the entire monsoon season as was done in the modeling experiments. Following the earlier work by Ellis (43), observational studies of possible relationships between SST over the Arabian Sea and Indian monsoon rainfall have been carried out by several investigators. Shukla and Misra (53) calculated the correlation coefficients between SST anomalies along a ship track at about 10°N between 60°E and 70°E using the data from Fieux and Stommel (54) and seasonal mean rainfall over Indian subdivisions. They found weak positive correlations for subdivisions along the west coast of India. Raghavan et al. (55) showed that during the 1964 monsoon season, the western Arabian Sea was 2–3°C colder during a short period of reduced rainfall over India compared to another short period of enhanced rainfall. Weare (56) calculated the empirical orthogonal functions (EOF) for seasonal mean rainfall of 53 stations over India, and sea surface temperature anomalies on a 5° × 5° grid over the Arabian Sea and Indian Ocean for six months (May–October). Correlation coefficients between the EOFs of monsoon rainfall and SST were of different sign. A recent analysis of SST anomalies along the ship tracks over the Arabian Sea suggests that the results of Shukla and Misra and Weare should be interpreted with caution. In Shukla and Misra's study the SST anomaly data have a significant negative bias for the period before 1940 and a positive bias after that (as will be discussed further on in this section). This bias was not removed before the calculation of correlation coefficients. The problem with Weare's calculations is that he grouped together the SST anomalies of pre-monsoon months and post-monsoon months. SST anomalies during April, May, and June are generally of opposite sign to those of August, September, and October and it is not meaningful to combine SST anomalies for all the six months together.

Analysis of Observed Sea Surface Temperature Anomalies. The monthly mean SST data for 75 years along the ship track shown in Figure 14.15 has been analyzed. The climatological monthly mean and the standard deviation of the SST were first calculated for 2° × 2° regions along the ship tracks. Anomalies with absolute value greater than four times the standard deviation or 3°C, whichever was larger, were not considered. (Such cases were rare and the discarded values were generally about 10°C or more). Spatial averages over the regions R1, R2, R3, R4, and R5 shown in Figure 14.15 were calculated by using appropriate weights based upon data density in each 2° × 2° box. Figure 14.16 shows the seasonal cycle which was subtracted to obtain the anomalies. Figure 14.17 shows the 10-year running mean for average SST anomalies for the regions R1, R2, R3, R4, and R5. For all the regions the anomaly was colder by about 0.25°C before the 1940s and warmer by about the same amount after the 1940s. This bias was seen for several other areas in the Indian Ocean and equatorial Pacific. Similar bias was also noted by other investigators. We have removed this bias from the observed SST anomalies for further analysis of data.

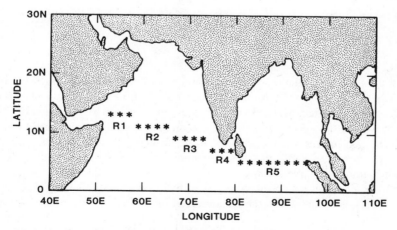

Figure 14.15. Ship track with highest sea surface temperature (SST) data density over the Arabian Sea and regions R1 through R5. The width of each region is 2° latitude.

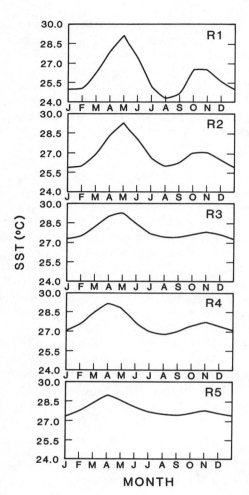

Figure 14.16. Long-term monthly mean sea surface temperature (SST) for the regions R1 through R5.

Figure 14.18 shows the time series of bimonthly [February, March (FM); April, May (AM); June, July (JJ); August, September (AS); October, November (ON); and December, January (DJ)] SST anomalies for 75 years for region R2. Figure 14.19 shows the June–July bimonthly SST anomalies for the five regions R1 through R5. These two figures show that most of the observed SST anomalies are within ±1°C. There is reasonable spatial and temporal consistency in the values of SST anomalies among the five locations.

We have averaged the SST anomalies for five heavy rainfall years (1916, 1917, 1933, 1961, 1975) for which the standardized rainfall anomaly (ratio of anomaly and its standard deviation) over India was 1.4, 2.7, 1.5, 2.0, 2.2, respectively, and for five deficient rainfall years (1905, 1911, 1918, 1951, 1972) for which the standardized rainfall anomaly was −1.6, −1.6, −2.6, −1.7, −1.8, respectively.

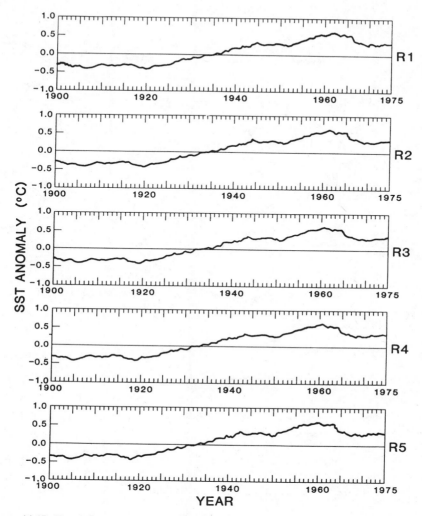

Figure 14.17. Ten-year running mean of sea surface temperature (SST) anomalies for the regions R1 through R5.

The composite time series for SST anomalies for these heavy rainfall and deficient rainfall years are shown in Figure 14.20. The SST anomaly for the heavy rain years is relatively warmer than deficient rain years during March, April, and May, and is colder during August, September, and October. This distribution is consistent with the inverse relationship between wind speed and SST anomaly earlier noted by Shukla and Misra (52) because stronger winds associated with above average monsoon rainfall tend to cool down the ocean surface in the post-monsoon months. This anomalous cooling effect is superimposed upon the annual cycle for which

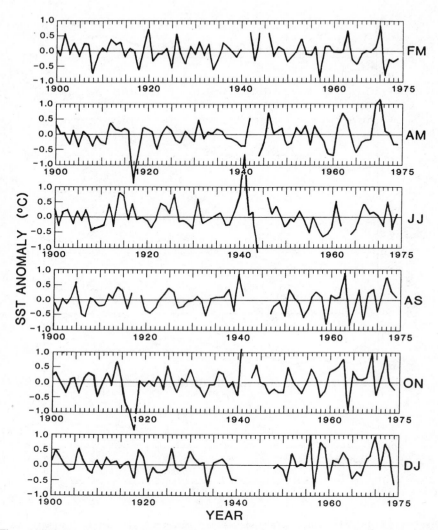

Figure 14.18. Time series of bimonthly [February, March (FM); April, May (AM); June, July (JJ); August, September (AS); October, November (ON); and December, January (DJ)] sea surface temperature (SST) anomalies for the region R2.

454

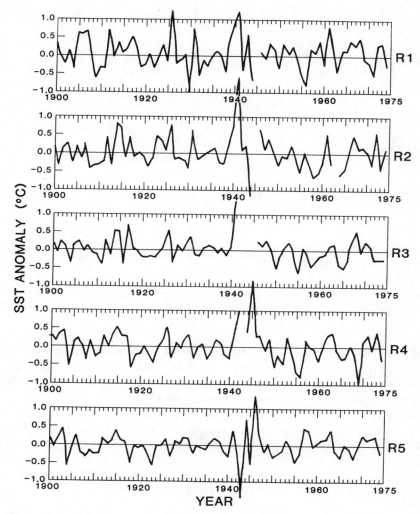

Figure 14.19. Time series of June, July (JJ) SST anomalies for the regions R1 through R5.

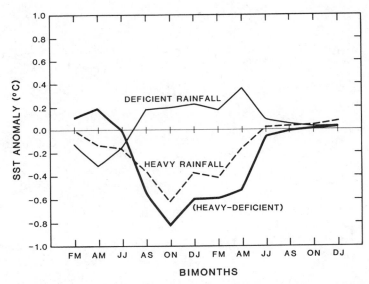

Figure 14.20. Composite sea surface temperature (SST) anomalies for heavy (dashed line) and deficient (solid line) monsoon rainfall. The thick solid line shows the difference (heavy-deficient) of SST anomalies.

cooling is primarily due to advective effects (see Chapter 13, Section 2.4). The magnitudes of the SST anomalies before the peak monsoon months are smaller than those for the post-monsoon months and therefore an average for the whole monsoon season is dominated by the post-monsoon months (56).

The general picture that emerges is as follows: a warm (cool) SST anomaly during April, May, and June is not necessarily indicative of above (below) average monsoon rainfall; however, heavy monsoon rainfall is followed by negative SST anomalies. The predictive value of this relationship is rather limited because the magnitude of SST anomalies during the pre-monsoon months is within the range of observational error (−0.1°C to −0.5°C).

Equatorial Pacific Sea Surface Temperature Anomalies. Subsequent to the pioneering works of Walker [(10), who discovered the "Southern Oscillation" while looking for correlations between the Indian monsoon rainfall and other atmospheric parameters over the globe] and Walker and Bliss (57), and the works of Bjerknes [(58), who coined the term "Walker circulation"], several observational papers have documented the relationship between the occurrence of warm SST anomalies in the equatorial Pacific and a shift of the heavy precipitation regime from the extreme western Pacific to the central Pacific near the international dateline. Aperiodic occurrences of warm equatorial Pacific SST anomalies, referred to as El Niño events, have been found to be associated with below normal summer monsoon rainfall over India. Sikka (15) showed a general association between El Niño events and deficient summer monsoon rainfall. Angell (59) showed that SST anomalies in the equatorial

Pacific were highly correlated (correlation coefficient 0.62) with rainfall over India during the preceding summer monsoon season. Neither Sikka nor Angell recognized the predictive value of their findings because of incomplete knowledge of the life cycle of the warm SST events later described by Rasmusson and Carpenter (60). The positive (warm) El Niño SST anomalies usually appear along the Ecuador–Peru coast several months before the monsoon season, so the association between El Niño events and deficient monsoon rainfall can be a useful forecasting tool. Rasmusson and Carpenter (11) have identified 25 El Niño events during the period 1875–1979, and have shown that the area averaged summer monsoon rainfall over India was below the median value in 21 of the 25 events. The solid black bars in Figures 14.12 and 14.13 denote these El Niño years. This relationship will be discussed in Chapter 16.

Fu and Fletcher (61) have examined the large-scale thermal contrast between the soil temperature over the Tibetan Plateau (given by the mean value for five representative stations), and sea surface temperature over the eastern equatorial Pacific averaged between 5°N–10°S and 120°W–160°W, and found that the higher (lower) values of the land–ocean contrast (i.e., warmer Tibetan Plateau and colder SST) are associated with higher (lower) monsoon rainfall. The correlation between an index of the Indian summer monsoon rainfall and the land–ocean temperature contrast is higher than that with either land or ocean temperature alone.

These results pertain only to a possible relationship between the summer monsoon rainfall and equatorial Pacific SST. Owing to the lack of data, relationships to Southern Hemisphere SST have not been investigated.

3.2.3 Soil Moisture. The annual net rainfall for the global continents is estimated to be about 764 mm, and runoff to the oceans about 266 mm (62). If there were no secular trends in the annual mean global soil moisture this would suggest that the annual global mean evaporation from the land surface alone is more than 60% of the annual and global mean precipitation over the land and is a very important component of the global water budget and hydrologic cycle. It does not necessarily follow that water evaporated locally from the land is important in determining the local rainfall over the land because the total rainfall is determined by the combined effects of available and precipitable moisture and the nature and intensity of the dynamical circulation.

The role of soil moisture in determining the interannual variability of atmospheric circulations is two-fold. First, it strongly influences the rate of evaporation and therefore, the moisture supply to the atmosphere. Second, it influences the heating of the ground which affects the sensible heat flux and ground temperature. In a set of idealized numerical experiments with a GCM, Shukla and Mintz (63) examined the role of soil moisture. They found that when the land surface is dry, and no evaporation is allowed from it, a very intense surface low develops over India during the summer monsoon season. Despite the lack of evaporation from the land surface in the model experiments, monsoon rainfall is greater than when the soil is wet and evaporation over land can take place. This is because the reduction of moisture caused by the lack of evaporation is more than compensated by moisture flux convergence from the surrounding oceans caused by the intensified monsoon low. If this result were valid for the real atmosphere, it implies that a very dry pre-monsoon season would be followed by enhanced rainfall during the monsoon season. No observational study has been carried out to verify this hypothesis.

Charney et al. (64) suggested that significant changes in precipitation over subtropical desert margin regions can occur by changes in albedo at the earth's surface. An increase in albedo reduces the absorption of the incoming solar radiation, and hence evaporation and cloudiness. The increase in the solar radiation reaching the ground due to reduced cloudiness is more than compensated by the reduction in the long wave radiation reaching the ground from the cloud base; therefore, there is a net reduction in solar radiative heating of the ground, evaporation, and precipitation. These factors appear to be of some importance in producing changes over northwest India near the desert region, but a more quantitative evaluation of their influence on observed interannual variability of monsoons has not been carried out.

Bavadekar and Mooley (65) computed the evapotranspiration and the moisture flux convergence of the atmosphere for a triangular volume of Peninsula India and found that the interannual variability as well as the intraseasonal variability during the monsoon season is negligible. This suggests that the land surface processes are not important in determining the rainfall variability over this region. However, the accuracy of evapotranspiration data used in this study is not known. From the limited number of observational and modeling studies, it is not possible to determine the contribution of soil moisture toward the observed interannual variability of monsoon circulation and rainfall.

4 CONCLUDING REMARKS

In the context of the annual variations of the global circulation the Asiatic monsoon appears to be a highly periodic phenomenon primarily determined by the seasonal variations of solar radiation and asymmetric continentality with respect to the equator. However, although the onset and duration of the monsoon is quite regular, the precise date of onset at a given location is highly variable (up to 1 month). Once the onset has taken place, the nature of the day-to-day fluctuations and seasonal mean for one year can be very different from another. The day-to-day changes during a monsoon season seem to be associated with a variety of factors, the most notable of which are:

1. the formation, growth, decay and propagation of weather-scale disturbances (lows and depressions, etc.) which are manifestations of the hydrodynamic and moist-convective instabilities of the large-scale flow; and
2. changes in the locations and intensity of east–west oriented convergence zones (the near-equatorial trough, the monsoon trough, etc.) which can occur either due to quasi-periodic (15–40 days) fluctuations in the tropics or due to influences of the traveling and quasi-stationary waves in the mid-latitudes of the Northern and Southern Hemispheres.

The role of these short period (i.e., shorter than a season) fluctuations in determining the seasonal mean is not clearly understood. It is also not clear how these transient (high frequency) fluctuations are influenced by global- or planetary-scale low frequency changes.

Since space and time averaged precipitation is determined largely by the amount of moisture that converges in a region, it is reasonable to suggest that large-scale— larger than the scale of lows and depressions—convergence is the primary determinant of the rainfall intensity. Monsoon lows and depressions are manifestations of dynamical instabilities which organize precipitation at preferred scales and, therefore, it may be more profitable to study the large and planetary-scale circulation features that produce a suitable environment for these instabilities to grow. If intraseasonal and interannual variability of rainfall were determined solely by the intensity and frequency of disturbances, the prospects for long-range prediction of time averaged rainfall would be hopeless, but there seems to be sufficient reason to believe that these instabilities are strongly controlled by planetary-scale circulations which are perhaps more predictable than the instabilities themselves. A part of the intraseasonal variability of rainfall is indeed accounted for by the fluctuations associated with these instabilities, and that part would be difficult to predict. However, the remaining part could be more predictable, at least in principle, if the planetary-scale circulations responsible for the rainfall variability were forced by slowing varying boundary conditions at the earth's surface.

We can pose the following questions:

1. Is the behavior of the seasonal mean monsoon primarily determined by a statistical average of a variety of independent short period fluctuations that are not related to any seasonal or other low frequency forcings?
2. Are there global- and planetary-scale "forcing functions" (either due to slowly varying boundary conditions at the earth's surface or due to very low frequency changes like the Southern Oscillation) that determine the interannual behavior of the seasonal mean monsoon circulation and rainfall, and is the interannual variability of the short period fluctuations controlled by such large-scale low frequency forcings?

These questions are similar to those raised in the beginning of the chapter about the relative importance of the internal dynamics and boundary conditions. Clearly the questions represent the two extreme possibilites and the reality must lie between. This author is reluctant to accept the premise of the first question, and is inclined to accept the hypothesis implicit in the second question. The role of the unpredictable day-to-day changes due to the instabilities and the nonlinear interactions can not be ignored; however, it is unlikely that they can account for the total observed variability without considering the influence of the boundary conditions. The significant correlations between the Southern Oscillation index and seasonal mean monsoon rainfall clearly suggest that the behavior of the seasonal mean monsoon is strongly influenced by planetary-scale low frequency changes. It is shown in Chapter 16 that 9 out of 12 heavy monsoon rainfall seasons were preceded by a negative Darwin pressure trend (a measure of the Southern Oscillation), and 12 out of 14 deficient monsoon rainfall seasons were preceded by a positive Darwin pressure trend. Moreover, there are significant correlations between the monsoon rainfall and the equatorial Pacific SST anomalies, and snow cover over Eurasia and the Himalayas. It would be unreasonable to discard these relationships, attributing them to random chance.

These observational facts combined with their physical plausability support the hypothesis that the interannual variability of monsoons is significantly influenced by slowly varying boundary conditions and very low frequency planetary-scale atmospheric fluctuations. In order to advance our understanding of these phenomena it will be necessary to examine the behavior of the three-dimensional flow at planetary and global scales rather than over limited regions as has been mostly the case during the past.

REFERENCES

1. J. Shukla, Dynamical predictability of monthly means, *J. Atmos. Sci.*, **38**, 2547–2572 (1981).
2. J. G. Charney and J. Shukla, "Predictability of Monsoons," in Sir J. Lighthill and R. P. Pearce, Eds., *Monsoon Dynamics*, Cambridge University Press, Cambridge, 1981, pp. 99–109.
3. B. Parthasarathy and O. N. Dhar, Secular variations of regional rainfall over India, *Quart. J. Roy. Meteor. Soc.*, **100**, 245–257 (1974).
4. O. N. Dhar, B. Parthasarathy, and G. C. Ghosh, A study of mean monthly and annual rainfall of contiguous India area, *Vayu Mandal*, **4**, 49–53 (1974).
5. R. S. Pareek and C. Ramaswamy, "Climatology of Droughts in Burma during the Southwest Monsoon Period," *Proceedings of the Indian National Science Academy*, Vol. 42, Part A, New Delhi, 1976, pp. 44–50.
6. B. Parthasarathy and D. A. Mooley, Some features of a long homogeneous series of Indian summer monsoon rainfall, *Mon. Wea. Rev.*, **106**, 771–781 (1978).
7. H. N. Bhalme and D. A. Mooley, Large-scale droughts/floods and monsoon circulation, *Mon. Wea. Rev.*, **108**, 1197–1211 (1980).
8. D. A. Mooley, B. Parthasarathy, N. A. Sontakke, and A. A. Munot, Annual rain-water over India, its variability and impact on the economy, *J. Clim.*, **1**, 167–186 (1981).
9. D. A. Mooley and G. B. Pant, "Droughts in India over the last 200 Years, Their Socio-Economic Impacts and Remedial Measures for Them," Proceedings of the Symposium on Climate and History, Norwich, Cambridge University Press, Cambridge, 1981, pp. 465–478.
10. G. T. Walker, Correlations in seasonal variations of weather, *Mem. India Meteor. Dept.*, **24**, 333–345 (1924).
11. E. Rasmusson and T. Carpenter, The relationship between eastern equatorial Pacific sea surface temperatures and rainfall over India and Sri Lanka, *Mon. Wea. Rev.*, **111**, 517–528 (1983).
12. O. N. Dhar, P. R. Rakhecha, and B. N. Mandal, Does the early or late onset of monsoon provide any clue to subsequent rainfall during the monsoon season?, *Mon. Wea. Rev.*, **108**, 1069–1072 (1980).

13. K. Ramamurthy, "Some Aspects of the 'Break' in the Indian Southwest Monsoon during July and August," Forecasting Manual (available from India Meteorological Department, Lodi Road, New Delhi), Part IV-18.3, 1969.

14. O. N. Dhar and P. Rakhecha, "Does Absence of Tropical Disturbances Cause Deficient Rainfall of Contiguous Indian Area," Proceedings of Indian National Science Academy, Vol. 42, Part A, New Delhi, 1976, pp. 81–89.

15. D. R. Sikka, "Some Aspects of the Large-Scale Fluctuations of Summer Monsoon Rainfall over India in Relation to Fluctuations in the Planetary and Regional Scale Circulation Parameters," Proceedings of the Indian Academy of Science (Earth and Planetary Science), Vol. 89, New Delhi, 1980, pp. 179–195.

16. T. Yasunari, Cloudiness fluctuations associated with the Northern Hemisphere summer monsoon, J. Meteor. Soc. Japan, 57, 227–262 (1979).

17. D. R. Sikka and S. Gadgil, On the maximum cloud zone and the ITCZ over Indian longitudes during the southwest monsoon, Mon. Wea. Rev., 108, 1840–1853 (1980).

18. T. N. Krishnamurti and D. Subrahmanyam, The 30–50 day mode at 850 mb during MONEX, J. Atmos. Sci., 39, 2088–2095 (1982).

19. T. N. Krishnamurti and P. Ardanuy, The 10 to 20 day westward propagating mode and 'breaks' in the monsoons, Tellus, 32, 15–26 (1980).

20. T. Yasunari, Quasi-stationary appearance of 30–40 day period in the cloudiness fluctuations during summer monsoon over India, J. Meteor. Soc. Japan, 58, 225–229 (1980).

21. T. Yasunari, Structure of an Indian summer monsoon system with around 40-day period, J. Meteor. Soc. Japan, 59, 336–354 (1981).

22. T. N. Krishnamurti and H. N. Bhalme, Oscillations of a monsoon system. Part I: Observational aspects, J. Atmos. Sci., 33, 1937–1954 (1976).

23. J. G. Charney and D. M. Straus, Form-drag instability, multiple equilibria and propagating planetary waves in baroclinic orographically forced, planetary wave systems, J. Atmos. Sci., 37, 1157–1176 (1980).

24. R. A. Madden and P. R. Julian, Description of global scale circulation cells in the tropics with 40–50 day period, J. Atmos. Sci., 29, 1109–1123 (1972).

25. P. J. Webster and L. C. Chou, Low-frequency transitions of a simple monsoon system, J. Atmos. Sci., 37, 368–382 (1980).

26. B. N. Goswami and J. Shukla, Quasi-periodic oscillations in a symmetric general circulation model, J. Atmos. Sci., 41, 20–37 (1984).

27. C. Ramaswamy, Breaks in the Indian summer monsoon as a phenomenon of interaction between the easterly and subtropical westerly jet-streams, Tellus, 14, 337–349 (1962).

28. A. K. Bannerjee, P. N. Sen, and C. R. V. Raman, On fore-shadowing southwest monsoon rainfall over India with mid-tropospheric circulation anomaly of April, Indian J. Meteor. Hydrol. Geophy., 29, 425–431 (1978).

29. Y. P. Rao. "The Monsoon as Reflected in the Behavior of the Tropical High-Pressure Belt," in Sir J. Lighthill and R. P. Pearce, Eds., Monsoon Dynamics, Cambridge University Press, Cambridge, 1981, pp. 209–212.

30. R. K. Verma, Importance of upper tropospheric thermal anomalies for long-range forecasting in Indian summer monsoon activity, Mon. Wea. Rev., 108, 1072–1075 (1980).

31. P. V. Joseph, R. K. Mukhopadhyaya, W. V. Dixit, and D. V. Vaidya, Meridional wind index for long range forecasting of Indian summer monsoon rainfall, Mausam, 32, 31–34 (1981).

32. M. Tanaka, Interannual fluctuations of the tropical easterly jet and the summer monsoon in the Asian region, J. Meteor. Soc. Japan, 60, 865–875 (1982).

33. C. R. V. Raman and Y. P. Rao, Blocking highs over Asia and droughts over India, *Nature*, **289**, 271–273 (1981).

34. D. R. Wiesnet and D. Matson, A possible forecasting technique for winter snow cover in the Northern Hemisphere and Eurasia, *Mon. Wea. Rev.*, **104**, 828–835 (1976).

35. H. F. Blanford, On the connexion of the Himalaya snowfall with dry winds and seasons of droughts in India, *Proc. Roy. Soc. London*, **37**, 3 (1884).

36. D. Hahn and J. Shukla, An apparent relationship between Eurasian snow cover and Indian monsoon rainfall, *J. Atmos. Sci.*, **33**, 2461–2463 (1976).

37. R. R. Dickson, Eurasian snow cover vs. Indian monsoon rainfall—An extension of the Hahn-Shukla results, *J. Clim. and Appl. Meteor.*, **23**, 171–173 (1984).

38. B. Dey and O. S. R. U. Bhanukumar, An apparent relationship between Eurasian spring snow cover and the advance period of the Indian summer monsoon, *J. Appl. Meteor.*, **21**, 1929–1932 (1982).

39. T. C. Yeh, R. T. Wetherald, and S. Manabe, A model study of the short-term climatic and hydrologic effects of sudden snow-cover removal, *Mon. Wea. Rev.*, **111**, 1013–1024 (1983).

40. L. T. Chen and Z. X. Yan, "The Statistical Analysis of the Influence of Anomalous Snow Cover over Qinghai-Tibetan Plateau in Winter and Spring on the Summer Monsoon" (in Chinese), Proceedings of the Symposium on Qinghai-Xizang (Tibet) Plateau, Kumming, China, Science Press, Beijing, 1978, pp. 151–161.

41. J. Shukla, "Predictability of Time Averages: Part II. The Influence of the Boundary Forcing," in D. M. Burridge and E. Kallen, Eds., *Problems and Prospects in Long and Medium Range Weather Forecasting*, Springler-Verlag, London, 1984, pp. 155–206.

42. B. J. Hoskins and D. Karoly, The steady linear response of a spherical atmosphere to thermal and orographic forcing, *J. Atmos. Sci.*, **38**, 1179–1196 (1981).

43. R. S. Ellis, "A Preliminary Study of a Relation between Surface Temperature of the North Indian Ocean and Precipitation over India," unpublished master's thesis, Department of Meteorology, Florida State University, Tallahassee, 1952.

44. P. R. Pisharoty, "Forecasting Droughts in the Subcontinent of India," Proceedings of the Indian National Science Academy, Vol. 42, Part A, New Delhi, 1976, pp. 220–223.

45. K. R. Saha, Some aspects of the Arabian Sea summer monsoon, *Tellus*, **26**, 464–476 (1974).

46. K. R. Saha and S. N. Bavadekar, Water vapor budget and precipitation over the Arabian Sea during the northern summer, *Quart. J. Roy. Meteor. Soc.*, **99**, 273–278 (1973).

47. S. K. Ghosh, M. C. Pant, and B. N. Devan, Influence of the Arabian Sea on the Indian summer monsoon, *Tellus*, **30**, 117–125 (1978).

48. D. Cadet and G. Reverdin, Water vapor transport over the Indian Ocean during summer 1975, *Tellus*, **33**, 476–487 (1981).

49. J. Shukla, Effect of Arabian Sea surface temperature anomaly on Indian summer monsoon: A numerical experiment with the GFDL model, *J. Atmos. Sci.*, **32**, 503–511 (1975).

50. D. R. Sikka and K. Raghavan, Comments on effects of Arabian sea-surface temperature anomaly on Indian summer monsoon: A numerical experiment with the GFDL model, *J. Atmos. Sci.*, **33**, 2252–2253 (1976).

51. J. Shukla, Reply, *J. Atmos. Sci.*, **33**, 2253–2255 (1976).

52. W. M. Washington, R. M. Chervin, and G. V. Rao, Effects of a variety of Indian ocean surface temperature anomaly patterns on the summer monsoon circulation: Experiments with the NCAR general circulation model, *Pageoph*, **115**, 1335–1356 (1977).

53. J. Shukla and B. M. Misra, Relationships between sea surface temperature and wind speed over the central Arabian sea and monsoon rainfall over India, *Mon. Wea. Rev.*, **105**, 998–1002 (1977).

54. M. Fieux and H. Stommel, Historical sea-surface temperature in the Arabian Sea, *Ann. Inst. Oceanogr. Paris*, **52**, 5–15 (1976).

55. K. Raghavan, P. V. Puranik, V. R. Mujumdar, P. M. M. Ismail, and D. K. Paul, Interaction between the west Arabian Sea and the Indian monsoon, *Mon. Wea. Rev.*, **106**, 719–274 (1978).

56. B. C. Weare, A statistical study of the relationship between ocean surface temperatures and the Indian monsoon, *J. Atmos. Sci.*, **36**, 2279–2291 (1979).

57. G. T. Walker and E. W. Bliss, "World Weather V," *Mem. Roy. Meteor. Soc.*, **4**, 119–139 (1932).

58. J. Bjerknes, Atmospheric teleconnections from the equatorial Pacific, *Mon. Wea. Rev.*, **97**, 163–172 (1969).

59. J. K. Angell, Comparison of variations in atmospheric quantities with sea surface temperature variations in the equatorial eastern Pacific, *Mon. Wea. Rev.*, **109**, 230–243 (1981).

60. E. Rasmusson and T. Carpenter, Variations in tropical sea surface temperature and surface wind fields associated with the Southern Oscillation/El Niño, *Mon. Wea. Rev.*, **110**, 354–384 (1982).

61. C. Fu and J. O. Fletcher, "The Role of the Surface Heat Source over Tibet in Interannual Variability of the Indian Summer Monsoon," unpublished manuscript, 1983.

62. H. Baumgartner and E. Reichel, *The World Water Balance: Mean Annual Global Continental and Maritime Precipitation, Evaporation and Runoff*, Elsevier, Amsterdam, 1975, pp. 179.

63. J. Shukla and Y. Mintz, Influence of land-surface evapotranspiration on the earth's climate, *Science*, **214**, 1498–1501 (1982).

64. J. G. Charney, W. J. Quirk, S. Chow, and J. Kornfield, A comparative study of the effects of albedo change on drought in semi-arid regions, *J. Atmos. Sci.*, **34**, 1366–1385 (1977).

65. S. N. Bavadekar and D. A. Mooley, "Use of the Equation of Continuity of Water Vapor for Computation of Average Precipitation over Peninsular India during the Summer Monsoon," in Sir J. Lighthill and R. P. Pearce, Eds., *Monsoon Dynamics*, Cambridge University Press, Cambridge, 1981, pp. 261–268.

PART VI

PREDICTION AND GOVERNMENT ACTION

How, and how well, are monsoons predicted? In a sense, prediction is the ultimate goal of the monsoon scientist. An accurate prediction of when the monsoon will come and how much rain it will bring can mean the difference between agricultural success and failure in a part of the world where agriculture is vital. When to plant, what to plant, where to plant, whether to ration water and power, whether to move people from one region to another, whether to irrigate, all depend critically on accurate forecasts with sufficient lead times for government action and public response.

These questions are viewed from various perspectives by the authors in Part VI. T. N. Krishnamurti and J. Shukla have dedicated a major part of their professional lives to the problem of monsoon prediction. As academics, they enjoy the luxury of studying monsoon prediction in a research environment. P. K. Das and Boon-Khean Cheang are also research scientists. But in addition, they have been long associated with operational monsoon forecasting at two of the most advanced meteorological services in Asia. It is from an operational viewpoint that they discuss the problems of monsoon prediction.

In Chapter 15 Krishnamurti provides background on the modern numerical prediction models—the types of models, how they work, and their shortcomings. Where Krishnamurti concerns himself with short to medium range predictions of 1 to 10 days, Shukla discusses monsoon prediction on monthly, seasonal, and interannual time scales in Chapter 16. Both discuss possibilities for improvement and directions for future research.

In contrast, in Chapters 17 and 18, Das and Cheang discuss day-to-day operational forecasting in Asia. Here we learn how forecasters faced with the problems of sparse weather observations, marginal computer power, and less than adequate models meet the demand for daily and longer range forecasts.

The subject of the final chapter by M. S. Swaminathan is how governments deal with the consequences of abnormal monsoons. Except perhaps for the meteorologist, a monsoon prediction is not an end unto itself. If it is to be of any practical use, it must be communicated, its implications must be understood, disaster management strategies must be developed, and, finally, the public must respond. In order to make intelligent decisions, it is clear that governments must have the best possible monsoon predictions as far in advance as possible.

Of course, it is virtually certain that from time to time monsoons will be abnormal. And, as Swaminathan points out, "although abnormal monsoons may be forecast, no technique has been developed to avert them." Governments, therefore, also have an obligation to prepare for the worst.

Swaminathan discusses a variety of preparatory and reactive strategies and programs that have been developed for India in response to abnormal monsoons. He recommends ways by which governments can prepare for the challenges of a future in which it is likely that the climate will differ radically from today's climate.

15

Monsoon Models

T. N. Krishnamurti
Department of Meteorology
Florida State University
Tallahassee, Florida

INTRODUCTION

One way to study monsoons is to build models of them, study the behavior of the models, and to the extent that the models represent reality, learn more about how monsoons work. A model is a conceptual or mathematical description of the physical processes that produce and influence monsoons.

1 SIMULATION MODELS

A number of physical models will be discussed that have provided major insights into the mechanisms of the monsoon. The starting point of this review is a "zonally symmetric" model in which the monsoon is treated as a simple, differentially heated system. This is followed by a description of two-dimensional and planetary boundary layer models which allow us to consider the monsoon circulation over the Arabian Sea. Finally, the more complex three-dimensional general circulation models will be described which are essential for simulating the planetary-scale monsoon circulation.

1.1 The Zonally Symmetric Monsoon Model

The zonally symmetric model is the simplest atmospheric model in which all three wind components, the temperature field, and the humidity field are represented. The only restriction imposed is that the variation of all variables in the zonal or east–west direction is zero, hence zonally symmetric. The Asian monsoon circulations in this model are driven by differential heating between land (to the north) and oceans (to the south). The components of heating will be described later in Section 1.4. Smoothed mountains over the land areas represent the Himalayas and the oceans to the south have a prescribed sea surface temperature. The important effect of condensation or latent heating is included; thus the zonally symmetric model contains the basic "ingredients" of a monsoon discussed in Chapter 1.

This model appears quite realistic for the simulation of the gross features of the regional monsoon circulation. A convincing illustration of this comes from the work of Murakami and his colleagues (1). Their model is started from a state of rest and is run with a constant solar insolation (July) modified by summer mean cloud cover. Slowly, due to the surface heating and attendant sensible heat flux, a shallow low-pressure area forms over the land area. This low gradually generates low-level wind inflow that draws moisture from the ocean to the south. The moisture is carried upward in the convection, condenses, and warms the upper troposphere where a strong anticyclone develops. After about 80 (model) days of integration, a statistical steady state circulation is reached. South of the upper anticyclone the zonal wind is strong and easterly while in the lower troposphere south of the low-pressure area, the wind is westerly. The meridional circulation is southerly at lower levels with northerly return flow in the upper troposphere. Thus this model is, in fact, able to simulate the observed winds of the summer monsoon; southwesterly at low levels and northeasterly at upper levels (see Fig. 15.1).

The monsoon simulation described here is similar to the simulation of a sea breeze, with a major difference. The scale of the sea breeze is horizontally about 100 km and 3 km vertically. The scale of the monsoon circulation, on the other hand, is horizontally about 3000 km (north–south) and 14 km vertically.

NUMERICAL SIMULATION OF INDIAN SUMMER MONSOON
ZONAL WIND

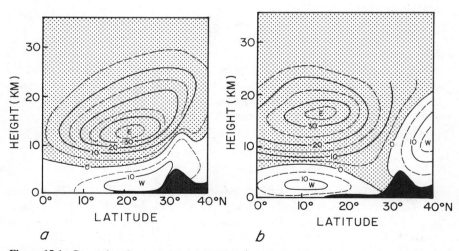

Figure 15.1. Comparison between the simulated wind (*a*) and the observed wind (*b*). From the study of Murakami and his colleagues (1). This height–latitude cross section shows the strength of the west to east wind along 80°E in m/sec. The heavy darkened area denotes mountains, that is, the Himalayas. Regions of easterly winds are stippled.

On this large scale, the effects of the earth's rotation became quite important. In the zonally symmetric model the transfer of kinetic energy from the meridional winds to zonal winds can occur only via the Coriolis force, that is, the effect of the earth's rotation. The manner by which energy is generated, transferred, and dissipated in the zonally symmetric monsoon is extremely illustrative of the monsoon as a differentially heated system. The mathematical description of the energy processes is:

$$\partial P/\partial t = \langle K_m \rightarrow P \rangle + G \tag{1}$$

$$\partial K_m/\partial t = -\langle K_m \rightarrow P \rangle + \langle K_z \rightarrow K_m \rangle - D_m \tag{2}$$

$$\partial K_z/\partial t = -\langle K_z \rightarrow K_m \rangle - D_z \tag{3}$$

where P, K_m, and K_z respectively denote the basic energy quantities (potential, meridional kinetic, and zonal kinetic) of the model. The symbol $\langle A \rightarrow B \rangle$ denotes an energy transformation or exchange from A to B (positive) or B to A (negative). D_z and D_m denote the dissipation terms for the zonal and the meridional kinetic energies, and G denotes the generation of potential energy. The symbol $\partial/\partial t$ denotes (in the calculus) the change of a quantity (e.g., P or K_m or K_z) with respect to time at a fixed location.

Note that by adding Eqs. (1), (2), and (3), and in the absence of dissipation and generation, the model system is governed by the conservation of total energy:

$$\partial/\partial t (P + K_m + K_z) = 0 \tag{4}$$

Indeed one of the many interesting properties of this model is that when a statistical steady state is reached after some 80 days of integration, the system does conserve total energy.

The zonal kinetic energy at this stage can be maintained against dissipation only by receiving energy from meridional motions [see Eq. (3)]. Put another way, any rotational component of the wind must be zonal and any divergent component must be meridional. Thus in counteracting against frictional dissipation the divergent wind component must steadily supply energy to the nondivergent component in a near steady state, zonally symmetric system. This is, in fact, a theorem that holds true even when the requirement for zonal symmetry is relaxed and the system is fully three-dimensional (2). The divergent kinetic energy is a measure of the strength of the tropical Hadley cell. However, the Hadley cell does not decrease in strength as it continuously provides energy to the nondivergent motions. Thus we must look for other sources in our model that supply energy to maintain the divergent motions in near steady state. The only source available is from the conversion from the available potential energy [Eq. (2)]. This happens when the ascending air of the meridional circulation is relatively warmer than the descending air (i.e., in a direct circulation such as the Hadley cell). In this manner, potential energy is lost to the divergent (meridional) kinetic energy. Since the potential energy also remains a near invariant, once a steady state is reached a source that generates potential energy must be present. The only generation term in this simple monsoon model is identified schematically by G in Eq. (1) and represents the covariance of the heating and the temperature; generation occurs where relatively warm areas are heated and/or relatively cold areas are cooled. This is the key to how a differentially heated system works. Continuous warming of the troposphere over the land area by sensible and latent heating through deep cumulus convection acts as the principle energy source for the potential energy necessary to produce the monsoon circulation and maintain it against frictional dissipation. The energy cycle is illustrated in Figure 15.2.

There are some obvious limitations to the zonally symmetric model. The zonal winds can lose kinetic energy only by frictional dissipation. Of course, the real atmosphere is not zonally symmetric; the zonal wind does vary in the zonal direction and this variation can be represented by the sum of many discrete zonal waves. Observational studies, for example (3), show that these zonal waves interact among

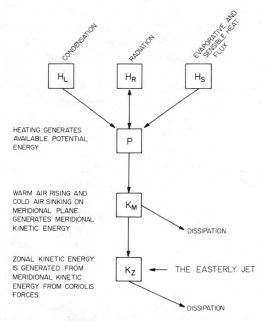

Figure 15.2. Energy exchanges in a zonally symmetric model simulation of the monsoon. The three main forms of heating [H_L, the condensation heating; H_R, the radiative heating (which is negative when it is cooling); and H_S, the sensible heating from the ocean] contribute to a net generation of available energy P. K_m denotes the meridional kinetic energy produced by vertical circulations, that is, by the ascent of warm air over land area and the descent of relatively colder air over the ocean to the south. The direction of the arrow from P to K_m represents the conversion from available potential to kinetic energy. The exchange of energy from K_m to K_z (K_z denotes zonal kinetic energy, i.e., kinetic energy of west to east motions) occurs as air traveling northward or southward is deflected by the Coriolis forces. Finally, there is dissipation that depletes both zonal and meridional energy.

themselves and gain energy (from longer waves) or lose energy (to shorter waves). Such interactions cannot be represented within a zonally symmetric monsoon model. In particular, because of zonal symmetry no baroclinic waves (weather disturbances) over the higher latitudes can be simulated. Their absence precludes the possibility of realistic middle latitude interactions with the tropical monsoon. This limitation could be overcome by representing the effects of these interactions by an explicit parameterization of heat and momentum fluxes in the thermal and momentum equations of the zonally symmetric model. This approach has been employed by climate modelers, although no studies on the evolution of zonally symmetric monsoons have been carried out in this context. This would be an endeavor of considerable interest.

1.2 Two-Dimensional Arabian Sea Circulation Models

The two-dimensional models described here are on a meridional–vertical plane. They contain the three components of the winds (zonal, meridional, and vertical) but they do not allow east–west variations; hence they are two-dimensional. They are dynamical models where the Coriolis force, advective accelerations and the frictional forces play a role in defining the boundary layer.

Many of the gross features of the Arabian Sea circulations over the lower troposphere have been simulated using simple two-dimensional models. The major observational features are illustrated in Figure 15.3 (4) and include the southern trades, the cross-equatorial flows, the Somali jet and its possible split into two branches over the northern Arabian Sea.

Most of the Arabian Sea simulations have been done with models that are formulated in terms of potential vorticity—a measure of circulation characteristics. In Section 2 some of the important properties of the two-dimensional potential vorticity conserving model are discussed.

Once a reasonable simulation of the aforementioned features of the Arabian Sea circulation are accomplished one must interpret the results:

1. How is the redistribution of potential vorticity accomplished?
2. How sensitive are the results of the simulation to the specification of mountains, to the lateral boundary forcing, to the effect of earth's rotation (beta effect) and to the domain size?

It is important to recognize that in the absence of friction, heating, and boundary forcing, air parcels conserve potential vorticity, even in the presence of mountains. During spring, as the Somali jet strengthens and the monsoon onset occurs, a major evolution of potential vorticity is observed with large positive values (north of the jet) and large negative values (south of the jet). To model these observed changes one needs to include the effects of heating, friction, and lateral import. If one considers a model domain between 30°S and 30°N and between 40°E and 80°E, the import of negative potential vorticity into this domain by the trade wind systems of the Southern Hemisphere and the export of positive potential vorticity by the

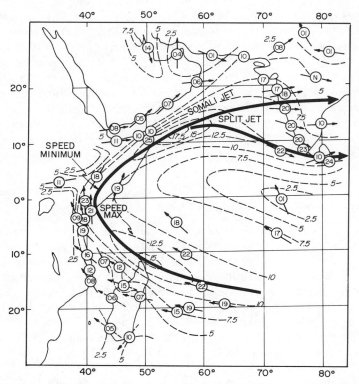

Figure 15.3. The monthly mean flow field for July at 1 km above the sea level, based on the work of Findlater (4). The heavy black line shows the main axis of the low-level jet which is called the Somali jet. Note the strong winds downwind from Somalia with a maximum speed around 17.5 m/sec. The lines of constant wind speed are shown by dashed lines. The circles indicate locations of weather stations and the numbers within the circles denote the number of observations within July. Also shown are the speed minimum and the splitting of the Somali jet into two branches.

southwest monsoon wind systems (north of the trades) need to be specified. In addition, a model that allows for parcel invariance of potential vorticity within the domain should have adequate interaction with the mountains in the form of lateral boundary conditions, such as a no-slip condition that allows for wind stress at the boundaries.

Figure 15.4 illustrates the major mountain systems relevant to this domain. The important ones are the East African highlands, western Ghats, Kenya Highlands, the Madagascar mountains, and especially those over Ethiopia. In some two-dimensional models the East African highlands appear as a sidewall boundary where the role of lateral friction becomes quite important in modifying the strong southerly flow, especially near the East African coast. The modification, producing very strong lateral wind shear, contributes to the generation of strong cyclonic shear, which is advected into the interior of the Arabian Sea. This is evidently a possible mechanism for the generation of the observed strong positive potential vorticity on the cyclonic shear side of the Somali jet. An axis of large negative potential vorticity is also observed on the anticyclonic shear side of the Somali jet north of the equator. Its presence is largely explained by the advective mechanism—its source of origin can be traced to the Southern Hemisphere trades. The split of the Somali jet into two flanks over the northern Arabian Sea may be a consequence of barotropic instability of the strong Somali jet near the Horn of Africa. Such splitting of jets has been examined in the barotropic context by Wiin-Nielsen (5) and in the present modeling context by Krishnamurti et al. (6), whose results are shown in Figure 15.5.

The question of sensitivity of the simulation to various parameters has been addressed by Hart (7), Bannon (8), and Krishnamurti et al. (6). There is general agreement on the importance of the lateral forcing across the eastern boundary in determining the intensity of the Somali jet; numerical experiments with weak lateral forcing along this wall exhibit a weak circulation system over the Arabian Sea. In particular, the role of the Ethiopian mountains has been examined and the results show that the strength of the Somali jet near the Socotra Islands off the Horn of Africa is diminished considerably unless the full impact of these mountains is included. Studies by Krishnamurti et al. (6) also emphasize the major role of the beta effect (β) in the two-dimensional dynamics. β is a measure of the change of the Coriolis parameter with latitude. When the model used a constant Coriolis parameter (i.e., $\beta = 0$) the simulations proved to be unrealistic, leading to the conclusion that the north–south variation of the Coriolis parameter is an important aspect of the circulation.

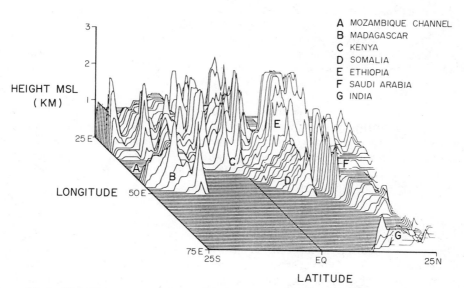

Figure 15.4. The mountains around the Arabian Sea are identified by letters A through G.

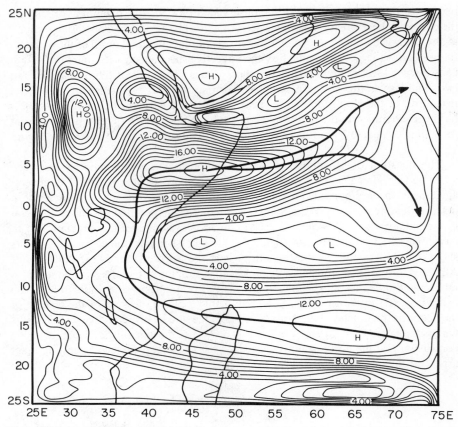

Figure 15.5. Results of numerical simulations of the Arabian Sea wind and the low-level jet at 1 km above sea level (from reference 6). The thin lines are lines of constant wind speed in m/sec. The heavy lines show the axis of maximum wind that includes the Somali jet.

One recognizes some of the major limitations of these simple two-dimensional models. Principally these include the absence of a planetary boundary layer (to be discussed in the following text) and the influence of heat sources and sinks. These can be important in the generation and redistribution of potential vorticity over the open oceans. These effects can only be included in multi-level, comprehensive two- and three-dimensional models.

1.3 Planetary Boundary Layer Models

The planetary boundary layer (PBL) is the lowest kilometer or so of the atmosphere where the influence of surface friction is important. There are two areas of interest in PBL modeling efforts. The first is the dynamics of the cross-equatorial, low-level monsoonal flows over the Arabian Sea and the second is air–sea interaction, where specifically one is interested in modeling the transports between the atmosphere and ocean of heat, moisture, and momentum, and their effects on monsoon convection.

The usual Ekman dynamics (see Chapter 9) describes the extratropical PBL quite adequately provided that the wind does not vary rapidly with height. The essential dynamical balance is among the Coriolis, pressure gradient, and frictional forces. Under these circumstances one expects a veering (clockwise turning) of wind with height in the Ekman layer over the Northern Hemisphere and a backing of wind (counterclockwise turning) over the Southern Hemisphere. Departures from Ekman winds are well known near the equator where the Coriolis force approaches zero and the advective effects can become important. In the tropical PBL the balance is among the advective accelerations, pressure gradient, and frictional forces. As one moves away from low latitudes the gradual increase of the Coriolis force diminishes the influence of the advective accelerations. The PBL dynamics of the Somali jet and, in fact, the entire Arabian Sea cross-equatorial low-level circulation raises a number of important questions. These include:

1. What are the transitions in the balance of forces across the equator?
2. What is the balance of forces across the Somali jet?
3. What is the balance of forces across the intertropical convergence zone which is located over the northern Arabian Sea north of the Somali jet?

In order to address these types of questions both two- and three-dimensional PBL models have been used. In these models one invokes a pressure distribution prescribed from detailed observations and integrates the Navier–Stokes equations with time to obtain a steady-state or equilibrium field of motion. Since, in general, the pressure gradient force varies with height, the simulations include thermal wind influences (which are non-Ekman). Frictional diffusion or mixing is modeled using the so-called K-theory which defines the vertical distribution of the momentum mixing coefficient as a simple parabolic profile. The lateral boundary conditions are defined from prescribed but varying upstream conditions, while at the meridional boundaries the motion field is time invariant and is obtained from a solution of the linearized Ekman equation. The unknowns are the three velocity components; their initial state is obtained from the Ekman balance at all latitudes except at the equator where it is linearly interpolated from the Ekman solution at the adjacent latitudes.

The numerical aspects of this type of model are fairly simple; one uses standard finite differences for the advective terms and the vertical velocity is obtained from a vertical integration of the mass continuity equation with the boundary condition that the vertical velocity is zero at the earth's surface.

Figures 15.6 and 15.4, respectively, illustrate the grid lattice and the mountains of a PBL model that was designed to study the Arabian Sea circulations. The model uses a vertical resolution of 200 m with higher resolution in the first layer above the surface. Figure 15.7 shows an example of a simulation of the Arabian Sea circulations obtained from the model study. The relevant balance of forces in this simulation is illustrated for the 1-km level. The intertropical convergence zone (ITCZ) is located along the asymptote of flow convergence (see Fig. 15.3). The

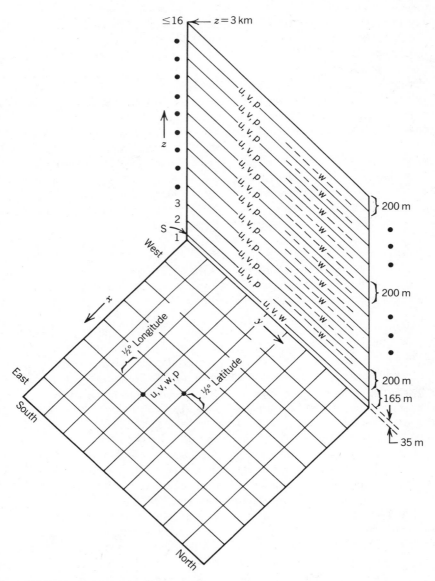

Figure 15.6. The computational grid for a planetary boundary layer model along x (west to east), y (south to north), and z (bottom to top). Along the horizontal axes the grid size is ½ degree latitude by ½ degree longitude. In the vertical, the grid size varies from 35 m near the ground to 200 m higher up. There are at least 16 levels in the vertical below 3 km. The top level for model computations is 3 km. The horizontal velocity components u, v and the pressure p are at the vertical levels shown by solid lines while the vertical velocity w is determined between these levels and is shown by a dashed line.

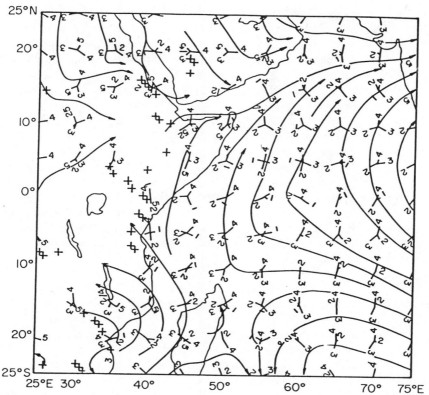

Figure 15.7. The flow field at the 1 km above sea level over the Arabian Sea based on a numerical simulation. The balance of forces is shown by the small line segments with numbers at each of the computational points. These numbers denote: (1), vertical advection; (2), horizontal advection; (3), Coriolis force; (4), pressure gradient force; and (5), frictional force. Thus a point near the Somali jet showing a balance among (2), (4), and (5) signifies a balance between horizontal advection, Coriolis and pressure gradient forces. Note that the 1-km level is well above the ocean so that the role of the frictional force (5) is less important compared to the other forces.

following approximate balance of forces and advective processes at the 1-km level, denoted by numbers (1)–(5) in Figure 15.7, is in general true over the following regions:

1. North of the asymptote of convergence there is a predominant balance between the Coriolis force (3) and the pressure gradient force (4); that is, a quasi-geostrophic balance. [The horizontal advection (2) appears over this region but it is relatively weak.] Below the 1-km level over this region (not shown) the intensity of the frictional force (5) increases rapidly and the planetary boundary layer is controlled by an Ekman balance with a marked veering of wind with height.

2. As we proceed south toward the equator we note a predominant balance between horizontal advection (2) and pressure gradient forces (4). The role of the Coriolis force (3) diminishes. In the PBL below the 1-km level over this region, the role of the frictional force (5) increases and an advective boundary layer (i.e., a balance among 2, 4, 5) exists.

3. Along the equatorial east coast of Africa sinking motions are strong and the role of vertical advection (1) is important. Over this region below the 1-km level (not shown) the balance in the PBL is among terms (1, 4 and 5). This is an advective boundary layer with a dominant role for the vertical (rather than the horizontal) advection.

4. As we proceed southward towards 20°S the geostrophic balance between Coriolis (3) and pressure gradient (4) forces again becomes dominant . As in the northern subtropics, below the 1-km level (not shown) the role of the frictional force (5) is strong and an Ekman balance exists with a veering of winds with height. It should be noted that pure Ekman, geostrophic or advective balance is rarely the case, since a small contribution from the other forces is almost always present.

The air–sea interactions and thermodynamics of the monsoon PBL have not been adequately modeled. Instead, the surface fluxes have been estimated from bulk aerodynamic principles applied to the scant observations that exist in these regions. The most exhaustive work, by Hastenrath and Lamb (9), shows that the southern trades and the Somali jet over the northern Arabian Sea are the most active regions in transporting sensible heat and moisture between the atmosphere and ocean during the summer monsoon season (Fig. 15.8a and b).

During undisturbed periods, the cloud transitions from East Africa to western India are quite analogous to those one finds in the trade wind belts from the west coast of continents to the near equatorial ITCZ. This downstream transition goes from a subsidence region near the East African coast with a marked temperature inversion to a gradual buildup of stratocumulus to fair weather convection and eventually to deep convection. Thus the numerical modeling aspects over the Arabian Sea are somewhat analogous to that of the trade wind belt. There are also some major differences. These are mostly in the environmental large-scale motion and thermal fields. The presence of the strong low-level westerly jet (with winds in excess of 30 m/sec) is quite unlike anything one sees in the trade wind regions. Another major difference is the presence of desert heat-lows to the north of the Arabian Sea. The current numerical models that are being used to study low-level thermodynamics, that is, cloud topped mixed layers, non-precipitating cumuli, and the ITCZ, are relevant to our understanding of the Arabian Sea PBL. Also important is the problem of the strong diurnal changes that have been observed over the western Arabian Sea and over Somalia. In this region the most pronounced signal is in the amplitude of the wind field and the strength of the coastal temperature inversion. Wind speeds in the early morning hours are about twice as large as those during the late afternoon hours. The marked coastal inversion weakens considerably during this period of wind decay. Models should include the effects of diurnal heating and momentum mixing in this environment of general descending motions over the western Arabian Sea; efforts to do so are under way.

1.4 General Circulation Models

A general circulation model is a hemispheric or global weather and climate simulation model. In principle one can start numerical integrations with the earth's atmosphere at a state of rest, impose the earth's rotation rate on the atmosphere, turn on the sun, and watch the weather and climate evolve in the results of very long-term integrations. However, the design of a general circulation model is a little more complicated. A number of important ingredients are common to the general circulation models that are used to simulate atmospheric behavior. Within such a model one does not explicitly treat the physics of one phenomenon any differently from another. The monsoon characteristics, such as the onset and the active and inactive phases, are no different in this regard. Starting from a state of rest, a general circulation model must describe the evolution of such events.

The source of all atmospheric motions is ultimately related to solar energy. The earth-ocean-atmosphere system absorbs, reflects, and scatters solar (short wave) radiation, and at the same time, transmits radiation at longer wave lengths. The total radiative forcing in the general circulation models requires calculations of both radiative warming and cooling rates within the atmosphere for the short and long wavelengths. Scattered as well as absorbed solar radiation and the emission and absorption of long wave radiation must all be included.

In the calculations, the emissivity of principal atmospheric constituents such as water vapor, ozone, and carbon dioxide has to be considered. Another complicating factor is what is called "cloud radiation," for the presence or absence of clouds greatly alters the atmosphere's radiative properties. At the starting point of any model simulation, the areas where there are clouds must be defined. Most general circulation models do not carry clouds as a large-scale dependent variable. Thus one must depend on some empirical definition of clouds. This is most often done using what are called critical or threshold relative humidity criteria. Climatological humidity information is a useful guide for defining certain threshold values of critical relative humidity. If the model calculated humidity exceeds the threshold value at any level along the vertical coordinate, then the model assumes the presence of a cloud, the prescribed radiative properties of which determine the estimates of the heating and cooling rates at that level.

A distinction is made between land and ocean in the radiation calculations. Most general circulation models are atmospheric models only; the state of the ocean is prescribed. (This is not the case for the few existing coupled models where the evolution of the atmosphere and the ocean are both calculated.) In the passive ocean models the state of the sea is defined in terms of a fixed sea surface temperature for each month. The ocean provides a variable structure in space and time via the air–sea interactions which are manifested as the fluxes of momentum, heat, and moisture at the air–sea interface. They vary because the atmospheric variables at the sea level—the winds, temperature, and the humidity—vary in space and time;

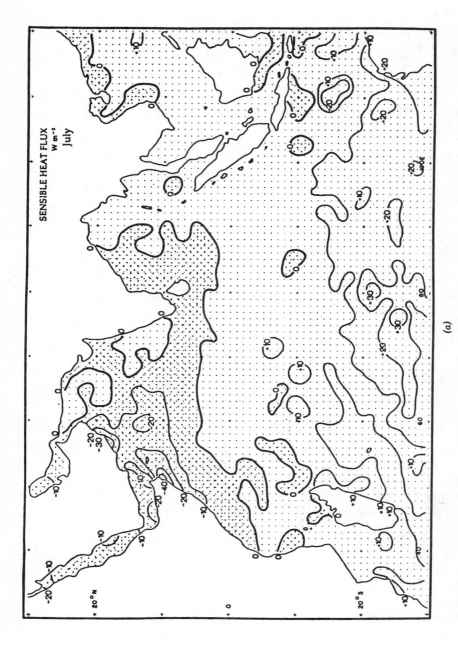

Figure 15.8. (*a*) The monthly mean sensible heat flux over the Indian Ocean, the Arabian Sea, and the Bay of Bengal based on the work of Hastenrath and Lamb (9). The units of fluxes are W/m². Shaded regions are those where the fluxes are from the atmosphere to the ocean. Over the regions with positive values (unshaded) the sensible heat flux is from the ocean to the atmosphere.

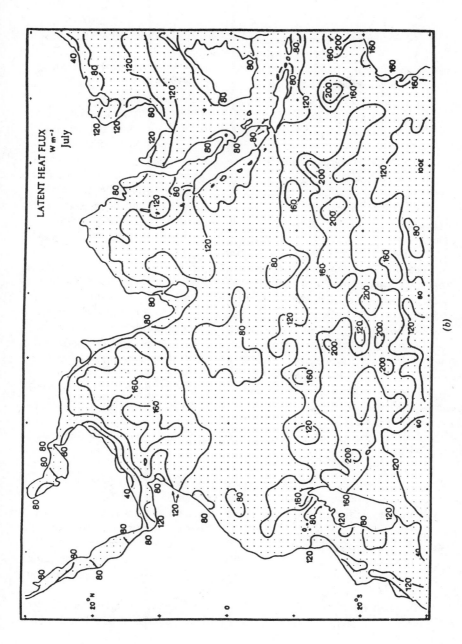

Figure 15.8. (*continued*) (*b*) is the same as (*a*) but for latent heat.

thus a fixed sea surface temperature model does not imply an altogether passive ocean. Specifying the characteristics of the land area in the radiative calculations is vital for the monsoon modeling effort (see Chapter 1). Here one has to consider the radiative heat balance of the earth's surface as it relates to the hydrological cycle. The elements include the calculation of the downward and upward short and long wave radiation, the fluxes of sensible and latent heat from the land surface (or vice versa) and the flux of heat into the deeper soil. It is necessary to specify the albedo of the soil in order to estimate the reflected short wave radiation. Using data sets from direct surface radiation measurements of upward and downward fluxes over various parts of the globe under different seasonal and surface conditions, Kondratyev (10), among others [e.g., (11)], has provided global albedo estimates. The soil heat flux estimates include the fluxes of sensible and latent heat. These estimates are obtained using theories of similarity analysis that have been developed in many boundary layer studies, for example (12). The analysis differentiates among the stable, neutral, and unstable stratifications of the model atmosphere.

The heat flux into the ground is based on empirical evidence and depends on having a reasonable model of soil temperature. Numerically, the radiative heat balance calculation is an elegant exercise where the soil temperature is usually solved by an iterative process for given measures of incoming solar radiation and downward long-wave radiation. A transcendental equation for the soil temperature is usually solved by standard numerical techniques such as the Newton–Raphson method and as a by-product it provides consistent measures of soil heat flux, assuming a small heat capacity of the soil.

It is important that general circulation models be capable of simulating the observed diurnal variability of the monsoon. This can be accomplished by a variable zenith angle of the sun. It is interesting to note that all of the diurnal variability in a general circulation model of the soil heat flux, long wave radiation, large-scale variables (wind, temperature, etc.), and cloud cover arises from this single and simple specification.

Figure 15.9 provides a schematic outline of the various components of the radiative forcing in a comprehensive general circulation model.

In addition to radiative heating and cooling, cumulus convection is a major element in the monsoon energetics. Convection does not contribute to the generation of strong winds directly. Usually the process takes a circuitous route: convection, if it occurs in relatively warm areas, and not as much in relatively colder areas, provides a strong covariance of convective heating and positive temperature departures. This is how *eddy* available potential energy is generated. Moreover, the large-scale mass and moisture convergence in regions of deep convection are associated with large-scale upward motions. Thus a strong positive covariance of upward motion and warm temperature is also found in these situations. Energetically, this implies a conversion of available potential energy to kinetic energy. In this manner, convection contributes to a strengthening of winds. It is not, a priori, evident that the energy released as a result of convection in a disturbance, for example a monsoon depression, would be used to drive only that disturbance. Studies suggest that most of the energy released by this process maintains the planetary-scale monsoon circulation. Pasch

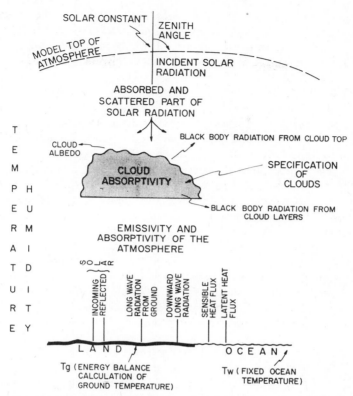

Figure 15.9. A schematic diagram of the radiative components used to calculate the net atmospheric heating or cooling in a general circulation model.

(13) has examined the maintenance of planetary-scale monsoons and concludes that convection in many disturbances is in fact organized in such a way as to provide most of its energy to larger planetary-scale motions.

Modelers are faced with the problem of how to represent the heat released by convection. There are two well-known methods, called parameterizations, currently in use. They were developed by Kuo (14) and Arakawa and Schubert (15). Both of these methods have undergone some modifications and have found wide application in large-scale modeling efforts.

The most widely used version of the Kuo scheme assumes that over regions of the atmosphere where certain convective instability criteria are met all of the available moisture supply goes into precipitation. The vertical distribution for the resultant latent heating is obtained by a simple formula based on thermodynamic principles. Kuo's method has been widely tested in general circulation simulations in the tropics and has been found to be quite reasonable when compared with observed convective heating distributions (16). Various versions of this method have been proposed; they all have several limitations. The most severe is that those versions that define the heating and rainfall rates adequately seem to underestimate the moistening of the atmosphere by the cumulus scale motion since most of the available large-scale water supply is used up for condensation. This can be rectified by the use of strong vertical diffusion, but this is, at best, an artificiality. New methods must be developed to overcome this deficiency.

The Arakawa–Schubert method is a complex scheme for the interaction of clouds and their environment. The generation of moist convective instability by large-scale processes is assumed to be in near-equilibrium with the destruction of this instability by the vertical flux of heat and moisture within the clouds. This is called a quasi-equilibrium hypothesis. The clouds modify the temperature and humidity of their (large-scale) environment through detrainment of saturated air and liquid water. At the same time the environment is warmed by subsidence around clouds. The subsidence compensates locally for net upward mass flux within the clouds. This upward mass flux obtained by solving an integral equation is based on the aforementioned quasi-equilibrium hypothesis (17). From the mass flux within the clouds it is possible to determine, sequentially, the heating by cumulus clouds and the rainfall rate. These are the three important ingredients of a cumulus parameterization procedure.

If the evolution of a phenomenon such as the monsoon crucially depends on convection, then vast differences in simulation would be expected depending upon how the convection is treated by the modeler. Results from a model that includes a good cumulus scheme can be expected to produce more realistic simulations and forecasts.

1.4.1 The Onset of the Southwest Monsoon.

Although a rigorous definition of the date of the onset of the southwest monsoon is not a simple matter, one can resort to a definition that is based on the rapid rise of area averaged rainfall or the arrival of moist air from the south (see Chapter 17). Recent observations have made it clear that one can also use the date of the rapid rise of area averaged kinetic energy over the Arabian Sea as a guide for this definition. Figure 15.10 shows the time evolution of the area averaged kinetic energy of the low-level winds over the Arabian Sea during 1979. The area averaged kinetic energy rises about one week prior to the rapid rise in the area averaged rainfall over India. This rise in the kinetic energy is one of the most spectacular aspects of atmospheric flows during the onset of the monsoon and has drawn the interest of general circulation modelers.

The onset is, of course, a part of the annual cycle of the large-scale circulation over the monsoon region and has been handled quite reasonably by many modeling groups. One in particular, the general circulation modeling group of the United Kingdom Meteorological Office, examined in some detail the rapid transition from the springtime to the summertime circulation. The group used a five-layer model with a grid spacing of 330 km over the entire globe. Their model contains detailed physical parameterizations including a cloud feedback cycle in their radiation calculations (18). It also includes an annual cycle of the sea surface temperature as well as a time varying zenith angle. Three-month integrations were made from an initial state based on real data for each of three years. Figure 15.11 illustrates the excellent simulation of the rapid increase in the winds over the lower troposphere

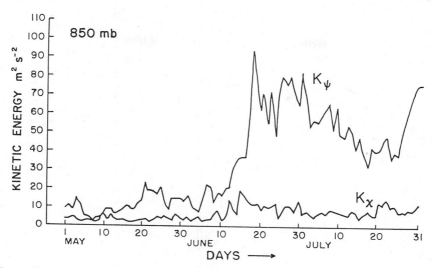

Figure 15.10. The explosive increase of winds over the Arabian Sea during the onset is seen in the time evolution of the rotational (ψ) and the divergent (χ) kinetic energy at 1.5 km above the sea level or the 850-mb level from May through June 1979. It is interesting and worth noting that the total kinetic energy (sum of K_ψ and K_χ) started to increase rapidly around June 11 over the Arabian Sea while the rainfall over central India commenced around June 18.

observed during the 1979 monsoon onset (cf. Fig. 15.10). The simulations are quite similar to observations for other years as well.

Although the three-month-long model run was not intended to provide a day-to-day prediction of the onset during 1977, 1978, and 1979, it served the purpose of illustrating that the physical processes responsible for the drastic rise of kinetic energy are contained within the prescription of the general circulation model. Thus these processes can be studied by means of detailed diagnoses of the output of the United Kingdom Meteorological Office's general circulation model or any other that demonstrates similar skill. Such a study was carried out by Pasch (13). Based on results from an 11-level, global spectral model, Pasch was able to determine the cause of the simulated rapid increase of kinetic energy of the low-level flows over the Arabian Sea. He noted that on the planetary-scale (zonal wave numbers 1–3) cumulus convection contributed to a significant generation and transfer of eddy available potential to eddy kinetic energy. Thus instead of why there is a rapid rise of kinetic energy during the onset, one must ask why there is a significant generation of available potential energy from cumulus convection.

This question was also addressed by Krishnamurti and Ramanathan (2). A gradual buildup and northward motion of the heat source east of India provides a field of differential heating which extends from the southwestern Indian Ocean to the foothills of the Himalayas. As this heating develops, major energy exchanges occur over the monsoon region. A sequence of energy exchanges, from convective heating to the generation of available potential energy to the release of divergent kinetic energy

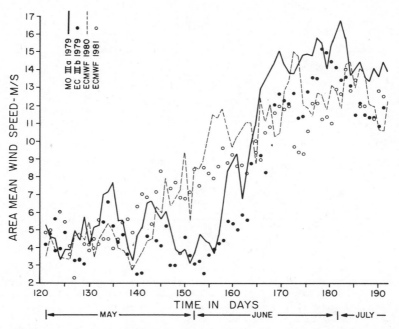

Figure 15.11. Area averaged wind speed during the onset of monsoon as seen from analyses of observations and from the results of numerical forecasts. Shown is the daily average low-level wind speed over the Arabian Sea. The heavy solid line shows the evolution of monsoon during 1979 as obtained from an operational analysis of observations by the Meteorological Office (MO) of the United Kingdom. The dashed line and the open and closed circles show the time variation of the area averaged wind speed obtained from forecasts (for 1980, 1979, and 1981, respectively) made by the European Center for Medium Range Weather Forecasts. All of the forecasts were initialized with analyses of May 1 data for their respective years. The dramatic increase of wind occurs in late May or early June.

and finally to the production of rotational kinetic energy, was shown to be of major importance during the evolution of strong low-level flows and eventually leads to the onset of the southwest monsoon. In this sequence, the first step, namely the generation of available potential energy from convective heating, depends on the covariance of convective heating and the air temperature. The magnitude of this covariance increases dramatically as the regions of convection move northward toward the regions of warm tropospheric temperatures over the heated Asian continent.

Prior to the onset of the monsoon, most of eastern India experiences hot, dry conditions and, although a surface low pressure trough exists in response to the heating, the area is relatively cloud-free because it is so dry. These conditions are akin to a heat-low with shallow lower-tropospheric convergence. However, shallow cumulus clouds gradually move into the region and, through condensation processes, the low-level heat content of the air gradually increases. When a threshold value of heating is reached, a sudden buildup of deep convection ensues and is accompanied by an increase in the generation of eddy available potential energy. This energy is rapidly transferred to the kinetic energy of divergent motions via the ascent of relatively warm air over the convective area and a compensating descent over relatively colder areas. The direction of the divergent motions suggests that this large-scale descent occurs over the western Indian Ocean, western Arabian Sea, South Africa, and the desert regions of North Africa and the Arabian peninsula.

The explosive growth of low-level winds over the Arabian Sea appears to depend crucially on the rate of transfer of the divergent kinetic energy to the rotational kinetic energy. The intensity of this transfer is determined by the relative orientations of the streamfunction (a measure of the rotational wind) and the velocity potential (a measure of the divergent wind). An orientation favoring the rapid transfer from divergent to rotational kinetic energy was noted in the 1979 observations collected during the summer monsoon onset period.

The Florida State University modeling group attempted to simulate the details of the observed energetics during the onset of the southwest monsoon using a grid point model. From observations they prescribed an initial state for pressure, temperature, and the rotational part of the wind field. In addition, a number of data sets for the divergent wind and the humidity distribution were extracted from different epochs of the pre- and post-onset periods. Each of these sets constituted initial conditions for separate experiments. Since the initial mass and moisture convergence was slightly different in each of these experiments, the initial convective heating was also slightly different.

The response of a model atmosphere with a pre-onset initial state to different configurations of differential heating between land and ocean was explored. First, it was found that, among the various components of diabatic heating (sensible, radiative, and convective), the convective heating had by far the largest effect in determining the structure and strength of the divergent wind. Second, the simulation experiments confirmed the observation that an explosive increase of the kinetic energy occurs when large heating (i.e., generation of eddy available potential energy) is found east of India around 20°N. This setting seems to provide an ideal geometry for the rapid transfer of energy, starting from convection and eventually reaching the rotational wind. Figure 15.12 illustrates this sequence of energy exchanges in schematic form.

1.4.2 The Upper Tropospheric Circulation.

During the Northern Hemisphere summer, the winds in the upper troposphere have well-known characteristic features which in the aggregate are referred to as the planetary summer monsoon circulation. They include an upper-level anticyclone which is usually located around 30 to 35°N over the Tibetan Plateau, an easterly jet which is found near 5 to 10°N to the south of this anticyclone and mid-oceanic troughs which are located over the central Atlantic and Pacific Oceans. These are shown in Figure 15.13a which is based on July observations (19). Figures 15.13b and c show the results of model simulations of the 200-mb July mean flows by Manabe et al. (20) and Hayashi (21). It is evident that the features of the planetary-scale monsoon circulations are essentially reproduced in this modeling effort. The success depends on having a nearly correct definition of the large-scale heat sources and sinks over the monsoon region.

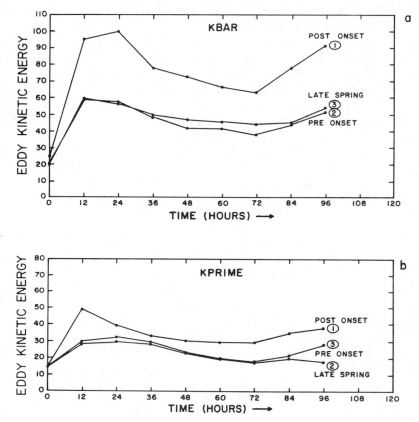

Figure 15.12. The change in various energy quantities during a period of the onset of the monsoon in 1979. The results of three separate numerical forecast experiments are shown. Different initial configurations of heating between ocean and land are determined by the choice of the moisture convergence. The curve labeled ''late spring'' represents heating from early May; ''pre onset'' represents heating conditions during a period just prior to the onset during 1979; and the ''post onset'' represents heating from late June, just after the onset of monsoon rains. The results are daily average values over a box contained between 30°E to 150°E longitude, 30°S to 40°N latitude and the entire troposphere. Panels (*a*), (*b*), and (*c*) illustrate, respectively, the time evolution of zonal kinetic energy (KBAR), eddy kinetic energy (KPRIME) and the conversion of eddy available potential energy into eddy kinetic energy.

When one compares these results (published in 1979) with those reported in the literature some 10 years earlier [see a review by Krishnamurti et al. (22)], we note that much progress has indeed been made. The improvements have come with better models (a spectral model versus a grid model), better vertical resolution, improved fields of sea surface temperature, improved physical parameterizations and a better representation of mountains.

Longer integrations will be necessary to study the interannual variability of the upper tropospheric summer monsoon circulation. In their long-term (18 years) general circulation integrations, Manabe and Hahn (23) noted considerable variation

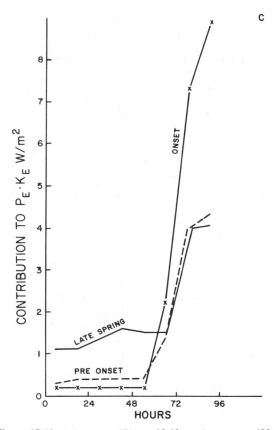

Figure 15.12. (*continued*) (*Figure 15.12 continues on p. 490.*)

in the monsoon circulation from year to year. A detailed analysis of their results on the monsoon has yet to be carried out.

Observations (24) suggest that during some years blocking situations (abnormal weather patterns that can persist for weeks) over western Europe can cause the middle latitude westerly current to move to lower latitudes. This can then lead to the establishment of an upper-tropospheric westerly jet south of the Himalayas which may often result in a delayed onset of, or a break in, the monsoon. Manabe and Hahn's long-term integrations include situations where the middle latitude westerlies indeed behaved in this manner.

1.4.3 Mountains. The Himalayas are the tallest mountains on earth, with individual peaks extending close to the upper limits of the troposphere. The *average* elevation of the massive Tibetan Plateau is over 4 km. Because of their dimensions and position, it has long been argued that the Himalayas must strongly influence the Asian monsoon. Simulation studies of the idealized zonally symmetric monsoon,

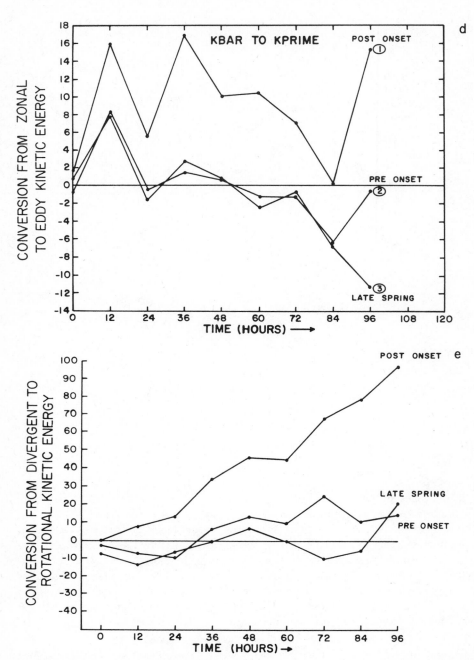

Figure 15.12. (*continued*) Panel (*d*) shows the time evolution of the barotropic energy conversion from zonal to eddy kinetic energy. Panel (*e*) shows the conversion from the divergent to the rotational kinetic energy. The units of energy quantities are m²/sec² while the energy conversion is expressed in W/m².

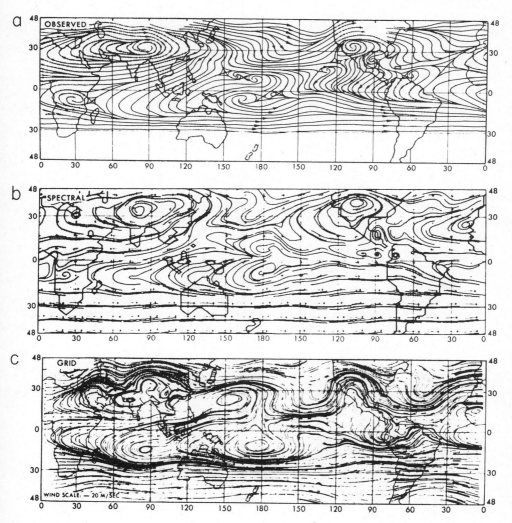

Figure 15.13. The observed and predicted July monthly mean flow field at 200 mb (close to 12 km above sea level). Panel (*a*) shows the observed circulation based on the analysis of Sadler (19). Panels (*b*) and (*c*) show the flow fields produced by a global spectral model (20) and a global grid point model (21).

such as those by Murakami et al. (1), gave the first direct evidence of the importance of the Himalayas in the Asian monsoon circulation. Such studies examine the results of a pair of general circulation model simulations: one with and one without mountains, keeping all other features of the model identical. Further complexity can be added when different representations (maximum height and smoothing) of the mountains are used. One hopes to learn about the effect of the mountains on the monsoon circulation and also how to most efficiently represent the mountains in the model.

Murakami and his coworkers noted that the inclusion of the mountains was essential for the simulations of (a) realistic southwesterly flows in summer in the lower troposphere over India, (b) the upper-level easterly jet, and (c) a realistic warm upper-tropospheric thermal field. The inclusion of mountains altered the meridional distribution of heat sources and sinks in the model, providing an elevated source for latent heat release and sensible heating. Of the two, the latent heat release associated with the orographic component of rain over the mountains was the major component of the heat source and contributed strongly to the model's ability to simulate a warm tropospheric thermal field that was consistent with the observed pressure distribution and circulation over the monsoon domain. In the absence of mountains the simulated monsoon was unrealistically shallow and confined below the 3-km level.

As mentioned earlier there are some inherent limitations in the zonally symmetric monsoon model, such as the one used by Murakami and his colleagues (1). We shall next discuss the results of a fully three-dimensional model that does not assume zonal symmetry. It should be noted that three-dimensional models allow more realistic interaction on many scales. For example, interactions of the monsoon and disturbances over the Pacific Ocean are possible. Over the middle and higher latitudes, frontal cyclones and associated upper-wave disturbances are better described as are interactions between the middle latitudes and tropics. Mountains are fully three dimensional, thus the influence of the Himalayas and the other mountain chains of the monsoon region can be described more realistically.

Hahn and Manabe (25) carried out the first promising three-dimensional simulations to investigate the role of mountains. They used an 11-level grid point, general circulation model with fairly complete physical parameterizations. The model included an annually varying field of sea surface temperature which was fixed for each month. It was integrated to 3.5 years from an initial state of rest and a uniform pressure of 988 mb. The simulation that included mountains successfully reproduced the observed large-scale flow features at low-levels except for an unusual frequency of storms over the western Pacific Ocean. This affected the pattern of tropical depressions over the Bay of Bengal and, in turn, the rainfall distributions over the monsoon region. Most of central and eastern India was dry; the majority of the precipitation occurred over southern India and over the Tibetan Plateau. This is similar to what we call a break pattern (see Chapter 11). More recently, Manabe et al. (20) repeated these experiments with a spectral general circulation model and major improvements were found in the rainfall patterns and in the representation of the tropical depressions over the Bay of Bengal. Since the only difference between these and the earlier experiments was the change from a grid point to a spectral model, it appears that the improvement was due to the use of a spectral model.

In their simulations without the mountains, Hahn and Manabe noted many peculiarities. The lower-tropospheric monsoon trough moved from 20°N, 90°E (its position with mountains and roughly its observed position) to 50°N and 125°E. The entire Indian subcontinent experienced northwesterly flows which were dry and the moist Southern Hemisphere air did not penetrate north of the equator. The warmest tropospheric temperatures occurred near 5 to 10°N, compared to 30°N when the mountains were included (and as is observed). In the absence of mountains, air temperatures over the Tibetan Plateau were about 12°C cooler than observed values at the 500-mb surface. On the whole, when comparing the observed and the simulated monsoon for experiments with and without mountains, the importance of mountains is quite evident.

Many issues concerning the role of orography cannot be handled by general circulation models because of their relatively gross resolution, but must be studied by high-resolution regional models. The importance of the East African highlands and the Ethiopian mountains for the simulation of the Somali jet was mentioned in Section 1.2. The importance of the western Ghats in rainfall distributions for western India near shore and offshore has been studied by Sarker (26). He used a simple linear mountain wave model to estimate orographically induced vertical motions which were, in turn, used to determine the rainfall rates using simple thermodynamic principles. Figure 15.14 from his study shows reasonable agreement between observed

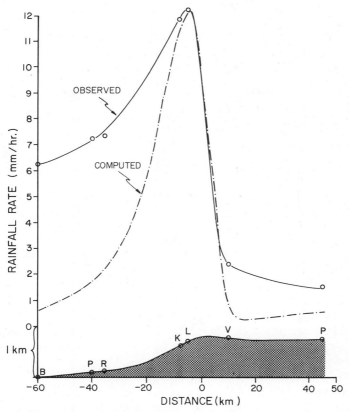

Figure 15.14. The orographic rainfall as computed and observed across the western Ghats of India. The units of rainfall are millimeters per hour. The distance along the abscissa is in kilometers. The weather stations from left to right are shown by the following letters: B (Bombay), P (Pen), R (Roha), K (Khandala), L (Lonavla), V (Vadgaon), P (Poona). The computed values are based on the work of Sarker (26).

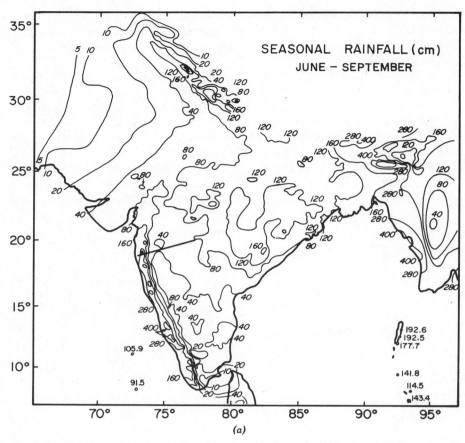

Figure 15.15. (*a*) The observed seasonal mean rainfall for the months of June–September. The lines are drawn at an interval of every 40 cm/season. Note the large orographic component of rain along the west coast of India and along the foothills of the Himalayas, especially north of the Bay of Bengal.

and calculated estimates across the Ghats from Bombay to Poona. The summer monthly mean patterns of rainfall over the Ghats, shown in Figure 15.15*a*, clearly illustrates the important role of orographic monsoon precipitation. (A detailed picture of the orography is presented in Fig. 15.15*b*.) Regional orographic effects also tend to produce stationary troughs over the Bay of Bengal (27); they may have a strong influence on monsoon disturbances.

1.4.4 Sea Surface Temperature Anomalies. A most promising area of research is the study of how sea surface temperature (SST) anomalies (departures from normal) over the western Arabian Sea and over the Pacific Ocean affect planetary-scale monsoon circulations. From a modeling perspective, the procedure usually consists of making two long-term model integrations. The first uses normal SST

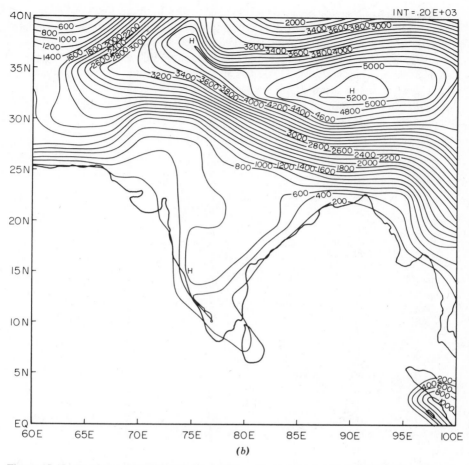

Figure 15.15. (*b*) The smoothed height of the earth's surface above sea level at different locations of the Asian monsoon region. The units are meters. Along their western slope, the Ghats rise from sea level to about 800 m. The Himalayas rise to above 5200 m. It is of interest to note that regions of heavy rain roughly correspond to mountain regions with large slope.

to obtain a hopefully realistic simulation of the monsoon, that is, the monsoon trough, monsoon westerlies, rainfall regions, tropical easterly jet, and the warm troposphere. The Tibetan anticyclone and the depth of the moist layer are to some degree faithfully simulated. This is essential if the second simulation, the anomaly experiment, is to be physically meaningful. However, the simulation is not always good and under such circumstances SST anomaly experiments are not very meaningful. The anomaly experiment is run with an identical model, the only difference is that an SST anomaly of a few degrees Celsius is introduced over a region known to exhibit such features climatologically.

Figure 15.16 illustrates some typical SST anomalies from the study of Washington et al. (28) who were interested in the question of local versus remote atmospheric response to SST anomalies. Among the three anomaly experiments shown (anomalies of −3°C, +1°C and +3°C), they found the maximum remote atmospheric response when a positive anomaly of +3°C was introduced over the central Arabian Sea (Fig. 15.16c). In this instance, the response consisted of suppressed vertical velocity and a statistically significant reduction of precipitation over Malaysia and the equatorial western Pacific Ocean.

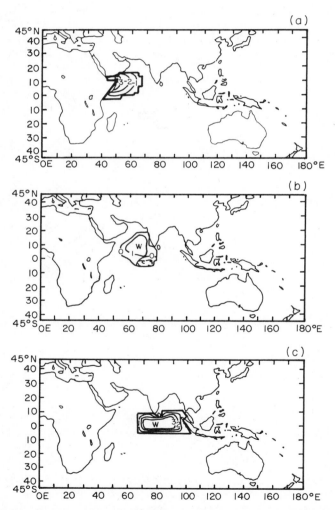

Figure 15.16. Locations of the sea surface temperature anomalies introduced by Washington, Chervin, and Rao (28) in their three general circulation experiments. The maximum values of the sea surface temperature anomaly in the three experiments are (*a*) −3°C, (*b*) +1°C, and (*c*) +3°C.

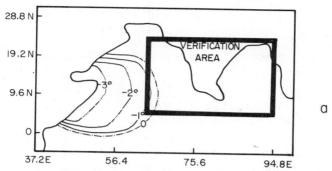

SEA-SURFACE TEMPERATURE ANOMALY (°C) OVER THE WESTERN
ARABIAN SEA AND LOCATION OF VERIFICATION AREA.

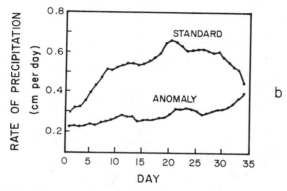

RATE OF PRECIPITATION (CM PER DAY) AVERAGED OVER THE
VERIFICATION AREA FOR THE STANDARD AND ANOMALY RUNS.

Figure 15.17. (*a*) The sea surface temperature anomalies introduced by Shukla in his experiments (29).
The cold anomaly has a magnitude of 3°C. The box shows the verification area for the two experiments,
standard and anomaly. (*b*) The area averaged rainfall rates in cm/day for each experiment.

Warm SST anomalies of up to a few degrees centigrade have recently been
observed over the subtropical latitudes of the southern Indian Ocean. Their impact
remains to be explored.

Shukla (29) used a general circulation model to study the impact of a cold
anomaly over the western Arabian Sea. Figure 15.17 shows the location and intensity
of this anomaly and the differences in the area averaged precipitation for the control
as well as the anomaly experiments. It appears from this experiment that the area-
averaged precipitation is reduced somewhat in the anomaly experiment. This leads
to the hypothesis that colder water temperatures result in a reduction of evaporation
over the Arabian Sea which in turn reduces the moisture supply for the precipitation
over India. Since the difference between the control and the anomaly experiment
decreases, especially after day 30, the question arises whether the result represents
a transient response related to the spin-up of the model during the first month.

Washington et al. (28) found a negligible response in a similar experiment using a different general circulation model (cf., Fig. 15.16*a*).

The impact of warm sea surface temperature anomalies over the equatorial Pacific Ocean is discussed later in conjunction with the Southern Oscillation.

1.4.5 Snow Cover in High Latitudes. One would expect the areal coverage of snow cover, especially during winter and spring months, to have some influence on the monsoons during the following summer months. If the snow cover, especially over the Himalayas, is large, the heating of the Plateau will be slower (due to the high albedo of snow) and, as a consequence, the establishment of the warm upper tropospheric Tibetan high-pressure cell can be slowed down. This can result in a delayed onset of the southwest monsoon and a below normal rainfall season as well. There are no published results on the modeling of this aspect of the monsoon variability. In order to study this problem, a climate model would have to be integrated over several decades.

1.4.6 Monsoons and the Southern Oscillation. The Southern Oscillation is an alternation in the sea level pressure difference between the eastern and western Pacific Ocean. This topic has been addressed by Webster (Chapter 11) and Shukla (Chapter 16). As stated earlier it was first documented from surface pressure records at Darwin and the Easter Islands. The period of this oscillation is between two and six years.

The normal long-term average sea level pressure is lower over the western Pacific Ocean than it is over the eastern ocean. In response, an east–west circulation flows on a zonal plane near the equator, characterized by ascent over the western Pacific and general descent over the eastern Pacific; it is called the Walker circulation and has been discussed earlier in this volume (see Chapter 11, Section 1.2.1). During the Southern Oscillation cycle the sea surface pressure over the eastern Pacific Ocean becomes lower than normal and the Walker circulation weakens or may even reverse its direction. A typical sequence of events is as follows:

1. the areal extent and the strength of the trade winds increase;
2. some time later, the trades diminish and the westerly low-level winds increase in strength over the western Pacific;
3. warm sea surface temperature anomalies occur over the eastern and central Pacific Ocean;
4. the Walker circulation weakens or reverses direction;
5. with further lowering of pressure over the eastern Pacific the west–east pressure differential weakens;
6. the warm sea surface temperature anomaly expands somewhat in extent and starts spreading westward along the near-equatorial latitudes;
7. the region of upward motion and convection spreads westward following the warm sea surface temperature anomalies; and
8. large-scale changes in the upper-tropospheric circulations are observed (30).

During the 1972 and 1983 episodes of El Niño as the warm SST anomaly advanced towards the central Pacific Ocean, westerly wind anomalies at 200 mb were observed to propagate eastward across Central America towards Africa and South Asia. They accounted for a diminution of high-level easterlies over the regions of the summer monsoons (30, 31). Currently, there is much interest in this entire problem from a modeling perspective since it has some important implications for global climate.

A numerical simulation of the atmospheric behavior during westward propagation of warm SST anomalies was modeled by Keshavamurty (32). In his study he placed a warm SST anomaly of 3°C amplitude over the eastern, central, and western Pacific Ocean (Fig. 15.18) in three different simulation experiments with the general circulation model described in Manabe et al. (20). This global spectral model has 12 waves in both the north–south and east–west directions and nine vertical levels. It includes

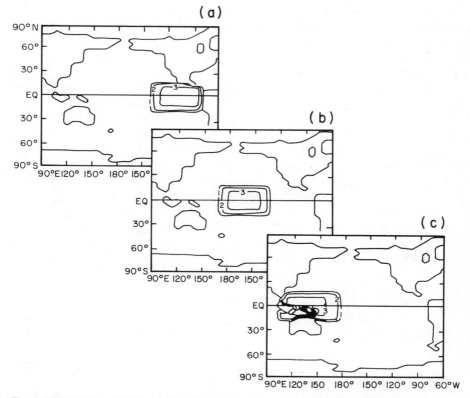

Figure 15.18. The configurations of sea surface temperature anomalies that were used in three separate general circulation experiments by Keshavamurthy (32). These are (a) the eastern Pacific anomaly, (b) the central Pacific anomaly, and (c) the western Pacific anomaly. The amplitude of the anomalies were 3°C in each experiment.

a complete physical parameterization as well as an annual cycle of prescribed SST. The SST anomalies are added to the annual cycle and kept fixed in each experiment.

The most striking differences in the three experiments were over the tropical upper-troposphere. Here, for each experiment, the time averaged 200-mb winds for June, July, and August were compared with a control experiment. The control, as defined earlier, is a separate simulation experiment where no sea surface temperature anomalies are inserted. The time averaged vector wind differences with respect to the control for the three experiments are shown in Figure 15.19. The insertion of the SST anomaly in the eastern Pacific produces an interesting wind difference pattern in its vicinity which consists of a clockwise (anticyclone) gyre to the north of the equator and a counterclockwise (anticyclone) gyre south of the equator. In between these gyres there is an easterly wind maximum. To the north of the northern gyre and to the south of the southern gyre, the normal (control) westerly winds are enhanced by westerly wind anomalies. When the anomaly of SST is placed over the central and western Pacific Ocean, the anomalous gyres still occur, roughly in the same juxtaposition with the SST anomaly, but the westerly wind anomaly expands equatorward and to the west over the monsoon regions of Africa and Asia. These last simulations are quite similar to what was reported by Pan (30) from his analysis of observations during the 1972 El Niño.

It is also of considerable interest to note that a local anomaly in one region of the equatorial tropics seems to have a major effect on the circulations around the globe. This has been reported in observational studies reviewed by Hoskins and Karoly (33) and discussed at length by Webster in Chapter 11. Tropical heat sources, such as those used in these model anomaly studies, evidently excite a geostationary train of Rossby waves which propagate to the middle latitudes. In addition, Gill (34) found theoretically that the heat sources would excite a Kelvin-like wave to the east and a mixed Rossby–gravity wave to the west of their positions in the equatorial latitudes. The structure of the circulation anomalies, as simulated by Keshavamurty's model experiments, are indeed consistent with these findings.

Ideally, instead of *imposing* an SST anomaly to study the interaction between the Asian Monsoon and the Pacific Ocean circulations, one would like to have a sophisticated general circulation climate model from which the SST anomalies would arise as a result of internal dynamics. While examining the output of Manabe and Hahn's (23) general circulation model integrations, Lau (35) noted that some of the large-scale interannual features such as the Southern Oscillation were not detectable in the 18-year simulations. Lau concluded that realistic simulation of the Southern Oscillation requires a coupled ocean–atmosphere model to fully represent the feedback processes. Development of such coupled models seems to be several years away. Since observational evidence shows the Southern Oscillation to be an important modulator of the monsoon (see Chapter 14), its realistic modeling is extremely important. An urgent task facing atmospheric scientists is the development of coupled ocean–atmosphere general circulation models that are capable of realistic simulation of phenomena like the Southern Oscillation and its interaction with the monsoon.

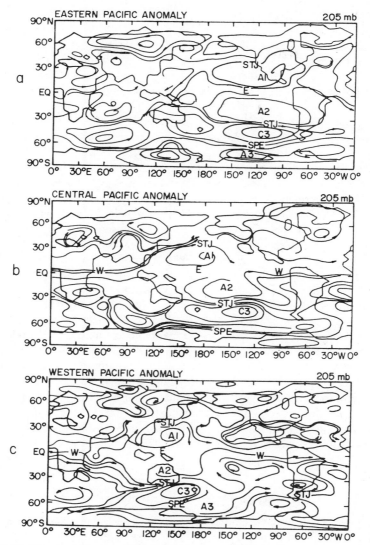

Figure 15.19. The circulation anomalies that evolved at the 205-mb level (about 12 km above sea level) in Keshavamurthy's modeling experiment (32) when the sea surface temperatures shown in Figure 15.18 were introduced over the eastern, central, and western Pacific Ocean. A feature to note is the 205-mb easterly flow above the sea surface temperature anomaly. On the north side of these easterlies is a clockwise circulation (an upper anticyclone). To its south a counterclockwise circulation (a Southern Hemisphere upper anticyclone) is found. These anticyclones are marked by the letters A1 and A2. STJ denotes a subtropical jet stream which is located to the north of the A1. Also seen along the equator are westerly wind anomalies (W) which extend over a large region. These anomalies emanate from the region of the warm sea surface temperature anomaly and extend eastward over a great distance. This represents an equatorial Kelvin wave.

2 PREDICTION MODELS

Numerical prediction is quite different from numerical simulation. Where simulation models are usually started with the atmosphere at rest, prediction models are "initialized" with current weather observations. Moreover, prediction models use current weather data to "update" or correct the forecast from time to time during the integration; simulation models do not.

Much progress has been made over the past several decades toward the application of statistical, empirical, and dynamical methods to predicting the behavior of the monsoon.

The statistical approach holds some promise for the seasonal forecasts of monsoonal activity. Since Shukla and Das discuss statistical long-range prediction in Chapters 16 and 17, respectively, this text will concentrate on empirical and dynamical methods. The empirical approach has been used for nearly a century for weather prediction over the Indian subcontinent. In recent years, with the availability of better data, digital computers, and weather satellites, this approach has advanced significantly. Dynamical weather prediction has benefited from the development of a hierarchy of models. Each type of model has provided an important contribution to overall model development. It is interesting that most major operational weather centers in the mid-latitude regions run three to four different dynamical weather prediction models. For example, the U.S. Weather Service makes daily forecasts using three different models. Why are so many models used? Each model serves a different practical purpose. A simple barotropic model provides a quick reference on the wind and vorticity evolution over the middle atmosphere. A limited-area fine-mesh (LFM) model (a multilevel baroclinic primitive equation model) is used for short-term forecasts of severe weather and rainfall forecasts. The global spectral model (also a baroclinic primitive equation model) provides short- to medium-range weather prediction. The point is that the type of model used depends on the kind of forecast that is required. We will see that a similar hierarchy of forecast models exists for the monsoon region.

2.1 Empirical Approaches to Monsoon Prediction

Over the years a considerable amount of experience has accumulated in the practice of weather prediction in monsoon regions. (See Chapter 8 for an historical perspective.) This has resulted in the evolution of a number of guiding rules, or models, that are not entirely based on dynamical principles. The wide application and usefulness of these models have made an impact on the monsoon prediction problem.

The axis of the monsoon trough over northern India oscillates north–south with a period of roughly two weeks. A well-established empirical rule relates active and break periods of the summer monsoons to this oscillation; the southerly position (around 15°N) represents an active period and the northerly position (around 25°N), a break period. Favorable conditions for the formation of a monsoon depression with its attendant heavy rains generally exist when the trough is located in its more southerly position. The position of the upper-tropospheric easterlies and the related easterly jet is also known to exhibit north–south variations in its mean latitude. When the jet axis moves closer to the equator, troughs in the westerlies tend to move eastward, bringing westerly winds south of the Himalayas. During these periods the axis of rainfall maxima shifts from around 20°N to double axes near 5°N and around 25°N, leaving central India with less than normal rainfall. Synoptic meteorologists watch carefully for signs of motion of the monsoon trough and the upper easterly wind belt to provide guidance for the development of dry or wet spells in India.

Another well-established empirical rule is based on the strength of the low-level southwesterly winds over the Arabian Sea. A dramatic increase in their speed signals the approach of the onset. In the past the practical use of this empirical relationship was limited because of the dearth of conventional wind information over the Arabian Sea. However, a major data source has emerged in the last decade, the cloud-drift wind estimates obtained from the geostationary satellites.

Cloud-drift can be estimated at two vertical levels, at low-levels and at the tropopause level. These estimates are obtained by tracing the position of the low-level stratocumulus cloud elements and the structures within the higher level cirrus clouds, respectively. These cloud "markers" maintain an identity for periods of about 30 minutes. The successive positions of these markers provide estimates of the wind speed. In 1983, the Government of India launched a geosynchronous weather and communications satellite (INSAT) which provides the images necessary to calculate low-level winds over the monsoon oceans. (See Fig. 15.20 for an

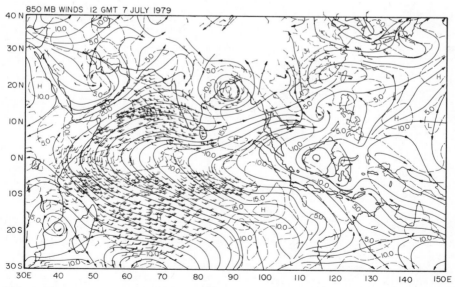

Figure 15.20. A typical flow field in the lower-troposphere during the summer monsoon determined from a variety of observing platforms including satellites. The solid lines denote streamlines while the dashed lines denote lines of constant wind speed. During this time there is a monsoon depression over the northern Bay of Bengal.

example of the remarkably high density of observations provided almost continuously by such satellites.)

The pressure differential between land and ocean builds in a dramatic manner during the onset period and provides a useful empirical forecasting model. Kuettner and Unninayar (36) have examined the sea level pressure difference between Diego Suarez Island (north of the Malagasy Republic in the western Indian Ocean) and Bombay (on the west coast of India). Figure 15.21 shows the time evolution of this parameter for the years 1976 through 1980. The dates of onset of monsoon rains over Trivandrum (8°N) and Bombay (19°N) are shown by vertical lines. There is an increase of the pressure differential from around 6 mb prior to the onset to around 9 mb at the time of the onset at Trivandrum. The transition period for the pressure change is very rapid and a similar pattern is noted for each year examined. In addition, except for 1976, the onset at Bombay occurred when the pressure difference reached 12 mb. The correlation coefficients between calendar day and

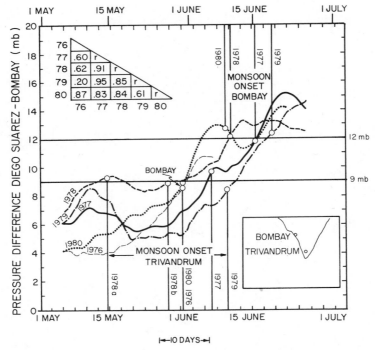

Figure 15.21. Pressure difference (in mb) between two points—one at Diego Suarez (near Madagascar) and the other at Bombay—for the years 1976–1980 based on the study of Kuettner and Unninayar (36). Also shown are the dates of onset of monsoon rains over Bombay and Trivandrum (see inset map for approximate locations). The inset on the top left shows the correlation coefficient of the pressure difference for the different years illustrating the nature of these curves is quite similar. In general when the pressure difference approached 9 mb, the onset of monsoon rain commenced over Trivandrum and when this difference approached 12 mb, the rainfall commenced over Bombay.

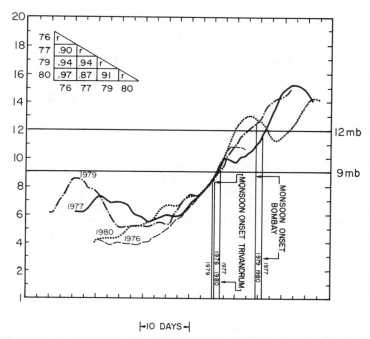

Figure 15.22. The pressure difference between Diego Suarez and Bombay is shown as a function of days (along abscissa). The curves of the previous diagram have been shifted so that they intersect 8 mb at the same point. This composite curve can be used as a nomogram. About two days after the 8-mb pressure difference is realized, the onset of rains over Trivandrum can be expected. The Bombay rains are expected about eight days after the 8-mb pressure difference is realized (36).

pressure difference for different years are shown at the upper left. Based on this result, Kuettner and Unninayar defined a threshold value of the pressure difference (9 mb) that determines the date of the onset of monsoon rains over Trivandrum during these years. In order to demonstrate the importance of the threshold pressure they shifted the date of onset for the late monsoon years (1977 and 1979) by a few days (Fig. 15.22), so that all curves intersect at around $p = 9$ mb. In this case the correlations are generally of the order of 0.9 or higher. This suggests the important predictive value for the threshold value of the pressure difference and the usefulness of the composite Figure 15.22 which may be used as a model for empirical prediction of the onset dates at Trivandrum and Bombay.

Rao (37) defined the date of withdrawal based on the southward displacement of the surface trough and the establishment of dry continental air over northern and central India. Figure 17.3 shows the normal dates of withdrawal of the monsoon. The interannual variability of this date is quite large and is controlled by the planetary-scale circulation as well as the oceanic thermal state. Although the interannual variability is large, the southward phase propagation is usually quite regular once this pattern of progression is established; thus the prospects for forecasting the withdrawal via empirical methods appear promising.

Forecasters have noted that during this period the subtropical high-pressure belt which is usually over the western Pacific Ocean tends to move westward and the associated easterly trade winds penetrate westward as well. As a result, the monsoon trough tends to weaken and move farther south. The movement of the western Pacific high-pressure belt is, therefore, a good empirical indicator of the withdrawal of the monsoon.

2.2 Dynamical Approaches to Monsoon Prediction

Major advances have been made over the past several decades in the area of dynamical or deterministic forecasting, often referred to as numerical weather prediction, using many kinds of models, from simple two-dimensional to multilevel, global spectral models. Of course, only the sophisticated global models are capable of carrying out medium range (3–10 day) dynamical predictions. Two areas of current emphasis which show much promise are the prediction of monsoon depressions and the prediction of the onset of monsoon rains a week in advance.

In this section we shall discuss these dynamical models. Here one considers a closed system of hydrothermodynamical equations with appropriate boundary and initial conditions which is used to make forecasts of the winds, temperature, humidity, and pressure. Forecasts are carried out for 2 to 5 days, but may be extended up to 10 days with global models. The forecasted variables are used at frequent intervals (e.g., every three hours) to provide estimates of quantities such as cloud cover and rainfall rates which are then used as input for the next cycle of the forecast.

2.2.1 Short-Range Forecasting with Simple Conservation Principles. There are two well-known conservation laws of atmospheric physics which have been used successfully for 2- to 3-day forecasts of the tropical motion field. The first of these is called the law of conservation of absolute vorticity. This law states that the absolute vorticity of a parcel of air does not change as it moves from place to place provided no external forces (friction and heating) act upon it. The absolute vorticity is the parcel's relative vorticity or spin plus the spin of the earth at the location of the parcel. There is an exchange among the shear, curvature, and the earth's vorticity within a parcel of air so that their sum is invariant. The forecast model uses the conservative vorticity equation written in terms of the rotational component of the wind. It is the evolution of the rotational component (which accounts for most of the total variance of the wind) that is predicted. This model is two-dimensional, that is, purely horizontal with no variations in the vertical direction. Although the model is deceptively simple, it is capable of forecasting weather developments that often occur when the shear vorticity of the low-level southwesterly (Somali) jet and/or upper-tropospheric easterly jet is transformed into curvature vorticity, or the cyclonic spin of storms.

The other conservation law is for the potential vorticity of a parcel. It requires that in the absence of external forces, as a parcel moves, the product of its absolute

vorticity and stability remains invariant. This model is (barely) three dimensional—
it has two layers in the vertical. The model contains three equations: the zonal and
the meridional equations of motion and the mass continuity equation. The three
basic variables of the model are the two horizontal velocity components and the
height of the free surface or a pressure surface. Motion and development of storms
can occur as follows: as parcels of air move through a region of horizontal convergence,
their stability decreases while their absolute vorticity (a measure of the storm's
strength) increases so that the product of the two is conserved. The converse, of
course, is the case for motion through regions of horizontal divergence.

The construction of these numerical weather prediction models requires careful
design of computational algorithms. These are described in standard textbooks on
numerical weather prediction, such as Haltiner and Williams (38), as well as in a
number of excellent papers (39–42).

The performance of a model can be assessed by a number of methods. For the
tropics the most commonly used is the root mean square (rms) correlation. The rms
error is the square root of the mean of the squares of the differences between the
forecast and the observations. The model climate (time averaged forecast) and the
observed climate are usually quite different and this could lead to large rms errors.
Often, the prediction of transient motions is most important. In this sense, forecast
skill can also be assessed from anomaly correlations where the anomalies are
departures from the climatic states.

Few forecast experiments have been carried out in the research mode for the
Indian subcontinent. The operational forecast centers in India and the global operations
modeling groups of the United States, Japan, the United Kingdom, and the European
Centre for Medium Range Weather Forecasts routinely carry out operational (as
contrasted to research) predictions over the monsoon region. Currently, few error
statistics for this region are available from any of these groups. (However, statistics
for a limited period are discussed by Das in Chapter 17.)

Because of the available observational research data from the Global Atmospheric
Research Program, the largest number of experimental predictions have been carried
out over the region of the West African monsoon. The summary of the error statistics
for 86 experiments, carried out with the special data sets of the GARP Atlantic
Tropical Experiment of 1974 and the Global Weather Experiment of 1979 indicate
that the performance of the absolute vorticity conservation model is an improvement
over persistence (albeit limited) in the first 48 hours of forecasts. The performance
of the potential vorticity conservation model is better than persistence for forecasts
up to 96 hours. The difference in performance can be attributed to the inclusion of
the effects of divergence and mountains in the latter case. We noted earlier that
the mountains over Cameroon, Guinea, and the North African Mountains over
Mauritania, the Ahaggar, and Tibesti do have a downstream influence. Furthermore
the simpler model uses the purely nondivergent wind which only slightly underestimates
the total wind initially and subsequently.

Figure 15.23 illustrates an impressive 96-hour forecast over the West African
monsoon region from a single level (700 mb) absolute vorticity conservation model.
Shown are the observed and forecast locations of two African disturbances up to

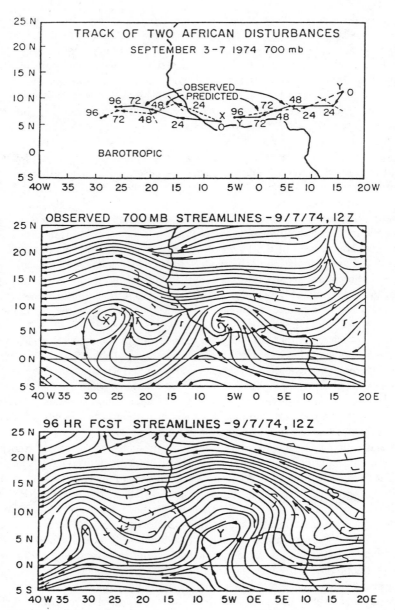

Figure 15.23. Top panel shows the observed (solid lines) and predicted (dashed lines) tracks of two African wave disturbances labelled X and Y. The positions of vorticity maxima every 24 hours are indicated. Middle panel shows the observed streamlines (on day 4) at 700 mb (roughly 3 km above sea level). Bottom panel shows the forecast wind field on day 4. These forecasts were made utilizing a simple model (the barotropic model with conserved absolute vorticity). For both the positions of the two storms are marked by the letters X and Y. A comparison shows that the African waves were predicted reasonably well by this simple model.

hour 96. Figure 15.23 (middle and bottom panels) also shows the details of the 96-hour observed and forecast streamlines of the circulation. Note the excellent agreement between observations and forecast for both the position and structure of the two disturbances (marked X and Y). In general, forecasts of this quality with this model are possible only for the first 48 hours.

Figure 15.24 shows an example of forecasts from a single-level potential vorticity conserving model. The effect of bottom topography (mountains) on the westward motion of a monsoon depression was examined in this series of experiments. In the absence of mountains the monsoon depression remained nearly stationary over northeastern India. The diagnosis of the forecast revealed that this was due to a balance between two terms in the model's governing (absolute vorticity) equation: the beta effect (a measure of the advection of the gradient of the Coriolis parameter) and the negative vorticity tendency in that region. However, when mountains were included, the northerly flow northwest of the storm center became downslope off the southern edge of the Himalayas. This resulted in the static stability in that region decreasing as the mean convergence increased, and an increase of the absolute vorticity followed under the conservation of potential vorticity. This increase of absolute vorticity made the storm move northwestward.

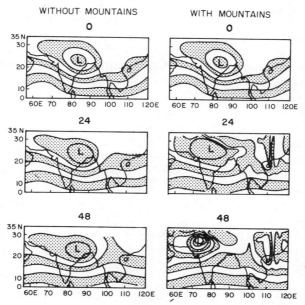

Figure 15.24. Two 48-hour forecasts of the 700-mb height fields, without and with mountains, showing the motion of a monsoon depression. The slope of the mountains has a strong influence on the northwestward motion of the monsoon depression. In the absence of mountains the depression remains nearly stationary (left) for the entire 48 hours. However, when the mountains are included (right) the storm shows a very reasonable northwestward motion. The model used for these studies is called the single-level primitive equation model and includes smoothed mountains. The contours are those of the height of the free surface of the single-level model. The interval between isopleths is 25 m.

Although this example is a fine illustration of the importance of including topographic effects in models of storm movement and development, it is not the only important effect. The role of heating by cumulus convection may play a major role in the westward motion of these storms. However, simple conservation models do not have the relevant physics and more complex models are required.

2.2.2 Short-Range Forecasting with Multilevel Models.

Short-range prediction is most important during periods of rapid changes of weather, for example, the transitions between dry and wet spells or the monsoon onset. Usually one is concerned with intense systems such as tropical storms that cause considerable loss of life and property and damage to crops. Tropical storm prediction requires a dense data network to define the initial state and a high resolution numerical model to carry out the prediction. Current prediction efforts are hampered by the paucity of data available in the tropics. The models must handle convective processes well. The monsoonal large-scale flow fields are very sensitive to cumulus convection and different ways of specifying the cumulus scale heating can lead to a somewhat different evolution of the circulation. The models must also be sophisticated enough to resolve energy generation processes from horizontal and vertical shear of the local mean flows.

Physical processes in multilevel forecast models are similar to those in the general circulation simulation models discussed in Section 1.4. The major difference is that, for the former, current weather observations are used as the starting point or initial fields. Weather observations are not uniformly distributed in time and space, therefore it is impossible to describe the state of the atmosphere perfectly at any given time. Thus a detailed initialization procedure is necessary. If the analysis of the observations of pressure and wind has large errors due to the sparsity of data, then a large, erroneous imbalance between the pressure gradient force and Coriolis force will likely be present. Forecasts, carried out from such (imbalanced) initial states, would contain large, erroneous accelerations of winds. Initialization procedures are based on physical laws that impose certain restrictions on the wind–pressure balance relationship. The purpose of initialization is to prevent these excessive initial accelerations of the wind which usually manifest themselves as large-amplitude fast-moving gravitational (not Rossby waves) modes. The presence of these spurious modes usually causes a forecast to deteriorate in a matter of 24 hours.

There are three procedures that are currently used by operational and research scientists, namely (a) static initialization, (b) dynamic initialization, and (c) normal mode initialization.

In the static initialization procedure one simply makes use of the geostrophic or the gradient wind laws to establish a relationship between the wind and the pressure field. In the tropical static initialization problem the wind analysis is usually given and the pressure field is deduced from the aforementioned laws.

The dynamic initialization does not restrict the balance between wind and pressure to be either geostrophic or gradient. This method entails a forward–backward integration of the dynamical equations around the initial time. The forward–backward integration results in a more realistic wind pressure balance that is valid for the tropical latitudes.

The normal mode initialization procedure examines the observational data in terms of the normal modes of a prediction model. If large pressure–wind imbalances are present then one finds in the spectrum of normal modes a large family of fast-moving high-frequency modes. In the initialization procedure, in principle, one can discard unwanted modes.

A number of operational and research groups participated in an intercomparison experiment on modeling the onset of the 1979 Asian summer monsoon. Global forecasts were carried out with each modeling group using identical initialized data sets and three different types of cumulus parameterization schemes: soft convective adjustment, Kuo's schemes, and a plume model with entrainment.

Most of the models were spectral. Each group carried out a 7-day forecast. The evaluation included the forecast of the formation of the onset vortex, the establishment of deep moist westerlies and the commencement of the rains over central India. The intercomparisons are described by Temperton et al. (43).

Primarily, it was noted that models that used the convective adjustment scheme or the classical Kuo's scheme failed to simulate the observed features of the onset; the convective heating rates were underestimated considerably. It was also noted that the forecast in which a plume convection model with entrainment was used produced unrealistically large heating rates. That model produced an explosive onset cyclone but it was too strong compared to the observed. The inflow into this cyclone occurred as a belt of low-level easterlies from India which persisted throughout the seven days. The model thus failed to predict the evolution of low-level moist westerlies and rain over India. Those models that used the modified version of the Kuo scheme (14) realized a reasonable heating rate (as evidenced by rainfall rates). In these instances, the observed features of the onset were predicted remarkably well. Figure 15.25 shows the observed and the predicted flow fields at 850 mb from a 7-day forecast using the modified Kuo scheme in the global spectral model. The predicted onset vortex and the monsoon westerlies compare very well with observations after seven days, a remarkably good forecast in a region where, even today, operational forecasters claim little forecast skill beyond two days.

Recent theoretical advances in the dynamics of the formation of monsoon depressions emphasize the effects of wind shear. Lindzen (44) speculates that the horizontal wind shear is most important whereas Arakawa and Moorthi (45) emphasize the importance of vertical wind shear. These two theories represent opposing views of the energy sources responsible for the formation of the monsoon depression— Lindzen ascribes the growth to barotropic processes which involve only kinetic energy exchanges, while Arakawa and Moorthi view it as a baroclinic process with potential to kinetic energy exchanges in organized convection, forced by what has been described by Charney and Eliassen (46) as conditional instability of the second kind or CISK. Both of these theories predict reasonable horizontal scales of a few thousand kilometers and growth rates of a few days for the depressions. In addition, Arakawa and Moorthi predict a reasonable phase for the vertical motion, that is,

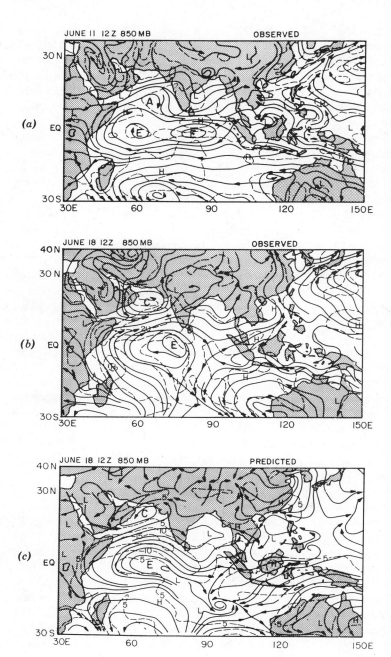

ahead of and west of the wave trough, and a reasonable thermal structure of a weak, cold core below 800 mb and a weak, warm core around 600 mb. Some of the relevant observational structures of monsoon depressions have been discussed in the literature (47–50).

Nitta and Masuda (50) noted that during the formative stage of the depression a deep layer of the tropical lower troposphere satisfies the conditions necessary for the existence of combined barotropic–baroclinic instability. This implies that both the aforementioned mechanisms proposed by Lindzen and by Arakawa and Moorthi may be relevant during the formative stage.

A successful forecast on the formation and landfall of a monsoon depression was recently carried out using the Florida State University's multilevel, global spectral model (43). The initial state at 850 mb, shown in Figure 15.26*a*, was characterized by strong zonal flows over the Bay of Bengal. Figures 15.26*b* and *c* illustrate the observed and the predicted fields on day 6 of the forecast.

This depression was first observed on July 4 in the middle troposphere and on July 5 it appeared as a closed circulation at sea level. The model forecasts of this development were accurate to within 12 hours.

The model output fields were used to diagnose the mechanisms responsible for the formation and growth of the depression. The depression grew initially from the (purely mechanical) barotropic conversion of zonal kinetic energy. As the cyclonic circulation of the depression formed, the vertical shear and cumulus convection dominated and led to the generation of available potential energy and finally kinetic energy of the depression via the baroclinic process of vertical overturnings of relatively warm air upward and relatively cold air downward.

Another important aspect of this model is its ability to handle large rainfall rates. Monsoon depressions have been known to produce as much as 15 to 20 cm of rain per day over India. This rainfall is mostly convective but about 40% can result from the large-scale ascent of saturated stable air. The present spectral models are designed to perform separate calculations for the convective and the nonconvective precipitation components.

In addition to the forecasting problem itself, the determination of the observed rainfall rates, especially over oceans, is a difficult problem. Several methods have been devised to do this. The Florida State University analysis scheme uses a mix of multiple regression and objective analysis to assimilate satellite and rain-gauge data (51).

Figure 15.25. A medium range (seven day) prediction of the onset of monsoon rain. The top panel shows the flow field at the initial time of the forecast on June 11, 1979, 12 GMT. These flow fields are at the 850-mb level (about 1.5 km above sea level). At the initial time the flow over India is from the northwest and is very dry. Over the Arabian Sea the flow is anticyclonic (marked by letter A). The middle panel shows the flow field seven days later as the onset of monsoon rain occurred over southwestern India. During this time there is westerly flow over the northern Arabian Sea and an equatorial eddy (marked by letter E) is present over the equatorial Indian Ocean. The bottom panel shows the corresponding field for day 7 of the forecast. The prediction appears quite reasonable with respect to several of the aforementioned features, that is, the strengthening of the westerlies, the storm over the northern Arabian Sea, and the prediction of the equatorial eddy. The solid lines denote the flow lines, that is, streamlines, and the lines of constant wind speed (m/sec) are shown by dotted lines.

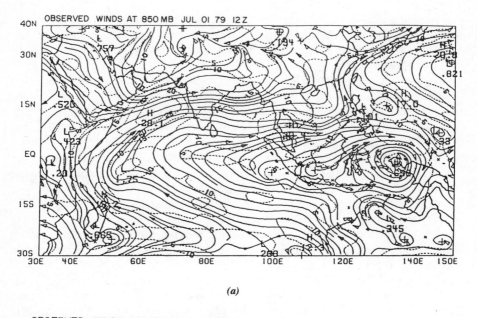

(a)

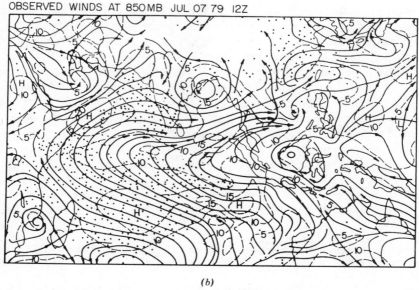

(b)

Figure 15.26. (*a*) The flow field four days before a monsoon depression forms over the northern Bay of Bengal. Note that there is no closed circulation over the Bay of Bengal. The solid lines in this diagram are streamlines and the dashed lines are the lines of constant wind speed in m/sec. These flows are for July 1, 1979, 12 GMT at the 850-mb level, roughly 1.5 km above sea level. The (*b*) observed and the (*c*) predicted flow fields on July 7, 1979, 12 GMT, six days after (*a*). Note that a monsoon depression has formed over the Bay of Bengal and is just making landfall over the northeast coast of India. In this six-day forecast the storm's formation and movement were predicted reasonably well.

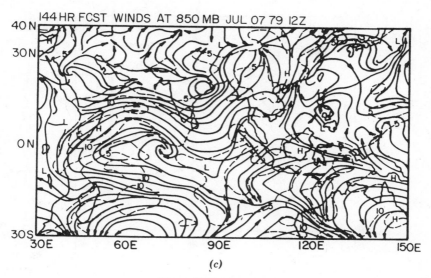

(c)

Figure 15.26. (*continued*)

Figure 15.27 shows the observed and predicted fields of 24-hourly rainfall totals for the depression described in Figures 15.26*a* and *b*. Overall the model was able to handle precipitation reasonably well for this particular storm. Although only a few such experiments have thus far been carried out by the research community, it appears that, given good data sets, the numerical weather prediction of tropical depressions including precipitation rates is quite promising.

2.2.3 Improvements in Short-Range Numerical Weather Prediction. There
are three requirements for improved short-range prediction: better data sets, better telecommunication systems, and improvement of numerical models, especially in the areas of initialization and parameterization of physical processes. As stated, an important aspect of numerical weather prediction is the proper specification of the initial state which depends, in part, on the availability and quality of observations. A number of investigators have carried out a series of numerical experiments to test the impact of the data by using different combinations of data types and density of observations.

Low-Nam (52) carried out such an intercomparison over the monsoon region. Using the multilevel primitive equations with a fine resolution of 100 km and 11 vertical levels, he integrated a physical model to 96 hours for each experiment. He investigated the formation and landfall of a tropical storm that caused considerable damage over eastern India during May 1979.

Low-Nam compared: (a) a current operational system that has only those data that are normally available on the Global Telecommunication System (GTS) within 12 hours after observation time; (b) a possible future observing system which

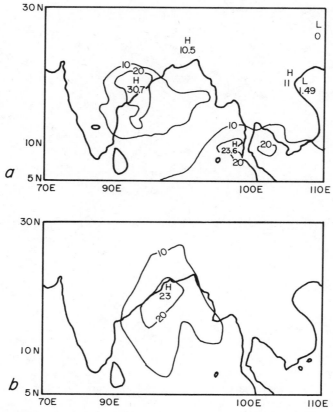

Figure 15.27. The 24-hour total (*a*) observed and (*b*) predicted rainfall for the period between days 5 and 6 of the monsoon depression discussed in Figure 15.26, in mm/day. The rainfall over the northeast coast of India is largely from the landfall of the monsoon depression and is mostly convective rain.

includes the GTS system observations plus winds derived from satellite cloud tracking over the Indian Ocean; and (c) a complete system which includes (a) and (b) above, plus the additional data available during the Global Weather and Monsoon Experiments, that is, the complete observing system that was deployed during 1979 including additional observations from research ships, constant level drifting balloons, drifting oceanic buoys, and wind and temperature observations from dropwindsondes deployed from high-flying research aircraft.

Figures 15.28*a*, *b*, and *c* illustrate respectively, the 850-mb wind forecast at 96 hours made with the three data sets as initial conditions. The results show that the forecasts made with the GTS plus the cloud winds [system (b)] are able to describe the formation of this major tropical storm that produced considerable loss of life and property over southeastern India. The current system with the GTS data alone [system (a)] was found to be quite inadequate for tropical numerical weather pre-

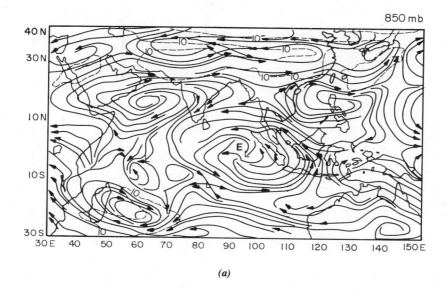

(a)

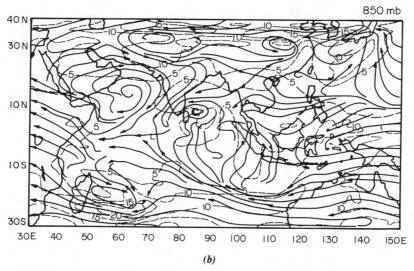

(b)

Figure 15.28. Three separate, four-day 850-mb circulation (streamlines) forecasts made during May 1979. Panel (*a*) shows a forecast made using only operational data sets for initial conditions. Panel (*b*) shows a forecast made using, in addition to (*a*) above, the cloud winds from a geostationary satellite. Panel (*c*) shows the forecast made with the complete observing system that was available during the Global Weather Experiment in 1979. The lack of a forecasted cyclone in (*a*) illustrates the deficiency of the current operational observing system. Panels (*b*) and (*c*) show a forecasted cyclone near southern India. They illustrate the benefits of including cloud winds from geostationary satellites for numerical weather prediction. (*Figure 15.28 continued on p. 518.*)

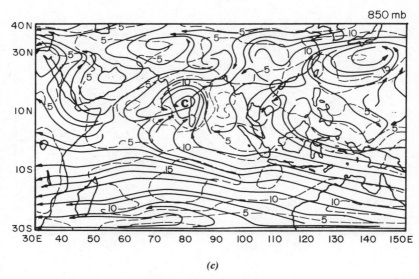

(c)

Figure 15.28. (*continued*)

diction. On the other hand the forecasts made by the complete system (c) are only slightly superior to those made from the GTS plus the cloud winds. It is apparent that the model is indeed quite sensitive to the data sets provided at the initial time.

It is important to note that the best forecasts were made when high resolution cloud winds from the geostationary satellite had been included [cases (b) and (c)]. Both of these forecasts were good to 96 hours. On the other hand, in the relatively data poor case (a), the model failed to predict the formation or the westward motion of the tropical storm.

3 CONCLUSIONS

Elsewhere in this volume Sah and Swaminathan have described the potential benefits that would accompany an improvement in monsoon forecasting. This chapter has suggested ways that this improvement might be accomplished. It is up to the nations of the world to unite and to cooperate to achieve this worthwhile goal. Numerical weather prediction in low latitudes is presently at an experimental stage. Paucity of atmospheric observations over vast data-void regions such as tropical oceans and parts of Africa and South America make numerical weather forecasts less than satisfactory. There are, however, three regions over the tropics where numerical models have shown some promise. These are located over the Caribbean, the Indian subcontinent to southeast Asia, and the western Pacific Ocean. In these three regions the use of conventional meteorological observations, supplemented by cloud winds from geostationary satellites, provides data density sufficient for useful forecasts up to a few days. Experimental forecasts over these tropical domains have shown that the physical processes (especially deep cumulus convection) and initialization are major areas of model sensitivity. In these areas considerable research is currently being carried out by many modeling groups in an effort to improve and extend still further, the range of useful numerical forecasts.

Some progress in predicting the onset of the monsoon and monsoon depressions is evident from these efforts. Time-averaged precipitation forecasts, made with high resolution experimental global models, have shown promising results for periods up to one week. This skill signifies much promise for operational prediction of precipitation for the monsoon countries. The availability of observations, suitable computers, and further improvement in modeling efforts are essential ingredients for further progress.

REFERENCES

1. T. Murakami, R. V. Godbole, and R. R. Kelkar, "Numerical Simulation of the Monsoon along 80°E," in C. S. Ramage, Ed., "*Proceedings of the Conference on the Summer Monsoon of Southeast Asia*," Navy Weather Research Facility, Norfolk, Virginia, 1970, pp. 39–51.
2. T. N. Krishnamurti and Y. Ramanathan, Sensitivity of the monsoon onset to differential heating, *J. Atmos. Sci.*, **39**, 1290–1306 (1982).
3. T. N. Krishnamurti, M. Kanamitsu, W. Koss, and J. Lee, Tropical east–west circulations during the northern winter, *J. Atmos. Sci.*, **30**, 780–787 (1973).
4. J. Findlater, A major low-level air current near the Indian Ocean during the northern summer, *Quart. J. Roy. Meteor. Soc.*, **95**, 362–380 (1969).
5. A. Wiin-Nielsen, On short- and long-term variations in quasi-barotropic flow, *Mon. Wea. Rev.*, **89**, 461–476 (1961).
6. T. N. Krishnamurti, J. Molinari, and H. L. Pan, Numerical simulation of the Somali jet, *J. Atmos. Sci.*, **33**, 2350–2362 (1976).
7. J. E. Hart, On the theory of the East African low-level jet stream, *Pure Appl. Geophys.*, **115**, 1263–1282 (1977).
8. P. R. Bannon, On the dynamics of the East African jet. III: Arabian Sea branch, *J. Atmos. Sci.*, **39**, 2267–2278 (1982).
9. S. Hastenrath and P. J. Lamb, *Climatic Atlas of the Indian Ocean*, Vol. I–II. University of Wisconsin Press, Madison, 1979.
10. K. Y. Kondratyev, *Radiative Processes in the Atmosphere*, World Meteorological Organization, Geneva, Switzerland, 1972, 214 pp.
11. J. W. Posey and P. F. Clapp, Global distribution of normal surface albedo, *Geofisica International* (Mexico), **4**, 33–48 (1964).
12. J. A. Businger, J. C. Wyngaard, Y. Izumi, and E. F. Bradley, Flux-profile relationships in the atmospheric surface layer, *J. Atmos. Sci.*, **28**, 181–189 (1971).
13. R. J. Pasch, "On the Onset of the Planetary Scale Monsoon," Report No. 83-9, Department of Meteorology, Florida State University, Tallahassee, Florida, 1983, 220 pp.
14. H. L. Kuo, Further studies of the parameterization of the influence of cumulus convection on large-scale flow, *J. Atmos. Sci.*, **31**, 1232–1240 (1974).

15. A. Arakawa and W. Schubert, Interaction of a cumulus cloud ensemble with the large-scale environment, Part I, *J. Atmos. Sci.*, **31**, 674–701 (1974).

16. T. N. Krishnamurti, S. Low-Nam, and R. Pasch, Cumulus parameterization and rainfall rates II, *Mon. Wea. Rev.*, **111**, 815–828 (1983).

17. S. J. Lord, Interaction of a cumulus cloud ensemble with the large scale environment, Part III: Semi-prognostic test of the Arakawa–Schubert cumulus parameterization, *J. Atmos. Sci.*, **39**, 88–103 (1982).

18. M. R. Newman and A. J. Gadd, "Variability in the Character of the Onset of the SW Monsoon in Global Objective Analyses and GCM Simulations," Proceedings of the International Conference on the Scientific Results of the Monsoon Experiment, Denpasar, Bali, Indonesia, October 26-30, 1981, World Meteorological Organization, Geneva, Switzerland, 1982.

19. J. C. Sadler, "The Upper Troposopheric Circulation of the Global Tropics," Technical Report No. 75-05, Meteorology Department, University of Hawaii, Honolulu, Hawaii, 1975.

20. S. Manabe, D. G. Hahn, and L. Holloway, Jr., "Climate Simulations with GFDL Spectral Models of the Atmosphere: Effects of Spectral Truncation," Proceedings of the Joint Organizing Committee Study Conference on Climate Models, Washington, D.C., Vol. 1, Global Atmospheric Research Program Publication Series No. 22, World Meteorological Organization, Geneva, Switzerland, 1979, pp. 41–94.

21. Y. Hayashi, "Studies of the Tropical General Circulation with a Global Model of the Atmosphere," Proceedings of the Seminar on the Impact of GATE, Woodshole, Mass. August 20–29, 1979, United States National Academy of Science, Washington, D.C., 1980.

22. T. N. Krishnamurti, S. M. Daggupaty, J. Fein, M. Kanamitsu, and J. D. Lee, Tibetan high and upper tropospheric tropical circulations during northern summer," *Bull. Amer. Meteor. Soc.*, **54**, 1234–1249 (1973).

23. S. Manabe and D. Hahn, Simulation of atmospheric variability, *Mon. Wea. Rev.*, **109**, 2260–2286 (1981).

24. T. N. Krishnamurti and M. Kanamitsu, "Northern Summer Planetary Scale Monsoons during Drought and Normal Rainfall Months," in Sir J. Lighthill and R. P. Pearce, Eds., *Monsoon Dynamics*, Cambridge University Press, Cambridge, 1981, pp. 19–48.

25. D. G. Hahn and S. Manabe, The role of mountains in the south Asian monsoon circulation, *J. Atmos. Sci.*, **32**, 1515–1541 (1975).

26. R. P. Sarker, Some modifications in a dynamic model of orographic rainfall, *Mon. Wea. Rev.*, **95**, 673–684 (1967).

27. S. Gadgil, Orographic effects on the southwest monsoon: A review, *Pure Appl. Geophys.* **115**, 1413–1430 (1977).

28. W. M. Washington, R. M. Chervin, and G. V. Rao, Effects of a variety of Indian Ocean surface temperature anomaly patterns on the summer monsoon circulation: Experiments with the NCAR general circulation model, *Pure Appl. Geophys.*, **115**, 1335–1356 (1977).

29. J. Shukla, Effect of Arabian sea-surface temperature anomaly on Indian summer monsoon: A numerical experiment with the GFDL model, *J. Atmos. Sci.*, **32**, 503–511 (1975).

30. H. L. Pan, "Upper Tropospheric Tropical Circulations during a Recent Decade," Report No. 79-1, Department of Meteorology, Florida State University, Tallahassee, Florida, 1979.

31. E. M. Rasmusson and T. H. Carpenter, Variations in tropical sea surface temperature and surface wind fields associated with the Southern Oscillations/El Niño," *Mon. Wea. Rev.*, **110**, 354–384 (1982).

32. R. N. Keshavamurty, Response of the atmosphere to sea surface temperature anomalies over the equatorial Pacific and the teleconnections of the Southern Oscillation, *J. Atmos. Sci.*, **39**, 1241–1259 (1982).

33. B. J. Hoskins and D. J. Karoly, The steady linear response of a spherical atmosphere to thermal and orographic forcing, *J. Atmos. Sci.*, **38**, 1179–1196 (1981).

34. A. E. Gill, Some simple solutions for heat-induced tropical circulation, *Quar. J. Roy. Meteor. Soc.*, **106**, 447–462 (1980).

35. N. C. Lau, A diagnostic study of recurrent meteorological anomalies appearing in a 15-year simulation with a GFDL general circulation model, *Mon. Wea. Rev.*, **109**, 2287–2311 (1981).

36. J. P. Kuettner and S. Unninayar, "The Onset Mechanism of the Indian Monsoon," Proceedings of the International Conference on the Scientific Results of the Monsoon Experiment, Denpasar, Bali, Indonesia, October 26–30, 1981, World Meteorological Organization, Geneva, Switzerland, 1982.

37. Y. P. Rao, *Southwest Monsoon*, Meteorological Monographs, Synoptic Meteorology, Indian Meteorological Department, Delhi, India, 1976, 367 pp.

38. G. Haltiner and R. Williams, *Numerical Prediction and Dynamic Meteorology*, Wiley, New York, 1980, 443 pp.

39. J. J. Stephens and K. W. Johnson, Rotational and divergent wind potentials, *Mon. Wea. Rev.*, **106**, 1452–1457 (1978).

40. F. G. Shuman, Numerical methods in weather prediction, II: Smoothing & filtering, *Mon. Wea. Rev.*, **85**, 375–381 (1957).

41. M. B. Mathur, A note on an improved quasi-langragian advective scheme for primitive equations, *Mon. Wea. Rev.*, **98**, 214–219 (1970).

42. J. G. Charney, "The Intertropical Convergence Zone and the Hadley Circulation of the Atmosphere," Proceedings of the Conference on Numerical Weather Prediction, Japan Meteorological Agency, Tokyo, 1968, pp. III 73–III 79.

43. C. Temperton, T. N. Krishnamurti, T. Pasch, and T. Kitade, "WGNE Forecast Comparison Experiments," World Climate Reserach Program Report No. 6, World Meteorological Organization, Geneva, Switzerland, pp. 73–104, 1983.

44. R. S. Lindzen, B. Farrell, and A. J. Rosenthal, Absolute barotropic instability and monsoon depression, *J. Atmos. Sci.*, **40**, 1178–1184 (1983).

45. A. Arakawa and S. Moorthi, "Baroclinic (and Barotropic) Instability with Cumulus Heating," Proceedings of the WMO Program on Research in Tropical Meteorology, October 1982, Tsukuba, Japan, World Meteorological Organization, Geneva, Switzerland, 1982.

46. J. G. Charney and A. Eliassen, On the growth of the hurricane depression," *J. Atmos. Sci.*, **21**, 68–75 (1964).

47. T. N. Krishnamurti, M. Kanamitsu, R. Godbole, C. B. Chang, F. Carr, and J. H. Chow, Study of a monsoon depression (I): synoptic structure, *J. Meteor. Soc. Japan*, **54**, 208–226 (1975).

48. R. V. Godbole, On cumulus-scale transport of horizontal momentum in monsoon depression over India, *Pure Appl. Geophys.*, **115**, 1373–1381 (1977).

49. D. R. Sikka, Some aspects of the life history, structure and movement of monsoon depressions, *Pure Appl. Geophys.*, **115**, 1501–1529 (1977).

50. T. Nitta and K. Masuda, Observational study of a monsoon depression developed over the Bay of Bengal during summer MONEX, *J. Meteor. Soc. Japan*, **59**, 227–240 (1981).

51. T. N. Krishnamurti, S. Cocke, R. Pasch, and S. Low-Nam, ''Precipitation Estimates for Raingauge and Satellite Observations,'' Report No. 83-7, Department of Meteorology, Florida State University, Tallahassee, Florida, 1983, 373 pp.

52. S. Low-Nam, 'On the Impact of Data Sets in the Prediction of a Tropical Storm during MONEX,'' Report No. 82-9, Department of Meteorology, Florida State University, Tallahassee, Florida, 1982, 122 pp.

16

Long-Range Forecasting of Monsoons

J. Shukla
Center for Ocean–Land–Atmosphere Interactions
Department of Meteorology
University of Maryland
College Park, Maryland

INTRODUCTION

As discussed in Chapter 14, the day-to-day changes in weather are mainly due to the growth, propagation, and decay of the synoptic-scale disturbances that owe their origin to the instabilities of the large-scale flow. The growth rate for the dominant instabilities and their nonlinear interactions with other scales and the mean flow are such that these synoptic-scale disturbances become totally unpredictable within a couple of weeks. This limit of deterministic predictability has been the greatest stumbling block for progress in dynamical, long-range forecasting, and therefore, it is no surprise that in the past most of the attempts at long-range forecasting have been either synoptic or statistical in nature.

Recent observational and modeling studies have suggested that although the synoptic scales lose their predictability within two weeks, the low-frequency planetary scales remain predictable up to a month (1). It has also been suggested that there may be additional predictability due to the influence of the slowly varying boundary conditions at the earth's surface (2). The situation is especially promising in the low latitudes where synoptic scale instabilities are too weak to degrade the predictability of the large scales, and the influence of the changes in the boundary conditions is large enough to be clearly distinguishable from the unpredictable day-to-day fluctuations. These results, collectively, have suggested a physical basis for dynamical prediction of average monthly and seasonal atmospheric conditions.

Section 1 describes the current status of our knowledge of the long-range predictability of monsoons. Section 2 covers the operational statistical forecasting of seasonal monsoon rainfall over India and Section 3 presents a new and simple technique for predicting summer monsoon rainfall over India.

1 PREDICTABILITY OF MONSOONS

Charney and Shukla (3) have suggested that low-latitude atmospheric flows in general, and the monsoon circulations in particular, are potentially more predictable than mid-latitude circulations. This suggestion was based on the results of observational, theoretical, and numerical studies which show that: the day-to-day fluctuations in the tropics are small, implying weak flow instabilities; the interannual fluctuations of seasonal averages are large and are related to the changes in the boundary conditions of such slowly varying parameters as sea surface temperature, soil moisture, albedo, and vegetation; and numerical simulations by general circulation models with prescribed boundary conditions underestimate the observed low-latitude variability. The planetary-scale tropical circulations are dominated by thermally forced Hadley and Walker type circulations (see Chapter 11), which are intrinsically stable because weak tropical disturbances cannot change them, and also because they are linked with slowly varying boundary conditions. Since the space and time averages in the tropics are dominated by the large-scale, low-frequency components, monthly and seasonal predictions are possible, even if the weak, synoptic-scale instabilities are not predictable on the shorter time scales.

Charney and Shukla (3) examined the variability among four July simulations with a global general circulation model (4) in which the global boundary conditions were identical but the initial conditions different. The observed initial conditions in the middle of June were altered at all the model grid points by superimposing random error fields of wind, temperature, and pressure. The spatial structure of the random error fields corresponded to Gaussian distributions with zero means and standard deviations of 1°C in temperature, 3 m/sec in horizontal wind components and 1 mb in surface pressure. The initial perturbation fields were quite large, at least for the tropics. The simulated fields were quite different at the end of 15 model days of integration. The first two weeks of integrations were ignored and monthly averages for July (days 16–46) sea level pressure and rainfall were calculated for each of the four simulations. Figure 16.1 shows the model (σ_M) and observed (σ_O) zonally averaged standard deviations and their ratios for July sea level pressure and rainfall as functions of latitude. The observed standard deviations were calculated for about 380 Northern Hemisphere stations for 10 years (1966–1975). It can be seen that the observed and the model simulated variabilities are comparable in mid-latitudes, but the variability in the simulations is considerably less than that observed in the low latitudes. The ratios of the observed and the model standard deviations are less than 1.5 between the latitudes 25°N and 55°N for pressure and to the north of 30°N for rainfall. For low latitudes, the ratio is more than 2.0. Similar results were obtained for the averages over limited areas in low and middle latitudes.

Manabe and Hahn (5) have carried out an 18-year integration of the GFDL (Geophysical Fluid Dynamics Laboratory) spectral model with prescribed seasonally varying but interannually fixed boundary conditions of sea surface temperature. Figure 16.2 shows the zonally averaged standard deviation of winter season 1000-mb geopotential height for the last 15 years of their simulation (σ_M), the observations (σ_O), the ratio (σ_O/σ_M). The ratio of observed and simulated standard deviations

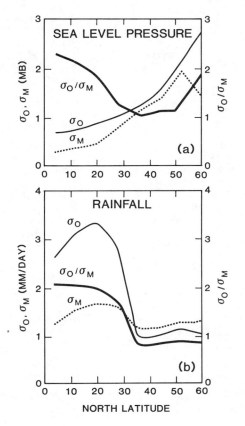

Figure 16.1. Model (σ_M, dotted line) and observed (σ_O, thin solid line) zonally averaged standard deviations as functions of latitude, and their ratio (σ_O/σ_M, thick solid line), for mean July: (*a*) sea level pressure and (*b*) rainfall. Observed values are for land stations and model values are for grid points over land. Reprinted with permission from the Cambridge University Press from Charney and Shukla, *Monsoon Dynamics* (reference 3).

is about 2.0 in the near-equatorial regions and about 1.0 in the middle and high latitudes. This ratio was found to be more than 3.0 for tropical upper tropospheric variables. If we can assume that this discrepancy in observed and simulated variability at low latitudes was not due to model deficiencies, Manabe and Hahn's results from a completely different model and based on long-term simulations provide further evidence for Charney and Shukla's (3) results.

Based on these results, Charney and Shukla proposed that the cause of the discrepancy between model and observed variability in the low latitudes could be the boundary conditions that were fixed and constant for all the simulations. This hypothesis is supported by observed correlations between boundary conditions and circulation patterns, and sensitivity experiments with global models which show that tropical boundary anomalies produce significant changes in the model simulated circulation and rainfall. There is a serious limitation to studies that compare model and observed variability because model deficiencies can also produce an underestimate of the interannual variability. Although in the Charney and Shukla study the day-to-day model fluctuations were quite realistic when compared to the observations, it would have been more appropriate to compare the variability of the

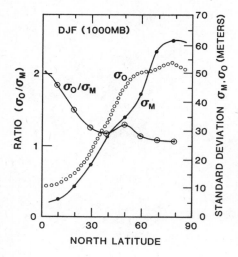

Figure 16.2. Zonally averaged means of standard deviation of seasonal mean 1000-mb geopotential height in meters; observations (σ_O, broken line of circles) and model (σ_M, solid line with black dots). On the left side is the ratio (σ_O/σ_M, solid line with circles) of observed and model standard deviations. Reprinted with permission from the American Meteorological Society from Manabe and Hahn, *Monthly Weather Review* (reference 5).

same model with and without changes in the boundary conditions. The following results describe such an experiment which was carried out with the GLAS (Goddard Laboratory for Atmospheric Sciences) general circulation model (6).

A 45-day integration was carried out starting from observed initial conditions in the middle of June, and long-term climatological mean boundary conditions of sea surface temperature (SST). This integration is referred to as the control run (C). For the identical boundary conditions, three additional integrations of 45 days each were carried out by randomly changing the initial conditions of the horizontal wind components u and v at each of the nine levels of the model. The spatial structure of the random errors corresponded to a Gaussian distribution with zero mean and standard deviation of 3 m/sec for each of u and v. These three integrations are referred to as predictability runs (P_1, P_2, and P_3). Although the statistical properties of the random errors were the same for each predictability run, the actual grid point values were randomly different. Three additional integrations were carried out for which, in addition to the randomly perturbed initial conditions, the climatological boundary conditions of SST between the equator and 30°N were replaced by each year's observed monthly mean SST during July of 1972, 1973, and 1974. These three integrations are referred to as boundary forcing runs (B_1, B_2, and B_3).

The variance (σ_P)^2 among C, P_1, P_2, and P_3 is a measure of the natural variability of the model; the variance (σ_B)^2 among C, B_1, B_2, and B_3 is a measure of the variability due to changes in the boundary conditions of the tropical SST.

$$(\sigma_P)_{i,j}^2 = \frac{1}{3} [(C - \overline{P})^2 + (P_1 - \overline{P})^2$$

$$+ (P_2 - \overline{P})^2 + (P_3 - \overline{P})^2]_{i,j} \qquad (1)$$

$$(\sigma_B)^2_{i,j} = 1/3 \; [(C - \overline{B})^2 + (B_1 - \overline{B})^2$$

$$+ (B_2 - \overline{B})^2 + (B_3 - \overline{B})^2]_{i,j} \qquad (2)$$

where $\overline{P} = (C + P_1 + P_2 + P_3)/4$ and $\overline{B} = (C + B_1 + B_2 + B_3)/4$, and C_{ij}, P_{ij}, B_{ij} denote the July mean at grid point i,j.

Figure 16.3 shows the plots of zonally averaged values of standard deviations σ_P, σ_B, σ_O and the ratios σ_O/σ_P and σ_O/σ_B where σ_O is the standard deviation for 10 years of observed monthly means. In agreement with the results of Charney and Shukla (3), it is seen that the ratio σ_O/σ_P is more than 2.0 in the tropical latitudes and close to 1.0 in the middle latitudes. The new result is that the curve σ_B lies nearly halfway between the curves σ_O and σ_P. This suggests that for this model about half of the "unexplained" variability is accounted for by changes in sea surface temperature between the equator and 30°N.

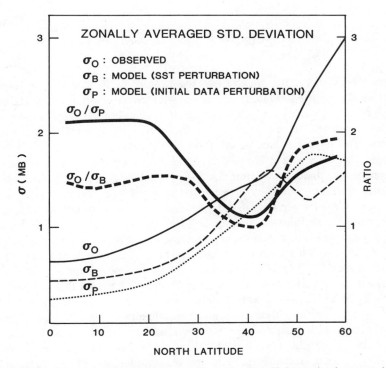

Figure 16.3. Zonally averaged standard deviation of monthly mean (July) sea level pressure (mb) for 10 years of observations (σ_O, thin solid line), four model runs with variable boundary and initial conditions (σ_B, thin dashed line), and four model runs with identical boundary conditions (σ_P, thin dotted line). Thick solid line and thick dashed line show the ratio σ_O/σ_P and σ_O/σ_B, respectively (from reference 6).

This supports Charney and Shukla's hypothesis that the slowly varying boundary conditions play an important role in determining the interannual variability of time averages for the tropical atmosphere. Additional effects of soil moisture or the Eurasian snow cover could possibly bring the σ_O and σ_B curves still closer. However, since the long period internal dynamical changes (e.g., tropical–extratropical interactions) can also contribute to the interannual variability of time averages, it would never be possible to explain the total σ_O by boundary conditions alone.

The model variability for the predictability and boundary forcing integrations have also been compared. Since the SST anomalies for B_1, B_2, and B_3 have many common features, it is more appropriate to calculate the changes in the monthly means due to boundary conditions (E_B) and due to random perturbations (E_P) as follows:

$$E_B{}^2 = \text{1/3} \sum_{k=1}^{3} (C - B_k)^2 \qquad \text{(at each grid point } i, j) \qquad (3)$$

and

$$E_P{}^2 = \text{1/3} \sum_{k=1}^{3} (C - P_k)^2 \qquad \text{(at each grid point } i, j) \qquad (4)$$

Figure 16.4 shows the zonally averaged values of E_B and E_P for July mean geopotential height at 300 mb. In agreement with the observations in the atmosphere the values of E_B and E_P are small for the low latitudes and large for the middle latitudes. However, in this experiment, the ratio E_B/E_P is more than 2.0 for low latitudes. The largest values of the ratio E_B/E_P occur between 20°N–20°S. This result also suggests a possible role of tropical SST anomalies for changes in the atmospheric circulation away from the SST anomalies. Although the SST boundary anomalies were imposed only between the latitudes 0–30°N, their effects on the circulation at 300 mb are seen in the Southern Hemisphere tropics also. This could be either due to meridionally propagating Rossby waves forced by heating due to the SST anomalies, or interhemispheric interactions associated with the fluctuations of Hadley cells which are strongly influenced by SST anomalies and their attendant convection (see Chapter 11). The results for geopotential height at 500 mb (not shown) are very similar to those shown in Figure 16.4 except that the peak of the ratio E_B/E_P at the equator is not as high.

Lau [7] has further analyzed the model simulations of Manabe and Hahn (who, we will recall, used interannually fixed boundary conditions of SST), and found that the model simulated mass field did not show any evidence of the Southern Oscillation, which is clearly seen in the observations as pressure anomalies of opposite sign in the Indian Ocean and the eastern equatorial Pacific. This gives indirect evidence for the importance of tropical SST anomalies for the Southern Oscillation.

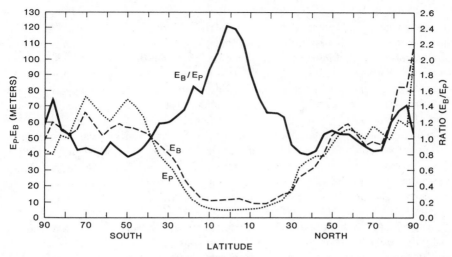

Figure 16.4. Zonally averaged standard deviation for predictability runs (E_P, dotted line), boundary forced runs (E_B, dashed line) and the ratio of boundary forced and predictability runs (E_B/E_P, thick solid line) for geopotential height at 300 mb (from reference 6).

The numerical results of Charney and Shukla (3), Shukla (6), Manabe and Hahn (5), and Lau (7) provide strong, albeit indirect, evidence that SST anomalies in the tropics are one of the most important determinants of the interannual variability of the tropical and monsoonal circulations. These results are also in agreement with Madden (8) who examined the predictability of monthly sea level pressure over the Northern Hemisphere. He compared the variances of the observed monthly means with the "natural variability" of monthly means that occur in the absence of boundary forcing and showed that the long period changes in low latitudes are potentially more predictable compared to those in the mid-latitudes. These conclusions are further confirmed by the results of Shukla and Gutzler (9).

In addition to the above mentioned "indirect" numerical and observational results, there is a large body of literature on numerical experiments with global general circulation models which show, more directly, the sensitivity of model climate to changes in the boundary conditions. The following brief description is presented of the physical mechanisms through which properly specified boundary conditions can enhance the predictability of monthly and seasonal atmospheric averages.

1.1 Boundary Forced Predictability

Simple arguments suggest that enhanced atmospheric predictability on monthly and seasonal time scales is possible with the proper specifications of boundary effects. The mechanisms involved are different for each boundary type.

1. Changes in boundary forcing directly influences the location and intensity of the heat sources and sinks that drive the atmospheric circulation. The forcing at the boundary itself is generally not sufficient to produce significant changes in the atmospheric circulation; however, under favorable conditions of large-scale convergence and divergence, the boundary effect is transmitted through the atmosphere, transforming a shallow boundary forcing into a three-dimensional change in heating distributions which can be quite effective in influencing the atmospheric circulation. The effectiveness of a boundary forcing in changing the atmospheric circulation therefore strongly depends upon its ability to produce a deep heat source and a means by which this influence can propagate away from the source. Since both of these factors are determined by the structure of the large-scale dynamical circulation itself, the response of a given boundary forcing can be very different depending upon its size and geographical location, and on the structure of the large-scale circulation.

2. Boundary forcings of SST, soil moisture, surface albedo, and snow and ice not only affect the sources and sinks of heat, but they also affect the sources and sinks of moisture, and in turn, the latent heat sources.

3. The strongly nonlinear character of the atmosphere is such that even a weak anomaly in a boundary forcing, under favorable conditions, can produce significant anomalies in atmospheric circulation, and therefore the actual response may be stronger than one estimated from linear theories.

1.1.1 Snow Cover. Large-scale anomalies of snow cover have the potential to influence the atmospheric circulation by several physical processes:

1. An increase in snow cover increases the albedo, and therefore reduces the incoming solar radiation. If there were no other feedbacks, this would produce colder temperatures (and also result in the snow cover anomalies persisting for a longer time).

2. Excessive snow anomalies in the mid-latitudes, by cooling the overlying air, can act as anomalous heat sinks which in turn can produce anomalous stationary wave patterns that can alter storm tracks and their frequency.

3. Persistent snow anomalies can change the components of the heat balance of the earth's surface. Even after the snow has melted completely, wet soil will maintain colder surface temperatures for longer periods of time than dry soil.

4. Persistent snow anomalies can produce anomalous meridional temperature gradients and associated anomalous vertical wind shears which in turn can change the instability characteristics of synoptic-scale weather disturbances.

1.1.2 Sea Surface Temperature. The factors that determine the influence of SST anomalies on the atmospheric circulation are rather complex. The immediate effect of an SST anomaly is to change the sensible heat flux and evaporation in its immediate vicinity. However, this type of thermal forcing, confined to the ocean's surface, is too weak to produce significant changes in the atmospheric circulation. The crucial factor that determines whether an SST anomaly can affect the atmospheric circulation is its ability to produce a vertically deep heat source in the atmosphere. It is the deep atmospheric heat source which produces anomalous atmospheric circulations. Based on several numerical model experiments, it has been suggested (2) that the creation of a strong and deep heat source associated with an SST anomaly depends upon the structure and magnitude of the anomaly, the structure and magnitude of the total sea surface temperature field, the structure of the large-scale flow of the atmosphere, the latitude of the anomaly, and the most dominant atmospheric instability mechanisms in the region.

The magnitude of the anomaly is important because it determines the amount of change in sensible heat and evaporation; however, due to the nonlinearity of the thermodynamic processes, the change in evaporation over the warm waters near the equator will be much larger in cooler waters than that at higher latitudes for the same incremental change in the SST. An even more important factor is the structure of the SST field and its effect on the horizontal convergence of the air and water vapor directly above. Small gradients of SST can produce large convergence in the tropics compared to the mid-latitudes and, therefore, can produce large heating fields.

If a warm SST anomaly occurs under the ascending branches of Hadley and Walker type circulations, it is more effective in increasing the intensity of these circulations than the same anomaly under the descending branches. Similarly, the location of an SST anomaly with respect to the planetary waves in the mid-latitudes may be crucial in determining the atmospheric response.

In addition to the local influences, significant changes can occur in the circulation away from the heating anomalies if the intervening medium is suitable for propagation of these influences. Possible mechanisms for the influence of tropical SST anomalies on extratropical circulation are through propagation of Rossby waves (10), and changes in the intensity of the Hadley circulation. The latter change affects the zonal flow in the mid-latitudes which, in turn, interacts with the quasi-stationary thermal and orographic conditions to produce anomalous atmospheric circulations. The influence of SST anomalies on the tropical circulation is effected mostly through changes in the moisture convergence and precipitation. A warm anomaly could increase the sensible heating, evaporation and moisture convergence, giving rise to enhanced precipitation. If the convergence and evaporation associated with the anomaly are strong enough, they can alter the position and intensity of planetary-scale tropical circulations which determine possible areas of excessive rain and drought.

If the convective instability of the second kind (see Chapter 9, Section 3.2) is the dominant instability mechanism in the region of the anomaly, an enhanced convergence due to horizontal gradients in SST could produce enhanced latent heating in the troposphere which would further enhance the moisture convergence in the boundary layer. This feedback mechanism helps to establish a vertically deep heat source in the atmosphere in association with the SST anomaly at the surface.

1.1.3 Soil Moisture. Since the maximum net heating of a vertical atmospheric column occurs over the tropical land masses (2), it is quite likely that changes in the heat sources over these areas could produce considerable change in the planetary-scale circulations of the tropical as well as the extratropical atmosphere. Despite the small (relative to the oceans) earth surface area covered by land, the soil moisture effects over land could be as important as SST anomaly effects. The soil moisture effects depend strongly upon season and latitude. During the winter season in high latitudes, the amount of solar radiation reaching the ground is not large enough to significantly alter the surface energy budget. Changes in soil moisture influence evaporation and heating of the land surface. The degree to which the total radiative energy impinging on the ground goes into latent rather than sensible heating is determined by the wetness of the ground. If the soil is wet, most of the radiative energy goes to evaporate the water, if the soil is dry and there is no vegetation, most of the radiative energy is used to heat the ground and the overlying air.

This discussion of the mechanisms through which boundary forcing influences the atmospheric circulation suggests that there is a physical basis for prediction of monthly and seasonal atmospheric anomalies due to the influence of boundary conditions. Since the boundary conditions change slowly, they can be prescribed and their effects on the atmosphere can be calculated using a dynamical model. Recognition of the importance of the boundary conditions for extended range predictability is an important step toward dynamical long-range forecasting that can now be attempted using current models and global observations of the boundary conditions.

2 FORECASTING OF SEASONAL RAINFALL OVER INDIA

After India experienced a great famine in 1877, a year with highly deficient summer monsoon rainfall over most of the country, the British Government called upon Henry F. Blanford to make monsoon forecasts. Blanford was a geologist and the first Meteorological Reporter for Bengal. He established the India Meteorological Department in 1875 and served as its director. Blanford (11) noted an association between large winter and spring snowfalls in the Himalayas and monsoon droughts over India in summer and he used this association to make preliminary forecasts during 1882–1885. He met with some success and official forecasts were issued beginning in 1886. Sir John Eliot, who succeeded Blanford, used weather conditions over the whole of India and the surrounding regions to prepare elaborate (perhaps too elaborate—as long as 30 pages) forecasts of monsoon rainfall. India experienced another great famine in 1899 and the newspapers were so critical of the long-range forecasts of the monsoon rainfall that for some time the forecasts were issued only to the Provincial Governments as confidential documents. Sir Gilbert T. Walker, a Senior Wrangler at Cambridge, succeeded Eliot and started objective methods of monsoon rainfall forecasting based on correlation. While searching for the potential predictors of Indian monsoon rainfall, and benefiting from the earlier work of Eliot, who had noted an association between high pressure over Mauritius and Australia, and droughts over India, Walker described and coined the words the "Southern Oscillation," as well as the two "Northern Oscillations" (North Atlantic and North

Pacific). Walker's search for global predictors at large distances away from India was motivated by the already published papers of Hildebrandsson (12) who had noted an opposite polarity of pressure at Sydney and Buenos Aires, and the Lockyers (13) who had further confirmed the pressure seesaw between the Indian Ocean and Argentina (14). In Chapter 8, G. Kutzbach provides a detailed description of this productive period in India's history.

Walker (15) developed several regression formulas to predict seasonal monsoon rainfall averaged over homogeneous subdivisions of India. A comprehensive review of the method, the factors used as predictors, and the performance of these methods has been documented by Jagannathan (16). The regression formulas used by Walker in 1924 for forecasting seasonal (June–September) rainfall over Peninsula and Northwest India (which are defined in Chapter 14, Fig. 14.5) are given below as an example:

Peninsula rainfall departure in inches
$$= 1.61 \, F_1 - 0.29 \, F_2 - 0.02 \, F_3 - 77.3 \, F_4 - 0.21 \, F_5 - 0.35 \, F_6 \quad (5)$$

Northwest rainfall departure in inches
$$= 0.29 \, F_1 - 44.5 \, F_4 - 0.36 \, F_5 - 0.95 \, F_7 - 0.53 \, F_8 - 17.0 \, F_9 \quad (6)$$

where F_1 = average of April and May departure from normal pressure averaged for Santiago, Buenos Aires, and Cordoba (mm of mercury),

F_2 = May Zanzibar percentage rainfall departure from normal,

F_3 = October through February Java percentage rainfall departure from normal,

F_4 = average of September, October, and November Capetown pressure departure (inches of mercury),

F_5 = October through April South Rhodesian rainfall departure from normal (inches),

F_6 = average of March and April Dutch Harbor temperature (degrees F),

F_7 = snow accumulation in Himalayas by end of May (tabulated on a numerical scale of departure),

F_8 = average of December through April Dutch Harbor temperature (degrees F), and

F_9 = average of: average February and March Seychelles pressure, average January through April Batavia pressure, and average March through May Port Darwin pressure departures (inches of mercury).

These regression equations were revised periodically to include new predictors and modified values of regression coefficients for old predictors. The following regression equation was used by the India Meteorological Department to predict Peninsula rainfall for the summer monsoon season of 1954 (17).

Peninsula rainfall departure in inches
$$= 1.825 \, X_1 + 0.912 \, X_2 - 0.067 \, X_3 - 0.0183 \, X_4 - 0.559 \, X_5 - 4.307 \quad (7)$$

where X_1 = average of April and May departure from normal pressure for Santiago,
Buenos Aires, and Cordoba (in mm of mercury),

X_2 = average April northerly wind speed (m/sec) over Bangalore at 6 km,

X_3 = October through April South Rhodesia rainfall departure from normal
(in inches),

X_4 = October through February Java percentage rainfall departure from
normal, and

X_5 = average May easterly wind speed (m/sec) over Calcutta at 4 km.

Eq. (7) shows that by the mid-1950s, upper air predictors had been included in the forecast schemes. The equations currently in use in India have been modified still further. They are discussed by Das (see Section 3.2.1 of Chapter 17).

From a close examination of the various predictors used during the past 60 years, with the possible exception of the Southern Oscillation and related circulation features, a physical basis for these statistical relationships is not clear. It is likely that these apparent relationships are due simply to random sampling; however, a definitive conclusion cannot be drawn without a detailed examination of the long-term variability of these predictors. A superficial analysis of available data suggests that most of the predictors chosen by Walker are indirect descriptors of the Southern Oscillation phenomenon.

Normand (14) verified 18 years (1931–1948) of forecasts of monsoon rainfall. Thirty-two cases were for the summer season of June through September (16 for Northwest and 16 for Peninsula India), 29 were for only August and September (15 for Northwest and 14 for Peninsula India), and nine cases were for the Northwest for the winter season of January through March. He showed that forecasts were considerably better than those based on pure chance, but only slightly better than those based on probability tables for odds of 4 to 1 against being wrong. Verification for the deficient rainfall years alone showed that 66% of these forecasts were wrong. For August and September rainfall over Northwest India, forecasts for all the seven years of deficient rainfall were wrong. Normand wondered whether these forecasts were of any use at all; however, he favored continuation, "if only to keep the subject alive and in the hope that ideas for progress will emerge." Unfortunately there is no available documentation for the performance of the regression equations for the last 20 years.

3 THE RELATIONSHIP BETWEEN THE SOUTHERN OSCILLATION AND MONSOON RAINFALL

As discussed in the preceding section, relationships between the Southern Oscillation and the Indian monsoon rainfall were established by Walker in the beginning of this century and versions of them have been used since for operational forecasting of monsoon rainfall. This section re-examines the relationship between the Southern Oscillation and Indian monsoon rainfall using the Darwin sea level pressures for the period 1901–1981, and the summer monsoon rainfall data described in Chapter 14. Darwin pressure was chosen because its long-term record is considered to be more accurate and more complete than that for any other station in that region. Although Tahiti minus Darwin pressure is considered to be a better index of the Southern Oscillation, Tahiti pressure is available only for the period 1935–1981, and for this period the correlation coefficient between the spring Tahiti pressure and Indian monsoon rainfall is only 0.01. The summer monsoon rainfall data used in this study is the area weighted average of the percentage departures for each of the 31 subdivisions of India, and is referred to as the whole Indian monsoon rainfall anomaly in Chapter 14.

3.1 Influence of the Southern Oscillation on Summer Monsoon Rainfall over India

Figure 16.5 shows the data used: the thin line denotes the 12-month running mean of the normalized Darwin pressure anomaly and the bars denote the normalized Indian monsoon rainfall anomaly. For normalization, the anomaly is divided by its

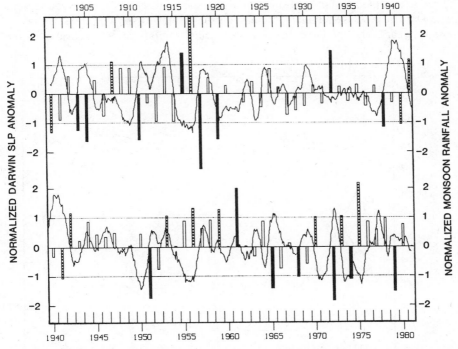

Figure 16.5. Twelve-month running mean of normalized monthly Darwin pressure anomaly (thin line) and normalized Indian monsoon rainfall anomaly (bars). Years with normalized rainfall anomaly of more than 1.0 or less than −1.0 standard deviation are shown by solid black bars for positive, and hatched bars for negative trend of the Darwin pressure anomaly. Reprinted with permission from the American Meteorological Society from Shukla and Paolino, *Monthly Weather Review*, **111**, 1830–1837 (1983).

standard deviation. The normalized rainfall anomaly is more than one standard deviation for the years 1908, 1916, 1917, 1933, 1942, 1953, 1956, 1959, 1961, 1970, 1973, and 1975, and is less than minus one standard deviation for the years 1901, 1904, 1905, 1911, 1918, 1920, 1939, 1941, 1951, 1965, 1968, 1972, 1974, and 1979. The former group of years will be referred to as the heavy monsoon rainfall years and the latter as the deficient monsoon rainfall years. The composite normalized seasonal mean Darwin pressure anomalies averaged for all the heavy rainfall years, and the deficient rainfall years, are shown in Figure 16.6. The rectangle on the graph denotes the summer season for which the monsoon rainfall was considered, and the following and the preceding months are represented along the abscissa to the right and to the left. Along the ordinate are the values of the composite, 3-month running mean, pressure anomaly.

The most remarkable feature of this figure is the simultaneous occurrence of high (low) Darwin pressure anomalies and deficient (heavy) monsoon rainfall anomalies that persist for two seasons after the monsoon. This association, however, has little usefulness for the long-range forecasting of the monsoon rainfall. For the purpose of predicting the monsoon rainfall, the most useful antecedent parameter appears to be the *trend* of the Darwin pressure anomaly before the monsoon season. The Darwin pressure anomaly decreases from winter to spring before the heavy rainfall years, and increases before the deficient rainfall years. The value of the

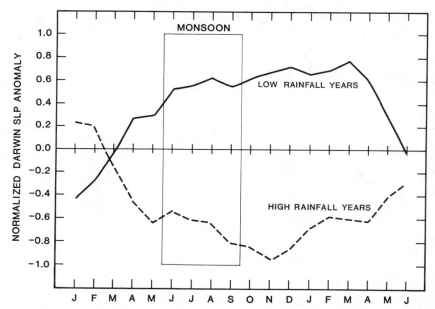

Figure 16.6. Composite of normalized Darwin pressure anomaly (3-month running mean) for heavy monsoon (high) rainfall years and deficient monsoon (low) rainfall years. Reprinted with permission from the American Meteorological Society from Shukla and Paolino, *Monthly Weather Review*, **111**, 1830–1837 (1983).

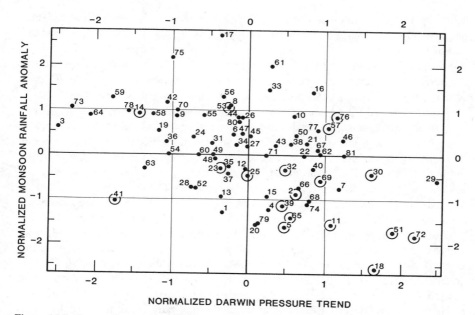

Figure 16.7. Scatter diagram between the normalized Darwin pressure trend (spring–winter) along the abscissa, and normalized Indian monsoon rainfall anomaly along the ordinate. The numbers denote the year (minus 1900). Open circles represent El Niño years. Reprinted with permission from the American Meteorological Society from Shukla and Paolino, *Monthly Weather Review*, **111**, 1830–1837 (1983).

Darwin pressure anomaly itself during the preceding winter and spring does not appear to be a useful parameter because it fluctuates around zero. We have therefore examined the association between the Darwin pressure trend and summer monsoon rainfall over India. The Darwin pressure trend is defined as the spring (March, April, May) minus the winter (December, January, February) pressure anomaly. The correlation coefficient between the normalized monsoon rainfall anomaly and the difference of the normalized spring and winter Darwin pressure anomaly is -0.46, which, in absolute value, is higher than that for the normalized spring Darwin pressure anomaly (0.32). The correlation coefficient between the normalized Darwin pressure trend and the Indian rainfall anomaly is -0.42. I am not aware of any other antecedent parameter with a correlation coefficient with the monsoon rainfall as high as -0.42 for a time series as long as 81 years. Pant and Parthasarthy (18) have shown that the correlation coefficient between a Southern Oscillation index [as defined by Wright (19) which consists of surface pressure averaged for several stations] and monsoon rainfall is 0.34.

Figure 16.7 shows a scatter diagram between the normalized Darwin pressure trend and the normalized Indian monsoon rainfall anomaly. Most of the severe drought years are in the lower right quadrant, and most of the excessive rainfall years are in the upper left quadrant of the scatter diagram. During the 81-year period examined here, there were only two occasions (1901 and 1941) when a negative Darwin pressure trend was followed by a negative normalized rainfall anomaly greater than −1.0. The near absence of points in the lower left corner of this scatter diagram suggests that a negative Darwin pressure trend is a very useful predictor for the nonoccurrence of droughts over India. Similarly a positive Darwin pressure trend can be a good predictor for the nonoccurrence of excessive rain.

Rasmusson and Carpenter (20) have identified 18 El Niño years during the 81-year period examined here, and these years have been denoted by open circles in Figure 16.7. For 8 out of 18 El Niño events, the normalized rainfall anomaly is less than −1.0, and for 14 out of 18 events the anomaly is negative. However, the predictive value of this relationship is limited only to El Niño years. During the 81-year period examined here, there were 16 instances when the normalized rainfall anomaly was close to or less than −1.0, and 8 out of these 16 cases were not associated with El Niño. If an El Niño event has already been observed in the previous winter and spring, a prediction for drought over India can be made with a high degree of confidence. The relationship between El Niño and monsoon rainfall is applicable for a limited number of cases, namely, the ones when El Niño is observed, whereas the relationship between the Southern Oscillation and monsoon rainfall is applicable in general. Monitoring of both the Southern Oscillation and El Niño can provide very useful guidance for the long-range forecasting of monsoon rainfall over India.

It can be argued that a negative trend in the winter to spring pressure is an indicator of below normal pressure at Darwin during the spring. If so, a combination of Darwin's winter–spring trend and its spring pressure anomaly should provide better guidance for forecasting the anomaly of monsoon rainfall. Figure 16.8 shows a scatter diagram between winter–spring Darwin pressure trend, along the abscissa, and its normalized spring pressure anomaly, along the ordinate. The numbers in the diagram represent the normalized monsoon rainfall anomaly for each of the 81 years. Nine out of 12 years with a normalized rainfall anomaly equal to or greater than 1.0 occur on the left half of the diagram for a negative Darwin pressure trend and 12 out of 14 years with a normalized rainfall anomaly less than −1.0 occur on the right side for positive values of the trend. It is rather remarkable that none of the 17 years with standardized rainfall anomaly less than −1.0 occur in the lower left quadrant of the diagram. In fact, out of 26 years in the lower left quadrant, there are no instances when the magnitude of the normalized rainfall anomaly is greater than 1 standard deviation in the negative sense. Most of the large negative values fall in the upper right quadrant. Out of 13 years with a standardized rainfall anomaly larger than 1.0, only one year (1961) with a value of 2.0, is in the upper right quadrant, and out of 24 years in the upper right quadrant, the rainfall anomaly is greater than or equal to 1.0 only in one year (1961). This scatter diagram suggests that if spring Darwin pressure is lower than its normal value, and if winter to spring trend shows that the Darwin pressure is falling, a prediction of nonoccurrence of drought over India in the subsequent monsoon season would be almost always right; similarly, a positive anomaly in spring Darwin pressure and a positive trend from

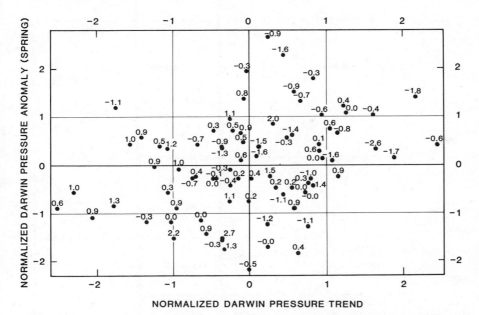

Figure 16.8. Scatter diagram between the normalized Darwin pressure trend (spring–winter) along the abscissa, and normalized spring Darwin pressure anomaly along the ordinate. The numbers denote the normalized Indian monsoon rainfall anomaly. Reprinted with permission from the American Meteorological Society from Shukla and Paolino, *Monthly Weather Review*, **111**, 1830–1837 (1983).

winter to spring would provide a highly reliable forecast of nonoccurrence of excessive monsoon rain.

3.2 Influence of Monsoon Rainfall on the Southern Oscillation

If one is to predict monsoon rain it is necessary to examine the Southern Oscillation before the monsoon season. However, it should be recalled that one of Walker's most important findings was that monsoon rainfall has very significant correlations with the subsequent global circulation. Normand (14) aptly wrote:

> To my mind the most remarkable of Walker's results was his discovery of the control that the Southern Oscillation seemingly exerted upon subsequent events and in particular of the fact that the index for the Southern Oscillation as a whole for the summer quarter June–August, had a correlation coefficient of 0.8 with the same index for the following winter quarter, though only of −0.2 with the previous winter quarter. It is quite in keeping with this that the Indian monsoon rainfall has its connections with later rather than with earlier events. The Indian monsoon therefore stands out as an active, not a passive feature in world weather, more efficient as a broadcasting tool than as an event to be forecast.

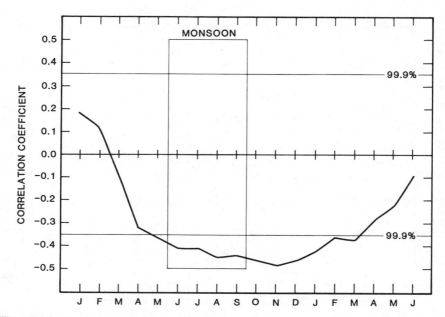

Figure 16.9. Correlation coefficient between the Indian monsoon rainfall anomaly for monsoon season (shown by rectangle) and 3-month mean Darwin pressure anomaly centered at the months shown along abscissa before and after the monsoon season. Reprinted with permission from the American Meteorological Society from Shukla and Paolino, *Monthly Weather Review*, **111**, 1830–1837 (1983).

Figure 16.9 shows the correlation coefficient for 81 years of data of the normalized monsoon rainfall anomaly and the Darwin pressure anomalies from six months before to six months after the monsoon season. The largest negative correlations are found in November following the monsoon season. This suggests a possible role of monsoon rainfall fluctuations (and the associated changes in the location and intensity of heating fields) in affecting the subsequent global circulation. The correlation coefficient between Indian monsoon rainfall anomalies and Darwin minus Tahiti pressure (not shown) is very similar to the one shown in Figure 16.9 with opposite sign.

Figure 16.10 shows the autocorrelation of seasonal mean Darwin pressure anomalies at different seasonal lags. In agreement with the earlier results of Walker, the largest correlation between adjacent seasons is found between summer and fall, and between fall and winter pressure anomalies. The slow decay of autocorrelations from summer to fall and from fall to winter and the largest correlations between monsoon rainfall and Darwin pressure following the monsoon season (Fig. 16.9) suggest a possible role of monsoons in modulating the Southern Oscillation. The smallest correlations between adjacent seasons are found between winter and spring, and spring and summer. This is seen further in Figure 16.11 which shows a scatter diagram between normalized Darwin winter–spring pressure trend and normalized

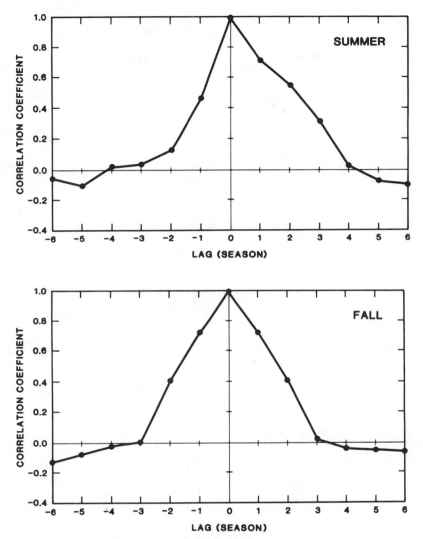

Figure 16.10. Autocorrelation of summer, fall, winter, and spring seasonal mean Darwin pressure anomaly with preceding (negative lags) and succeeding (positive lags) seasons. Reprinted with permission from the American Meteorological Society from Shukla and Paolino, *Monthly Weather Review*, 111, 1830–1837 (1983). (*Figure 16.10 continues on p. 542.*)

winter Darwin pressure anomaly. The most prominent feature is a strong inverse relationship (correlation coefficient = -0.74) between the winter pressure anomaly and winter to spring trend; if the winter pressure is higher than normal, the (spring–winter) tendency is negative and vice versa. This suggests that Darwin pressure undergoes a marked transition from winter to spring. However, the winter Darwin

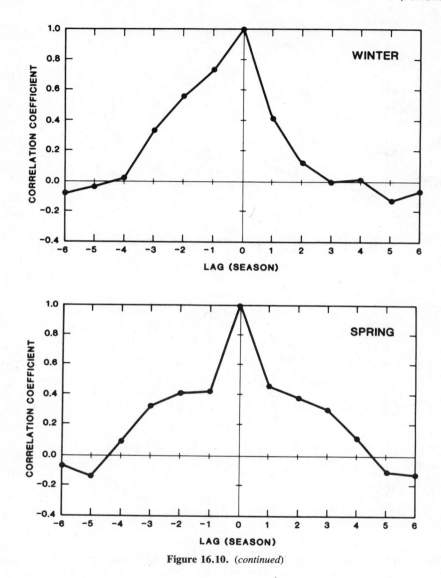

Figure 16.10. (*continued*)

pressure anomaly itself is not found to be of any particular importance for long-range forecasting of monsoon rainfall.

4 SUGGESTIONS FOR FURTHER RESEARCH

Considering the great socioeconomic importance of the fluctuations of monsoon rainfall, a more concerted research effort is required to understand its variability

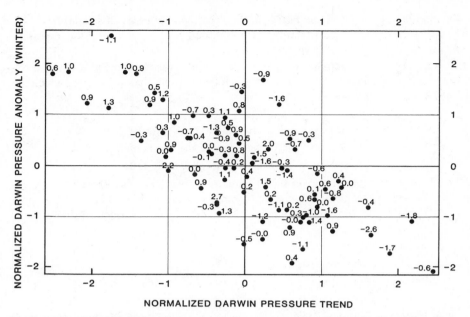

Figure 16.11. Scatter diagram between the normalized Darwin pressure trend (spring–winter) along the abscissa, and normalized winter Darwin pressure anomaly along the ordinate. The numbers denote the normalized Indian monsoon rainfall anomaly. Reprinted with permission from the American Meteorological Society from Shukla and Paolino, *Monthly Weather Review*, **111**, 1830–1837 (1983).

and predictability. Modeling and observational studies should be carried out to determine the influence of global boundary conditions and interrelationships with global circulation features. I believe that global general circulation models are already realistic enough to be used for dynamical prediction of monthly and seasonal anomalies, and that actual forecasts should be carried out with observed global initial and boundary conditions. I also believe that, parallel to this approach, it is also necessary that comprehensive synoptic and statistical studies be carried out to document the nature of monsoon variability and its relation to other tropical and mid-latitude circulation features. We have more data than Walker had 60 years ago, and we also have a better understanding of the mean circulation of the earth's atmosphere. There is reason to believe that a Walker-like effort of massive data analysis and diagnostic studies would provide valuable insights into the nature and causes of monsoon variability. During the recent years we have gained a better understanding of large-scale atmospheric phenomena such as El Niño and the Southern Oscillation, the quasi-biennial oscillation, and atmospheric blocking. This new knowledge provides a better synoptic and dynamical framework to examine the interannual and the long-term variability of monsoons. The El Niño–Southern Oscillation seems to be the single most important feature of the ocean–atmosphere system. Its period is quite large (2–5 years) and therefore it can be of practical value for predicting fluctuations of a seasonal phenomenon like the monsoon. It is therefore necessary to document the main features of the global circulation, including the monsoon, during different phases of the Southern Oscillation. It should, however, be recognized that the fluctuations of the monsoon can also be one of the important factors affecting the Southern Oscillation.

Based on the findings of several recent studies, we recommend a detailed examination of the long-term records for the following circulation features, which are not necessarily independent of each other:

1. sea level pressure over India, Australia, and Southeast Asia;
2. snow cover and snow depth over Eurasia and the Himalayas (21, 22);
3. sea surface temperature over the equatorial Pacific (20);
4. upper air circulation over India, latitudinal position of the 500-mb ridge (23), wind speed and direction over Indian upper air stations (17, 24, 25), upper troposphere thickness anomalies (26), and trough and ridge positions at 50 mb (27);
5. quasi-biennial oscillation (28);
6. blocking in the mid-latitudes (29);
7. convection over Indonesia and the equatorial Pacific;
8. typhoon activity over the western Pacific (30); and
9. boundary conditions and circulation in the Southern Hemisphere.

Combining global circulation parameters, such as those listed above, for different phases of the Southern Oscillation, and for years of droughts and excessive monsoon rainfall, might provide useful insight into the nature of monsoon variability and value of these parameters as potential predictors of monsoon rainfall.

Abilities and limitations of stochastic and dynamic models (31) for long-range forecasting of monsoon rainfall should also be further examined.

5 SUMMARY AND CONCLUDING REMARKS

The prospects for long-range forecasting of large-scale, seasonal mean monsoon rainfall appear to be good. There are significant correlations between large-scale seasonal mean Indian rainfall anomalies and low-frequency changes in the Southern Oscillation. There are also significant correlations between seasonal Indian rainfall anomalies and slowly varying boundary conditions of sea surface temperature and snow cover. Collectively, these observed associations support the notion that the seasonal mean Indian rainfall anomalies are not merely a consequence of random statistical variations in the atmosphere, but are associated with low-frequency, large-scale changes in the global circulation.

Tropical and monsoon flows are dominated by the thermally forced planetary-scale Hadley and Walker type circulations for which the primary energy source is the latent heat of condensation. The large-scale moisture convergence required for the release of the latent energy is organized by gradients of temperature at the earth's surface. Solar heating can produce thermal low-pressure areas over the land which can further deepen due to latent heating if the dynamical circulation is favorable for moisture convergence. Therefore fluctuations of soil moisture can influence the intensity of the tropical heat sources over the land. Similarly, the tropical heat sources over the oceans can be influenced by the anomalies of sea surface temperature. It is therefore reasonable to expect that the changes in the large-scale tropical flows would be related to the changes in the slowly varying boundary conditions at the earth's surface. Since dynamical instabilities are not too strong in the tropics, it is also reasonable to hypothesize that the changes in the large-scale flows are dominated by the changes in the boundary conditions. These arguments collectively suggest that there is a physical basis for predictability of the large-scale, seasonally averaged monsoon flow and rainfall.

If the daily rainfall patterns related to the monsoon's high frequency, synoptic-scale disturbances were the consequence of dynamical instabilities of the large-scale flow, and if the changes of the large-scale flow itself were caused mainly by its interaction with such unstable disturbances, the prospects for long-range forecasting, beyond the limits of deterministic prediction, would not be very good. Fortunately, this does not appear to be the case. While it is indeed true that the rain producing disturbances form only when the structure of the large-scale flow (i.e., horizontal and vertical gradients of wind, temperature, and moisture) is favorable, the changes in the large-scale flow itself appear to be primarily related to planetary-scale boundary forcing manifested as tropical heat sources and to orographic barriers. This provides a physical basis as well as hope for long-range forecasting of monsoon rainfall. It is also of interest that during the monsoon season, even the biweekly and monthly anomalies have significant spatial coherence, which further suggests that the prospects for predicting biweekly and monthly anomalies are also quite good.

REFERENCES

1. J. Shukla, Dynamical predictability of monthly means, *J. Atmos. Sci.*, **38**, 2547–2572 (1981).
2. J. Shukla, "Predictability of Time Averages: Part II. The Influence of the Boundary Forcing," in D. M. Burridge and E. Kallen, Eds., *Problems and Prospects in Long and Medium Range Weather Forecasting*, Springler-Verlag, London, 1984, pp. 155–206.
3. J. G. Charney and J. Shukla, "Predictability of Monsoons," in Sir James Lighthill, and R. P. Pearce, Eds., *Monsoon Dynamics*, Cambridge University Press, Cambridge, 1981, pp. 99–110.
4. J. G. Charney, W. J. Quirk, S. Chow, and J. Kornfield, A comparative study of the effects of albedo change on drought in semi-arid regions, *J. Atmos. Sci.*, **34**, 1366–1385 (1977).
5. S. Manabe and D. G. Hahn, Simulation of atmospheric variability, *Mon. Wea. Rev.*, **109**, 2260–2286 (1981).
6. J. Shukla, "Predictability of the Tropical Atmosphere," NASA Tech. Memo. 83829, NASA/Goddard Space Flight Center, Greenbelt, MD, 1981.

7. N.-C. Lau, A diagnostic study of recurrent meteorological anomalies appearing in a 15-year simulation with a GFDL general circulation model, *Mon. Wea. Rev.*, **109**, 2287–2311 (1981).

8. R. A. Madden, Estimates of the natural variability of time-averaged sea level pressure, *Mon. Wea. Rev.*, **104**, 942–952 (1976).

9. J. Shukla and D. S. Gutzler, Interannual variability and predictability of 500 mb geopotential heights over the Northern Hemisphere, *Mon. Wea. Rev.*, **111**, 1273–1279 (1983).

10. B. J. Hoskins and D. Karoly, The steady linear response of a spherical atmosphere to thermal and orographic forcing, *J. Atmos. Sci.*, **38**, 1179–1196 (1981).

11. H. F. Blanford, On the connexion of the Himalaya snowfall with dry winds and seasons of droughts in India, *Proc. Roy. Soc. London*, **37**, 3 (1884).

12. H. H. Hildebrandsson, Quelques recherches sur les entres d'Action de l'Atmosphere, *Kon. Svenska Vetens.-Akad. Handl.*, **29**, 33 (1897).

13. N. Lockyer and W. J. S. Lockyer, On the similarity of the short-period pressure variation over large areas, *Proc. Roy. Soc. London*, **73**, 457–470 (1902).

14. C. Normand, Monsoon seasonal forecasting, *Quart. J. Roy. Meteor. Soc.*, **79**, 463–473 (1953).

15. G. T. Walker, Correlations in seasonal variations of weather, X, *Mem. India. Meteor. Dept.*, **24**, 333–345 (1924).

16. P. Jagannathan, "Seasonal Forecasting in India, A Review," Special Publication No. DGO.82/650, India Meteorology Department, New Delhi, 1960.

17. P. Jagannathan and M. L. Khandekar, Predisposition of upper air structure in March to May over India to the subsequent monsoon rainfall of the peninsula, *Indian J. Meteor. Geophys.*, **13**, 305–316 (1962).

18. G. B. Pant and B. Parthasarathy, Some aspects of an association between the Southern Oscillation and Indian summer monsoon, *Arch. Meteor. Geoph. Biokl.*, *Ser. B.*, **29**, 245–252 (1981).

19. P. B. Wright, "An Index of the Southern Oscillation," Climatic Research Unit Report No. CRU RP4, University of East Anglia, Norwich, U.K., 1975.

20. E. M. Rasmusson and T. H. Carpenter, The relationship between eastern equatorial Pacific sea surface temperature and rainfall over India and Sri Lanka, *Mon. Wea. Rev.*, **111**, 517–528 (1983).

21. D. Hahn and J. Shukla, An apparent relationship between Eurasian snow cover and Indian monsoon rainfall, *J. Atmos. Sci.*, **33**, 2461–2463 (1976).

22. R. R. Dickson, Eurasian snow cover vs. Indian monsoon rainfall—An extension of the Hahn-Shukla results, *J. Clim. and Appl. Meteor.*, **23**, 171–173 (1984).

23. A. K. Bannerjee, P. N. Sen, and C. R. V. Raman, On foreshadowing southwest monsoon rainfall over India with mid-tropospheric circulation anomaly of April, *Indian J. Meteor. Hydrol. Geophys.*, **29**, 425–431 (1978).

24. P. V. Joseph, Sub-tropical westerlies in relation to large scale failure of Indian monsoon, *Indian J. Meteor. Hydrol. Geophys.*, **29**, 412–418 (1978).

25. P. V. Joseph, R. K. Mukhopadhyaya, W. V. Dixit, and D. V. Vaidya, Meridional wind index for long range forecasting of Indian summer monsoon rainfall, *Mausam*, **32**, 31–34 (1981).

26. R. K. Verma, Importance of upper tropospheric thermal anomalies for long range forecasting of Indian summer monsoon, *Mon. Wea. Rev.*, **108**, 1072–1075 (1980).

27. V. Thapliyal, "Stratospheric Circulations in Relation to Summer Monsoon over India," Proc. of Hydrological Aspects of Droughts I, India Meteorology Department, New Delhi, 1979, pp. 347–362.

28. K. S. Rao Raja and N. J. Lakhole, Quasi-biennial oscillation and summer southwest monsoon, *Indian J. Meteor. Hydrol. Geophys.*, **29**, 403–411 (1978).
29. M. Tanaka, Interannual fluctuations of the tropical easterly jet and the summer monsoon in the Asian region, *J. Meteor. Soc. Japan*, **60**, 865–875 (1982).
30. M. Kanamitsu and T. N. Krishnamurti, Northern summer tropical circulations during drought and normal rainfall months, *Mon. Wea. Rev.*, **106**, 331–347 (1978).
31. V. Thapliyal, Stochastic dynamic model for long range prediction of monsoon rainfall in peninsular India, *Mausam*, **33**, 399–404 (1982).

17

Short- and Long-Range Monsoon Prediction in India

P. K. Das
India Meteorological Department*
New Delhi, India

INTRODUCTION

India experiences two monsoons every year. The first is a "hundred day" summer monsoon from around the first of June to the end of September, while the second is a winter monsoon of shorter duration from mid-October to the end of December. The summer monsoon is generally considered to be *the* monsoon because of its larger sphere of influence—it covers the entire country—while the winter monsoon is largely confined to the southern half of the Indian peninsula. This chapter is concerned with the summer monsoon.

Summer rains are important to India because of their impact on agriculture. Every year the country faces the problem of maximizing the output of food grains from an unknown rainfall pattern. The variability of rainfall from one year to another adds to the difficulties of agriculture.

Agricultural planning depends strongly on three monsoon predictions:

1. The dates of onset or arrival of monsoon rains over different parts of India.
2. Short-period fluctuations in rainfall of 5 to 7 days duration within the monsoon season which represent the "active" and "break" phases of the monsoon.
3. The total amount of rainfall from the beginning of June to the end of September.

1 MONSOON FEATURES

Improvements in our understanding of the monsoon and the mechanisms responsible for it, as well as improvements in prediction, depend largely on the availability of high quality observations, especially over the normally data-sparse regions such as the Indian Ocean. Encouraging developments have taken place in recent years. Space technology in India and the meteorologist's access to satellite observations have helped prediction. To this, one must add a growing spirit of cooperation

---

* Present affiliation, Centre for Atmospheric Sciences, Indian Institute of Technology, Delhi.

through international data collection exercises. The Monsoon Experiment (MONEX) is the best known example. As precursors to MONEX, an International Indian Ocean Expedition (IIOE) was mounted in the mid-1960s. Subsequently, there were two other multinational expeditions in 1973 and 1977. These experiments have added to our knowledge about regions over which little or no data were available in the past. The data have helped to provide better insight into the dynamics of monsoonal winds and rain.

1.1 Dates of Onset and Retreat

Prediction of the dates of onset and retreat of rains help farmers to select the most suitable crops to plant and to determine when to prepare the land. Times of sowing, the initial supply of fertilizers, and adjustments in the density of seedlings are examples of operations that depend on a prediction of the date of onset. A similar situation prevails at the time of the monsoon's retreat. Heavy rains, or damp conditions after the harvest could lead to a substantial reduction in output.

Opinions differ on what really constitutes an onset of the monsoon. Some define it by a change in wind direction. On the other hand, it is often more convenient to delineate an onset by the commencement of rains. The latter is the practice in India because the main impact of the monsoon is generated by rainfall. Farmers associate the summer monsoon with the arrival of rains.

It must be emphasized that neither the onset nor the withdrawal of the monsoon can be defined with mathematical precision. Ambiguity on this aspect has often led to divergence of opinion on the duration of the monsoon. Some consider the monsoon to be a 120-day phenomenon because long-range forecasts are prepared for a period beginning with June and ending with September. But the duration of rains is not the same over all parts of the country. It is shorter over northern India and longer over the southern parts of the country. For the country as a whole, the average duration is 100 days.

One cannot overlook the fact that the onset and the cessation of rains are gradual processes. The onset starts with a transition period when the atmosphere changes from a state of extreme heat to one of high humidity and light rain. Many parts of northeastern and eastern India experience violent thunderstorms in the premonsoon months of April and May. They are well known as Nor'westers because the winds associated with them blow from the northwest. In some years, when premonsoon thundershowers are severe and frequent, it is difficult to distinguish between a premonsoon thunderstorm and genuine monsoon rain; but on most occasions there is a fairly well-defined period of transition with gradual changes.

The onset of the monsoon is better defined than its withdrawal, especially over southern India. In some years there is no clear demarcation between the retreat of the summer monsoon and the onset of the winter monsoon over the Indian peninsula. For this reason one can not be too precise about the duration of the summer monsoon. The question is often asked: Is winter rain over the Indian peninsula influenced by the preceding summer monsoon? Patnaik et al. (1) found that the two monsoons are independent of each other; the correlation coefficient between summer and winter rainfall was only around 0.3 over the southern half of the Indian peninsula and much less over other parts of the country.

Figure 17.1 depicts the meteorological subdivisions of India. A meteorological subdivision does not of necessity coincide with the boundaries of different Indian states. Thus the plains of Bihar and Bihar plateau represent two meteorological subdivisions, but both belong to the single state of Bihar.

Figure 17.1. The meteorological subdivisions of India. The material presented in this map as well as all subsequent maps or figures in this chapter does not imply the expression of an opinion by the author on the frontiers or boundaries of countries. The maps shown in Figures 17.1–17.5 and Figure 17.7 were provided by the Indian Meteorological Department.

Monsoon rains first set in over Kerala, a state located on the southern tip of the Indian peninsula. A survey of nearly 70 years of rainfall data reveals that the onset is between May 11 and June 15 each year. Rainfall prior to May 11 is consequently not regarded as part of the monsoon by the Indian Meteorological Department.

Empirical rules are used to define the onset over different parts of the country. These rules are designed to ensure that the observed rain is uniformly spread and is not merely a transient phenomenon. As an illustration we reproduce the working rules that are used by the Indian Meteorological Department to define the onset over Kerala:

1. Beginning with May 10 at least five out of the seven meteorological stations in Kerala should record 24-hour rainfall amounts of 1 mm or more for two consecutive days. The monsoon's arrival is announced on the second day.
2. If three or more of the seven observing stations report no rain for three consecutive days, the monsoon is considered to have receded from Kerala on the third day. A recession implies weak monsoon conditions on the preceding two days. As the monsoon is an oscillating current, it is not unusual for it to set in over Kerala and then withdraw temporarily. But an onset followed by a temporary withdrawal is rarely observed after the rainbelt has advanced north of 13°N.
3. Temporary onsets are not considered for preparing the climatology of the onsets.

Similar working rules exist over other parts of the country. The onset is discerned by an increase in the moisture content of the atmosphere and a sustained increase in rainfall. For this purpose, moving averages of rainfall over five days, or pentads as they are called, are considered. The midpoint of a pentad in which a sudden increase in rainfall is recorded and subsequently sustained is considered to be the date of onset. In fixing the date of onset, other supporting evidence, such as an increase in moisture, a rise in dew point and changes in wind direction are also considered. Guided by these considerations, the normal dates of onset are shown in Figure 17.2. It is observed that the standard deviation of the onset date is about a week for most parts of the country. For Kerala it is about 7.7 days. Distress situations occur if the onset is delayed by more than a week.

The normal dates of retreat are shown in Figure 17.3. While monsoon rains set in over southern India by the beginning of June and gradually move northward until the entire country is covered by mid-July, the withdrawal proceeds in the opposite direction. This has an interesting impact on agriculture.

A late onset is more harmful over northern India than a similar delay over the south. A delay in the onset by a week or more over the northern and central parts of the country means little or no rain for June because the rains normally arrive in the second half of June. A similar situation occurs if the monsoon retreats earlier than expected. A beginning of the retreat toward the end of August may well mean little or no rain in the last three or four weeks of the monsoon over northern India, while the southern states are still receiving rain.

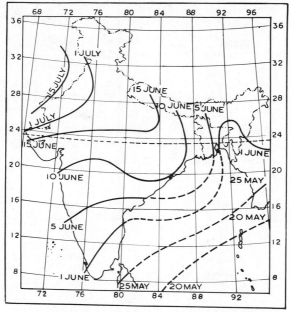

Figure 17.2. Normal dates of onset of the summer monsoon.

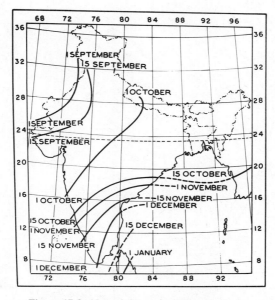

Figure 17.3. Normal dates of withdrawal of rains.

Figure 17.4 shows the distribution of the coefficient of rainfall variability. This expresses the variance as a percentage of the mean rainfall. The normal rainfall distribution is shown in Figure 17.5. It is worthwhile to note that the largest variance in rainfall occurs in regions that receive the least amount of rain. Rainfall variability is the outcome of several diverse features, such as the proximity of a region to the normal path of rain-bearing depressions, or the prevailing winds over the region, but the shorter duration of monsoon rains is also a factor that contributes to the greater variability of monsoon rains over northern India.

A late onset coupled with an early withdrawal of the monsoon over most parts of India can cause severe droughts as occurred, for example, in 1972 and 1979. The planetary circulation features that lead to good monsoons in some years and

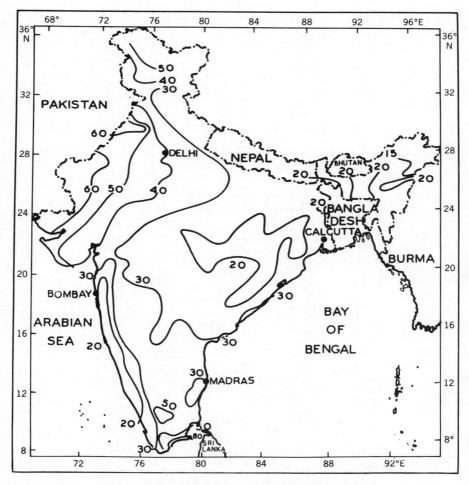

Figure 17.4. Coefficient of rainfall variability in percentages.

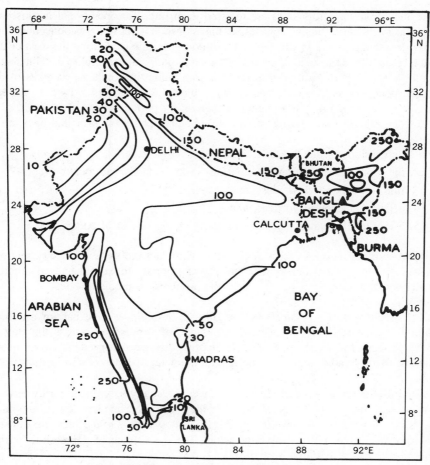

Figure 17.5. Normal rainfall distribution in cm.

deficient ones in others are not yet well understood. This topic is explored in other parts of this volume; this chapter will be confined to the regional features of the summer monsoon.

1.2 Short Duration Rainfall Fluctuations

Rainfall fluctuations of 5 to 7 days duration are frequently observed throughout the monsoon. In recent years, some studies have provided evidence to suggest cycles of 15 and 40 days. Opinion is divided on the influence of such cycles on monsoon rainfall; consequently, periodicities of this nature have not been taken into account for rainfall prediction in India. Even if the suspected periodicities were true, a considerable amount of variability would still remain in monsoon rainfall.

Fluctuations of 5 to 7 days are referred to in India for operational purposes as the "active" and "break" phases of the monsoon. Active monsoon conditions provide copious rainfall over the plains of northern India and along the west coast, but a break moves the zone of maximum rainfall, often dramatically, to the Himalayan foothills of northern India. An increase in rain along the southeastern coast of the Indian peninsula is also observed during a break. Meanwhile, as long as a break persists, the remaining parts of India experience a period of lean rainfall.

Of the several rain-bearing systems of an active phase the principal ones are:

1. depressions in the Bay of Bengal;
2. the monsoon trough and fluctuations in its intensity and location;
3. mid-tropospheric low-pressure systems;
4. offshore and onset vortices; and
5. a low-level equatorial jet stream along the east coast of Africa.

The coupling between these regional systems and the larger planetary motions is still obscure. Nevertheless it is entirely appropriate to examine the distinctive features of these systems to see where the problems in prediction lie.

1.2.1 Monsoon Depressions in the Bay of Bengal.

A good part of monsoon rainfall is generated by the westward passage of depressions across the Bay of Bengal. On an average, two or three depressions are observed in the monsoon months of July and August. The horizontal diameter of a depression is about 1000 km.

The lifetime of a depression is about a week. Usually they move northwestward for the first three or four days; thereafter, they will either recurve toward the north or continue to move westward. Figure 17.6 shows the surface pressure and the rainfall associated with a Bay depression. The mean track of depressions during the monsoon is also shown in Figure 17.6.

A peculiar feature of the monsoon depression is the concentration of rainfall in its southwestern sector. After the formation of a depression, forecasters predict heavy rain along the southern and southwestern sectors of its anticipated track. The vertical structure of a monsoon depression has a slight tilt toward its southwestern sector (2). Joseph and Chakravarty (3) observed horizontal temperature gradients of 2–4°C within the field of a depression in the lower troposphere. Small temperature gradients of this magnitude do not support the existence of fronts, as is the case in mid-latitude depressions. Monsoon depressions usually have a cold core; consequently, their intensification into tropical cyclones is not frequent, but there are occasional exceptions as in the first half of June 1982, when an incipient depression intensified to form a tropical cyclone.

Studies by V. D. Iyer (4, 5) suggested that the remnants of many tropical cyclones over the South China Sea move into the Bay of Bengal as residual low-pressure systems. A much more recent study by Saha et al. (6) supports Iyer's hypothesis made over 50 years ago. While it is possible that Bay depressions are activated by

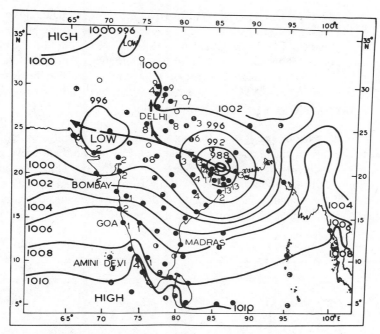

Figure 17.6. Surface pressure in mb and 24-hour rainfall in cm for a Bay depression. Circles show locations of weather observing stations with relative cloudiness indicated. Thick line with arrow indicates normal path of Bay depressions (from reference 53).

westward moving systems from the east, their development and movement is controlled by:

1. horizontal and vertical wind shears;
2. vertical transport of mass, heat, and momentum by convection; and
3. the transfer of sensible and latent heat from the sea surface.

Unfortunately, quantitative information on these factors is difficult to acquire because of a sparse network of upper-air stations. For short-range prediction of depression development, forecasters generally look for weak vertical wind shears accompanied by an increase in the cyclonic curvature of winds in the lower troposphere, but no threshold value of the shear (horizontal or vertical) has been found that clearly indicates cyclogenesis over the Bay of Bengal.

Dewan et al. (7) find a sharp increase (by a factor of three) in the eastward flux of moisture across the Bay of Bengal between a weak and an active phase of the monsoon. Pearce and Mohanty (8) suggest that a threshold value of 4 cm of precipitable water is needed for convective instability. Although Pearce and Mohanty confined their study to the Arabian Sea, Dewan et al. found that this value of precipitable water is present on most occasions over the Bay of Bengal. Observations by Ghosh et al. (9) support this view. Thus convective instability is present in the atmosphere over the Bay of Bengal and its release is associated with the changes in wind direction in the formative stage of a Bay depression. This is a useful guide for predicting the formation of a Bay depression, especially now that the new Indian geostationary satellite (INSAT) has considerably improved India's capability for observing clouds and deriving upper-level winds.

1.2.2 The Monsoon Trough.

After approaching the Indian coastline as a southwesterly current the monsoon air is deflected to the northeastern region of India and the northern parts of Burma. Subsequently, it flows along the plains of northern India and the southern periphery of the Himalayas as an easterly current. The flow of air is around a quasi-permanent zone of low pressure over the plains of northern India. This is referred to as the monsoon trough (Fig. 17.7). The axis

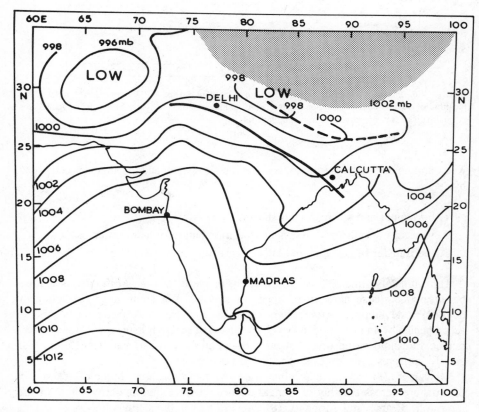

Figure 17.7. The monsoon trough. The dashed line and isobars show the trough axis and typical pressure pattern for a break situation. Full line shows the trough axis in its normal position. The shaded area shows the approximate location of the Himalayas.

of this trough is oriented in a northwest–southeast direction. It runs parallel to the southern edge of the Himalayas.

Although the trough is quasi-permanent in nature, its axis exhibits wide variations in location during the course of the monsoon. The normal position of the axis of the trough is shown on Figure 17.7. The normal position runs approximately from Delhi (28.5°N, 77°E) to Calcutta (22.5°N, 88°E) through the city of Allahabad (25.5°N, 82°E). It must be made clear that the normal position is not a line which can be drawn with mathematical precision. It represents a line of symmetry between westerly or southwesterly winds to its south and easterly or southeasterly winds to its north. Active phases of the monsoon occur when the axis of the trough is to the south of its normal position and its eastern end dips into the Bay of Bengal. In such a situation there is heavy rain over the plains of northern India, the central parts of the country and along the west coast. An extension of the axis into the Bay of Bengal is often the precursor of a Bay depression. Once a depression forms, it is followed by an increase in rainfall intensity over those parts of central India that lie to the south of the axis. But when the axis of the trough moves to the north and lies close to the Himalayas the result is a break in monsoon activity. During this phase the rainbelt moves north to a new location very close to the foothills of the Himalayas. One of the major rivers of northeastern India, the Brahmaputra, flows through a narrow valley between the Himalayas to the north and the mountains of northeastern India. This valley receives very heavy rain during a break phase. As the origin of several Indian rivers and their tributaries lies over the eastern Himalayas, a northward location of the trough often leads to heavy discharge and resultant flows over the northeastern states of the country. As a consequence of deforestation, the rate of sedimentation on these river beds has increased over the years. A decrease in their holding capacity has made the land in their vicinity more vulnerable to floods.

As pointed out, the duration of a break normally varies from 5 to 7 days, but abnormal breaks lasting up to three weeks can occur at times. The duration and frequency of breaks tends to increase during the second half of the monsoon, that is, in August and September.

The mechanism of fluctuations in the position of the monsoon trough is not yet properly understood. For many years it was believed that the monsoon trough was a mechanical effect brought about by the orientation of mountains. However, recent model experiments by Das and Bedi (10) suggest that the mountains by themselves are not sufficient to generate a quasi-permanent trough. Fluctuations in the earth–atmosphere radiation balance and changes in the reflectivity of the soil are thought to have an important bearing on the movement of the trough axis (11). While these ideas are in an exploratory state at present, they tend to stress the importance of feedback mechanisms generated by changes in cloud distribution and conditions at the earth's surface.

1.2.3 Mid-Tropospheric Disturbances.
Mid-tropospheric disturbances represent cyclonic vortices that are confined to the middle troposphere. They are observed over the northeastern sector of the Arabian Sea and the adjoining parts of India.

The vortices are most prominent at altitudes between 3 and 6 km, with their largest amplitude near 600 mb (4 km). The horizontal radius of a mid-tropospheric vortex is about 500 km.

Observations suggest that the vortices have a warm core above the middle level and a slightly colder core below. Unlike the depressions of the Bay of Bengal, these low-pressure systems exhibit little movement. They tend to remain quasi-stationary for many days, but are capable of generating as much as 20 cm of rain in 24 hours. Similar vortices are also observed over the South China Sea but not as frequently as over the Arabian Sea. The cyclonic circulation associated with the Indian systems lies between westerly winds at the surface and easterly winds in the upper troposphere. Large horizontal wind shear in the north–south direction is believed to lead to the development of mid-tropospheric systems, but the precise mechanism is still under investigation.

1.2.4 Offshore and Onset Vortices.
In addition to mid-tropospheric vortices, offshore vortices bring spells of heavy rain to the west coast of India.

The western Ghats are an important orographic feature of the west coast of India. These mountains run in a north–south direction along the western coast line and are approximately 1000 km in length and 250 km in breadth. The average altitude of the mountains is between 1 and 1.5 km. When the southwesterly monsoon winds strike the Ghats, they tend to be deflected around the mountains and the return current often forms an offshore vortex. Despite their small dimensions (roughly 100 km in diameter) these vortices generate large amounts of rain, about 10 cm in 24 hours lasting for 2–3 days.

Recently there has been a resurgence of interest in the process of the formation of a vortex off the coast of Kerala a few days prior to the monsoon's onset. This is usually referred to as an *onset vortex*. Its horizontal dimension is small but, like offshore vortices, it is capable of causing heavy rain. Interest in the onset vortex has been stimulated by the discovery of a rapid increase in the kinetic energy of the lower atmosphere over the Arabian Sea about a fortnight before the monsoon's onset (12). Some regard this vortex as an indispensable condition for the monsoon's onset, but opinion is divided on this question. In some years the vortex can be discerned on weather charts, but in other years its existence is doubtful.

1.2.5 Low-Level Cross-Equatorial Jet and the Somali Current.
During the mid-1960s and the late 1970s J. Findlater (13), a British meteorologist who was stationed in Kenya, observed a narrow current of very strong winds off the coast of East Africa. The jet stream was most pronounced at low levels between 1 and 1.5 km (see Chapter 15, Figure 15.3). It is now known to flow from a location east of Madagascar before reaching the coast of Kenya at around 3°S. Later it flows northward before it is deflected by the highlands of Kenya and Ethiopia toward India.

An important feature of the low-level jet is that its path near 9°N is associated with a zone of coastal upwelling. The coastal upwelling, caused by the low-level jet stream, generates a narrow strip of cold sea surface temperature off the coast of Somalia. On some occasions the temperature of the sea surface is as low as 15°C while it is close to 30°C off the west coast of India. A strong sea surface temperature gradient is thus set up over the Arabian Sea.

2 SHORT-TERM PREDICTION

Numerical methods for operational short-term prediction have been introduced in India recently. Model experiments to simulate the wind, temperature, and precipitation for the monsoon region contribute to our understanding of the models' capabilities for prediction. They provide an opportunity for evaluating the impact of external forces and boundary conditions on the evolution of the summer monsoon. Indian work in this area so far has been confined to regional models with a limited amount of physics, but work on a general circulation model with a variable grid has just commenced. For the present we will confine ourselves to the simple regional model.

2.1 Numerical Prediction of the Monsoon

Forecasts of winds at five levels in the troposphere are made on an operational basis in India with a quasi-geostrophic model. The details of the model are provided by Ramanathan and Bansal (14).

The governing model equations represent conservation of the vertical component of vorticity and the conservation of entropy through the first law of thermodynamics. By elimination, a governing equation is obtained for the vertical wind. This is solved by a numerical algorithm with suitable boundary conditions (15).

The equations of the model are, strictly speaking, valid where the Coriolis force is relatively large, that is, in the subtropical and higher latitudes. The model's performance is less satisfactory over the region between the equator and 10°N. The energetics and a scale analysis for the relevant equations of the model have been provided by Lorenz (16) and Phillips (17).

There are three forcing functions that determine the vertical wind:

1. the advection of vorticity;
2. the advection of the temperature; and
3. the diabatic heating.

When the advection of vorticity increases with height, the model provides for ascending motion. Similarly, subsidence occurs whenever the vorticity decreases with height. All other things being equal, this leads to upward motion ahead of and descending motion to the rear of vorticity maxima.

The advection of the temperature, which also generates vertical motion, is computed at the levels marked $Z_1, Z_2, \ldots, Z_5$ (Fig. 17.8). When there is advection of warm air into a region, there is ascending motion; similarly the advection of colder air generates descending motion.

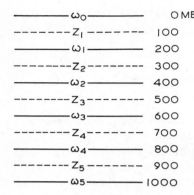

Figure 17.8. The vertical structure of the Indian model. The vertical wind is denoted by omega (ω). $Z_1, Z_2 \ldots$ denote the heights of pressure surfaces (from reference 14).

Diabatic heating is incorporated into the model by a simple empirical method. The convergence of moisture is computed for a vertical column of the atmosphere. It is then assumed that a small fraction of the moisture is used to moisten the atmosphere (instead of releasing latent heat). This fraction is assumed to remain constant throughout the depth of the atmosphere. The remaining part of the moisture is assumed to form clouds and release latent heat. Diabatic warming takes place only over regions of ascending motion. This method of parameterizing the effect of cumulus clouds was designed by Kuo (18, 19).

The model assumes that the vertical wind vanishes at the top of the atmosphere. Near the earth's surface the vertical wind is affected by both terrain and surface friction. The influence of mountains is introduced by a terrain generated vertical wind which is approximated by the equation

$$(\omega_s)_1 = \mathbf{V} \cdot \boldsymbol{\nabla} p_s \tag{1}$$

where the subscript s refers to the surface, $\mathbf{V}$ is the wind, ω stands for the vertical wind, and $\boldsymbol{\nabla} p$ is the horizontal pressure gradient across the terrain.

The influence of surface friction on the vertical wind is introduced at the lower boundary by the empirical relation

$$(\omega_s)_2 = \frac{g\rho_s}{f} \left[\frac{\partial}{\partial y} (C_D u_s V_s) - \frac{\partial}{\partial x} (C_D v_s V_s) \right] \tag{2}$$

Where g is the acceleration due to gravity, ρ_s is the density of air at the surface, f stands for the Coriolis parameter and u_s, v_s are the eastward (x) and northward (y) components of the surface wind speed $V_s = (u_s^2 + v_s^2)^{1/2}$.

The following empirical values are used for the drag coefficient (C_D):

$$C_D = 0.0005 \, |V_s|^{1/2} \, ; \qquad V_s \leq 15 \text{ m/sec} \tag{3a}$$

$$C_D = 2.6 \times 10^{-3} \, ; \qquad V_s > 15 \text{ m/sec} \tag{3b}$$

Frictional dissipation thus decreases the vertical wind in areas of negative vorticity. The vertical wind at the lower boundary (1000 mb) is the sum of $(\omega_s)_1$ and $(\omega_s)_2$.

The vertical profile of the model is shown in Figure 17.8. The region over which computations are made extends from the equator to 60°N and from 0°E to 140°E (Fig. 17.9). This region is covered by a rectangular grid and the mathematical equations are solved by the computer at each grid point. This is done for five layers in the model and the numerical method conserves the kinetic energy and the mean vorticity of the model atmosphere during integration. Along the lateral boundaries of the model domain the tendency of the height field $(\partial z/\partial t)$ vanishes.

A coarse grid at intervals of 5° of longitude and latitude first provides a forecast over the entire region. The computed values of the height tendency are later used as boundary conditions for the high resolution region (Fig. 17.9), which is embedded in the coarse grid and extends from the equator to 42.5°N and from 40°E to 120°E. The spacing of grid points in the finer mesh is 2.5°. A mercator map projection is used. The input data consist of contour heights for 850, 700, 500, 300, and 100 mb, along with the moisture content of the atmosphere at these levels. The contour heights for 900 mb (Fig. 17.9) are obtained by interpolation.

Upper-level winds, forecast in real time by the European Centre for Medium Range Weather Forecasts (ECMWF), are distributed on the Global Telecommunication System of the World Weather Watch. The operational global model of the ECMWF is considerably more sophisticated than the regional model described here. We tried to see how the performance of the simple regional model compared with that of ECMWF. The comparison of the root mean square (rms) errors for two levels is shown in Table 17.1 for the period August 1982 to January 1983.

The figures for the ECMWF forecasts in Table 17.1 are for a region extending from 6°N to 33°N and from 72°E to 120°E (19). The verification for the Indian

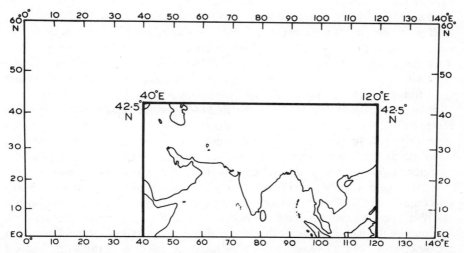

Figure 17.9. The region covered by the Indian model. Coarse (outer rectangle) and fine (inner rectangle) grid domains are shown (from reference 14).

TABLE 17.1 Upper-Level Wind Forecasts (24 hours) Average RMS Vector Wind Errors (m/sec)

		Pressure (mb)			
		850		200	
		ECMWF	India	ECMWF	India
1982	August	4.2	8.0	7.2	10.0
	September	3.5	7.2	5.7	10.4
	October	3.0	5.9	5.8	10.9
	November	3.5	6.9	7.7	12.5
	December	3.8	6.9	7.8	12.5
1983	January	3.6	6.8	8.2	18.0

model is for the inner region in Figure 17.9, which is larger. The differences are partly due to this difference in the verification areas, and partly because of different standards of comparison. The ECMWF forecasts are measured against ECMWF analyses of observations, whereas the Indian model forecasts are measured against the Indian analyses.

The ECMWF forecasts are better because they are made with a more sophisticated model. The Indian model does not perform very well near the equator and over the mountainous regions to the east and west of India. According to the analyses by the Indian Meteorological Department, rms errors that would result from forecasts of "no change" or persistence, are approximately 7 m/sec for the lower troposphere (850 mb) and 14 m/sec for the upper troposphere (200 mb). At this stage we can only state that, even with limited computer facilities, it is possible to prepare forecasts of upper-level winds which perform a little better on the average than persistence forecasts, especially in the upper troposphere. January 1983 was an exception because the data over India were sparse in that month.

It is encouraging to note that after January 1983 further improvements have been made in the ECMWF model (20). The rms errors are now within 3 m/sec for the lower troposphere and 6 m/sec for the upper troposphere. This bodes well for the meteorology of the tropics, but need not preclude national meteorological services from trying to develop models for their own regions.

It is interesting to note that both the ECMWF and the Indian models show an increase in errors during winter. This is the season of a strong westerly jet stream across northern India. It is possible that both models tend to smoothen the high wind speeds that occur within the jet and thus underestimate its maximum winds.

Attempts have been made in the past—in a research mode—to compute the vertical motion associated with monsoonal circulations with a multi-level quasi-geostrophic model. But for operational use the prediction of vertical motion has not been very successful because our understanding of convective systems and how to represent them in larger-scale forecast models is not yet very clear. For example, the observations of the International Monsoon Experiment of 1979 (MONEX) suggest the existence of mesoscale vortices within a monsoon depression in the Bay of Bengal. These small-scale systems can not be represented explicitly by the relatively coarse grid of existing forecast models.

2.2 Prediction of Bay of Bengal Phenomena

Short-range predictions are made operationally in India for two other phenomena which we wish to describe briefly. These are tropical cyclones and storm surges. Although these systems are not very frequent during the summer monsoon, we mention them because they provide examples of objective methods of prediction in India.

2.2.1 Tropical Cyclones.
Tropical cyclones are cyclonic storms with surface winds in excess of 63 km/hr. They are most frequently observed in the Bay of Bengal in the premonsoon months of April and May and the postmonsoon months of October and November. The precise reason for this frequency distribution is not well understood. Research has shown that diverse features, such as a warm sea surface and advection of warm air into an incipient vortex, lead to conditions that favor tropical cyclone formation, but firm prediction rules do not yet exist.

Despite their low frequency during the monsoon, a few tropical cyclones do form over the Bay of Bengal and the Arabian Sea. In recent years, a tropical cyclone struck the Indian coastal state of Orissa on June 3, 1982 and caused much devastation. The port of Paradeep (20°N, 86.5°E) was badly damaged, and 245 lives were lost (21).

Tropical cyclones of the monsoon season are usually observed in the region between 16°N and 21°N and 90°E to 95°E. Initially they are observed as Bay depressions (discussed earlier), some of which intensify into deep depressions or tropical cyclones. Firm guidelines for deciding which depressions will intensify and which will not have yet to be established, but satellite observations indicate greater organization of clouds in those depressions that are likely to intensify. Progress toward better prediction is hampered by the paucity of upper-level winds, but this difficulty is being overcome, to some extent, by weather satellites.

The initial movement of a cyclone in the Bay of Bengal is toward the west or northwest. Some continue to move in the same direction and eventually strike the Indian coast, while some recurve and move toward the north or northeast. The recurvature of a tropical cyclone is a difficult prediction problem.

Regression equations are used to predict the future movement of storms. A number of possible predictors are first screened. Those which explain the largest percentage of variance of the predictand, namely the storm track, are combined to form a regression equation for its future movement. The predictors are:

1. the previous 12- and 24-hour positions of the cyclone;
2. estimated pressure at the storm center 24 hour earlier; and
3. estimated pressure near the outer periphery of the cyclone 24 hours earlier.

Various regression equations have been developed from time to time. The most recent one is by Neumann and Mandal (22). The rms error for a one-day forecast of the position of a cyclone by this technique is about 145 km (23).

The performance of regression equations is sensitive to the accuracy with which the current location of a cyclone is fixed. Unless the observed cyclone has a well defined "eye" which is visible in satellite observed clouds or on the radar screen, the location of its center is difficult to determine. This limits the accuracy of such predictions.

Apart from regression equations, prediction has been tried with the help of analogues (24, 25). The history of all cyclones for the 90-year period from 1891 to 1982 have been put on magnetic tape and stored in a computer. Given the location of a tropical cyclone, the computer is programmed to select from its stored history the nearest analogue to the observed cyclone. The best analogue is selected by comparing the path, the size and other distinguishing features of the observed cyclone with those of the past. Sen Sharma (23) found that the average error in a 24-hour analogue forecast of the cyclone's position is about 157 km.

Prediction by a numerical model was tried by Ramanathan and Bansal (26), but only for a small number of cyclones. Their model results have not been compared with predictions by other means.

2.2.2 Storm Surges.

Much of the damage caused by tropical cyclones is through coastal inundation. A sudden rise in sea level is often associated with the cyclone's landfall. This is known as a "storm surge." It is caused by strong winds with an onshore component and a reduction in atmospheric pressure. The cyclone which hit the coast of Orissa on June 3, 1982 generated a surge of 3.9 m.

Objective methods have been developed in India to predict a surge with the help of models. These models determine the response of the sea near coastal regions when it is disturbed by strong cyclonic winds (27–29).

For prediction purposes an estimate is first made of the speed of the storm as it approaches the coast, its time of likely landfall, its maximum wind speed, and its central pressure. In addition, characteristics such as the coastal geometry, variations in the depth of the sea and sea bed friction are specified. This information is incorporated in a mathematical model, the governing equations of which are solved on a computer. The computer output provides an estimate of the peak surge (rise in sea level) at landfall for storms of different size and speed. The intensity of a storm is measured by the pressure difference between its outer periphery and its center (Δp), while its size is determined by the radial extent of strong winds from the storm center.

The peak surge (h_L) in meters is related to the pressure difference in mb and the storm's speed in kilometers per hour (c) by the relation

$$h_L = a_0 \Delta p + a_1 (\Delta p)^2 + a_2 c \tag{4}$$

where a_0, a_1, and a_2 are constants that are determined by the model. These constants assume different values for different storm tracks. As an illustration, Table 17.2

TABLE 17.2 Numerical Values of Constants

Track	a_0 $(\times 10^2)$	a_1 $(\times 10^4)$	a_2 $(\times 10^2)$
(a) Northeast	9.59	−0.91	−4.60
(b) North	2.88	3.08	−1.20
(c) Northwest	8.24	−1.60	−5.15

provides numerical values of the constants for three commonly observed tracks in the Bay of Bengal. They represent a storm moving from the central parts of the Bay of Bengal (a) northeastward to the coast of Bangladesh, (b) north to the coast of West Bengal in India, and (c) toward the northwest to the coastal state of Orissa.

Nomograms are available to a forecaster for rapid estimation of the peak surge. Similar models for predicting the storm surge have been designed for other parts of the Indian coastline.

In the past the operational forecast model did not represent the coastal geometry of the Bay of Bengal very well. It also needed improvement in treating storms that tended to strike the coast at an angle. Improvements on these aspects have been made by Johns et al. (28). Dube et al. (29) suggest that better forecasts are obtained if the winds associated with a cyclone are specified by an empirical dependence on radial distance from the storm center. This suggestion requires further study.

The validation of the model forecasts has not been very satisfactory so far, because reliable data on surge amplitude and the time of the maximum surge are difficult to acquire. It is important to note that the maximum surge need not occur when the storm strikes the coast. Sometimes it occurs a little later. Therefore it is very important to improve the network of tide gauges on the coastline. But, we should point out that the data that have been acquired so far suggest reasonable agreement between the predicted and observed surge amplitudes.

3 LONG-RANGE PREDICTION OF MONSOON RAINFALL

India's annual food production is now around 155 million metric tons. This annual output is achieved in two main seasons, which are known as the Kharif and the Rabi. The duration of the Kharif season coincides, roughly, with the summer monsoon, while the Rabi crop is grown during the winter rains. Of the total output of food grains, nearly two-thirds is generated by the Kharif crop, and the remainder comes from the Rabi season.

The average rainfall over India during the summer monsoon is around 85 cm. In a good monsoon the average exceeds 100 cm while a poor monsoon brings it down to about 60 cm. A variation of about 25% makes a large difference to the output of the Kharif crop. The purpose of long-range prediction is to try to anticipate the likely deviation from normal, so that action could be taken to maximize the output from the Kharif crop.

Long-range forecasts help planners to select those varieties of crops that are likely to do best in a given rainfall pattern. As an example, if the rains are likely to be delayed, this information could be used to delay sowing and use the intervening period for growing crops with a shorter gestation period, such as millet.

The forecasts are also useful for planning how much of the existing storage of food grains should be released to meet the estimates of demand and supply. The support that might be needed by areas that are likely to face rainfall deficiency is based on a long-range forecast.

3.1 Rainfall Distribution

As indicated earlier, regions that receive small amounts of rain generally show high variability. The coefficient of variability (Fig. 17.4) is highest over the semi-arid tracts of northwestern India (greater than 50%) where the annual rainfall is only 25 cm. In contrast, the eastern parts of India and the southwestern parts of the peninsula are regions of small variability and large rainfall (100–250 cm). Exceptions to this general rule occur in some states. The annual rainfall of Bombay on the western coast of Maharashtra is 250 cm, which is more than twice the rainfall of Pune (formerly known as Poona) which is located in the western part of the state, just 150 km to the east of Bombay, yet there is hardly any difference in the variability of rain.

Rainfall variability helps one decide which parts of India will benefit from long-range prediction. Long-range forecasts are not so important over areas of small variability. Thus over northeast India one can be reasonably confident that the rainfall will not deviate much from the normal climatological expectation, but large deviations, by as much as 30 to 60%, usually occur over northwestern parts of the country.

It is often convenient to examine rainfall distribution in terms of its principal components. This determines how the rainfall varies both in space and time and helps one to locate regions of similar patterns. Bedi and Bindra (30) examined the rainfall of 70 evenly distributed stations for a 60-year period (1911–1970). They found that:

1. The first four principal components were sufficient to explain 47% of the total variance.

2. There was an opposition in rainfall pattern between the northeastern and western sectors of India, that is, a spell of heavy rain over northwest India was accompanied by sparse rainfall over west India. This principal component was the dominant feature which accounted for 26.9% of the variance. Contributions by the other three components were relatively small.

3. There was a certain amount of similarity in rainfall between the eastern coastline of India near the head of the Bay of Bengal and the northern sector of the Arabian Sea. This suggests a tendency for mid-tropospheric troughs to be activated in consonance with monsoon depressions in the north Bay.

3.2 Long-Range Prediction Techniques

Long-range predictions in India are based on statistical associations and meteorological teleconnections. The best known work in this area is that of Sir Gilbert Walker in the early years of this century. As his outstanding contributions have been discussed in other parts of this volume they will not be repeated in this chapter, but we wish to refer the reader to excellent reviews on the Indian work by Jagannathan (31) and Rao (32, 33).

3.2.1 Multiple Regression Equations. Over the years it was realized that the original correlation coefficients introduced by Sir Gilbert were changing with time. Statistical associations between monsoon rain and antecedent features, which appeared strong at one stage, turned out to be not as strong as originally thought as new observations were added. Many correlation coefficients even changed sign. As a consequence, Indian meteorologists have tried to find better predictors.

Regression equations based on different predictors are now used for predicting the date of onset of the monsoon over the southern tip of India and also the total rainfall from the first of June to the end of September. The predictors are selected through a screening procedure from a large number of potential candidates.

To predict the onset date, the following predictors are used:

(a) direction (in degrees) of the mean preceding January wind at 300 mb over Delhi (X_1);

(b) direction (in degrees) of the mean preceding January wind at 200 mb over Darwin in Australia (X_2);

(c) difference in the mean February wind direction (in degrees) at 200 mb between Trivandrum and Madras (X_3); and

(d) mean meridional wind component (in m/sec) at 200 mb over Calcutta in December of the previous year (X_4).

These predictors, when taken jointly in the form of the following multiple regression equation, serve as the basis for predicting the onset date:

$$Y = -0.0767X_1 - 0.1220X_2 - 0.0794X_3 - 0.5107X_4 + 32.1500 \qquad (5)$$

where Y is the departure from the normal date of onset over Trivandrum (in days).

A similar technique is used for forecasting the total rainfall from June through September for (i) the Peninsula and (ii) Northwest India. These regions are defined in Chapter 14, Figure 14.2 of this volume.

The predictors for the Indian peninsula are:

(a) South American pressure measured by departures from normal, in mm, averaged for the months of April and May for Buenos Aires, Cordoba, and Santiago (Y_1);

(b)   the mean position, measured in degrees of latitude, of the axis of a ridge
       at 500 mb in April located along 75°E (Y_2); and
(c)   the mean minimum temperature in March for three Indian stations (Jaisalmer,
       Jaipur and Calcutta) in degrees Celsius (Y_3).

For Northwest India the predictors for the total rainfall from June through September are:

(a)   South American pressure measured by departures from normal, in mm,
       averaged for Buenos Aires and Cordoba in April (Z_1);
(b)   equatorial pressure measured as departures from normal, in inches, averaged
       for Seychelles, Jakarta, and Darwin from January to May (Z_2); and
(c)   mean April temperature at a north Indian station (Ludhiana) in degrees
       Celsius (Z_3).

The prediction equations that are currently in use are:

(i)   Peninsula

$$R = -0.296Y_1 + 1.127Y_2 + 2.242Y_3 - 58.1595 \qquad (6)$$

(ii) Northwest India

$$R = +0.759Y_2 - 1.433Z_1 + 25.863Z_2 - 0.129Z_3 - 11.6821 \qquad (7)$$

where R stands for the predicted departure in monsoon rainfall (June–September)
from its normal value in inches.

A similar prediction (formula not shown here) is made with the help of a regression
equation for the second half of the monsoon (August–September).

More recently a similar search for statistical associations for predicting the date
of onset was made by Kung and Sharif (34, 35). Their predictors are:

(a)   1200 GMT geopotentials and temperatures over the Indian region at 100
       and 700 mb during April;
(b)   daily geopotential, temperature, and winds at 200 and 700 mb at six Australian
       stations from January 1 to April 30;
(c)   mean sea surface temperature over the Indian region, divided into grids of
       10 degrees by 10 degrees; and
(d)   monthly precipitation (snow) from December to March for 21 stations over
       the Eurasian region.

Kung and Sharif reported encouraging results, but they used a limited data set.
For the period 1958–1977 they found that the date of onset could be predicted with
an accuracy of 3.2 days, which is one-third of the variance. Considering a 15-year
period 1966–1980, the Indian regression equations predicted the onset with an

accuracy of 1.7 days,* but clearly it is premature at this stage to compare the results because different data sets have been used. On some occasions regression equations have not been very successful in predicting a long delay in onset. The Indian model did not capture the delayed onset of 1979 very well, and Kung and Sharif did not have access to data beyond 1977. What is needed now are tests on independent data sets for longer periods.

Regression equations have been less successful, especially in years of rainfall deficiency, for predicting the total amount of rainfall from June through September. But one of the Indian predictors, namely, the mean position of the ridge at 500 mb in April [first suggested by Banerji et al. (36)] has shown remarkable stability. The correlation coefficient between this predictor and the predictand, June through September rainfall, has never fallen below 0.5 in the last 40 years. For most of this period it was a little more than 0.6.

3.2.2 Trends and Periodicities.
Many scientists have searched for trends and periodicities in monsoon rainfall (37). Opinions differ but the search has not yet yielded positive results for long-range forecasts.

By analyzing the power spectrum of monsoon rainfall Raghavendra (38) found some association with the well-documented quasi-biennial oscillation, in addition to a cycle of 3.3 years for northwest India. The quasi-biennial oscillation is a major feature of the equatorial stratosphere. At pressure altitudes between 60 and 10 mb, alternate spells of easterly and westerly winds are observed with periods of two years or slightly longer. Theoretical studies suggest that the cause of this oscillation is an interaction between vertically propagating internal waves from the troposphere and the mean flow of the stratosphere (39). Thapliyal (40) found that the phase of the oscillation at 50 mb in January over the Northern Hemisphere is related to the performance of the subsequent monsoon. If the oscillation is in its westerly phase, the summer monsoon rains are good, but the easterly phase is followed by poor or indifferent monsoons.

The coupling between fluctuations in the stratosphere and the troposphere is not yet well understood, but some find a relation between the phase of the oscillation and the date of the monsoon's onset (41–43).

Raja Rao et al. (44) found that the winter preceding a poor monsoon is marked by large northward transports of momentum and sensible heat away from the monsoon region, and larger conversions from the zonal to the eddy components of available potential energy in the equatorial stratosphere between 50 and 30 mb.

There is another oscillation which appears to influence the summer monsoon. This is the Southern Oscillation, which was discovered by Sir Gilbert Walker at the beginning of the present century. As this is discussed in greater detail in other parts of this volume, we will make only a passing reference here. The Southern Oscillation is represented to some extent by South American pressures which, as we have seen, are used as predictors for monsoon rain. The Southern Oscillation

---

* Personal communication, V. Thapliyal, India Meteorological Department.

has been shown to be associated with El Niño, a phenomenon related to upwelling off the coast of Peru. Recently, it has been argued that El Niño in the preceding winter or spring will lead to deficient monsoon rainfall. This conjecture is now the subject of considerable research and firm results are likely to emerge in the next few years.

Bryson and Starr (45, 46) assume that small fluctuations of the earth's axis of rotation influence monsoon rain and may be used for forecasts of rainfall a year ahead. Wilson (47), on the other hand, believes that the atmospheric response to such fluctuations is too small to influence weather. In a later publication, Bryson and Campbell (48) note that cold periods in the history of climate have been associated with periods of weak monsoons, while warm periods were marked by strong monsoons. Therefore, they argue that since fluctuations in the earth's axis should be associated with global temperature changes, the fluctuations should also be related to monsoon rain. Although rainfall forecasts on this basis have been prepared by Bryson and Campbell, it is difficult to assess their success at this stage.

3.2.3 Autoregressive Models.
Although autoregressive models have been used by economists for many years, their application to meteorological problems is comparatively recent.

When observations of rainfall are available at equal intervals of time they form a time series. A stationary time series is one in which the statistical properties, such as its mean and its variance, do not change with time. In reality, no time series is stationary. Autoregressive models seek to convert the observed series into one which is as stationary as possible by applying weights to the observations. A new series is thus generated in which the value of the predictand, in our case summer rainfall, is a function of the influence of its past values and a series of independent random "shocks." The word "regress" implies moving backwards, so a model in which the current value of the predictand is expressed by an aggregate of its past values and a set of random deviations is an "autoregressive" model.

The forecast made using a time series is often improved by information coming from another associated series. This is especially the case if changes in the original series are strongly influenced by the second series. Economists refer to the second series as a "leading indicator" for the first. For predicting monsoon rainfall, the April position of the 500-mb ridge [item (b) under predictors for June–September rainfall over the Indian Peninsula in Section 3.2.1] is a useful leading indicator for the time series of rainfall.

Thapliyal (49) designed an autoregressive model for forecasting summer monsoon rain. A brief summary of the technical details of this model is provided in Appendix A. Figure 17.10 depicts the performance of the model for the period 1973–1980 in terms of a comparison of the observed and predicted rainfall over the Indian peninsula. Figure 17.10 shows that the model was able to capture the poor monsoons of 1974 and 1979 fairly well. The root mean square prediction error for rainfall over the peninsula was found to be 4.4 cm, but when multiple regression equations were used, the error was 8.0 cm for the same period. Therefore it appears there is higher skill in the autoregressive model than in multiple regression equations.

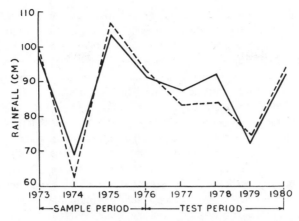

Figure 17.10. Performance of an autoregressive model by Thapliyal (49). The thick line shows the observed rainfall over the Indian peninsula for the period 1973–1980. Predicted rainfall is shown by the dotted line.

To further test the model's performance, Thapliyal split the period 1973–1980 into two parts. Forecasts for the first part (1973–1976) were prepared by utilizing all past rainfall data for the model's design. For the next four years (1977–1980) the model was not updated with more recent rainfall data. Despite this, the performance of the model was good.

An advantage of autoregressive models is that they provide a mix between deterministic and stochastic models. The output from a deterministic general circulation model could be used as input for an autoregressive model, for example, model generated sea surface temperatures.

Despite encouraging signs a word of caution is in order. An autoregressive model has been tried for forecasting rainfall a month ahead for each meteorological subdivision of India (Fig. 17.1). The results have not been as satisfactory as those shown in Figure 17.10. This is because no suitable leading indicator has been discovered yet for each individual subdivision. It appears at this stage that much of the success of autoregressive models depends on the discovery of the right leading indicators for which an intensive search seems necessary. There is yet another aspect of autoregressive models that merits consideration. While multiple regression equations relate the predictand with several predictors, an autoregressive model tends to rely on one predictor, namely the leading indicator. It is not clear at present whether this is a limitation because the contribution from a good leading indicator may be more significant than a number of other predictors, but an extension of model design to include more than one leading indicator would clearly be helpful.

3.3 Predictability of Monsoon Rain

The general question of predictability was discussed in Chapter 16; consequently we will confine ourselves to a few aspects that are related to operational prediction in India.

It is clear that an assessment of what is predictable, and to what extent, is needed for the summer monsoon. Some experience has been gained with limited area models. Further improvements will undoubtedly take place. Forecasts of upper-level winds, tropical cyclones, and surges are specific areas where the advent of satellite data will help to improve performance. The period of short-range forecasts in monsoon regions could well be extended to 72 hours.

But several problem areas still remain. Inadequate definition of terrain irregularities and parameterization of convective clouds are examples of areas that need improvement. The paucity of data over oceanic regions, especially data on upper-level winds and vertical profiles of temperature and moisture are other factors that limit the accuracy of short-range prediction. (Although recently, satellites have helped in these areas.) Notwithstanding these difficulties, the principal rain bearing systems of the summer monsoon are now reasonably well defined and their main characteristics are known. This is of considerable help to an operational forecaster.

The situation is less satisfactory for long-range prediction. Measured in terms of the ratio of interannual to natural variability, Charney and Shukla (50) find there is greater potential for long-range prediction of the monsoon's performance, especially monthly means of rainfall. But the techniques for realizing this potential are not clear. For the present, the emphasis on statistical methods may have to continue, with a more vigorous search for better predictors and leading indicators.

Experiments with general circulation models are just beginning in India. Long-term integration of this model will require more computing facilities. It is not clear at present whether a monthly forecast will show more skill than what is currently achieved by statistical methods. But Shukla's (51) suggestion concerning the importance of boundary forcings certainly could be tested by the model.

The full potential of general circulation models for long-range prediction has not yet been realized, but interesting results are beginning to emerge which are encouraging. Long-term integrations of a model developed by the Meteorological Office of the United Kingdom (52, 53) show interesting contrasts between the rainfall predicted for the July of 1979 and 1980. The 1980 July simulation shows greater convective activity over the Bay of Bengal and more monsoon rain. By way of contrast, that of July 1979—a year of delayed onset and rainfall deficiency— showed less convection over the Bay of Bengal. It is not clear how much the results were influenced by the different initial conditions for the two years as opposed to the differences in the convective activity during the two Julys, but experiments of this kind are needed to understand the interannual variability of monsoons.

4 SUMMARY AND CONCLUSIONS

In this chapter we have summarized the problems that confront meteorologists making short- and long-range predictions for the Indian summer monsoon. International expeditions, such as the Monsoon Experiment of 1979, have provided much valuable data on monsoon systems, but while there has been substantial progress in observational techniques, a number of basic problems still remain. It is still not very clear what features of the monsoon should be monitored on a year-to-year basis for improving prediction. Some scientists (i.e., Shukla, see Chapter 16) have emphasized lower boundary fluctuations, such as sea surface temperature, the reflective power of the earth's surface, ground hydrology, and snow cover over the Himalayas. However, few general circulation models have been applied to study the impact of fluctuations of boundary forcings on monthly forecasts. Much research needs to be done to combine the statistical approach with prediction by general circulation models and it is our firm conviction that research on these aspects should continue.

ACKNOWLEDGMENTS

Monsoons have attracted a considerable volume of research in the past few years. In this growing field, it is natural that differing views will emerge. I have tried to put across my views and where divergences of opinion occur, this has been stated in the text. I am indebted to all my colleagues in India and elsewhere who helped me with their time and expertise to prepare this article.

The permission of the Director General of the Indian Meteorological Service to reproduce the maps and diagrams from different departmental publications is gratefully acknowledged.

APPENDIX A.   AUTOREGRESSIVE MODELS

An autoregressive model relates the input X_t of the time series of a physical system to its output Y_t by

$$Y_t = \sum_{k=0}^{\infty} h_k X_{t-k} + N_t \tag{A.1}$$

The coefficients h_k represent the response of the system at each time index k and N_t is a system of random shocks which represents white noise. The values of h_k may be estimated by multiplying each term of equation (A.1) by X_{t-m}, where m is another time index. Thus we have

$$\gamma_{XY}(m) = h_0 \gamma_{XX}(m) + h_1 \gamma_{XX}(m-1) + \cdots \tag{A.2}$$

where $\gamma_{XY}(m)$, $\gamma_{XX}(m)$ are covariances between X and Y. But this is difficult to achieve because we do not know where the series (A.2) should be terminated.

It was shown by Box and Jenkins (54) that this difficulty can be overcome by making X_t and Y_t stationary. This is achieved by differencing the series a finite number of times. By this process a new series

$$X_t = (1 - B)^d X_t = \nabla^d X_t \tag{A.3}$$

is generated. B stands for a backward shift operator in time and d is a positive integer. We have

$$BX_t = X_{t-1} \tag{A.4}$$

$$B^2X_t = X_{t-2} \tag{A.5}$$

and so on. The new series (A.3) is now stationary. This enables one to express equation (A.1) as

$$\Phi(p)Y_t = \Theta(q)X_{t-b} \tag{A.6}$$

where $\Phi(p)$, $\Theta(q)$ represent the polynomials

$$\Phi(p) = 1 - \Phi_1B - \Phi_2B^2 - \cdots - \Phi_pB^p \tag{A.7}$$

$$\Theta(q) = \Theta_0 - \Theta_1B - \Theta_2B^2 - \cdots - \Theta_qB^q \tag{A.8}$$

and b represents a lag or delay in the response of the system. Equation (A.6) represents an autoregressive model of order (p, d, q). To design a model we have to determine p, d, and q. This is done by starting with a first guess and then improving upon the first guess by iteration. This enables one to obtain least square estimates of the weights $\Phi_1, \Phi_2 \ldots, \Phi_p$ and $\Theta_0, \Theta_1 \ldots, \Theta_q$.

REFERENCES

1. J. K. Patnaik, R. R. Rao, and R. Ramanadham, Some characteristics of Indian monsoon rains, *Indian Geog. J.*, **52**(1), 23–30 (1977).
2. S. M. Dagupatty and D. R. Sikka, On the vorticity budget and vertical velocity distribution of a monsoon depression, *J. Atmos. Sci.*, **33**, 773–792 (1977).
3. P. V. Joseph and K. K. Chakravorty, "Lower Tropospheric Temperature Structure of a Monsoon Depression," Summer Monsoon Field Phase Research (Part B), World Meteorological Organization (WMO), Geneva, No. 9, 1980, pp. 257–265.
4. V. D. Iyer, "Typhoons of the Pacific Ocean and South China Sea," Indian Meteorological Department, Scientific Notes, Vol. 3, No. 29, New Delhi, 1931, 25 pp.
5. V. D. Iyer, "Typhoons and Indian Weather," Mem. Indian Meteorological Department, New Delhi, Vol. 3, Part VI, 1935, pp. 93–130.
6. K. R. Saha, F. Sanders, and J. Shukla, Westward propagating predecessors of monsoon depressions, *Mon. Wea. Rev.*, **109**, 330–343 (1981).
7. B. N. Dewan, S. Jaipal, and S. Kumar, Flux of moisture into the Bay of Bengal during the summer monsoon, *Mausam*, **35**(4), 493–498 (1984).
8. R. P. Pearce and U. C. Mohanty, Onsets of the Asian summer monsoon 1979–1982, *J. Atmos. Sci.*, **41**(9), 1620–1639 (1984).
9. S. K. Ghosh, M. C. Pant, and B. N. Dewan, Influence of the Arabian Sea on the Indian monsoon, *Tellus*, **30**, 117–125 (1978).

10. P. K. Das and H. S. Bedi, "A Numerical Model of the Monsoon Trough," in Sir James Lighthill and R. P. Pearce, Eds., *Monsoon Dynamics*, Cambridge University Press, Cambridge, 1981, pp. 351–364.

11. P. J. Webster, Mechanisms of monsoon low frequency variability: Surface hydrological effects, *J. Atmos. Sci.*, **40**(9), 2110–2124 (1983).

12. U. C. Mohanty, R. P. Pearce, and M. Tiedtke, "Numerical Experiments on the Simulation of the 1979 Asian Summer Monsoon," Technical Report Number 44, European Centre for Medium Range Weather Forecasts (ECMWF), Reading, U.K, 1984, 45 pp.

13. J. Findlater, Aerial explorations of the low level cross-equatorial current over eastern Africa, *Quart. J. Roy. Meteor. Soc.*, **98**, 274–289 (1972).

14. Y. Ramanathan and R. K. Bansal, "The Northern Hemisphere Analysis Centre (NHAC) Quasi-Geostrophic Model," Indian Meteor. Dept. Sci. Report No. 76/1, New Delhi, 1976, 20 pp.

15. G. J. Haltiner, *Numerical Weather Prediction*, Wiley, New York, 1971, 371 pp.

16. E. N. Lorenz, Energy and numerical weather prediction, *Tellus*, **12**, 364–373 (1960).

17. N. A. Phillips, Geostrophic motion, *Rev. Geophys.* **1**, 123–176 (1963).

18. H. L. Kuo, On the formation and intensification of tropical cyclones through latent heat released by cumulus convection, *J. Atmos. Sci.*, **22**, 40–63 (1965).

19. H. L. Kuo, Further studies on the parameterization of cumulus convection on large scale flow, *J. Atmos. Sci.*, **31**, 1232–1240 (1974).

20. L. Bengtsson, personal communication, 1984.

21. A. A. Ramasastry, A. K. Chowdhury, and N. C. Biswas, Cyclones and depressions over the Indian seas in 1982, *Mausam*, **35**, 1–10 (1984).

22. C. J. Neumann and G. S. Mandal, Prediction of the path of tropical cyclones, *Ind. J. Meteor. Hydrol. Geophys.*, **29**, 40–58 (1978).

23. A. K. Sen Sharma, Tropical cyclone movement, *Vayu-Mandal*, **13**, 30–34 (1983).

24. D. R. Sikka and R. Suryanarayana, An objective method of predicting tropical cyclones, *Ind. J. Meteor. Hydrol. Geophys.*, **23**, 35–40 (1972).

25. R. K. Datta and R. N. Gupta, "Tropical Cyclone Prediction by the Method of Analogues," Indian Meteorological Department, Scientific Notes No. 157, New Delhi, 1975, 24 pp.

26. Y. Ramanathan and R. K. Bansal, Forecasting tropical cyclone path with a barotropic model, *Ind. J. Meteor. Hydrol. Geophys.*, **28**, 50–62 (1977).

27. P. K. Das, M. C. Sinha, and V. Balasubramanyam, Storm surges in the Bay of Bengal, *Quart. J. Roy. Meteor. Soc.*, **100**, 437–447 (1974).

28. B. Johns, P. C. Sinha, S. K. Dube, U. C. Mohanty, and A. D. Rao, On the effect of bathymetry on storm surge simulation, *Computers and Fluids*, **11**(3), 161–174 (1983).

29. S. K. Dube, P. C. Sinha, and A. D. Rao, Effect of coastal geometry on location of the peak surge, *Mausam*, **33**, 445–450 (1982).

30. H. S. Bedi and M. M. S. Rindra, Principal components of monsoon rainfall, *Tellus*, **32**, 296–298 (1980).

31. P. Jagannathan, "Seasonal Forecasting in India—A Review," Indian Meteorological Department, Scientific Pub. No. DGO 82/650, New Delhi, 1960.

32. K. N. Rao, "Seasonal Forecasting in India," Tech. Rept. No. 66, World Meteorological Organization (WMO), Geneva, 1964, pp. 17–30.

33. K. N. Rao, "Agroclimatic Classification of India", Indian Meteorological Department Monograph on Agroclimatology, No. 4, New Delhi, 1972, 51 pp.

34. E. C. Kung and T. A. Sharif, Regression forecasting of the onset of the Indian summer monsoon with antecedent upper air conditions, *J. Appl. Meteor.*, **19**, 370–380 (1980).

35. E. C. Kung and T. A. Sharif, Long range forecasting of the Indian summer monsoon onset and rainfall, *J. Meteor. Soc. Japan*, **104**, 635–647 (1982).

36. A. K. Banerji, P. N. Sen, and C. R. V. Raman, The 500 mb sub-tropical ridge in April, *Ind. J. Meteor. Hydrol. Geophys.*, **29**, 425–431 (1978).

37. P. Koteswaram and S. M. A. Alvi, Secular trend and variations in rainfall over India, *Idojarus*, (Budapest), **3–4**, 175–183 (1970).

38. V. K. Raghvendra, "A Statistical Study of Southwest Monsoon Rainfall," Indian Meteorological Department Meteorological Monograph on Climatology, No. 6, New Delhi, 1973, 14 pp.

39. R. S. Lindzen and J. R. Holton, A theory of the quasi-biennial oscillation, *J. Atmos. Sci.*, **25**, 1095–1107 (1969).

40. V. Thapliyal, "Stratospheric Circulations in Relation to the Summer Monsoon over India," Proceedings Symposium Hydrological Aspects of Droughts, Indian Institute Technology, New Delhi, 1979, 347–372 pp.

41. B. Parthasarathy and O. N. Dhar, A study of trends and periodicities in the seasonal and annual rainfall of India, *Ind. J. Meteor. Hydrol. Geophys.*, **27**, 23–28 (1976).

42. S. J. Reddy, Forecasting the onset of the southwest monsoon over Kerala, *Ind. J. Meteor. Hydrol. and Geophys*, **28**, 113–114 (1977).

43. B. S. Chuchkalov, "Relation between Variations in the Vertical Structure of the Equatorial Atmosphere and Planetary Circulations," Proceedings International Conference on Early Results of FGGE and MONEX, Tallahassee, Florida, World Meteorological Organization (WMO), Geneva, No. 5, 1981, pp. 48–49.

44. K. S. Raja Rao, S. T. Awade, and M. V. H. Nair, Dynamics of Stationary Eddies in the Lower Stratosphere," Proceedings Indian Academy Science, Earth and Planetary Science, Bangalore, **93**, 1–16 (1984).

45. R. A. Bryson and T. B. Starr, Chandler tides in the atmosphere, *J. Atmos. Sci.*, **34**, 1975–1986 (1977).

46. R. A. Bryson and T. B. Starr, "Indications of Chandler Compensation in the Atmosphere," in K. Takahashi and M. M. Yoshino, Eds., *Climatic Change and Food*, University of Tokyo Press, Tokyo, 1978, pp. 410–425.

47. C. R. Wilson, Comments and additional investigations concerning Chandler tides in the atmosphere, *J. Atmos. Sci.*, **35**, 2381–2387 (1978).

48. R. A. Bryson and W. H. Campbell, "Year in Advance Forecasting of the Indian Monsoon Rainfall," Center for Climatic Res., Scientific Report Number 5, University of Wisconsin, Madison, 1982, 17 pp.

49. V. Thapliyal, A stochastic dynamic model for long range prediction of monsoon rainfall, *Mausam*, **33**, 399–404 (1982).

50. J. G. Charney and J. Shukla, "Predictability of Monsoons," in Sir James Lighthill and R. P. Pearce, Eds., *Monsoon Dynamics*, Cambridge University Press, Cambridge, 1981, pp. 99–110.

51. J. Shukla, Dynamical prediction of monthly means, *J. Atmos. Sci.*, **38**, 2547–2572 (1981).

52. A. Gilchrist, personal communication, 1984.

53. P. K. Das, "The Monsoons—A Perspective," Perspective Report Series, No. 4, Indian National Science Academy, New Delhi, 1984, 52 pp.

54. G. E. P. Box and G. M. Jenkins, *Time Series Analysis, Forecasting and Control*, Revised Edition, Holden–Day Inc., San Francisco, 1976, 575 pp.

18

Short- and Long-Range Monsoon Prediction in Southeast Asia

Boon-Khean Cheang
Malaysian Meteorological Service
Selangor, Malaysia

INTRODUCTION

Southeast Asia experiences two monsoons, winter and summer,* every year. Locally the winter monsoon is known as the northeast monsoon and the summer monsoon as the southwest monsoon, except in Indonesia where they are known as the west and east monsoons, respectively. These local names are derived from the low-level prevailing winds of the two seasons, shown in Figures 18.1a and 18.1b. The winter monsoon arrives around November and retreats in late March while the summer monsoon starts in late May and retreats in late September. Between them are the brief transitional periods.

The winter monsoon is important to those Southeast Asian countries located south of about 10°N because it brings about 50% of the annual rain of these countries. For the same reason, the summer monsoon is important to the Southeast Asian countries situated north of about 10°N (Fig. 18.2). However, in the western part of Peninsular Malaysia and southwest Thailand, normally more rain is received during the two transitional periods than during the monsoon periods.

Like the people of India, the inhabitants of Southeast Asia depend heavily on the monsoon rains for agriculture and many other activities. As in India, the monsoons can also bring disaster. Loss of lives and damage to crops and property due to excessive monsoon rains or drought are common. Good monsoon predictions can help to mitigate these problems and are therefore essential for the planning of agricultural activities and disaster prevention.

This chapter focuses on day-to-day weather forecasting in Southeast Asia with emphasis on the heavy-rain spells in the South China Sea–Malaysia region. Only major problems pertaining to both short- and long-range prediction will be considered.

---

* Unless otherwise stated, winter and summer refer to Northern Hemisphere seasons.

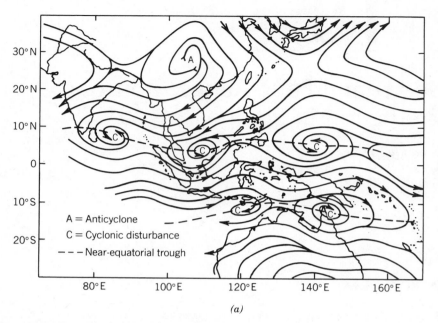

(a)

Figure 18.1. (*a*) Mean December 850-mb streamline analysis. Near-equatorial troughs are indicated by dashed lines. A and C stand for anticyclone and cyclone.

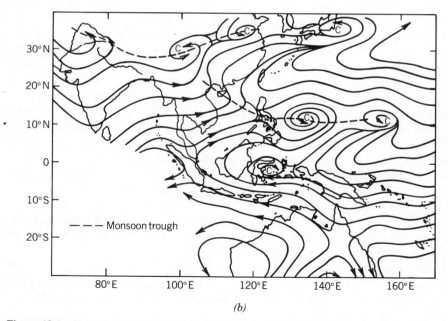

(b)

Figure 18.1. (*b*) Mean July 850-mb streamline analysis. Monsoon troughs are indicated by dashed lines. C stands for cyclone.

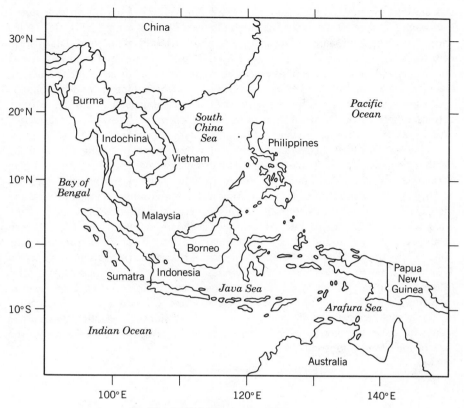

Figure 18.2. Map of Southeast Asia.

1 FEATURES OF THE WINTER MONSOON

The winter monsoon circulation is complex with many important features of different space and time scales. Many of the large- and planetary-scale aspects have been examined elsewhere in this volume (for example, Chapters 11 and 12). Attention will be devoted to the synoptic- or weather-scale features.

1.1 Onset and Retreat of the Winter Monsoon

To the laymen in the affected countries, the onset of a monsoon means the beginning of a rainy season. For the meteorologists, the definition is more involved. Cheang (1) used the wind steadiness at the 850-mb level and the total 24-hour rainfall at a station as the two criteria to determine the mean onset date of the winter monsoon in Malaysia based on data over a 16-year period. Wind steadiness is defined as the ratio of the magnitude of the mean vector wind over a time period to the mean wind speed over the same period. Using the 850-mb wind data, the wind steadiness at a station is computed for every successive 30-day period starting from October 1 through October 30, October 2 through October 31, and so forth (Fig. 18.3). To avoid the complications introduced by surface friction, 850-mb wind data instead of surface wind data are used. At a given station, the critical ratio (wind steadiness) and the critical 24-hour total rainfall, which define the onset of the winter monsoon, are 0.6 and 3 inches, respectively. It is found that beginning in mid-November, the onset normally starts along the northeast coast of Peninsular Malaysia, arriving at the southern part of the peninsula around early December. It then moves eastward, arriving at the northwestern coast of Borneo in late December.

In Peninsular Malaysia the onset of the winter monsoon is generally accompanied by three major features in the atmospheric circulation pattern. First beginning in September, a large-scale monsoon trough (low-pressure system) found over the northern part of the South China Sea during late summer moves southward. During the onset of the winter monsoon, it establishes a quasi-stationary position over the equatorial South China Sea and is then known as the northern near-equatorial trough (compare Figs. 18.1a and 18.1b). Steady northeasterly trade winds prevail to the north of this trough. The second feature is the cold surge (to be described in Section 1.2) which frequently reaches the equatorial South China Sea one or two days prior to the onset in Peninsular Malaysia. The third feature is the reversal from easterly to westerly winds at the 200-mb level over southern China due to the reversal of the north–south temperature gradient across the Asian continent. Operational forecasters watch for these weather patterns in daily weather analyses in order to predict the date of the winter monsoon onset.

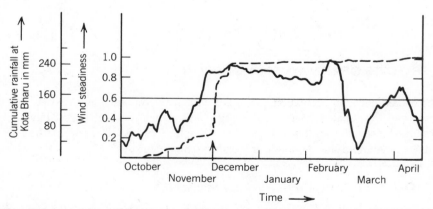

Figure 18.3. The 850-mb wind steadiness at Kota Bharu, Peninsular Malaysia (102°E, 6°N) for the period October 1, 1981 to April 15, 1982. Each point on the graph (full line) represents the 850-mb wind steadiness computed for every successive 30-day period starting from Oct. 1 to Oct. 30, Oct. 2 to Oct. 31, and so forth. Horizontal line in the graph represents the critical value of the wind steadiness (= 0.6). Cumulative daily rainfall (in mm) from October 1, 1981 to April 15, 1982 at Kota Bharu is represented by the dashed line. The vertical arrow along the time axis indicates the winter monsoon onset date when both the steadiness is equal to or greater than 0.6 and 3 inches of rainfall is recorded during 24 hours at any station along the east coast of Peninsular Malaysia.

Along the eastern Philippines, the onsets in the northern, central, and southern parts occur in November, December, and January, respectively; generally there is a sharp increase in rainfall during these months as the northeasterly winds become well established. In Indonesia, the onset takes place in the months of November and December when there is a distinct rise in rainfall and when the westerlies are established between the northern near-equatorial trough and the southern near-equatorial trough (Fig. 18.1a).

Cheang (1) uses the criterion that when the wind steadiness at Kota Bharu drops below 0.6 for more than a week the winter monsoon has retreated from Malaysia. He found that the dates of retreat vary considerably from year to year. It can occur as early as February 5 in some years and as late as March 31 in others.

1.2 Cold Surges

The cold surge is one of the main weather features over South China and the South China Sea during the winter season. At that time, the Siberian high pressure system is a persistent feature over the East Asia continental region. Radiative cooling and persistent cold air advection in this region maintain a layer of very cold air over the frozen land. The presence of the Himalayan Tibetan massif blocks the southward movement of this cold air mass and contributes to the build up of the surface high. A strong baroclinic zone exists between the cold continental air mass and the warm tropical air mass to its south. Under these circumstances, short waves passing through a long, quasi-stationary, upper-level westerly wave trough east of the Asian continent (to be discussed in Section 1.3) in the middle latitudes often trigger intense anticyclogenesis over China and cyclogenesis over the East China Sea. Coupled with these changes is a downstream acceleration of the upper-level westerly jet over East Asia. As the pressure gradient across the East China coast increases, a cold surge is initiated when low-level cold air bursts forth from the continent toward the South China Sea. Generally, cold fronts associated with the cold surge can be followed as far south as 25°N over China during most of the cold outbreaks; however, they are rarely observed over southeastern China. It is only during vigorous cold surges that the diffuse clouds associated with these fronts can move southeastward out of southeastern China to the northern part of the South China Sea, northern and central Philippines and the West Pacific.

Cold surges are most frequent during the period from November to February. In a normal year they occur at intervals of several days to about 20 days. (See Chapter 11 for a discussion of the variability of cold surges and their relationship to other circulation features.) In general, the cold surges arrive at the northern part of the South China Sea from two different directions (2). During the early winter they tend to flow directly from southeastern China to the South China Sea bringing north–northeasterly winds. In contrast, most of the cold surges during mid- and late winter arrive at the northern part of the South China Sea as easterlies from the Pacific. This variation in the direction has different impacts on the weather in the Malaysia–Southern South China Sea region and will be discussed further in Sections 1.3 and 1.4.

By the time a cold surge arrives at the equatorial South China Sea, it has been warmed and moistened over the long sea track and has lost its cold and dry continental air mass properties. Nevertheless, the northeasterly wind maintains considerable strength, with speeds as high as 40 knots. Some surges propagate so rapidly that they arrive at the southern part of the South China Sea in less than a day; others take somewhat longer. Chang et al. (3) observed gravity wave characteristics in rapidly propagating cold surges. Most surges persist for 2–4 days over the South China Sea, although some last for only one day while others last as long as five days (1).

Cold surges have strong effects on the weather in Southeast Asia. In the Northern Hemisphere, minimum surface temperatures as low as 0°C have been recorded at land stations that are located at about 17°N during intense cold surges (4). Over the South China Sea, north of about 10°N, cold surges bring low-level stratocumulus and cumulus clouds with occasional light rain. Over the equatorial South China Sea, south of about 10°N, convection associated with pre-existing near-equatorial disturbances is enhanced following cold surges (5–7). This results in heavy precipitation which very often causes severe flooding in southern Thailand and Malaysia. Ramage (5) attributed the heavy rain to the enhancement of the local Hadley cell by the cold surges. During periods of strong cold surges, sea conditions are also very rough and dangerous for shipping and fishing.

Cold surges affect not only the South China Sea and Southeast Asia, but also have a far-reaching effect on the Pacific. It has been observed that the northeasterly trades are freshened to as high as 40 knots over the West Pacific 3–4 days after the onset of a cold surge. Associated with this strengthening of the trade winds, convection associated with pre-existing near-equatorial disturbances over the West Pacific, and the central and southern Philippines are enhanced. Lau et al. (8) observed similar effects over the central Pacific during the international Winter Monsoon Experiment conducted in 1978–1979.

Following a vigorous cold surge, the Siberian high pressure expands southward and creates a west-to-east pressure gradient in the equatorial region of the South China Sea. Convective cloud bands, associated with convergence in the equatorial westerlies close to the southern near-equatorial trough, have been observed to intensify and move eastward (9). Lim and Chang (10) attributed the movement of these cloud bands to the propagation of equatorial Kelvin waves which are excited by the rapid intensification of the Siberian high pressure.

1.3 Upper-Level Westerly Wave Trough and Upper-Level Anticyclone

During the early winter there is a quasi-stationary, upper-level westerly long wave trough located to the east of the Asian continent between 120°E and 160°E (Fig. 18.4a). Below this long wave trough, a semi-permanent surface subtropical high pressure center prevails over the northwest Pacific with its east–west axis located north of 20°N (Fig. 18.4b). In this synoptic situation, cold surges tend to move directly from southeastern China to the South China Sea with north–northeasterly winds. The early winter cold surges tend to enhance convective activity in the Malaysia–South China Sea Region (1, 5–7).

During mid- and late winter, the upper level westerlies of the mid-latitudes expand southward and the quasi-stationary upper-level long wave trough shifts eastward (Fig. 18.4c). In response to these changes, the surface subtropical high-pressure center over the Pacific moves nearer to the equatorial region, with its east–west axis between latitudes 15°N and 20°N over the northern region of the South China Sea and northern Indochina. Cold surges can no longer move directly out from southeastern China. They have to traverse a long distance across South Korea and southern Japan and around the subtropical high in the West Pacific and they arrive at the northern South China Sea as easterlies (Fig. 18.4d). The easterlies finally back in direction to become the northeasterly, cross-equatorial flow over the

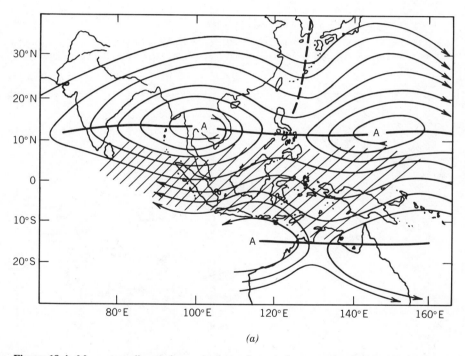

(a)

Figure 18.4. Mean streamline analyses. At the surface: A denotes anticyclone and C cyclone, the subtropical ridge is indicated by a solid line, the low-level easterly wind maximum is shown as a dashed arrow, and the cold surge is indicated by a bold arrow. At 200 mb: A denotes anticyclone, the subtropical ridge is indicated by a solid line, the upper-level westerly wave trough is shown as a dashed line, and areas favorable for disturbances to develop are indicated by hatching. (a) Mean 200-mb streamline analysis for early winter. In the Northern Hemisphere, the subtropical ridge extends westward from the Pacific to southern India. In the East Asia and West Pacific regions there are two upper-level anticyclones, one over the Bay of Bengal and Southeast Asia and one over the West Pacific. The early winter mean upper-level long wave trough is located around longitude 130°E. (*Figure 18.4 continued on p. 586.*)

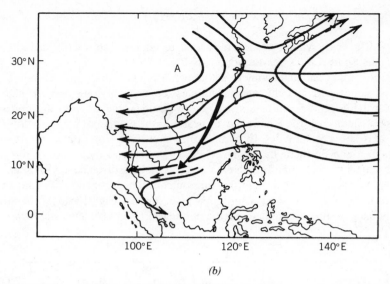

(b)

Figure 18.4. (b) Mean surface streamline analysis for early winter and a schematic representation of a surface north-northeasterly cold surge from southern China to the South China Sea during this period.

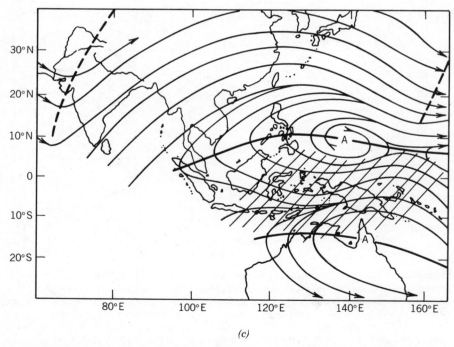

(c)

Figure 18.4. (c) Mean 200-mb streamline analysis for late winter. Note the appearance of the upper-level westerly wave trough south of the Tibetan Highlands, the southeastward shifting of the mean upper-level westerly wave trough and the eastward shifting of the areas of near-equatorial disturbances compared to the early winter positions.

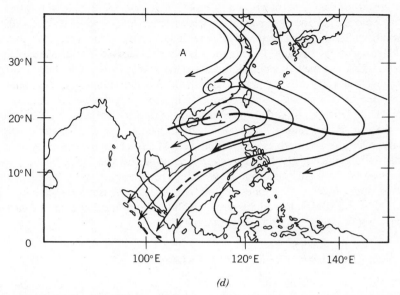

(d)

Figure 18.4. (*d*) Mean surface streamline analysis for late winter and a schematic representation of a surface easterly surge and cross-equatorial flow over the South China Sea during this period. Compare with panel *b* to follow the southeastward movement of the subtropical ridge and the low-level easterly wind maximum.

Malaysia–equatorial South China Sea region. These late winter cold surges are associated with dry weather in the Malaysia–South China Sea region (11).

Besides the quasi-stationary upper-level long wave trough east of the Asian continent, another large amplitude upper-level westerly wave trough tends to develop south of the Tibetan Highlands during the mid- and late winter (Fig. 18.4*c*) at intervals of 10–30 days. Its origin is still unknown, but it has a significant effect on the weather in Southeast Asia. In intense cases, this trough can be detected from the 700-mb to the 200-mb levels over the whole of India. Ramage (12) and Harris et al. (13) have observed that toward late winter it moves northeastward to southern China. Further observations show that it frequently moves southeastward to Malaysia before turning northeastward to southern China. Weather in Malaysia tends to become dry and stable for several days after the passage of this upper-level westerly trough (14).

Another important feature of the upper-level winter circulation is the upper-level anticyclone. During the summer monsoon, the upper-level anticyclone is generally located over the region of the Tibetan Highlands, an elevated heat source. During the winter Monsoon, due to the reversal in the meridional temperature gradient in Asia, the upper-level westerlies and anticyclone shift southward. There are two such anticyclones during the early winter in the Northern Hemisphere, one over the Southeast Asia–Bay of Bengal region and the other over the West Pacific (Fig. 18.4*a*). Near-equatorial disturbances tend to intensify below the divergent upper-level easterlies south of these anticyclones. Normally a few days following an intense cold surge, the upper-level easterlies tend to veer to southeasterlies over the southern South China Sea–Malaysia region, a result of the intensification of the local Hadley circulation (5). During the late winter, due to the southward expansion of the westerlies of the temperate latitudes and the development of the large amplitude, upper-level westerly wave trough south of the Tibetan Highlands, the upper-level anticyclone over Southeast Asia disappears leaving only one anticyclone over the West Pacific (Fig. 18.4c). At the same time, an upper-level anticyclone is present over the Australian region. Along with this change, the widespread convective activity over the southern South China Sea–Malaysia region of early winter is substantially reduced. By late winter it is confined to the West Pacific, eastern Indonesia, northern Australia, and New Guinea region below the easterlies between the two upper-level anticyclones in the West Pacific–Australia region (compare Figs. 18.4a and 18.4c). Recent research has shown significant interannual variations of these major equatorial convective areas (15).

1.4 Near-Equatorial Troughs and Disturbances

There are two near-equatorial troughs in the lower troposphere during the winter monsoon, one on either side of the equator (Fig. 18.1a). Both normally extend westward from the Pacific into the Indian Ocean. The regions under their influence are generally characterized by convective cloud bands and rainy weather. However, the convective cloud bands are more organized in the near-equatorial cyclonic disturbances that are embedded in the troughs and in the convergence zone in the equatorial westerlies between the troughs. The near-equatorial disturbances in the southern South China Sea and Malaysia region are known locally as monsoon disturbances. On weather maps, they often appear as cyclonic vortices and are normally associated with tropical waves in the easterlies. The tropical easterly waves can easily be detected over the West Pacific from the 700-mb up to the 500-mb level.

Many of the disturbances in both the northern and southern near-equatorial troughs move westward from the Pacific to the Indian Ocean. Disturbances originating north of New Guinea follow the northern near-equatorial trough across the South China Sea to the Bay of Bengal and bring torrential rains to Sri Lanka and southern India as they intensify into tropical cyclones. Some of the disturbances in the northern near-equatorial trough develop into typhoons in the West Pacific during the early winter. They normally move northwestward and recurve in the region of the northern Philippines and Taiwan and become extratropical storms. However, others move across the Philippines and dissipate over the South China Sea in the presence of cold surges. Only a small percentage of these early winter typhoons move westward to Indochina. Similarly, close to the southern near-equatorial trough, disturbances originating over the Arafura and its neighboring seas may intensify into tropical cyclones and traverse long distances into the Indian Ocean or they may recurve southward to northern or northwestern Australia.

Finally, weak cyclonic vortices may develop in situ, and remain quasi-stationary in both troughs. One of the regions favorable to the formation of these disturbances is the equatorial South China Sea. Because of the mountain ranges in Viet Nam, Peninsular Malaysia, and Borneo, the low-level trades are deflected cyclonically over the South China Sea causing a cyclonic vortex to form in the presence of the near-equatorial trough. The weather associated with this quasi-stationary disturbance appears to be strongly influenced by diurnal variation of land and sea breezes (16) as well as by cold surges (17).

The position of the northern near-equatorial trough oscillates between the equator and about 10°N while the southern near-equatorial trough oscillates between the equator and about 15°S. The northern trough also periodically retreats eastward to the West Pacific when northeasterly cross-equatorial flow develops near Peninsular Malaysia and Borneo (see Section 1.3 and Fig. 18.4d).

1.5 Winter Rainfall Fluctuations

Murakami (18) has shown that the cold surges over the South China Sea exhibit oscillations with periods of 4 to 5 days and 10 to 20 days. Surface pressure and rainfall (Fig. 18.5) over Malaysia, the southern South China Sea, and the West Pacific also exhibit similar oscillations (19–21). Cheang et al. (19) attributed the 4–5 day rainfall fluctuations to the interaction of cold surges and westward moving monsoon disturbances from the West Pacific. They found that the time series of rainfall in the West Pacific, southern South China Sea, and Malaysia were correlated with the surface pressure over China.

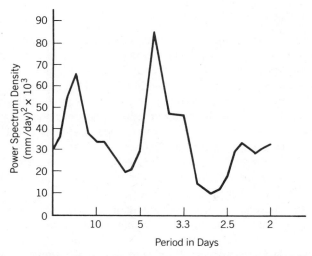

Figure 18.5. Power spectrum density of Malaysian winter (November 1978–March 1979) monsoon rainfall. Spectral peaks are found in two frequency domains, 4–5 days and 10–20 days (from reference 20).

Yap et al. (20) found that the 10–20 day wet and dry cycle over the South China Sea–Malaysia region during the winter monsoon is strongly influenced by the major cold surges which occur on a 10–20 day time scale (the weaker surges occur on a 4–5 day time scale (see Chapter 11, Section 1.2.4.). Dry spells over the South China Sea–Malaysia region are marked by the presence of low pressure over China, by an active southern near-equatorial trough and by two large-amplitude, upper-level westerly troughs, one over the India–Bay of Bengal region (see Section 1.3) and another over the central Pacific. Cheang and Krishnamurti (21) also found that the intensification of the 500-mb westerly wave trough over the Central Pacific is often followed about five days later by a dry spell in Malaysia. During the dry spell in Malaysia, disturbances in the West Pacific and the South China Sea become quasi-stationary. In contrast, rain spells over the South China Sea–Malaysia region are marked by the presence of high pressure over China, by an inactive southern near-equatorial trough and by the disappearance of the two upper-level westerly wave troughs described previously.

2 FEATURES OF THE SUMMER MONSOON

Although the summer monsoon in Southeast Asia is not as dramatic as that over the Indian subcontinent, it is an important component of the Asian monsoon system. As does the winter monsoon, the summer monsoon circulation in Southeast Asia has many important features. This section will be limited to discussion of only the synoptic-scale features of the summer monsoon in Southeast Asia.

2.1 Onset and Retreat of the Summer Monsoon

In general, the onset of the summer monsoon in Southeast Asia precedes that over the Indian subcontinent by several weeks. Orgill (22) studied the latitudinal variation of the equatorial westerlies at 700 mb (about 3 km) to examine the onset of the summer monsoon over Southeast Asia (mainly for Indochina). He defined the onset as the date when the equatorial westerlies at 700 mb extend north of 15°N over Southeast Asia. At this time, the Northern Hemisphere near-equatorial trough disappears and the summer monsoon trough forms over Indochina, allowing the equatorial westerlies to penetrate northward (Fig. 18.6). Orgill found that the mean date of onset in Indochina is May 17, with onsets in certain years as early as May 1 and some others as late as June 3. In his study, Orgill showed that: (a) the onset in Indochina coincides with an increase of rainfall but that the increase is typically irregular in both time and space; (b) single station rainfall statistics often reflect topographic influences; (c) in some years pre-monsoon rains mask the actual onset; (d) two-thirds of the summer monsoon onsets in Indochina (based on a 29-year data set, 1936–1964) occur concurrently with developing tropical cyclones over the Bay of Bengal; and (e) the upper-level subtropical high over Southeast Asia is displaced north of Indochina during the onset. Orgill's research has proved to be useful for operational meteorologists in Southeast Asia.

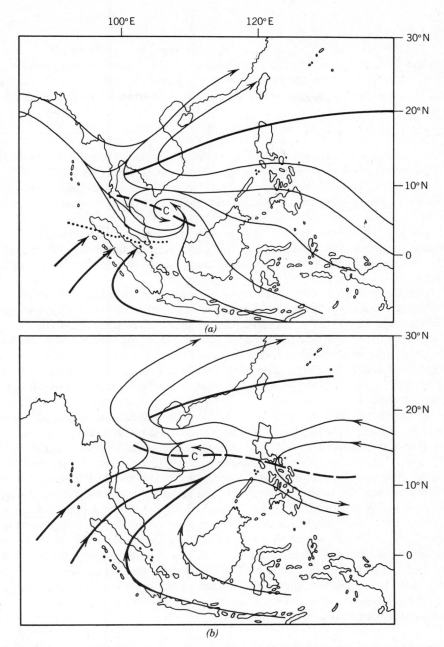

Figure 18.6. An example of the onset of the summer monsoon in Southeast Asia. (*a*) Low-level flow (700 mb) before the onset. The near-equatorial trough (dashed line) lies between northern Borneo and Peninsular Malaysia. The southwesterlies are denoted by the bold arrows south of the near-equatorial trough. The dots represent a convergence line of southwesterly and northwesterly winds. (*b*) Low-level (700 mb) flow after the onset. The near-equatorial trough has been replaced by the summer monsoon trough over Indochina and the subtropical ridge has moved to the northeast. Notations are the same as in Figure 18.4.

The summer monsoon begins its gradual retreat from the northern part of Southeast Asia in September. It takes more than a month for the monsoon to retreat completely from Indochina and the Philippines because the monsoon trough exhibits a low frequency north–south oscillation that occurs throughout the summer monsoon period.

2.2 Monsoon Trough

The monsoon trough is one of the main features of the summer monsoon in Southeast Asia. It extends westward from the West Pacific to Southeast Asia (Fig. 18.1b). At times it extends all the way from the West Pacific, across southern China and Southeast Asia to the Bay of Bengal and India. At other times, there are two troughs; one over the West Pacific and Southeast Asia separated from another over China and India. As a region of low pressure, the monsoon trough, which is also a heat source, is characterized by wet and windy weather and closely linked to the monsoon disturbances that produce significant precipitation during the summer in Southeast Asia.

According to Riehl (23), three types of monsoon disturbances that are related to significant precipitation can be identified in Southeast Asia. They are: (a) cyclones that are well established at the surface and in the lower troposphere, and decrease in intensity above those levels (assumed to be warm core); (b) cyclones well established in the middle troposphere and weak or non-existent at low levels (assumed to be cold core); and (c) an equatorial shearline (similar to a monsoon trough axis) without a definite center extending across all or most of Southeast Asia with its strongest intensity near the 700-mb level.

Harris (24) studied the structure of the monsoon trough in Indochina. He found that the converging westerlies, south of the monsoon trough axis, possess strong vertical shear between about 3 and 10 kms. Within the trough region there is synoptic-scale ascent, however, convective instability is weak. Extensive layer-clouds with rain prevail, and thunderstorms are infrequent. In the easterly wind regime north of the monsoon trough axis, the vertical shear of the horizontal wind is weak below 10 kms, and convective instability is strong although the weather is mainly fine. Cumulonimbus and thunderstorms are orographically controlled. Our observations also show that very often there is a strong wind belt in the lower tropospheric westerlies of the monsoon trough. It extends from the Indian Ocean through the Bay of Bengal to South Thailand during mid- and late summer and is associated with maximum (minimum) cloudiness north (south) of this strong westerly wind belt. This is one manifestation of the expected relationship between heavy rain and maximum low-level convergence.

2.3 Summer Monsoon Rainfall Fluctuations

Analyses of the temporal variations of the atmosphere have demonstrated that quasi-periodic behavior can be identified for short- and long-time scales. Oscillations of 4 to 5 days and 10 to 20 days exist both in the observed wind components in the troposphere and in the derived satellite brightness temperatures over the tropical

West Pacific and Southeast Asia during summer (25–29). In addition, Yasanari (30) found 40–50 day fluctuations in summer monsoon cloudiness.

The 4–5 day oscillation of summer rainfall has been associated with two types of tropical disturbances: on the synoptic scale, westward propagating waves in the troposphere with horizontal wavelength of about 3000 km and phase speeds of about 6–7 degrees per day, and on the planetary scale, westward propagating equatorial waves in the lower stratosphere with horizontal wavelengths of about 10,000 km. Tropospheric tropical waves can be identified as cloud clusters, which appear approximately every 4–5 days, moving westward from the Pacific to Southeast Asia in the daily satellite cloud pictures. Under favorable conditions (weak vertical shear of the horizontal winds in the troposphere, sufficiently large areas of warm sea surface temperature exceeding 26°C, and upper-level tropospheric outflow above a pre-existing wave disturbance) some of these tropical wave disturbances can develop into tropical depressions, tropical storms, and typhoons. These same upper-level wave disturbances often continue moving to the west. Krishnamurti et al. (31) and Saha et al. (32) found that a significant percentage of tropical depressions, storms and cyclones over the Bay of Bengal appear to originate over the West Pacific or South China Sea. The typhoon season starts as early as June and ends as late as December in Southeast Asia with the peak occurring in September, October and November. Most of the typhoons that reach the countries of Southeast Asia originate over the Pacific.

Zangvil (29) found 10–20 day fluctuations in satellite brightness (an indirect measure of the intensity of the convection activity, hence precipitation) over India and Southeast Asia. Krishnamurti and Bhalme (33) also studied the fluctuations in the summer monsoon trough in India. They suggested that the variations in the intensity of the trough on the 10–20 day time scale are the result of variations in solar radiation and the Hadley cell. Murakami (34) had found a remarkably high positive correlation between cloudiness fluctuations over India with those over Indochina. The explanation put forward by Krishnamurti and Bhalme for these 10–20 day fluctuations may possibly be applied to the 10–20 day fluctuations of the monsoon trough over Southeast Asia.

Yasanari (30) attributed the 40–50 day oscillation in summer monsoon cloudiness to the north–south fluctuation in the position of the monsoon trough. Cheang et al. (35) observed several interesting phenomena in the West Pacific and Southeast Asia related to this fluctuation which is also associated with the active–break cycle of the Asian summer monsoon. The active and break phases are defined in terms of conditions over northern Indochina and southern China. Thus, during the active phase, convective activity is strongest over northern Indochina and southern China, but rainfall in Malaysia is at a minimum. At this time the westerly winds in the lower troposphere (from the surface up to the 500-mb level) are at a maximum over Malaysia and southern Indochina. During the break phase the zone of maximum convective activity associated with the monsoon trough shifts northward to central China and the convective activity over southern China and Indochina is suppressed. However, the rainfall observed is at a maximum in Malaysia. The transition from active to break period is caused by the development of a subtropical ridge over

Indochina and southern China. Concurrently, two near-equatorial troughs, one north and one south of the equator, tend to develop. During the break phase, the low-level westerly winds over Indochina and Malaysia are weak and may even be replaced by easterlies. Disturbances in the double troughs bring along short but intense rain spells to Malaysia, Borneo, Sumatra, and the Java Islands.

3 SHORT-RANGE MONSOON PREDICTION

Local forecasting methods developed in Southeast Asia are based on experience. Most of the empirical rules and statistical methods used by the weather services in Southeast Asia are for specific events such as the occurrence of heavy rain spells, dry spells, and cold surges as well as typhoon intensification and movement.

3.1 Winter Monsoon Prediction

Predicting cold surges and heavy rain are the two major problems in Southeast Asia during the winter monsoon.

3.1.1 Predicting Cold Surges. In China and Hong Kong, synoptic methods have been developed to predict the arrival of cold surges. Chu (36) and Lam (37) of Hong Kong have discussed techniques that employ the 500-mb analysis. They watch for an intensifying 500-mb westerly trough east of Lake Baikal (90°E) with a ridge to the northwest. This trough may originate as far to the west as Greenland. Once the trough reaches Lake Baikal and intensifies, a cold surge is expected to arrive at Hong Kong one or two days later.

One of the main difficulties in predicting cold surges is forecasting how far south the surge will move and its intensity; several statistical objective methods have been developed to deal with this. In 1969, Riehl (38) developed a regression equation to predict the strength of the surge which is still being used operationally. Since there is little variation in the direction of the surges, it is necessary to predict only the wind speeds. Certain conditions must be met before Riehl's quantitative prediction technique can be applied: either the difference in the sea-level pressure between 30°N, 115°N and Hong Kong (22.2°N, 114.1°E) must exceed 10 mb or the pressure difference must be at least 8 mb with the difference having increased by not less than 7 mb in the preceding 24 hours. Riehl's equation gives the maximum surface wind speed $\bar{V}$ (in knots) over the South China Sea to be expected in the next 24 to 48 hours.

$$\bar{V} = -2.7 + 0.75\, P_{30} + 0.60\, \Delta P_{30,24\text{ hrs}} \tag{1}$$

where P_{30} is the existing sea-level pressure in millibars at 30°N, 115°E (minus 1000 mb) and $\Delta P_{30,24\text{ hrs}}$ is the forecast 24-hour sea-level pressure tendency in millibars at 30°N, 115°E.

To apply Riehl's equation, it is necessary to predict the sea-level pressure at 30°N, 115°E. Riehl found that when the necessary conditions for the occurrence of a surge are met, one could usually obtain satisfactory predictions of the sea-level pressure by simply extrapolating in time and space the previous 24-hour sea-level pressure changes.

3.1.2 Predicting Heavy Rain.
Since the winter monsoon is a wet season in Malaysia, prediction of rain in itself is not as important as the prediction of a heavy rain spell. Empirical rules have been developed to forecast the occurrence of heavy rain according to the prevailing synoptic weather situation.

Figure 18.7 illustrates the four common flow patterns at the 850-mb level (1):

1. Where neither a cold surge nor a monsoon disturbance is present in the region of the southern South China Sea, the main synoptic feature over the Malaysia–southern South China Sea region is the near-equatorial trough extending westward from a cyclonic vortex off the southeast coast of the Philippines. At times a shallow, quasi-stationary cyclonic vortex which develops in situ, is observed at the 850-mb level over the southern South China Sea. It is not shown here. This situation is not associated with either widespread moderate or with heavy rains in the Malaysia–southern South China Sea region.

2. When there is no cold surge over the South China Sea, and the vortex over the western Pacific moves to the southern South China Sea as a monsoon disturbance, it moves westward along the northern near-equatorial trough with a speed of 5–10 degrees of longitude per day and very often it moves across Peninsular Malaysia into the Bay of Bengal during the early winter. In the absence of a cold surge, this disturbance will bring widespread heavy rain in its path.

3. When a cold surge arrives at the southern South China Sea at about the same time as the monsoon disturbance from the West Pacific, convective activity intensifies because of enhanced low-level convergence. The disturbance may develop into a tropical depression which is associated with widespread torrential rain. During its intensification stage, the depression may become quasi-stationary over the southern South China Sea–Malaysia region for several days before moving further westward to the Bay of Bengal.

4. When a cold surge arrives at the southern South China Sea in the absence of a moving disturbance from the western Pacific, a local, quasi-stationary disturbance develops in situ over the southern South China Sea, intensifies and normally brings widespread torrential rain to the northern coast of Borneo. Immediately west of the disturbance, there is usually a clearing in the weather due to the stabilizing effect of the intrusion of cold air, especially if the surge is intense and lasts for several days.

The Malaysian Meteorological Service forecasters carefully monitor cold surges and the movement of tropical disturbances. Besides the routine surface weather maps and upper-air streamline analyses, forecasters use global longitude–time sections of 500-mb geopotential height along 30°N, graphs of the daily surface pressure of Hong Kong and five stations in central China, latitude–time sections of the surface

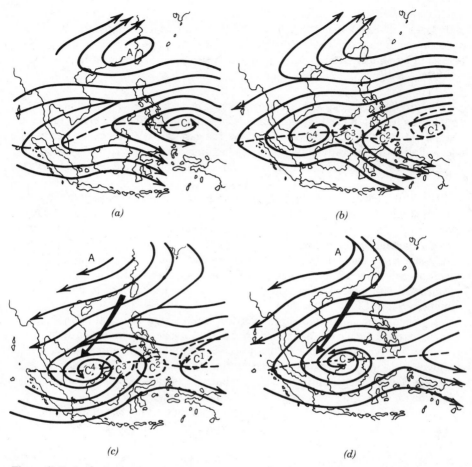

Figure 18.7. Streamline analyses at 850 mb of four synoptic situations associated with different weather patterns over the Malaysia–Borneo–South China Sea region during the winter monsoon. (*a*) No heavy rain over the equatorial South China Sea–Malaysia region. This situation is characterized mainly by the near-equatorial trough (indicated by the dashed lines) over the southern South China Sea when neither a cold surge nor a near-equatorial disturbance is present. There is, however, a near-equatorial disturbance located to the southeast of the Philippines. (*b*) Heavy rain over the southern South China Sea and Malaysia. This situation is characterized by a disturbance in the near-equatorial trough moving westward from the West Pacific to the southern South China Sea. The track of the disturbance is shown by C^1-C^2-C^3-C^4 which denote the positions of the disturbance on four consecutive days. A cold surge is absent over the South China Sea. (*c*) Widespread torrential rain over the Malaysia–southern South China Sea region. This situation occurs when a cold surge (indicated by the bold arrow) and a westward moving disturbance interact over the southern South China Sea. (*d*) Widespread torrential rain over the northern Borneo–southern South China Sea region. This situation occurs when a cold surge initiates the development of a near-equatorial disturbance over the southern South China Sea. There is no westward-moving disturbance from the West Pacific.

596

winds over the South China Sea, and longitude–time sections of satellite cloud pictures.

From experience it is known that the movement of monsoon disturbances is best monitored by using the 700- and 500-mb streamline analyses.

It is very important to predict whether or not a disturbance will move across Peninsular Malaysia from the southern South China Sea because of its associated heavy rain. Two synoptic patterns that are known to favor the westward movement of the disturbances are: (a) when the northern near-equatorial trough (Fig. 18.1a) extends across Peninsular Malaysia from the South China Sea to the Bay of Bengal at the 850- and 700-mb levels, and (b) a belt of easterlies of about 10 degrees latitude in width appear north of the trough all the way from Southeast Asia to southern India through the entire troposphere. Usually these synoptic patterns are observed during the early winter.

On the other hand, they are normally not present during late winter when the circulations in the lower- and upper-troposphere over Southeast Asia are characterized, respectively, by northeasterly cross-equatorial flow and an upper-level westerly wave trough south of the Tibetan Highland (see Sections 1.3 and 1.4). At this time monsoon disturbances tend to be quasi-stationary over the southern South China Sea and West Pacific, and Malaysia is not affected by heavy rain.

From analysis of satellite images and rainfall data, it is now known that heavy rain usually occurs in the area between the near-equatorial trough and the low-level (900–700 mb) easterly wind maximum where cyclonic shear is strongest (1, 6). This is the area to the forward right side of the monsoon disturbance. The horizontal width of the belt of strong easterly winds is about 2 degrees of latitude. The strongest easterly winds are normally observed at the 850-mb level (where the highest speed recorded at Kota Bharu, located along the east coast of Peninsular Malaysia, was 48 knots during a severe cold surge). During the mid- and late winter, as the near-equatorial trough moves eastward to Eastern Borneo and the West Pacific, the low-level easterly winds tend to back in direction and become a zone of maximum northeasterlies. At the same time the axis of maximum wind speed moves southeastward to southern Peninsular Malaysia and Northwestern Borneo. Operational forecasters monitor the movement of this strong wind belt together with the cold surges and monsoon disturbances to predict the occurrence of heavy rain.

Winter monsoon rainfall statistics show that heavy rains seldom occur in Peninsular Malaysia, Borneo, and Indonesia after mid-January, mid-February, and early March, respectively. Along the eastern parts of the Philippines, heavy rains seldom occur in the northern, central, and southern parts after January, February, and March, respectively. Operational forecasters compare daily the current surface and upper-air wind analyses with the long-term mean patterns to look for deviations that alert them to the possibility of unusual weather phenomena.

3.2 Summer Monsoon Prediction

During the summer monsoon, most heavy rain spells in Southeast Asia are associated with three types of monsoon disturbances, namely, tropical waves, mid-tropospheric cyclones over the Indochina and the South China Sea, and the convergence zone in the southwesterlies over Indochina (the convergence zone is not associated with any detectable cyclonic vortices). The prediction of the two latter disturbances is difficult since they usually develop in situ over the South China Sea and Indochina region. For this reason, no more will be said about them in the context of operational forecasting. In contrast, the rainfall associated with tropical waves is more easily predicted since the waves can be identified and tracked on surface weather maps and upper-air streamline analyses with the aid of geostationary satellite cloud pictures.

As in winter (see Section 2.3), tropical waves can sometimes develop into tropical depressions, tropical storms, and typhoons over the Pacific and occasionally over the South China Sea. Although these disturbances rarely hit Indochina directly, their presence in the West Pacific and the South China Sea strongly influences the distribution of dry and wet weather throughout Southeast Asia. Predicting disturbance behavior, particularly the typhoon, is a very important short-range prediction problem in the summer monsoon period in Southeast Asia.

When a typhoon is located over the northern part of the South China Sea, separated from the monsoon trough to its south, there is usually fine weather to the west of the typhoon where there is a high-pressure ridge associated with subsidence (Fig. 18.8a). Fine weather prevails over extensive areas of Indochina and Malaysia when the monsoon trough is located well to the north of Indochina and the ridge of high pressure expands to cover most of Southeast Asia (Fig. 18.8b). Krishnamurti et al. (31) attributed this large-scale ridging effect to the downstream amplication of tropical waves associated with the westward movement of typhoons.

In contrast, if a typhoon is located in the region of the West Pacific and the Philippines, and is embedded in a monsoon trough which extends from the West Pacific to Indochina, a mid-tropospheric cyclone tends to develop over the region of the South China Sea and Indochina. Rainy weather in Southeast Asia will occur particularly in the convergence zone between the low-level northwesterlies and southwesterlies to the left of the typhoon track. Rainy weather may occur as far south as Peninsular Malaysia and Singapore.

Much effort has been devoted to develop methods for forecasting typhoon behavior, its trajectory, intensity change, rainfall, wind speed, and the associated storm surges in Southeast Asia. There are five approaches: the extrapolation method, satellite image analysis, statistical methods, the dynamical or numerical weather prediction method, and the synoptic method.

The extrapolation method is just that: the central position of the typhoon is predicted by extrapolating its past movement. Often forecasters are concerned with only the speed and direction of movement of the typhoon; in some cases, particularly for longer range forecasts, they are also concerned with its acceleration.

Satellite image analysis is especially useful for determining typhoon intensity. Dvorak (39) has developed a technique that allows a 24-hour intensity forecast to be made from satellite observed cloud features. The technique combines meteorological analysis of satellite imagery with a conceptual model of tropical cyclone development. In the technique, the cyclone's (typhoon's) cloud characteristics are related to

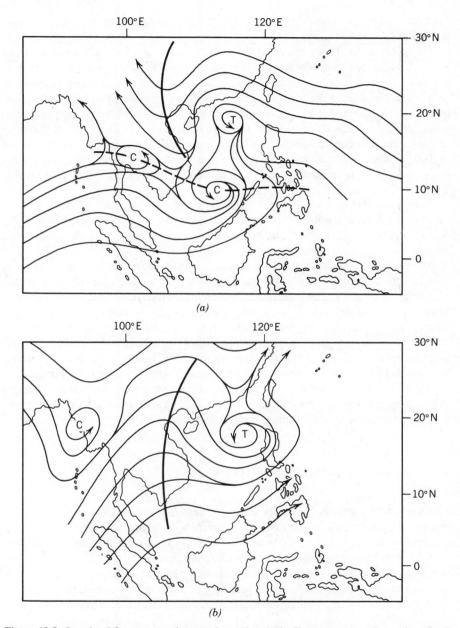

Figure 18.8. Low-level flow pattern when a typhoon (denoted by T) is present over the northern South China Sea during the summer monsoon season. (*a*) When a monsoon trough (dashed line) is present over southern Indochina, the ridging (high pressure), indicated by the bold line, brings clear dry weather to northern Indochina. (*b*) In the absence of a monsoon trough, the ridge (high pressure) and associated clear weather extends throughout most of Southeast Asia.

intensity based on models of "typical" slowly or rapidly developing cyclones (typhoons).

Statistical methods are effective when the typhoon movement is simple. Some of the best known methods are the persistence and climatology method (40) and Arakawa's method (41). Persistence is exactly what it implies; the typhoon is forecast to continue to do what it has done with no changes in its speed or direction of movement. The persistence and climatology method modifies the simple persistence forecast to include the climatological behavior of the typhoon. This method normally employs the stepwise screening procedure. Arakawa's method is based on a set of regression equations developed from data from past years. The equations predict the typhoon's geographical position and its central pressure at the surface for 12, 24, and 48 hours.

Principally because of the lack of large computers, dynamical or numerical weather prediction methods for typhoon forecasting have not been adopted for operational use in Southeast Asia. Nevertheless various research models have been tested and several show promise.

Three synoptic methods are widely used for forecasting. The steering method is based on the assumption that typhoons move with the velocity of a steering current. The velocity of the steering current is estimated by the movement of the cyclonic and/or anticyclonic circulation systems around the typhoon at the 500- and 300-mb levels. The time-change-of-pressure method is based on the observation that typhoons tend to move toward the area of maximum pressure fall. Thus by drawing a chart of pressure change over a specific time period (usually not more than 24 hours), forecasters can predict the typhoon's movement. Finally, if a typhoon's trajectory, intensity, and relationship to the large-scale weather situation is similar to typhoons in the past, the analogy method may be used. Here it is assumed that a typhoon will "repeat history," following the pattern of earlier storms.

4 LONG-RANGE MONSOON PREDICTION

Medium- and long-range forecasts are not issued on a routine basis by the weather services of most of the Southeast Asia countries. Nevertheless, requests for such predictions are increasing. Considerable damage is caused by monsoon-related weather every year (as demonstrated in Chapters 5 and 6). In particular, the large variability of monsoon rainfall makes economic planning difficult. The coefficient of variation of the seasonal rainfall (defined as the standard deviation divided by the mean) for most of Southeast Asia, for both the winter and summer monsoons, exceeds 50%.

To predict the behavior of the monsoon months, or even weeks, in advance requires a good knowledge of how planetary-scale forces influence the monsoon (as has been suggested in Chapters 11, 14, and 16). It would appear then, that global data are needed if we are to attempt long-range forecasts.

Nevertheless, there is some evidence that useful statistical long-range prediction of the monsoon can be done based on regional data alone. More than 60 years ago

Braak (42) asserted that Indonesian rainfall during the period of July to December could be predicted from observations of Indonesian surface pressure. Berlage (43) correlated total September–October rainfall in Java with the average July–August surface pressure at Darwin. He used data from 1877 to 1926 and found a correlation of -0.63, confirming Braak's suggestion that above-normal pressure was followed by below-normal rainfall. More recently, Nicholls (44) confirmed Berlage's results; using a different data set, he found a remarkable similarity in the correlation patterns. In another study, Nicholls et al. (45) also made use of local data to examine the onset and subsequent rainfall during the wet season (September to May) in northern Australia. They demonstrated that in the Darwin region of the Australian tropics, a quantitative index for the onset of the wet season and amount of precipitation can be defined in terms of rainfall at a single station, in this case Darwin. The index is the date after August 1 by which x (mm) of rainfall is accumulated. In their study five values of x were arbitrarily selected: 10, 50, 100, 250, and 500 mm. They correlated the Darwin June–August average surface pressure with the five values of the onset index. Significant correlations were found between the dates by which the first 50, 100, 250, and 500 mm rainfall were received and the Darwin June–August average surface pressure. Accumulated rainfall amounts thus appear to be predictable from observations of pressure several months ahead.

Although the above studies examine rainfall patterns outside of the winter monsoon season (November–March) they provide a good example of how regional data can be used to produce long-range statistical forecasts for seasonal rainfall.

The importance of the Southern Oscillation in influencing the winter monsoon rainfall in South Asia was studied by Tanaka and Yoshino (46) who showed that when the Southern Oscillation Index (SOI, see Chapter 11) is positive, the convergence zone (area of active convection) of the near-equatorial trough is found to shift toward the Indian Ocean. They also found that there is a tendency for two convergence zones to form, one on either side of the equator, which brings heavy precipitation to the South China Sea–Indonesian region. In contrast, when the SOI is negative, only one convergence zone is observed near New Guinea. In this case the winter monsoon is weak and there is below average rainfall over the South China Sea region.

Given these general results, the use of the SOI for long-range prediction of rainfall over Malaysia has been examined employing a 33-year rainfall data set (1951–1983) and a 16-year synoptic data set (1968–1983). It has been found that the correlation coefficient between the total November-to-March rainfall in Malaysia and the SOI for the same months is about 0.6. Out of 19 winter monsoons with a negative SOI, 14 had below-normal rainfall in Malaysia. Conversely, out of 14 winter monsoons with a positive SOI, 10 had above-normal rainfall. Hence, there is a high probability that the normal winter monsoon rainfall in Malaysia will be above (below) normal when the winter SOI is positive (negative). Since the aim of the investigation was to predict the normality of the seasonal rainfall and not the actual amount of rainfall, the above results are encouraging. If we can predict the SOI one month in advance, it will be possible to forecast one month in advance the normality of the winter monsoon rainfall in Malaysia (see Chapter 16 for a discussion of the SOI and the Indian summer monsoon).

To predict the winter season SOI one month in advance, synoptic conditions in East Asia and the West Pacific region have been examined during 16 Octobers (1968–1983). October was chosen because it is about one month before the onset of the winter monsoon in Peninsular Malaysia. The synoptic conditions examined were the north–south movement of the monsoon trough over the South China Sea, the number of days in October when typhoons were present over the West Pacific and the South China Sea, the 200-mb winds over Bombay (India) and Penang (Malaysia) and the frequency of the upper-level westerly wave troughs over the equatorial central Pacific. Each of these conditions was divided into two categories, one for those Octobers that preceded a winter season of positive SOI (henceforth known as category A) and another for those Octobers that preceded a winter season of negative SOI (henceforth known as category B). The average obtained for each synoptic condition in each category turned out to be fairly representative of conditions on any given day during the month because the deviations from the average were small.

In category A, the monsoon trough was found most of the time over the South China Sea and was always located north of Malaysia (7°N). Because of this, over Malaysia, both the low-level (850- and 700-mb) westerly winds and the upper-level (200-mb) easterly winds were stronger compared to the long-term mean. The monsoon trough was also found to move steadily southward from the northern to the central part of the South China Sea. In category B the monsoon trough very often disappeared from the South China Sea and the Southeast Asia region and was replaced by zonal easterlies in the lower troposphere. As a result, the low-level and the upper-level winds over Malaysia were just the opposite of those in category A.

Typhoons were found to be present more often in category A than in category B in which most typhoons developed over the Pacific recurved and moved northward before they reached the western Pacific or the South China Sea. The early recurving of the typhoons over the Pacific occurred at the same time that the monsoon trough (in category B) disappeared from the South China Sea and the Southeast Asia region. Early recurving also brought about a weakening of the trades over the western and central Pacific.

The average 200-mb winds over Bombay in category A were easterlies. An upper-level anticyclone was frequently detected over the region of eastern India and the northern Bay of Bengal. In contrast, the winds in category B were westerlies and the position of the upper-level anticyclone was frequently to the south of the category A position.

The 200-mb winds at Penang in category A and category B had easterly and westerly anomalies, respectively, compared to the long-term mean. The westerly anomaly in category B was associated with the frequent appearance of a north–south oriented upper-level westerly trough over the region of the South China Sea and Malaysia.

Finally, the Pacific upper-level westerly trough was found to penetrate to the equatorial central Pacific region more often in category A than in category B. It is not known whether this condition is related to the development of typhoons.

These preliminary results indicate that the sign of the SOI during the winter months (November–March) can be predicted by examining the October synoptic conditions in East Asia and the West Pacific. Hence, a qualitative long-range (about one month in advance) prediction of the normality of the winter season rainfall in Malaysia is possible.

5 CONCLUSIONS

Weather forecasting in Southeast Asia is difficult. Daily predictions for local areas are disappointingly inaccurate since even in the most "predictable" synoptic conditions the local weather is complicated by orography and land and sea breezes.

Forecasters have come a long way from the time when forecasts were based mainly on persistence. Much use now is made of data from the geostationary satellites. From the international Monsoon Experiments conducted in 1978–1979, forecasters have learned much about rainfall variations on synoptic time scales and how they relate to changes in the monsoon circulation in Southeast Asia. This knowledge enables them to predict monsoon phenomena with more confidence. The accuracy of 24-hour forecasts has improved. However, planners and decision makers require forecasts with longer lead times. To meet their demands, in the near future weather services will probably have to provide on a routine basis qualitative medium- and long-range forecasts of significant weather events. This undertaking appears feasible in view of the encouraging correlations between the Southern Oscillation and monsoon rainfall (see Chapters 11 and 16 and Section 4 of this chapter). However, since the Southern Oscillation Index alone cannot account for all the variance in monsoon rainfall, weather services in Southeast Asia will continue to look to research scientists for an improved understanding of the intraseasonal and interannual variability of the monsoon. With the implementation of the two international projects, the Long-Term Asian Monsoon Studies and the Tropical Ocean and Global Atmosphere Experiment, we are confident that in the very near future, research scientists will be able to achieve their aims in attaining a fundamental understanding of the behavior of the monsoon, and hopefully, this will help to improve the ability of the Southeast Asian weather services in short-, medium- and long-range monsoon predictions.

At this time, all the weather services in Southeast Asia are still using synoptic and empirical methods in day-to-day prediction. Nevertheless, preliminary steps have been taken by several Southeast Asian countries toward developing a numerical weather prediction capability. Experiments with simple models have begun. However, bear in mind that developing a numerical weather prediction capability is a long-term effort. It involves several stages of implementation, namely, the training of research meteorologists in numerical weather prediction and computer techniques, acquiring computers of adequate capacity, developing and testing various numerical models, establishing computer–telecommunication networks to allow for real-time data input to the computer for running models in an operational mode, and finally, the interpretation of the results of the model output. Even when numerical weather prediction is in operation, research needs to be continued to improve the physics of the models so that better forecasts can be obtained. All these stages assume that an effective and efficient data collection and dissemination system will be operating on a global scale. For the tropics and subtropics, this is far from the case at present, and therein lies an additional and very important problem to be overcome.

Although it may require a long time, development of numerical weather prediction capability in Southeast Asia holds a great deal of promise because the method is based on physical principles of the atmosphere. Day-to-day prediction in the future in all the Southeast Asia weather services will probably combine the two existing methods (synoptic and statistical) with numerical weather prediction.

REFERENCES

1. B. K. Cheang, "Some Aspects of Winter Monsoon and Its Characteristics in Malaysia," Research Publication No. 2, Malaysian Meteorological Service, Jalan Sultan, Petaling Jaya, Selangor, Malaysia, 1980.
2. P. C. Chin, "Cold Surges over South China," Technical Note No. 28, Royal Observatory of Hong Kong, 1969.
3. C. P. Chang, J. E. Millard, and G. T. Chen, Gravitational character of cold surges during winter MONEX, *Mon. Wea. Rev.,* **111**, 293–307 (1983).
4. "The ASEAN Compendium of Climatic Statistics," The ASEAN Committee on Science and Technology, ASEAN Secretariat, 70 Jalan Sisingamangaraja, Kebayoran Baru, Jakarta, Indonesia, 1982, 545 pp.
5. C. S. Ramage, *Monsoon Meteorology,* Academic Press, New York, 1971, pp. 158–160.
6. B. K. Cheang, Synoptic features and structures of some equatorial vortices over the South China Sea in the Malaysian region during the winter monsoon December 1973, *Pure Appl. Geophys.,* **115**, 1303–1333 (1977).
7. C. P. Chang, J. E. Erikson, and K. M. Lau, Northeasterly cold surges and near-equatorial disturbances over the winter MONEX area during December 1974, Part I: Synoptic aspects, *Mon. Wea. Rev.,* **107**, 812–829 (1979).
8. K. M. Lau, C. P. Chang, and P. H. Chan, Short-term planetary-scale interactions over the tropics and mid-latitudes, Part II: Winter MONEX period, *Mon. Wea. Rev.,* **111**, 1372–1388 (1983).
9. M. Williams, "Interhemispheric Interaction during Winter MONEX," in International Conference on Early Results of FGGE and Large-Scale Aspects of Its Monsoon Experiments, Tallahassee, Florida, January 1981, World Meteorological Organization, Geneva, 1981.
10. H. Lim and C. P. Chang, A theory for mid-latitude forcing of tropical motions during winter monsoon, *J. Atmos. Sci.,* **38**, 2377–2392 (1981).
11. J. T. Lim and L. C. Quah, Cross-equatorial flow over Southeast Asia during the northeast monsoon, *Indian J. Meteor. Hydrol. Geophys.,* **29**, 5–15 (1978).
12. C. S. Ramage, The cool season tropical disturbances of Southeast Asia," *J. Meteor.,* **12**, 252–262 (1955).
13. B. E. Harris, J. Sadler, I. Gordon, F. P. Ho, and W. R. Brett, "Synoptic Regimes which Affect the Indochina Peninsula during the Winter Monsoon," AFCRL-71-0232, UHMET-71-1, Department of Meteorology, University of Hawaii, Honolulu, 1971.
14. T. L. Gan, "The Circulation Pattern over Singapore and the East Coast of West Malaysia during January and February 1967 compared with that of January and February 1968," in Proceedings of Forecasting of Heavy Rains and Floods, Joint Training Seminar RAII

and V, Kuala Lumpur, 11–23 November 1968, World Meteorological Organization, Geneva, pp. 283–289.

15. J. S. Boyle and C. P. Chang, "Monthly and Seasonal Climatology over the Global Tropics and Subtropics for the Seasonal Decade 1973–1983. Vol. I: 200 mb Winds," Technical Report, NPS-63-84-006. Department of Meteorology, Naval Post Graduate School, Monterey, California, 1984.

16. R. A. Houze, S. G. Geotis, F. D. Marks, and A. K. West, Winter monsoon convection in the vicinity of North Borneo, Part I: Structure and time variation of the clouds and precipitation, *Mon. Wea. Rev.*, **109**, 1595–1614 (1981).

17. R. H. Johnson and D. L. Priegnitz, "Convection over the Southern South China Sea— Part II: Effects on Large-Scale Fields," in International Conference on Early Results of FGGE and Large Scale Aspects of Its Monsoon Experiments, Tallahassee, Florida, January 1981, World Meteorological Organization, Geneva, 1981.

18. T. Murakami, Winter monsoonal surges over East and Southeast Asia, *J. Meteor. Soc. Japan*, **57**, 133–158 (1979).

19. B. K. Cheang, K. G. Lum, and R. Radhakrishnan, "Oscillations of the Winter Monsoon System," International Conference on the Scientific Results of the Monsoon Experiments," Denpasar, Bali, Indonesia, October 1981, World Meteorological Organization, Geneva, 1982.

20. K. S. Yap, K. G. Lum, and B. K. Cheang, "Active and Break Cycles over the South China Sea–Malaysia Region during Winter Monsoon," Research Publication No. 6, Malaysian Meteorological Service, Jalan Sultan, Petaling Jaya, Selangor, Malaysia, 1983.

21. B. K. Cheang and T. N. Krishnamurti, Middle latitude interactions during the winter monsoon, *Papers in Meteor. Res.*, **4**(1), 38–61 (1982).

22. M. M. Orgill, "Some Aspects of the Onset of the Summer Monsoon over Southeast Asia," Technical Report, Department of Atmospheric Science, Colorado State University, Fort Collins, Colorado, 1967, pp. 11–20.

23. H. Riehl, "Southeast Asia Monsoon Study," Technical Report, Department of Atmospheric Science, Colorado State University, Fort Collins, Colorado, 1967, 22 pp.

24. B. E. Harris, "The Summer Monsoon over Southeast Asia," in Proc. Synoptic Analysis and Forecasting in the Tropics of Asia and the Southwest Pacific, Regional Training Seminar, Singapore, December 2–15, 1970, World Meteorological Organization, Geneva, 1972, pp. 182–214.

25. M. Yanai, Tropical meteorology, *Rev. Geophys. Space Phys.*, **13**(3), 685–710 (1975).

26. J. M. Wallace and C. P. Chang, Spectral analysis of large scale wave disturbances in the tropical lower troposphere, *J. Atmos. Sci.*, **26**, 1010–1025 (1969).

27. M. Yanai and T. Murakami, A further study of tropical wave disturbances by the use of spectral analysis, *J. Meteor. Soc. Japan*, **48**, 331–347 (1970).

28. C. P. Chang, V. F. Morris, and J. M. Wallace, A statistical study of easterly waves in the western Pacific, *J. Atmos. Sci.*, **27**, 195–201 (1970).

29. A. Zangvil, Temporal and spatial behavior of large-scale disturbances in tropical cloudiness deduced from satellite brightness data, *Mon. Wea. Rev.*, **103**, 904–920 (1975).

30. T. Yasanari, Cloudiness fluctuations associated with the Northern Hemisphere summer monsoon, *J. Meteor. Soc. Japan*, **57**, 227–241 (1979).

31. T. N. Krishnamurti, J. Molinari, H. L. Pan, and V. Wong, Downstream amplification and formation of monsoon disturbances, *Mon. Wea. Rev.*, **105**, 1281–1297 (1977).

32. K. R. Saha, F. Sanders, and J. Shukla, Westward propagating predecessors of monsoon depressions, *Mon. Wea. Rev.*, **109**, 330–343 (1981).

33. T. N. Krishnamurti and H. N. Bhalme, Oscillations of a monsoon system. Part I: Observational aspects, *J. Atmos. Sci.*, **33**, 1937–1954 (1976).

34. T. Murakami, Cloudiness fluctuations during the summer monsoon, *J. Meteor. Soc. Japan*, **54**, 175–181 (1976).

35. B. K. Cheang, K. S. Yap, K. G. Lum, and T. Y. Chang, "Variations of Rainfall in Malaysia in Response to the Oscillations of the Summer Monsoon Circulation during Winter MONEX," Research Publication No. 4, Malaysian Meteorological Service, Jalan Sultan, Petaling Jaya, Selangor, Malaysia, 1981.

36. E. W. K. Chu, "A Method of Forecasting the Arrival of Cold Surges in Hong Kong," Technical Note No. 43, Royal Observatory, Hong Kong, 1978.

37. C. Y. Lam, "Synoptic Patterns Associated with the Onset of Cold Surges reaching the South China Sea," in International Conference on the Scientific Results of the Monsoon Experiments, Denpasar, Bali, Indonesia, October 1981, World Meteorological Organization, Geneva, 1982.

38. H. Riehl, "The Diagnosis and Prediction of SEA Northeast Monsoon Weather," Navy Weather Research Facility, Naval Air Station, Norfolk, Virginia, 23511, 1969.

39. V. S. Dvorak, Tropical cyclone intensity analysis and forecasting from satellite imagery, *Mon. Wea. Rev.*, **103**, 420–430 (1975).

40. T. Aoki, A statistical prediction of the tropical position based on persistence and climatological factors in the western north Pacific (the PC method), *Geophys. Mag.*, **38**, 17–28 (1979).

41. A. Arakawa, "Statistical Method to Forecast the Movement and the Central Pressure of Typhoons in the Western North Pacific," in Proceedings of Symposium of Tropical Meteorology, Rotorua, New Zealand (1963), New Zealand Meteorological Service, Wellington, 1964.

42. C. Braak, "Atmospheric Variations of Short and Long Duration in the Malay Archipelago," Verhandelingen No. 5, Koninklijk Magnetisch en Meteorologisch Observatorium te Batevia, Indonesia, 1919, 57 pp.

43. H. P. Berlage, "East-Monsoon Forecasting in Java," Verhandelingen No. 20, Koninklijk Magnetisch en Meteorologisch Observatorium te Batevia, Indonesia, 1927, 42 pp.

44. N. Nicholls, Air–sea interaction and the possibility of long-range weather prediction in the Indonesian archipelago, *Mon. Wea. Rev.*, **109**, 2435–2443 (1981).

45. N. Nicholls, J. L. McBride, and R. J. Ormerod, On predicting the onset of the Australian wet season at Darwin, *Mon. Wea. Rev.*, **110**, 14–17 (1982).

46. M. Tanaka and M. M. Yoshino, "Synoptic Study of the Interannual Fluctuations of the Winter and Summer Monsoon in the Asian and Australian Regions," International Conference on the Scientific Results of the Monsoon Experiments, Denpasar, Bali, Indonesia, October 1981, World Meteorological Organization, Geneva, 1982.

19

Prediction and Warning Systems and International, Government, and Public Response: A Problem for the Future

M. S. Swaminathan
International Rice Research Institute
Manila, Philippines

INTRODUCTION

History has taught us that the fate of many past civilizations has been determined by the ability of governments to maintain adequate food supplies. The decline of the Roman Empire and the disappearance of the ancient civilizations of Mohenjodaro and Harrappa of the Indus valley can be attributed to inadequate food supplies. This chapter illustrates how governments in India have dealt with the consequences of famine caused by abnormal monsoons, be it floods that destroyed crops or droughts that reduced food grain production. In recent times, food shortages have been minimized by agricultural technologies developed to reduce risks related to abnormal monsoons. Although abnormal monsoons may be forecast, no techniques have been developed to avert them. To avert widespread famines, various action programs at both the national government and the international level are being geared toward providing relief to regions and countries struck by abnormal monsoon behavior. Government concern and public response to disaster management brightens the prospect of success in the battle against famine and disaster caused by abnormal monsoons.

1 GOVERNMENT AND PUBLIC RESPONSES TO ABNORMAL MONSOONS

The evolution of government policies to manage human hardships during famine caused by deviations of the monsoon from the normal pattern can be studied with examples from India, which suffers periodically from abnormal monsoons. The famines referred to in ancient Indian records were localized, and relief was generally provided by private charity rather than by the State. Government policies to combat famine, and food scarcity began to evolve only during the late nineteenth century.

Colonel Baird Smith was appointed by the Government of India to investigate the causes of the 1860–1861 famine in the northwestern provinces and to suggest remedial measures. He recommended extension of the irrigation system, improvement of communication, and remission of land tax (1).

Four years later the Orissa State in India was struck by one of the nineteenth century's worst famines (1865–1867). The Orissa Famine Commission, appointed by the Indian government and chaired by George Campbell, agreed with Smith's recommendations and suggested that the system of land tenure be improved, irrigation canals be constructed, postal communications be streamlined, and correct rainfall data and agricultural statistics be compiled (1). Both Smith's and the Commission's reports received little attention.

The Famine Inquiry Commission under General R. Stratchey, appointed in 1878, recognized "the paramount duty of the State" to offer "protection to the people of India from the effects of uncertainty of the seasons" (1). It identified drought-prone and relatively drought-immune areas of the country. It traced the history of different famines since 1770 and found that between 1770 and 1880 there were 21 famines, 8 of which were severe. One of the major recommendations of the Commission was to create a separate department to administer food relief and to appoint a full-time Famine Commission (as an extension to the Inquiry Commission) to coordinate relief activities during actual food scarcity. The Inquiry Commission also considered decentralization of financial responsibility to the provincial level. It prepared a provincial famine code recommending that:

1. a systematic record of agricultural and other statistics about crop failure be kept;
2. employment be provided to all able-bodied persons through public works projects such as construction of rural roads and drains, building percolation tanks, cleaning and deepening of wells;
3. gratuitous relief in the form of food and money be provided to destitutes through the village community;
4. a suitable system of village inspection of fields and homes and monitoring of famine relief operations be organized; and
5. suspending or remitting taxes and offering loans to peasants with small land holdings.

Although the Famine Commission of 1880 established a general framework for famine relief administration, subsequent famine commissions continually updated the provincial famine code to widen the provincial official's ability to assist in the recovery from natural disasters caused by abnormal monsoons, including drought (1).

The earliest recorded case of actual relief work in India was that which followed the famine of 1899–1900, when an estimated 1 million people died. The human disaster was of such dimension that the British government created a special agency to organize relief operations. The system required that relief work be organized on the basis of the anticipated reduction in harvest caused by drought, floods, and other natural calamities. As part of this program, the agency developed the *Annawari* system for estimating crop losses by visual judgment (before the current metric system of currency, the Indian rupee was divided into 16 annas; an 8-anna crop, therefore, meant the harvest would be 50% less than normal).

The Bengal famine of 1943–1944 was the biggest in India in this century. Extra mortality was officially estimated at 1.5 million, although Amartya Sen (2) placed the real death toll at about 3 million. Although the number of deaths could have been reduced by a more equitable distribution of the available food, this famine was in part triggered by a widespread devastation of the rice crop by a disease caused by the fungus *Helminthosporium oryzae*. Rice yields were reduced by 40–90% in different parts of Bengal. The disease epidemic was largely caused by the abnormally heavy rains that led to leaching of nutrients from the soil. It was made worse by unseasonal rainfall in winter, which favored the pathogen's release of spores. Had there been a good forecast, coupled with a suitable pest and disease surveillance system, the disease epidemic might have been avoided through timely control measures.

The democratic political system and the improved internal communication system established after India achieved independence in 1947 have helped government officials to identify potentially serious situations, allowing them to take action to prevent disasters (2). Since India's independence, no massive mortality from food shortages due to abnormal monsoons have occurred, thanks to prompt governmental and private action and international assistance.

A policy of building up of food reserves was initiated in 1967 and as a result, the Government of India had a stock of more than 22 million tons of food grains in 1984. A network of fair-price shops was established throughout the country for the distribution of food grains at reasonable prices. There are over 200,000 such shops scattered in different parts of the country. Both in the establishment of grain storage facilities and fair-price shops, the needs of drought and flood prone areas are given particular attention. This helps to instill confidence in the people, thereby preventing panic purchase by affluent consumers during seasons of crop failure caused by abnormal monsoons. This is a major factor in stabilizing prices in seasons when food production goes down due to drought and/or floods. It should be mentioned that success in building large grain reserves was the result of significant increases in grain production following the introduction of high yielding varieties of wheat and rice in 1966. Wheat production, for example, went up from 12 million tons in 1964 to 45 million tons in 1984.

2 FIGHTING THE FAMINE OF JOBS

As early as the middle of the last century, Colonel Smith stated "Indian famines are rather famines of work than of food; for, when work can be had and paid for, food is always forthcoming'' (1). Rural works programs have provided relief to and augmented income for victims of natural disasters. For example, Maharashtra State, one of the chronically drought-prone areas of India, has a long record of organizing extensive relief work programs. From that experience, a program termed "Employment Guarantee Scheme" (EGS) developed in recent years. It helped farmers increase their crop yields by 25%, increased food production by 2.5%, improved rural wages, and decreased rural indebtedness (3). Its effectiveness can be illustrated by a comparison of abnormal monsoon conditions in Bangladesh with those in Maharashtra (India), as reported by Bongaarts and Cain (4).

A severe flood in 1974 destroyed crops in the low-lying areas of Bangladesh. Even before the flood, however, Bangladesh was in the grip of high inflation that had seriously eroded the real income of the poorer population. Pockets of famine were reported in the northwest as early as spring 1974 due to lack of food in the market. The flood later in the year caused complete or partial crop failure over large areas of the country and led to a further rise in food-grain prices. In this comparison by Bongaarts and Cain (4) between India and Bangladesh, we must keep in mind that Bangladesh, unlike India, was politically unstable in the early 1970s. This may have seriously exacerbated the problems of food shortages. However, when government can and does take remedial action, the impacts of natural disasters can be substantially mitigated.

Even with this combination of disasters, famine need not have been a necessary consequence. Famine ensued because private, government, and international response in rushing food to the affected population was not timely. The occurrence of modern famine is evidence both of a high-risk environment and the absence or failure of risk insurance mechanisms—including public relief (5). Political stability helps to ensure that relief is adequate and timely.

Many parts of India have harsh natural environments similar to those of rural Bangladesh. The Maharashtra drought of 1970–1973 was comparable in its adverse impact on food production to the Bangladesh floods of 1974. But government responses to the disaster caused by abnormal monsoon weather differed probably due to the difference in the options open to the concerned governments. Using the frequency of "distress" land sales as a response to community disaster and an early sign of developing famine conditions, Bongaarts and Cain (Fig. 19.1) have compared the frequency of land sale transactions in Char Gopalpur (Bangladesh) and in three Indian villages in Maharashtra (India). They collected data from about 120 households in each village. The total transactions within a given time were standardized by the number of households during the period. The data show that the frequency of transactions in Char Gopalpur appears highly responsive to natural and manmade disasters. The sharp peak from 1970–1974 reflects a high volume of distress land sales. In contrast to the Bangladesh data, the frequency of land sales in the combined Indian samples has been low and constant over time. In the most drought-stricken villages, almost all agricultural activity stopped in 1972–1973, yet no household in the sample sold land under duress.

Bongaarts and Cain (4) explained the contrast in frequency of distress sales between India and Bangladesh on the basis of the provision of public relief. Government

Av no. of land sales per household

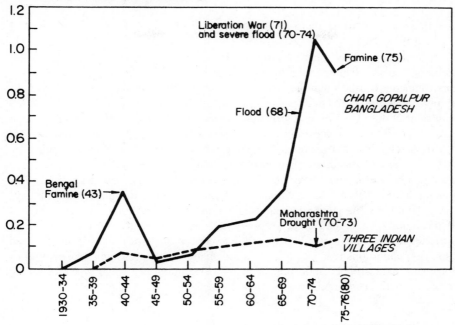

Figure 19.1. Frequency of land sale transactions in India and Bangladesh (from reference 4).

response to the Maharashtra drought was timely, appropriate, and consisted largely of a massive public works employment program. A survey of one of the villages during the worst part of the drought indicated that wages from public employment equalled 46% of total income whereas during normal periods it is insignificant. Since the drought ended, Maharashtra State has instituted a permanent public employment program (the Maharashtra Employment Guarantee Scheme) that provides a reliable source of employment and wages to all wishing to work. Funds for operating this scheme come from a special tax levy imposed on the more well-to-do sections of the community.

Similarly, the massive Food for Work and Food for Nutrition programs undertaken during the 1979 drought were designed to avoid widespread famine as a result of monsoon failure. The Food for Work program provided either wheat or rice to every person who could do physical work. The Food for Nutrition program provided food to old and infirm persons, pregnant and nursing mothers, and young children unable to do physical work. The Government of India released more than 2 million tons of food grains from its own reserves for these programs (6). These examples illustrate the pivotal role of political will in mitigating human distress during unfavorable monsoon years.

3 MITIGATING THE ADVERSE IMPACT OF ABNORMAL MONSOONS

Are there ways and means to avoid the adverse impact of abnormal monsoons? I contend there are.

A detailed strategy must be prepared for each major growing condition of all important crops. Using rainfed, upland rice as an example, the various rice-growing environments can be quantified in terms of their climatic and soil properties. As shown in Figure 19.2, most upland rice areas in South Asia receive 1200–1500 mm rainfall during a 5-month growing season (8). Because of its large interannual variability, it is more meaningful to express rainfall in probabilistic terms (Fig. 19.3). A scientific strategy for stabilizing production can be developed by selecting rice varieties with maturity characteristics based upon rainfall probability information. Fortunately rice breeders have developed early maturing varieties with good yield potential (Fig. 19.4). Old traditional varieties that mature in more than 135 days have about the same yield potential as new improved rices that mature in 110 days. If the monsoon's length could be predicted, farmers could be advised on which varieties to use. Thus it would be possible to choose rice types suitable for expected monsoon conditions (for example, rain-fed shallow water, drought-prone varieties for a short and irregular monsoon, or rain-fed medium-deep water types for a long monsoon with heavy rains).

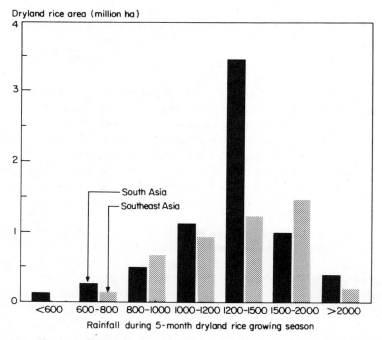

Figure 19.2. Distribution of dryland rice area by the amount of rainfall (mm) received during a 5-month growing period, South and Southeast Asia (from reference 7).

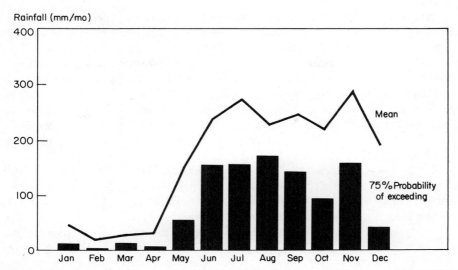

Figure 19.3. Mean monthly and 75% probable rainfall in Los Baños, Philippines (data from IRRI Climate Unit).

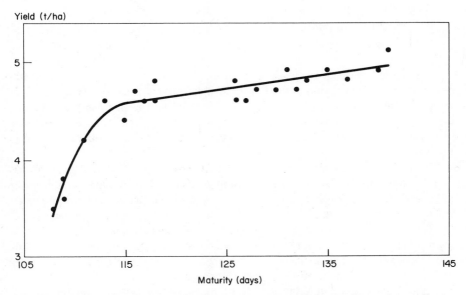

Figure 19.4. Yield of new rice varieties in relation to maturity period (from reference 9).

Scientists often have tried to predict the impact of weather on crop yields. Crop–weather simulation models provide a useful insight into the effect of weather variables on crop production (10). In the United States, scenarios for yield prediction based on expected weather conditions have been constructed for wheat, sorghum, soybeans, maize, and other crops, and models for rice are being developed. Although these models are still in an early stage of development, their value for monsoon regions is obvious: if the monsoon behavior can be predicted early, farming strategies can be adjusted and yields estimated. Planners, decision makers, and governments would then be able to estimate the size of food reserves that must be maintained to ensure food security.

Remote sensing techniques and continuous monitoring of monsoon behavior have helped in the development of crop production forecasts which have been used since 1975 by the Food and Agriculture Organization of the United Nations (FAO) in operating a global early warning system. This system provides information that helps to initiate suitable preventive and remedial measures for overcoming the problems that could arise from crop failure. Obviously, the early warning system has to be accompanied by a timely action capability. If an early warning on yield and production decline can be analyzed with reference to the quantities of food required to meet the needs of the affected population, in theory, it should not be difficult to arrange for the needed supply through coordinated national and international action. For example, according to FAO statistics, the world cereal stocks at the end of 1984-1985 are expected to be about 291 million tons. This volume of stocks represents 18% of expected world consumption. Cereal stocks by the close of 1984–1985 are forecast to be composed of the following commodities:

wheat: 140 million tons, 6% more than the year's beginning stocks;
coarse grains: 106 million tons, 18% more;
rice (milled):   45 million tons, 5% more.

Thus there are adequate food reserves in the world to prevent famine deaths provided the early warning system is utilized effectively for preparatory action in arranging for the delivery of food in the areas that will need them most. Recent experience in Africa, however, indicates the complexity of famine relief operations, particularly when the physical infrastructure such as rural roads and political framework for action are weak.

To capitalize on newly developed varieties and cropping technology, it would be advisable to organize in each agroclimatic zone a crop weather watch group including a meteorologist, an agronomist, a statistician, an agricultural engineer, and a development administrator. This group should regularly monitor weather conditions relative to crop growth potential and make recommendations to the appropriate agricultural extension and development agencies about crop selection and management strategies. Support from mass media will be particularly helpful in the dissemination of useful information and guidelines.

Besides steps that limit the effects of monsoon failure on crop production, administrative measures are needed to ensure a stable supply of food and drinking water in calamity-prone areas. It is generally good to encourage conservation and storage of water and to increase food grain reserves in areas vulnerable to abnormal monsoons. Several recent studies have evaluated natural disaster planning: Sarma (11) reviewed literature on contingency planning for famines and other acute food shortages. In 1983 the FAO of the United Nations (12) documented the impact of drought in the Sahel. Amartya Sen (2) wrote a penetrating analysis of the great Bengal famine of 1943, the Ethiopian famine, drought and famine in the Sahel, and famine in Bangladesh. All these analyses provide guidelines for future action to mitigate the adverse impact of abnormal monsoons.

Briefly stated, the drought and disaster management strategy will consist of the following major components:

1. *Relief and Rehabilitation Measures in the Most Seriously Affected (MSA) Areas.* MSA regions characterized by acute distress to the human and livestock populations have to be given the highest priority in relief and rehabilitation measures. Carefully designed *Food for Nutrition* programs sponsored by enlightened donors can provide much needed relief to old and infirm persons, children, and pregnant and nursing mothers.

Simultaneously, a *Food for Work* program has to be operated on an open-ended basis to provide opportunities for all who are in a position to work, to earn their daily bread, and to develop the infrastructure necessary for handling drought management procedures.

The kinds of work that can be undertaken will vary from place to place. Wherever possible, Food for Work programs should include measures like the establishment of drinking water sanctuaries, underground water harvesting, and organization of livestock camps near a water source where livestock are fed with a maintenance ration consisting of any available cellulosic material enriched by urea and an energy source like molasses.

2. *Compensatory Food Production Programs in the Most Favorable Areas (MFA).* MFA areas are defined as those where soil moisture will be adequate for sustaining a crop. Areas with assured irrigation and areas with moisture retentive soils will fall under this category. In such areas, more intensive production drives can be launched through the supply of inputs like improved seeds and fertilizer at reasonable prices. Steps should be taken in such areas to bridge the gap between potential and actual yields, using the best available technology, supported by effective packages of services, particularly of credit, and of government policies in input and output pricing and marketing.

3. *Good Weather Code for Drought Prone Areas.* It is only in the occasional good rainfall years that meaningful work can be done to contain desertification and build the ecological infrastructure necessary for stable agriculture. A *Good Weather Code* is hence necessary for such areas. The code would list all the steps that should be undertaken in normal rainfall years such as soil conservation, water harvesting, afforestation, aerial seeding, building seed reserves, and the other steps essential for building the ecological and production infrastructure essential for sustained agricultural advance.

Unfortunately, in normal rainfall years neither national nor international sources of additional funding become available. Consequently, unique opportunities for economic development are lost. While a "fire brigade" operation attracts funds, the less glamorous kind of programs essential to build productivity over the longer term tend to attract neither funds nor administration attention. This is why a good weather code is very important in any scientific monsoon management system.

4 PREPARING FOR THE FUTURE

Although we are already tackling complex problems caused by present-day abnormalities of climate, we must be prepared to deal with living in a changing world climate. There is a fast-growing literature on climate change. An example of a changing scenario from a computer model calculation by Kellogg and Schware (13) is shown in Figure 19.5. There are many other examples. According to Roberts and Friedman (14) who constructed a model of the climate for AD 2040, the world will be, on the average, 3°C warmer than now and 10°C or more warmer at the highest latitudes. Although these predictions are controversial and much more research is needed, we may—assuming that they are at least qualitatively correct—expect that in the next century, East Africa and the Sahara will receive twice as much precipitation, but most of the area will still remain semiarid to arid. Temperatures in this wetter belt will be about 2°C cooler. Parts of West Africa will be slightly drier. In India, monsoon rains will be more reliable and bring 50% more total precipitation than in the period 1960–1980.

How should we prepare to face these possible climatic changes? Fortunately there are possibilities for developing plant genotypes that can perform well in different environments. Investment in such anticipative research is as important as refining and taking advantage of the knowledge and material already available for facing the consequences of today's abnormal monsoons.

While preparing for the future is important, we should immediately use currently available technologies to stabilize production under varying monsoon scenarios. Strategies to do this have been described in Chapter 6. In addition, to assist policymakers in making decisions that affect national food security, monsoon region governments should compile for the crops of importance to their countries, data to answer the following questions:

1. what is the state of the art in the prediction of monsoon behavior in the countries where the concerned crop is a staple?;
2. what are the agricultural assets and liabilities based on the most frequent patterns of monsoon rainfall?;
3. what is the available know-how on risk-distribution agronomy and contingency planning and what are the major gaps in current knowledge?; and
4. what is a workable policy framework for ensuring adequacy of supplies as well as access to supplies particularly by the poorer section of the population?

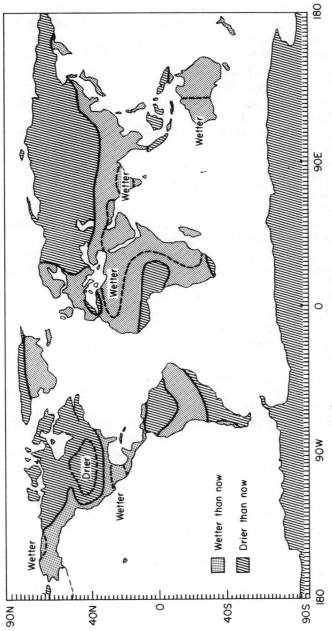

Figure 19.5. A scenario of possible soil moisture patterns on a warmer earth. It is based on paleoclimatic reconstructions of the Altithermal Period (4500–8000 years ago), comparisons of recent warm and cold years in the Northern Hemisphere, and a climate model experiment (from reference 13).

In relation to the last question, Raj Krishna and Chhibber (15) recently developed a model for wheat in India. Three scenarios are simulated in their projections. In the basic scenario, rainfall is 10–20% below normal in 6 of 14 years. In addition to drought, foreign exchange availability is used as a variable in the model. Thus the impact of drought and foreign exchange shortage on the Indian wheat grain system can be assessed. Using the model, Raj Krishna and Chhibber have projected the values of four policy variables—purchases from within the country, imports, concessional sales, and stocks—from 1979 to 1992.

By developing complementary computer models on the production and post-harvest aspects of each cropping system, on the basis of most likely patterns of monsoon behavior, every country can develop strategies for marketing and distribution conducive to achieving agricultural growth with stability of supplies and equity of food entitlements.

The question still remains as to what use these models could be to the political leaders and administrators who have to manage the food budget in the event of extensive crop failure. A political administrative system sensitive to its primary obligation of providing the people of the country physical and economic access to food at all times can use the indications provided by such models for the following purposes:

1. determining the minimum grain reserves that will be needed by the country to manage the food shortages that may occur as a result of damage to crops by deviations from normal monsoon behavior;
2. estimating the financial cost of maintaining such reserves and making appropriate funding arrangements for building the reserves through purchases of surplus produce within the country and/or importing food grains through commercial purchase or concessional aid arrangements;
3. developing a strategy for positioning the reserve food grain stocks at locations that are chronically prone to weather-induced crop failures;
4. improving the communication system including all-weather rural roads, and water and air transport systems that will enable the rapid movement of food grains to affected areas;
5. making the planning for adverse weather an integral part of the planning process so that there is no element of surprise or panic when in some seasons crops do fail due to monsoon abnormalities; and
6. developing compensatory production plans in areas with irrigation facilities or adequate soil moisture to sustain crop growth.

5 CONCLUDING REMARKS

There is an interplay of darkness and light in government and public response to disaster management. Darkness arises from human hardship due to natural calamities that are aggravated by technological backwardness stemming from illiteracy and poverty, by poor communications hampering the timely delivery of food grains to needy areas and people, and by sociopolitical instability. Light comes from the prospects of improved monsoon forecasts on both short and long time scales (see Krishnamurti, Shukla, Das and Cheang, and others in this volume), from the opportunities now open for achieving accelerated agricultural advances coupled with greater stability of production through codes of action designed to maximize the benefits of good monsoons and minimize the adverse impact of unfavorable monsoons, and from the continuously improving national, regional, and global food security and early warning systems.

REFERENCES

1. M. S. Swaminathan, *Science and Integrated Rural Development*, Concept Publishing Company, New Delhi, 1982, 354 pp.
2. A. Sen, *Poverty and Famines: An Essay on Entitlement and Deprivation*, Clarendon Press, Oxford, United Kingdom, 1981.
3. E. H. D'Silva, "The Effectiveness of Rural Works Programs in Labor-Surplus Economies: The Case of the Maharashtra Employment Guarantee Scheme," Cornell International Agriculture Mimeograph 97, Cornell University, Ithaca, 1983, 47 pp.
4. J. Bongaarts and M. Cain, "Demographic Response to Famine," in K. M. Cahill, Ed., *Famine*, Orbis Books, New York, 1982, pp. 44–59.
5. A. Sen, "Food Battles: Conflicts in the Access to Food," 12th Coromandel Lecture, Coromandel Fertilizers Ltd., New Delhi, 1982, 14 pp.
6. M. S. Swaminathan, "Climate and Agriculture," in A. K. Biswas, Ed., *Climate and Development*, Tycooly International Publishing Ltd., Dublin, Ireland, 1984, pp. 65–95.
7. D. P. Garrity, "Asian Upland Rice Environments," in *An Overview of Upland Rice Research*, International Rice Research Institute, Los Baños, Philippines, 1984, pp. 161–183.
8. D. P. Garrity, "Rice Environmental Classifications: A Comparative Review," in *Terminology for Rice Growing Environments*, International Rice Research Institute, Los Baños, Philippines, 1984, pp. 11–35.
9. IRRI, *Annual Report for 1978*, International Rice Research Institute, Los Baños, Philippines, 1978, 478 pp.
10. A. K. Biswas, "Crop-Climate Models: A Review of the State of the Art," in J. Ausebel and A. K. Biswas, Eds., *Climatic Constraints and Human Activities*, Pergamon Press, Oxford, 1980, pp. 75–92.
11. J. S. Sarma, "Contingency Planning for Famines and Other Acute Food Shortages: A Brief Review," International Food Policy Research Institute, Washington, D.C., 1983, 28 pp.
12. E. Saouma, "World Food Report 1983," Food and Agriculture Organization of the United Nations, Rome, 1984, 64 pp.
13. W. W. Kellogg and R. Schware, *Climate Change and Society—Consequences of Increasing Atmospheric Carbon Dioxide*, Westview Press, Boulder, CO, 1981, 178 pp.
14. W. O. Roberts and E. J. Friedman, "Living with the Changed World Climate," Aspen Institute for Humanistic Studies, New York, 1982, 37 pp.
15. R. Krishna and A. Chhibber, "Policy Modelling of a Dual Grain Market: The Case of Wheat in India," Research Report 38 of the International Food Policy Research Institute, Washington, D.C., 1983, 74 pp.

Index